Fruit Ripening
Physiology, Signalling And Genomics

Fruit Ripening

Physiology, Signalling and Genomics

Edited by

Pravendra Nath

Council of Scientific and Industrial Research-National Botanical Research Institute, Lucknow, India

Mondher Bouzayen

Institut National Polytechnique – ENSA Toulouse, France

Autar K. Mattoo

Beltsville Agricultural Research Center, USDA, Beltsville, USA

Jean Claude Pech

Institut National Polytechnique – ENSA Toulouse, France

www.cabi.org

CABI is a trading name of CAB International

CABI Head Office
Nosworthy Way
Wallingford
Oxfordshire OX10 8DE
UK

Tel: +44 (0)1491 832111
Fax: +44 (0)1491 833508
E-mail: info@cabi.org
Website: www.cabi.org

CABI
745 Atlantic Avenue
8th Floor
Boston, MA 02111
USA

Tel: +1 (617)682-9015

E-mail: cabi-nao@cabi.org

A catalogue record for this book is available from the British Library, London, UK.

The Library of Congress has cataloged the hardcover edition as follows:

Fruit ripening : physiology, signalling and genomics / editors: Pravendra Nath, Mondher Bouzayen, Autar K. Mattoo, Jean Claude Pech.
pages cm
ISBN 978-1-84593-962-5 (hbk : alk. paper) 1. Fruit--Ripening. I. Nath, P. (Pravendra) II. Bouzayen, Mondher. III. Mattoo, Autar. IV. Pech, J. C. (Jean Claude)

SB360.5.F78 2014
634'.0441--dc23

2014006976

ISBN-13: 978 1 78639 537 5

Commissioning editor: Sarah Hulbert/Vicki Bonham
Editorial assistant: Emma McCann
Production editor: Lauren Povey

Typeset by Columns Design XML Ltd, Reading, UK.
Printed and bound by CPI Group (UK) Ltd, Croydon, CR0 4YY
First printed in hardback in 2014. Transferred to POD paperback in 2019.

Contents

Contributors

Agnès Ageorges, INRA, UMR1083 Sciences pour l'œnologie 2, place Viala F-34 060 Montpellier Cedex 01, France. E-mail: ageorges@supagro.inra.fr

Asaph Aharoni, Department of Plant Sciences, Weizmann Institute of Science, PO Box 26, Rehovot, 76100, Israel. E-mail: asaph.aharoni@weizmann.ac.il

Raheel Anwar, Department of Horticulture and Landscape Architecture, Purdue University, West Lafayette, IN 47907-2010, USA. Email: ranwar@purdue.edu

Pierre Baldet, INRA, UMR 1332 Biologie du Fruit et Pathologie, 33140 Villenave d'Ornon, France, University of Bordeaux, UMR 1332, 33140 Villenave d'Ornon, France. E-mail: pierre.baldet@bordeaux.inra.fr

Kailash C. Bansal, National Bureau of Plant Genetic Resources (ICAR), Pusa Campus, New Delhi 110012, India. E-mail: kailashbansal@hotmail.com

Cornelius S. Barry, Department of Horticulture, Michigan State University, East Lansing, MI 48824, USA. E-mail: barrycs@msu.edu

Christine Böttcher, CSIRO Plant Industry, PO Box 350, Glen Osmond, SA 5064, Australia. E-mail: christine.bottcher@csiro.au

Mondher Bouzayen, Institut National Polytechnique – ENSA Toulouse, France. E-mail: mondher.bouzayen@ensat.fr

Mathilde Causse, INRA, UR1052 Génétique et Amélioration des Fruits et Légumes, F-84143 Montfavet, France. E-mail: mcausse@avignon.inra.fr

Kun-Song Chen, Laboratory of Fruit Quality Biology, Zhejiang University, Zijingang Campus, Hangzhou 310058, PR China. E-mail: akun@zju.edu.cn

Véronique Cheynier, INRA, UMR1083 Sciences pour l'œnologie 2, place Viala, F-34 060 Montpellier Cedex 01, France. E-mail: cheynier@supagro.inra.fr

Fabrizio Costa, Research and Innovation Centre, Fondazione Edmund Mach Via E. Mach 1, San Michele all'Adige, 38010, TN, Italy.

Abhaya Dandekar, Department of Plant Sciences, University of California, One Shields Avenue, Mail Stop 2, Davis, CA, 95616, USA. E-mail: amdandekar@ucdavis.edu

Christopher Davies, CSIRO Plant Industry, PO Box 350, Glen Osmond, SA 5064, Australia. E-mail: christopher.davies@csiro.au

Hiroshi Ezura, Faculty of Life and Environmental Sciences, University of Tsukuba, Tennodai 1-1-1, Tsukuba 305-8572, Japan. E-mail: ezura@gene.tsukuba.ac.jp

Alisdair R. Fernie, Max-Planck-Institut für Molekulare Pflanzenphysiologie, Am Mühlenberg 1, 14476 Potsdam-Golm, Germany. E-mail: fernie@mpimp-golm.mpg.de

Carine Ferrand, INRA, UMR 1332 Biologie du Fruit et Pathologie, 33140 Villenave d'Ornon, France, University of Bordeaux, UMR 1332, 33140 Villenave d'Ornon, France. E-mail: carine.ferrand@bordeaux.inra.fr

James Giovannoni, United States Department of Agriculture and Boyce Thompson Institute for Plant Research, Cornell University, Ithaca, NY 14853, USA. E-mail: jjg33@cornell.edu

Donald Grierson, Division of Plant & Crop Sciences, School of Biosciences, University of Nottingham, Sutton Bonington Campus, Loughborough, Leicestershire LE12 5RD, UK. E-mail: donald.grierson@nottingham.ac.uk

Avtar K. Handa, Department of Horticulture and Landscape Architecture, Purdue University, West Lafayette, IN 47907-2010, USA. E-mail: ahanda@purdue.edu

Kyoko Hiwasa-Tanase, Faculty of Life and Environmental Sciences, University of Tsukuba, Ten-nodai 1-1-1, Tsukuba 305-8572, Japan. E-mail: kytanase@gene.tsukuba.ac.jp

Angelos K. Kanellis, Aristotle University of Thessaloniki, Department of Pharmaceutical Sciences, 54124 Thessaloniki, Greece. E-mail: kanellis@pharm.auth.gr

Rahul Kumar, Repository of Tomato Genomic Resources, Department of Plant Sciences, School of Life Sciences, University of Hyderabad, Hyderabad 500046, India. E-mail: rahulpmb@gmail.com

Susheel Kumar, National Bureau of Plant Genetic Resources (ICAR), Pusa Campus, New Delhi 110012, India. E-mail: kumar.vishwakarma@gmail.com

Justin Lashbrooke, Institute for Wine Biotechnology, Stellenbosch University, Stellenbosch, South Africa. E-mail: jstnlshbrk@gmail.com

Alain Latche, Insitut National Polytechnique – ENSA Toulouse, France. E-mail: latche@ensat.fr

Surendra Kumar Malik, National Bureau of Plant Genetic Resources (ICAR), Pusa Campus, New Delhi 110012, India. E-mail: skm1909@gmail.com

George A. Manganaris, Cyprus University of Technology, Department of Agricultural Sciences, Biotechnology & Food Science, 3603 Lemesos, Cyprus. E-mail: george.manganaris@cut.ac.cy

Federico Martinelli, Dipartimento di Sistemi Agro-ambientali, Università degli Studi di Palermo, Viale delle Scienze, Palermo, 90128, Italy. E-mail: federico.martinelli@unipa.it

Autar K. Mattoo, Sustainable Agricultural Systems Laboratory, USDA-ARS, Beltsville Agricultural Research Center, Beltsville, MD 20705-2350, USA. E-mail: autar.mattoo@ars.usda.gov

Pravendra Nath, Plant Gene Expression Laboratory, CSIR-National Botanical Research Institute, Rana Pratap Marg, Lucknow 226001, India. E-mail: pravendranath@hotmail.com

Sonia Osorio, IHSM-UMA-CSIC, Departamento de Biología Molecular y Bioquímica, Universidad de Málaga, 29071 Málaga, Spain. E-mail: sosorio@uma.es

Jean Claude Pech, Institut National Polytechnique – ENSA Toulouse, France. E-mail: jean-claude.pech@ensat.fr

Bénédicte Quilot-Turion, INRA, UR1052 Génétique et Amélioration des Fruits et Légumes F-84143 Montfavet, France. E-mail: quilot@avignon.inra.fr

Christophe Rothan, INRA, UMR 1332 Biologie du Fruit et Pathologie, 33140 Villenave d'Ornon, France, University of Bordeaux, UMR 1332, 33140 Villenave d'Ornon, France. E-mail: christophe.rothan@bordeaux.inra.fr

Arun K. Sharma, Department of Plant Molecular Biology, University of Delhi South Campus, New Delhi 110021. India. E-mail: arun@genomeindia.org

Nancy Terrier, INRA, UMR1083 Sciences pour l'œnologie 2, place Viala, F-34 060 Montpellier Cedex 01, France. E-mail: terrier@supagro.inra.fr

Mark L. Tucker, Soybean Genomics and Improvement Lab, USDA/ARS, Building 006, BARC-West, 10300 Baltimore Avenue, Beltsville, MD 20705, USA. E-mail: mark.tucker@ars.usda.gov

Bo Zhang, Laboratory of Fruit Quality Biology, Zhejiang University, Zijingang Campus, Hangzhou 310058, PR China. E-mail: bozhang@zju.edu.cn

Preface

Fruit development and ripening represent the terminal phase of plant development. It is during this phase that fruits are enriched with sensory and nutritional quality attributes. The science of the basis of fruit ripening has progressed strikingly in recent years largely due to a number of breakthrough discoveries that have uncovered some of the key factors and signalling pathways by which ripening-related genes are set into motion. Fruits are a dietary source of vitamins, minerals and fibre, but, due to their short postharvest life, particularly in fleshy fruits, a large portion of the produce is lost. It is not, therefore, surprising that a major effort of public and private research in this area has been to find ways to control the perishability of fleshy fruit and thereby increasing their postharvest life span. In the past, such practices included fruit treatment with chemicals that inhibited ripening, use of plastic films to prevent loss of fruit moisture, and dip treatment with waxes and other adjuvants. Most of these treatments increased the shelf-life and therefore marketability of the fruit.

Advances in understanding the fruit-ripening process have led to the finding that a major effector of ripening in many fruits is a small, gaseous hydrocarbon, ethylene. Once the biosynthetic pathway of ethylene was elucidated, molecular genetics approaches established that suppression of ethylene biosynthesis genes prevented ripening of tomato fruit. Treatment of such genetically engineered fruit with exogenous ethylene made them ripen, providing definitive evidence that ethylene is a major catalyst of ripening in fruit. Since then, the field of fruit ripening has blossomed, and in recent years, the literature has been flooded with new information on the physiological, biochemical and genetic basis of ripening. These advances have been occurring rapidly, and there is no single treatise available that provides a one-stop source to gather them.

Although fruit softening and ethylene signalling have remained two major areas in fruit research, newer focal areas have generated considerable attention including nutritional value – particularly nutrients bearing potential health benefits. In this context, exploration of naturally occurring genetic variability and engineering key steps in metabolic pathways have allowed enhancement of specific nutrients to levels never before seen, thus making fruit research a new centre of interest for industries dealing in nutraceuticals and pharma products.

The primary aim of this book is to gather comprehensive and concise information on the advances in fruit research. It also aims to fill a long-felt need for a comprehensive treatise contributed to by world experts in each subdiscipline, covering various aspects of fruit development and ripening. Each chapter concisely describes the recent advances in our understanding of individual ripening pathways including those of cell walls, aroma volatiles and cuticle/waxes; micro-nutrients such as carotenoids, vitamins, other antioxidants and bioactive compounds; features of chromoplast development; and primary and secondary metabolism. Developments on the hormonal control of fruit ripening with a special focus on ethylene biosynthesis and responses are described. The potential role(s) of other phytohormones such as abscisic acid, auxin, gibberellins, brassinosteroids and the jasmonate family and their interaction with ethylene responses is also featured. Similarly, biogenic amines are shown to impact on fruit ripening. The involvement of key transcription factors and their interactions with the promoters of ripening-related genes provide a deeper insight into the molecular basis of fruit ripening.

The implementation of advanced genomics and post-genomics tools has opened up new perspectives in the elucidation of the complex network of interactions between the different signalling and regulatory elements, and has shown that the fruit-ripening process is under both genetic and epigenetic control. Building on recently available genome sequences for a number of fruit species and implementing 'omics' approaches offer new opportunities for addressing in an unprecedented way the mechanisms underlying fruit ripening and quality traits. Thus, chapters on ripening mutants, genomics/metagenomics and epigenetic control of ripening, emphasizing novel strategies, together with the subjects mentioned above and those on fruit biodiversity, the genetics of sensory quality and biotechnology, provide overall a present day meaning to diverse and complex processes that regulate the life of a fruit.

This book became a reality because of the willingness of experts to devote their valuable time to compiling recent advances in their fields and helped us bring out this treatise, together with the help of the publishers.

Pravendra Nath
Mondher Bouzyan
Autar K. Mattoo
Jean-Claude Pech
26 August 2013

1 Climacteric and Non-climacteric Ripening

Kyoko Hiwasa-Tanase and Hiroshi Ezura*
University of Tsukuba, Tsukuba, Japan

1.1 Introduction: The Role of Fruit Maturation and Ripening in Plants

Plants have developed a fruit architecture with the characteristics necessary to protect their seeds from the natural environment and disseminate those seeds. Seeds are frequently dispersed by animals as well as by wind and rain. Plants have made their fruit more attractive for surrounding animals by providing sources of energy and nutrition, thereby leading to successful seed spreading and propagation. Thus, fruit formation in plants is a reproductive strategy acquired during evolution, which has led to countless varieties of fruit types throughout the world.

Biologically, fruit is the framework that contains the seed in an angiosperm. Therefore, a fruit mainly consists of an ovary, including partial or whole carpel tissues. However, the edible part of a fleshy fruit also develops from various floral components, such as the receptacle, sepal and inflorescence. The fruit tissues usually ripen in concert with the maturation of the seed, regardless of their derivation, and attract animals by their colour and aroma. Not only do fruits stimulate the visual and olfactory senses of animals, but their sweet taste and juicy texture also induce repetitive eating. Differences in fruit shelf-lives that affect the area of seed dispersal may be an adaptive strategy. For example, fruit with a long shelf-life can be transported to more distant areas, thereby expanding the territory of the plant. In contrast, fruit with a short shelf-life are eaten in an area near the parent plants, and the seeds grow into buds that may create a dense community. Thus, fruit maturation and ripening are important phenomena that influence the dispersal of mature seeds.

Currently, consumable fruits are indispensable to the human diet from both a nutritional and an enjoyment standpoint. In terms of their economic importance, various studies of fruits have been performed, such as cultivation studies, breeding studies, environmental biology studies, morphological studies, physiological studies and nutritional studies. In particular, the dramatic change from immature to mature fruit has stimulated the interest of researchers, and studies have been performed extensively at the biochemical, physiological and molecular levels to understand better the ripening mechanism and to improve the quality and shelf-life of fruits, especially as it relates to their economic value.

* ezura@gene.tsukuba.ac.jp

In this chapter, we describe the differences and similarities among climacteric and non-climacteric fruits with respect to the different types of fruit maturation.

1.2 The Concept of Categories in Climacteric and Non-climacteric Fruit Ripening

Fruits generally display various biochemical and physiological modifications, including the loss of chlorophyll, the synthesis of pigments such as anthocyanins and carotenoids, increased aroma and flavour, alterations in the sugar and acid components, and softening, in association with fruit maturation. These phenomena vary significantly, depending on the ripening of different fruits. Classically, the types of fruit ripening are categorized into two groups: climacteric and non-climacteric. The classification of fruits depends on whether ripening-associated increased respiration occurs in the fruit (McMurchie *et al.*, 1972) (Fig. 1.1). Climacteric fruits are characterized by an increase in respiration with a concomitant and rapid production of ethylene at the initiation of ripening; the produced ethylene, which is a plant hormone, accelerates the ripening process (Leliévre *et al.*, 1997). In contrast, these changes do not occur in non-climacteric fruits, and maturation proceeds relatively slowly.

McMurchie *et al.* (1972) used the terms 'system I' and 'system II' to describe the differences in the ethylene biosynthesis pattern. Fruits continuously synthesize ethylene at a low level throughout development, even during the immature stage. The production of basal ethylene levels is called system I synthesis and occurs in both climacteric and non-climacteric fruits. The increased production of ethylene at the initiation of ripening is called system II synthesis and is observed only in climacteric fruits. System I ethylene production is inhibited by exogenous ethylene in an autoinhibitory manner; system I ethylene also functions when the plant responds to stress. In contrast, the production of system II ethylene is promoted by ethylene in an autocatalytic manner. Whether a transition from system I to system II ethylene biosynthesis is observed during the initiation of fruit ripening is one of the differences between climacteric and non-climacteric fruits (McMurchie *et al.*, 1972; Alexander and Grierson, 2002).

Typical climacteric fruits include apple, avocado, banana, pear, peach, melon and tomato, and typical non-climacteric fruits include strawberry, grape and citrus.

1.3 Characterization of Climacteric Fruit Ripening

The term 'climacteric' rise was first used by Kidd and West (1930) to describe the rapid increase in respiration that was observed at the end of maturation in apples, and 'climacteric' was originally defined as an augmentation of respiration. However, the definition of climacteric has changed over time and now usually refers not only to an increase in respiration but also to the rapid increase in ethylene production at the onset of fruit ripening. To date, the role of ethylene in the ripening mechanism has been studied extensively in climacteric fruits (Leliévre *et al.*, 1997; Alexander and Grierson, 2002; Barry and Giovannoni, 2007; Lin *et al.*, 2009).

1.3.1 Ethylene and climacteric fruit ripening

In climacteric fruits, the plant hormone ethylene plays an important role in the fruit-ripening process. Generally, a burst of ethylene production acts as an initiator of ripening and triggers the autocatalytic ethylene production process, thereby resulting in dramatic changes in colour, texture, aroma, flavour and other biochemical and physiological attributes of the fruit (Fig. 1.1). Moreover, ethylene production by the autocatalytic processes accelerates the ripening phenomena. Therefore, the shelf-life of climacteric fruits is shorter than that of non-climacteric fruits. A typical example is

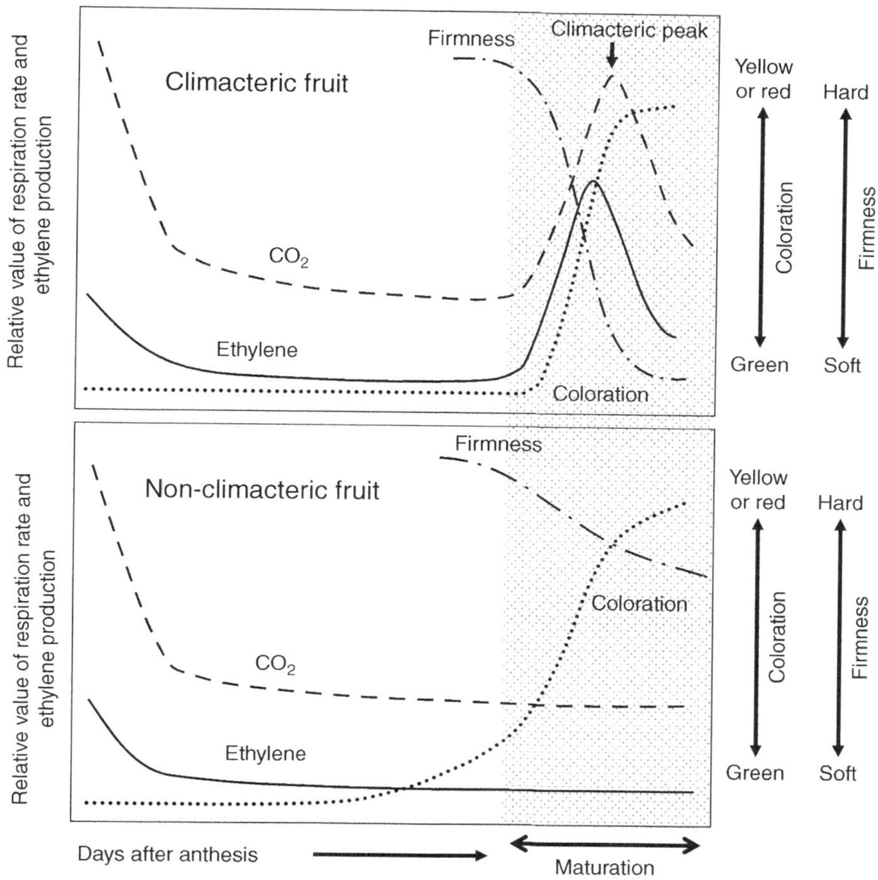

Fig. 1.1. Typical ripening pattern in climacteric and non-climacteric fruits.

demonstrated in the Charentais melon variety (*Cucumis melo* cv. *reticulatus* F1 Alpha), in which the ripening process from the pre-ripe to over-ripe stages occurs within 24–48 h (Rose *et al.*, 1998). In accordance with these dramatic changes, the expression of numerous ripening-related genes, including those involved in cell-wall breakdown and carotenoid biosynthesis, is induced and increased at the transcriptional and translational levels at the onset of ripening (Picton *et al.*, 1993; Gray *et al.*, 1994; Fei *et al.*, 2004).

Exposure to exogenous ethylene can also induce a rapid increase in autocatalytic ethylene production in climacteric fruits, even at the pre-climacteric stage, thereby accelerating the ripening process. In contrast, treatment with inhibitors of the action of ethylene can suppress and delay fruit maturation (Blankenship and Dole, 2003; Watkins, 2006, 2008). Similar results have been shown in a number of transgenic tomatoes in which ethylene production was reduced by the antisense-induced repression of the 1-aminocyclopropane-1-carboxylic acid (ACC) synthase (ACS) and ACC oxidase (ACO) genes, which encode ethylene biosynthesis enzymes; in addition, ethylene sensitivity was reduced by the modification of ethylene receptor genes (Hamilton *et al.*, 1990; Oeller *et al.*, 1991; Picton *et al.*, 1993; Wilkinson *et al.*, 1995).

These results confirm the pivotal role of ethylene in regulating the ripening of climacteric fruits.

To determine whether ethylene is necessary only for the initiation of ripening, the physiological role of ethylene in the initiation and subsequent progression of ripening has been investigated. When 1-methylcyclopropene (1-MCP), a gaseous inhibitor of ethylene action, was applied after the initiation of ethylene production in pear fruit, the subsequent flesh softening and ethylene production was suppressed (Hiwasa *et al.*, 2003). A similar suppression has been observed in melon fruit treated with 1-MCP during ripening (Nishiyama *et al.*, 2007). In the Charentais melon cultivar, ethylene production was inhibited by the introduction of the antisense gene of ACO, which encodes the ethylene synthesis enzyme. In this transgenic melon, the ripening parameters, including the degreening of the rind, flesh softening, detachment of the peduncle (Guis *et al.*, 1997) and aroma production (Bauchot *et al.*, 1998), were significantly suppressed by a reduction in ethylene production, and treatment with exogenous ethylene accelerated the ripening process (Flores *et al.*, 2001). These results indicate that ethylene is directly involved not only in the initiation of ripening but also in maintenance of the ripening process.

1.3.2 Ethylene-dependent and -independent ripening processes

The ripening processes in climacteric fruits are not necessarily regulated in an ethylene-dependent manner, although continuous ethylene action is necessary for adequate ripening. In melon fruit, flesh softening and membrane deterioration are not strictly ethylene-dependent processes and are regulated by both ethylene-dependent and -independent processes. In contrast, the detachment of the abscission zone and colour change of the rind, which includes chlorophyll degradation and an increase in yellow components such as carotenoids, are completely regulated by ethylene (Guis *et al.*, 1997; Flores *et al.*, 2001). In mountain papayas, the degreening of the skin and flesh softening are only partially dependent on ethylene, and the production of volatile aromatic compounds is strictly dependent on ethylene for certain esters, including ethyl acetate, and partially dependent on other esters, including butyl acetate (Moya-León *et al.*, 2004; Balbontín *et al.*, 2007). Moreover, pear fruit softening is completely suppressed by the absence of ethylene, whereas the softening process in melon and papaya shows only a partial dependence on ethylene (Hiwasa *et al.*, 2003). Thus, the ripening of climacteric fruits proceeds under both developmental and ethylene-related regulation, and the degree of dependence on ethylene varies with respect to the individual ripening parameter and the diversity of the species and cultivars.

The threshold values for the required ethylene levels are known to be different for each ripening parameter. For example, ethylene-dependent flesh softening is not induced by exogenous treatment with less than 1 ppm. of ethylene in an ethylene-suppressed transgenic melon but is induced by 2.5 ppm. of ethylene, which is sufficient to produce softening similar to that observed in wild-type melons (Flores *et al.*, 2001). The abscission zone detachment and rind degreening processes are induced by 2.5 and 1 ppm. of ethylene, respectively, and the extent of change is concentration dependent up to 5 ppm. Furthermore, at the molecular level, the expression level of each ethylene-inducible gene, such as *E4*, *E8*, *E17* and *J49*, has a unique dose–response curve that is dependent on the exogenous ethylene concentration from 0.11 to 72 ppm. (Lincoln and Fischer, 1988). The expression peak varies between 0.75 and 23 ppm. These data show that individual ethylene-dependent processes have distinct sensitivities to ethylene, and that the complicated ripening processes of climacteric fruits are intricately controlled by these differences in ethylene sensitivity.

1.3.3 Characterization of non-climacteric fruit ripening

In non-climacteric fruits, such as strawberry, grape, citrus, pineapple and cherry, the maturation and ripening process progresses without a burst of ethylene production or an increase in respiration, which is in contrast to that of climacteric fruits (Fig. 1.1). Generally, ethylene is not required for the ripening process (Given *et al.*, 1988; Pretel *et al.*, 1995). None the less, non-climacteric fruits have the capacity to produce a certain level of endogenous ethylene and to respond to exogenous ethylene. Recently, the possibility that ethylene is involved in the ripening process of non-climacteric fruit ripening has been proposed.

In mature strawberry, exogenous ethylene applied after harvesting does not accelerate fruit ripening (Perkins-Veazie *et al.*, 1996), and ethylene and 1-MCP treatments do not affect its storage life (Bower *et al.*, 2003). Recently, however, experiments have suggested the involvement of ethylene in strawberry fruit ripening. The respiration rate in strawberry fruit was stimulated by ethylene in a dose-dependent manner and resulted in a slight acceleration of the coloration and flesh softening (Tian *et al.*, 2000). Iannetta *et al.* (2006) investigated ethylene production and CO_2 levels from the flowering stage to fruit maturity in strawberry. An increase in ethylene levels was observed at the expanded pale-green and red-ripe stages, and an upsurge in the CO_2 level was also detected in association with the ethylene burst at the red-ripe stage. Moreover, ethylene production was regulated by negative feedback until the fruit was expanded and by positive feedback when it was red-ripened. Trainotti *et al.* (2005) showed that the expression level of the ethylene receptor increased in response to the increase in ethylene levels in strawberry fruit, thereby suggesting the possibility that a low level of ethylene is sufficient to trigger the ripening process. These findings support the possible role of ethylene in fruit ripening and/or senescence in strawberry fruit.

Grape has been classified as a non-climacteric fruit due to the lack of an obvious increase in ethylene production and concomitant increase in respiration at the veraison stage during which the berry size and concentrations of anthocyanin and sugars start to increase dramatically as the fruit grows (Coombe and Hale, 1973). However, the levels of endogenous ethylene display a relatively small peak prior to the veraison stage, and the application of ethylene leads to an increase in this low-level peak (Coombe and Hale, 1973; El-Kereamy *et al.*, 2003; Chervin *et al.*, 2004). Exogenous ethylene treatment at the veraison stage also hastens the fruit coloration and cell growth, increases the internal ethylene content (Coombe and Hale, 1973; El-Kereamy *et al.*, 2003) and stimulates the expression of genes related to anthocyanin biosynthesis as well as the alcohol dehydrogenase gene (El-Kereamy *et al.*, 2003; Tesniere *et al.*, 2004). Moreover, the application of 1-MCP prior to the veraison stage suppresses the acidity decrease, cell enlargement and anthocyanin accumulation (Chervin *et al.*, 2004).

The exposure of citrus fruits to exogenous ethylene accelerates the respiration rate and stimulates chlorophyll degradation (Purvis and Barmore, 1981) and carotenoid biosynthesis (Stewart and Wheaton, 1972). Furthermore, treatments with the ethylene antagonists 2,5-norbornadiene and silver nitrate prevented the degreening process (Goldschmidt *et al.* 1993). Katz *et al.* (2004) reported that young citrus fruitlets were able to synthesize ethylene in an autocatalytic manner similar to system II in climacteric fruits, although the mature fruit did not respond to exogenous ethylene with the exception of the degreening of the skin. In the Nijisseiki Japanese pear cultivar, a non-climacteric fruit, the continuous exposure to propylene, which is an analogue of ethylene, hastened skin degreening but did not induce ethylene production or climacteric respiration (Downs *et al.*, 1991). In conclusion, these aspects of non-climacteric fruit ripening may underscore the role of ethylene in certain non-climacteric fruits and various ripening processes.

1.4 Involvement of Other Phytohormones in Fruit Ripening

Abscisic acid (ABA) is a plant hormone that plays roles in the regulation of plant growth and development, seed dormancy and the adaptation to various stresses (e.g. cold and osmotic stress). However, its involvement in fruit ripening is also considered possible because an increase in ABA content is found during and/or preceding fruit ripening in both climacteric (Vendrell and Buesa, 1989; Chernys and Zeevaart, 2000; Zhang *et al.*, 2009a,b) and non-climacteric (Inaba *et al.*, 1976; Kondo and Tomiyama, 1998; Wheeler *et al.*, 2009; Jia *et al.*, 2011) fruits. In addition, the application of ABA accelerated fruit coloration and softening (Inaba *et al.* 1976; Jiang and Joyce, 2003; Wheeler *et al.*, 2009; Zhang *et al.*, 2009a,b).

In an ABA-deficient mutant in orange fruit, the coloration of the fruit skin was delayed compared with that of the wild type during maturation (Rodrigo *et al.*, 2003). This phenotype was overcome by the application of ABA. In contrast, the response to exogenous ethylene and the ethylene inhibitor 2,5-norbornadiene was weaker than that of the wild-type fruit (Alferez and Zacarias, 1999). These results suggest that both ethylene and ABA are involved in the coloration of citrus fruits and have a complex correlation.

The function of ABA as a ripening factor has recently been emphasized in several fruits. Zhang *et al.* (2009b) showed that endogenous ABA accumulated prior to the ethylene burst in tomato fruit, and exogenous ABA treatment promoted ethylene synthesis and fruit ripening. However, the application of fluridone or nordihydroguaiaretic acid (NDGA), which are inhibitors of ABA, suppressed the ripening processes, including fruit coloration and softening. In addition to the aforementioned results, they demonstrated four phenomena: (i) ABA treatment led to increases in both ABA content and ethylene production, although ethylene treatment did not produce an increase in ABA content; (ii) the application of NDGA suppressed the increase in ABA content at fruit ripening, although the suppression was not clearly correlated with ethylene production; (iii) treatment with a combination of ABA and 1-MCP inhibited ethylene production but did not affect ABA content; and (iv) the application of NDGA or 1-MCP inhibited fruit ripening and softening. Sun *et al.* (2012a) also demonstrated that expression of genes encoding cell-wall degradation enzymes was controlled by both ABA and ethylene. Taken together, these findings support the hypothesis that ABA might act as an upstream regulator prior to ethylene production in the fruit-ripening process, and suggest that the presence and perception of both ABA and ethylene is important for normal ripening in tomato fruit (Fig. 1.2). This model is supported by studies in other non-climacteric (grape) and climacteric (peach) fruits (Zhang *et al.*, 2009a).

Additional molecular mechanisms have been demonstrated based on evidence that 9-*cis*-epoxycarotenoid dioxygenase (NCED), a key enzyme in the biosynthesis of ABA, is involved in the ripening of several fruits, including avocado (Chernys and Zeevaart, 2000), orange (Rodrigo *et al.*, 2003), peach, grape, tomato (Zhang *et al.*, 2009a,b) and strawberry (Jia *et al.*, 2011). In tomato fruit (*Solanum lycopersicum*), the suppression of *SlNCED1* led to the inhibition of fruit softening and extended the shelf-life by up to four times compared with that of the wild type (Sun *et al.*, 2012a). Moreover, in strawberry fruit (*Fragaria × ananassa*), the downregulation of *FaNCED1* resulted in an uncoloured phenotype, which was rescued by exogenous ABA (Jia *et al.*, 2011). However, the suppression of *SlNCED1* was not sufficient to inhibit fruit ripening as measured by coloration and the ethylene production rate. In fact, coloration and ethylene production in the *SlNCED1*-suppressed fruit were enhanced relative to

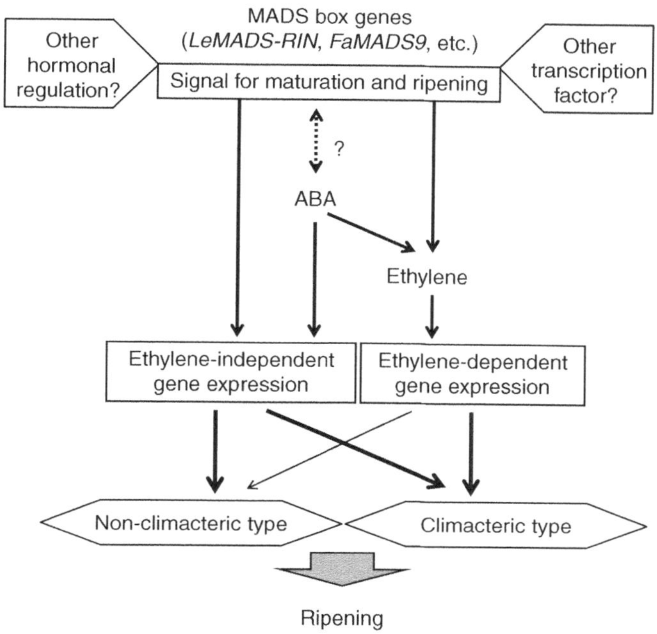

Fig. 1.2. Summary of the ripening regulation in climacteric and non-climacteric fruits.

the control fruit (Sun *et al.*, 2012a,b). A similar enhancement of ethylene production was observed in an ABA-deficient mutant in orange fruit (Alferez and Zacarias, 1999). These results suggest that ABA plays a partial role in fruit ripening and ethylene production or that the remaining ABA is sufficient to trigger ethylene synthesis.

Brassinosteroids (BRs) are steroidal plant hormones that are essential for normal plant development and affect processes such as cell elongation, cell division, vascular differentiation, reproductive development, and pathogen and abiotic tolerance (Clouse, 2002). The involvement of BRs in fruit ripening remains unclear. However, Symons *et al.* (2006) reported that endogenous BR levels increase in association with ripening in grape, and the ripening process is promoted by exogenous BRs and delayed by an inhibitor of BR biosynthesis. Exogenous BRs promote the ripening processes in the same manner in tomato (Vardhini and Rao, 2002). Although the evidence implicating a role of BRs in fruit ripening has been inconclusive, a possible role for BRs in fruit ripening may be revealed by further studies.

Auxin has been proposed to function as a negative regulator of fruit ripening in a number of climacteric and non-climacteric fruits, including avocado (Tingwa and Young, 1975), pear (Frenkel and Dyck, 1973), grape (Coombe and Hale, 1973; Davies *et al.*, 1997) and strawberry (Given *et al.*, 1988). For example, exogenous auxin treatments delay the onset of ripening in these fruits. In strawberry, the expression levels of genes for polygalacturonase and expansin, which are related to cell-wall degradation, were shown to be repressed by auxin (Villarreal *et al.*, 2008; Figueroa *et al.*, 2009). In grape, auxin treatment delays the expression of a number of ripening-related genes (Davies *et al.*, 1997).

Similar to auxin, gibberellin (GA) has been implicated to have a negative regulatory role in the ripening process. Certain studies have described the function of GA in fruit ripening. In strawberry, GA has inhibitory effects on fruit ripening, including decreases in respiratory activity, anthocyanin synthesis and chlorophyll degradation (Martínez *et al.*, 1994). Additionally, GA treatment delays softening in persimmon fruit (Ben-Arie *et al.*, 1996).

Historically, ethylene has been considered the most crucial factor for ripening initiation and the ripening process, particularly in climacteric fruits. Even today, this perception remains unchanged. However, fruit ripening would not proceed normally without the effect of other phytohormones, such as ABA, BR, auxin and GA. The evidence reviewed here indicates that the ripening process is regulated by the coordinated actions of multiple phytohormones in a complex relationship in both climacteric and non-climacteric fruits.

1.5 Distinctions Between Climacteric and Non-climacteric Ripening

The cumulative experimental evidence published to date has obscured the distinctions between climacteric and non-climacteric ripening. For example, there are a number of species in which the fruits of different varieties and cultivars exhibit both climacteric and non-climacteric behaviours (Barry and Giovannoni, 2007). In particular, apple (Wang *et al.*, 2009), Japanese pear (Downs *et al.*, 1991; Itai *et al.*, 1999), Chinese pear (Yamane *et al.*, 2007) and melon (Zheng and Wolff, 2000) have numerous natural variations related to ripening, ethylene production and shelf-life.

The storage properties of apple fruit (*Malus domestica*) vary substantially depending on the variety. Certain varieties can be stored for up to a year under optimal conditions, whereas others rapidly deteriorate. This shelf-life characteristic is closely related to the level of ethylene production (Gussman *et al.*, 1993). Several allelic forms of the *MdACS1* and *MdACS3* genes that encode the ripening-related ACS, which is a limiting enzyme for the production of ethylene, were identified in apple fruit (Sunako *et al.*, 1999; Wang *et al.*, 2009). Isolated alleles modified the function of this enzyme at the transcriptional level by a DNA insertion in the 5′ flanking region or null mutation, or at the activation level by a single-nucleotide polymorphism in the coding region. Wang *et al.* (2009) indicated that the ethylene production level and shelf-life in apple are determined by the combination of allelotypes of *MdACS1* and *MdACS3*.

In Chinese pear (*Pyrus bretschneideri*), the alleles *PbACS1A* and *PbACS1B* caused by a single-nucleotide polymorphism and sequence insertion in the promoter region were observed (Yamane *et al.*, 2007). The homozygous *PbACS1A* and heterozygous *PbACS1B* lines exhibited climacteric characteristics, whereas the homozygous *PbACS1B* line showed non-climacteric characteristics. Based on restriction fragment length polymorphism analysis in Japanese pears, Itai *et al.* (1999) indicated that the fruit cultivars that produce higher levels of ethylene possess an additional *PPACS1* allele, those that produce moderate levels of ethylene possess an additional *PPACS2* allele and non-climacteric-type fruit did not have additional versions of these alleles. The involvement of the *ACO* allele in controlling ethylene production levels was demonstrated in melon fruit using restriction fragment length polymorphism analysis: the A_0 allele was related to a high level of ethylene production, whereas the B_0 allele was related to a low level of ethylene production (Zheng *et al.*, 2002).

Périn *et al.* (2002) generated a melon with climacteric characteristics by crossing a typical climacteric-type Charentais melon (*C. melo* var. *cantalupensis* cv. Vedrantais) with a non-climacteric melon, PI161375 (*C. melo* var. *chinensis*). A genetic analysis indicated that the climacteric characteristics described by fruit abscission and ethylene production were controlled by two independent loci (*Al-3* and *Al-4*). Similarly, a cross between the climacteric-type Charentais and the non-climacteric-type Honeydew (*C. melo* var. *inodorus*) melon led to the generation of a climacteric melon in a different study (Ezura *et al.*, 2002). Climacteric-type melons have also been created from two non-climacteric melons, Piel de Sapo (var. *inodorus*) and PI161375 (var. *chinensis*) (Moreno *et al.*, 2008). However, the genetic position of the quantitative trait loci associated with ethylene production and respiration rates in this study did not coincide with the *Al* loci described by Périn *et al.* (2002). These results suggest, at a minimum, that the climacteric trait is dominant and not determined by only a single locus.

In conclusion, non-climacteric-type fruits derived from species that have both climacteric and non-climacteric varieties in the same genus may be generated by the loss or repression of function of fruit ripening-related gene(s). Conversely, fruits that are classically categorized as non-climacteric, such as strawberry, grape and citrus, have not been found to show these variations in ethylene production levels, although they do show variations at low ethylene levels. A concrete distinction between climacteric and non-climacteric types may be formulated based on these results, whereas novel data suggest the involvement of ethylene in classical non-climacteric fruits.

1.5.1 Common programmes of climacteric and non-climacteric ripening

Classically, fleshy fruits have been classified into two categories according to their dependence on ethylene for fruit ripening: climacteric and non-climacteric. However, the ripening phenomena of both classes, including changes in colour, texture and flavour, have much in common. As a common point, the ripening processes in climacteric fruits are controlled by an ethylene-independent pathway similar to that in non-climacteric fruits in parallel with an ethylene-dependent pathway (see Section 1.3.2 for further details). In certain non-ripening tomato mutants, such as *ripening-inhibitor* (*rin*), *non-ripening* (*nor*) and *Colourless non-ripening* (*Cnr*) mutants, the inhibited ripening phenomena are not restored by exogenous ethylene, although ethylene-regulated gene expression can be partially restored (Griffiths *et al.*, 1999; Thompson *et al.*, 1999; Barry *et al.*, 2000); i.e. these mutants do not lose functional ethylene perception and signalling. Conversely, these observations suggest that the mutations in *rin*, *nor* and *Cnr* affect the ripening cascade upstream of ethylene and are involved in both ethylene-dependent and -independent ripening pathways.

In the *rin* mutant tomato, *LeMADS-RIN*, a *SEPALLATA* (*SEP*)-like MADS-box transcription factor, was identified at the *rin* locus (Vrebalov *et al.*, 2002). Genetic complementation and antisense experiments confirmed that *LeMADS-RIN* acts both upstream of the ethylene cascade and in an ethylene-independent pathway. Data from chromatin immunoprecipitation studies have revealed that RIN interacts with the promoter of ripening-related genes – including the transcriptional control network involved in the overall regulation of ripening, ethylene biosynthesis, ethylene perception (downstream of the ethylene response), cell-wall metabolism and carotenoid biosynthesis – and directly controls these processes at the transcriptional level during ripening (Ito *et al.*, 2008; Fujisawa *et al.*, 2011; Martel *et al.*, 2011). Intriguingly, the *FaMADS9* transcription factor identified in strawberry had a similar biological function to that of *LeMADS-RIN*, as antisense strawberry fruit led to the inhibition of normal development and ripening of the petal, achene and receptacle tissues (Seymour *et al.*, 2011). Taken together, these results indicate that *RIN*-like genes play a central role as master regulators of the ripening cascade in both climacteric and non-climacteric fruits (Fig. 1.2).

1.6 Conclusions

The advancement of experimental techniques and tools has elucidated many fruit-ripening mechanisms at the physiological, genetic and molecular levels. The data summarized here provide novel insights into the ripening mechanism based on many new discoveries. The unknown factors and new queries that have emerged recently will help clarify the mechanism.

A relatively low-level peak in ethylene production in non-climacteric fruits was detected recently using continuous high-precision measurements in strawberry and grape. These discoveries suggest a relationship between ethylene and fruit ripening in fruits classically categorized as non-climacteric. However, the concrete function and involvement of ethylene in the ripening process remains unknown.

The transition from system I to system II at the onset of ripening is based on the participation of different *ACS* genes, although the regulation of *ACS* differs in the two systems (i.e. both negative- and positive-feedback mechanisms have been observed) (Barry *et al.*, 2000; Yokotani *et al.*, 2009). ABA may play a role in the transition stage because its accumulation is induced prior to the system II ethylene burst. Alternatively, changes in the accumulation or balance of ABA and/or other hormones may serve as a trigger for system II. Moreover, it is not clear whether the signal for the initiation of system II originates from the seed or is derived from another factor.

The discovery of RIN revealed the master regulator of the fruit-ripening process. The ripening cascade is considered to be a common pathway in the fruit-ripening process regardless of the fruit category. Additionally, various transcription factors have been identified by an exhaustive analysis, and their involvement in the ripening process has been discussed (Manning *et al.*, 2006; Lin *et al.*, 2008; Vrebalov *et al.*, 2009; Martel *et al.*, 2011). However, the key signal for the activation of these transcription factors has not been determined.

Many questions remain to be addressed within the field of fruit ripening, as described above. In the future, additional experimental approaches derived from various research areas will be required to answer these complex questions.

References

Alexander, L. and Grierson, D. (2002) Ethylene biosynthesis and action in tomato; a model for climacteric fruit ripening. *Journal of Experimental Botany* 53, 2039–2055.

Alferez, F. and Zacarias, L. (1999) Interaction between ethylene and abscisic acid in the regulation of citrus fruit maturation. In: Kanellis, A.K., Chang, C., Klee, H., Blecker, A.B., Pech, J.C. and Grierson, D. (eds) *Biology and Biotechnology of the Plant Hormone Ethylene II*. Kluwer Academic Publishers, Amsterdam, the Netherlands, pp. 183–184.

Balbontín, C., Gaete-Eastman, C., Vergara, M., Herrera, R. and Moya-León, M.A. (2007) Treatment with 1-MCP and the role of ethylene in aroma development of mountain papaya fruit. *Postharvest Biology and Technology* 43, 67–77.

Barry, C.S. and Giovannoni, J.J. (2007) Ethylene and fruit ripening. *Journal of Plant Growth Regulation* 26, 143–159.

Barry, C.S., Llop-Tous, M.I. and Grierson, D. (2000) The regulation of 1-aminocyclopropane-1-carboxylic acid synthase gene expression during the transition from system-1 to system-2 ethylene synthesis in tomato. *Plant Physiology* 123, 979–986.

Bauchot, A.D., Mottram, D.S., Dodson, A.T. and John, P. (1998) Effect of aminocyclopropane-1-carboxylic acid oxidase antisense gene on the formation of volatile esters in cantaloupe Charentais melon (cv. Vedrantais). *Journal of Agricultural and Food Chemistry* 46, 4787–4792.

Ben-Arie, R., Saks, Y., Sonego, L. and Frank, A. (1996) Cell wall metabolism in gibberellin-treated persimmon fruits. *Plant Growth Regulation* 19, 25–33.

Blankenship, S.M. and Dole, J.M. (2003) 1-Methylcyclopropene: a review. *Postharvest Biology and Technology* 28, 1–25.

Bower, J.H., Biasi, W.V. and Mitcham, E.J. (2003) Effects of ethylene and 1-MCP on the quality and storage life of strawberries. *Postharvest Biology and Technology* 28, 417–423.

Chernys, J.T. and Zeevaart, J.A.D. (2000) Characterization of the 9-*cis*-epoxycarotenoid dioxygenase gene family and the regulation of abcisic acid biosynthesis in avocado. *Plant Physiology* 124, 343–353.

Chervin, C., El-Kereamy, A., Roustan, J.P., Latché, A., Lamon, J. and Bouzayen, M. (2004) Ethylene seems required for the berry development and ripening in grape, a non-climacteric fruit. *Plant Science* 167, 1301–1305.

Clouse, S.D. (2002) Brassinosteroids: plant counterparts to animal steroids hormones? *Vitamins and Hormones* 65, 195–223.

Coombe, B.G. and Hale, C.R. (1973) The hormone content of ripening grape berries and the effect of growth substance treatments. *Plant Physiology* 51, 629–634.

Davies, C., Boss, P.K. and Robinson, S.P. (1997) Treatment of grape berries, a nonclimacteric fruit with a synthetic auxin, retards ripening and alters the expression of developmentally regulated genes. *Plant Physiology* 115, 1155–1161.

Downs, C.G., Brady, C.J., Campbell, J. and McGlasson, W.B. (1991) Normal ripening cultivars of *Pyrus serotina* are either climacteric or non-climacteric. *Scientia Horticulture* 48, 213–221.

El-Kereamy, A., Chervin, C., Roustan, J.P, Cheynier, V., Souquet, J.M., Moutounet, M., Raynal, J., Ford, C., Latché A., Pech, J.C. and Bouzayen, M. (2003) Exogenous ethylene stimulates the long-term expression of genes related to anthocyanin biosynthesis in grape berries. *Physiologia Plantarum* 119, 175–182.

Ezura, H., Akashi, Y., Kato, K. and Kuzuya, M. (2002) Genetic characterization of long shelf-life in honeydew (*Cucumis melo* var. *inodorus*) melon. *Acta Horticulture* 588, 369–372.

Fei, Z.J., Tang, X., Alba, R.M., White, J.A., Ronning, C.M., Martin, G.B., Tanksley, S.D. and Giovannoni, J.J. (2004) Comprehensive EST analysis of tomato and comparative genomics of fruit ripening. *Plant Journal* 40, 47–59.

Figueroa, C.R., Pimentel, P., Dotto, M.C., Civello, P.M., Martínez, G.A., Herrera, R. and Moya-León, M.A. (2009) Expression of five expansin genes during softening of *Fragaria chiloensis* fruit: effect of auxin treatment. *Postharvest Biology and Technology* 53, 51–57.

Flores, F., Ben Amor, M., Jones, B., Pech, J.C., Bouzayen, M., Latché, A. and Romojaro, F. (2001) The use of ethylene-suppressed lines to assess differential sensitivity to ethylene of the various ripening pathways in Cantaloupe melons. *Physiologia Plantarum* 113, 128–133.

Frenkel, C. and Dyck, R. (1973) Auxin inhibition of ripening in Bartlett pears. *Plant Physiology* 51, 6–9.

Fujisawa, M., Nakano, T. and Ito, Y. (2011) Identification of potential target genes for the tomato fruit-ripening regulator RIN by chromatin immunoprecipitation. *BMC Plant Biology* 11, 26.

Given, N.K., Venis, M.A. and Grierson, D. (1988) Hormonal regulation of ripening in the strawberry, a non-climacteric fruit. Planta 174, 402–406.

Goldschmidt, E.E., Huberman, M. and Goren, R. (1993) Probing the role of endogenous ethylene in the degreening of citrus fruit with ethylene antagonists. *Plant Growth Regulation* 12, 325–329.

Gray, J.E., Picton, S., Giovannoni, J.J. and Grierson, D. (1994) The use of transgenic and naturally occurring mutants to understand and manipulate tomato ripening. *Plant, Cell and Environment* 17, 557–571.

Griffiths, A., Barry, C., Alpuche-Solis, A. and Grierson, D. (1999) Ethylene and developmental signals regulate expression of lipoxygenase genes during tomato fruit ripening. *Journal of Experimental Botany* 50, 793–798.

Guis, M., Botondi, R., Ben-Amor, M., Ayub, R., Bouzayen, M., Pech, J.C. and Latché, A. (1997) Ripening-associated biochemical traits of cantaloupe Charentais melons expressing an antisense ACC oxidase transgene. *Journal of the American Society for Horticultural Science* 122, 748–751.

Gussman, C.D., Goffreda, J.C. and Gianfagna, T.J. (1993) Ethylene production and fruit-softening rates in several apple fruit ripening variants. *Hortscience* 28, 135–137.

Hamilton, A.J., Lycett, G.W. and Grierson, D. (1990) Antisense gene that inhibits synthesis of the hormone ethylene on transgenic plants. *Nature* 346, 284–287.

Hiwasa, K., Kinugasa, Y., Amano, S., Hashimoto, A., Nakano, R., Inaba, A. and Kubo, Y. (2003) Ethylene is required for both the initiation and progression of softening in pear (*Pyrus communis* L.) fruit. *Journal of Experimental Botany* 54, 771–779.

Iannetta, P.P.M., Laarhoven, L.J., Medina-Escobar, N., James, E.K., McManus, M.T., Davies, H.V. and Harren, F.J.M. (2006) Ethylene and carbon dioxide production by developing strawberries show a correlative pattern that is indicative of ripening climacteric fruit. *Physiologia Plantarum* 127, 247–259.

Inaba, A., Ishida, M. and Sobajima, Y. (1976) Changes in endogenous hormone concentration during berry development in relation to the ripening of Delaware grape. *Journal of the Japanese Society for Horticultural Science* 45, 245–252.

Itai, A., Kawata, T., Tanabe, K., Tamura, F., Uchiyama, M., Tomomitsu, M. and Shiraiwa, N. (1999) Identification of 1-aminocyclopropane-1-carboxylic acid synthase genes controlling the ethylene level of ripening fruit in Japanese pear (*Pyrus pyrifolia* Nakai). *Molecular and General Genetics* 261, 42–49.

Ito, Y., Kitagawa, M., Ihashi, N., Yabe, K., Kimbara, J., Yasuda, J., Ito, H., Inakuma, T., Hiroi, S. and Kasumi, T. (2008) DNA-binding specificity, transcriptional activation potential, and the *rin* mutation effect for the tomato fruit-ripening regulator RIN. *Plant Journal* 55, 212–223.

Jia, H.F., Chai, Y.M., Li, C.L., Lu, D., Luo, J.J., Qin, L. and Shen, Y.Y. (2011) Abscisic acid plays an important role in the regulation of strawberry fruit ripening. *Plant Physiology* 157, 188–199.

Jiang, Y. and Joyce, D.C. (2003) ABA effects on ethylene production, PAL activity, anthocyanin and phenolic contents of strawberry fruit. *Plant Growth Regulation* 39, 171–174.

Katz, E., Lagunes, P.M., Riov, J., Weiss, D. and Goldschmidt, E.E. (2004) Molecular and physiological evidence suggests the existence of a system II-like pathway of ethylene production in non-climacteric *Citrus* fruit. *Planta* 219, 243–252.

Kidd, F. and West, C. (1930) Physiology of fruit. I. Changes in the respiratory activity of apples during their senescence at different temperatures. *Proceedings of the Royal Society B* 106, 93–109.

Kondo, S. and Tomiyama, A. (1998) Changes of free and conjugated ABA in the fruit of 'Satohnishiki' sweet cherry and the ABA metabolism after application of (s)-(+)-ABA. *Journal of Horticultural Science and Biotechnology* 73, 467–472.

Leliévre, J.M., Latché, A., Jones, B., Bouzayen, M. and Pech, J.C. (1997) Ethylene and fruit ripening. *Physiologia Plantarum* 101, 727–739.

Lin, Z., Hong, Y., Yin, M., Li, C., Zhang, K. and Grierson, D. (2008) A tomato HD-Zip homeobox protein, LeHB-1, plays an important role in floral organogenesis and ripening. *Plant Journal* 55, 301–310.

Lin, Z., Zhong, S. and Grierson, D. (2009) Recent advances in ethylene research. *Journal of Experimental Botany* 60, 3311–3336.

Lincoln, J.E. and Fischer, R.L. (1988) Diverse mechanisms for the regulation of ethylene-inducible gene expression. *Molecular and General Genetics* 212, 71–75.

Manning, K., Tör, M., Poole, M., Hong, Y., Thompson, A.J., King, G.J., Giovannoni, J.J. and Seymour, G.B. (2006) A naturally occurring epigenetic mutation in a gene encoding an SBP-box transcription factor inhibits tomato fruit ripening. *Nature Genetics* 38, 948–952.

Martel, C., Vrebalov, J. and Giovannoni, J.J. (2011) The tomato MADS-box transcription factor RIPENING INHIBITOR interacts with promoters involved in numerous ripening processes in a COLORLESS NONRIPENING-dependent manner. *Plant Physiology* 157, 1568–1579.

Martínez, G.A., Chaves, A.R. and Añón, M.C. (1994) Effect of gibberellic acid on ripening of strawberry fruits (*Fragaria annanassa* Duch.). *Journal of Plant Growth Regulation* 13, 87–91.

McMurchie, E.J., McGlasson, W.B. and Eaks, I.L. (1972) Treatment of fruit with propylene gives information about the biogenesis of ethylene. *Nature* 237, 235–236.

Moreno, E., Obando, J.M., Dos-Santos, N., Fernández-Trujillo, J.P., Monforte, A.J. and Garcia-Mas, J. (2008) Candidate genes and QTLs for fruit ripening and softening in melon. *Theoretical and Applied Genetics* 116, 589–602.

Moya-León, M.A., Moya, M. and Herrera, R. (2004) Ripening of mountain papaya (*Vasconcellea pubescens*) and ethylene dependence of some ripening events, *Postharvest Biology and Technology* 34, 211–218.

Nishiyama, K., Guis, M., Rose, J.K.C., Kubo, Y., Bennett, K.A., Wangjin, L., Kato, K., Ushijima, K., Nakano, R., Inaba, A., Bouzayen, M., Latche, A., Pech, J.C. and Bennett, A.B. (2007) Ethylene regulation of fruit softening and cell wall disassembly in Charentais melon. *Journal of Experimental Botany* 58, 1281–1290.

Oeller, P.W., Lu, M.W., Taylor, L.P., Pike, D.A. and Theologis, A. (1991) Reversible inhibition of tomato fruit senescence by antisence RNA. *Science* 254, 437–439.

Périn, C., Gomez-Jimenez, M.C., Hagen, L., Dogimont, C., Pech, J.C., Latché, A., Pitrat, M. and Leliévre, J.M. (2002) Molecular and genetic characterisation of a non-climacteric phenotype in melon reveals two loci conferring altered ethylene response in fruit. *Plant Physiology* 129, 209–300.

Perkins-Veazie, P.M., Huber, D.J. and Brecht, J.K. (1996) *In vitro* growth and ripening of strawberry fruit in the presence of ACC, STS or propylene. *Annals of Applied Biology* 128, 105–116.

Picton, S., Barton, S.L., Bouzayen, M., Hamilton, A.J. and Grierson, D. (1993) Altered fruit ripening and leaf senescence in tomatoes expressing an antisense ethylene-forming enzyme transgene. *Plant Journal* 3, 469–481.

Pretel, M.T., Serrano, M., Amoros, A., Riquelme, F. and Romojaro, F. (1995) Non-involvement of ACC and ACC oxidase activity in pepper fruit ripening. *Postharvest Biology and Technology* 5, 295–302.

Purvis, A.C. and Barmore, C.R. (1981) Involvement of ethylene in chlorophyll degradation in peel of citrus fruit. *Plant Physiology* 68, 854–856.

Rodrigo, M.J., Marcos, J.F., Alferez, F., Mallent, M. and Zacarias, L. (2003) Characterization of Pinalate, a novel *Citrus sinensis* mutant with a fruit-specific alteration that results in yellow pigmentation and decreased ABA content. *Journal of Experimental Botany* 54, 727–738.

Rose, J.K.C., Hadfield, K.A., Labavitch, J.M. and Bennett, A.B. (1998) Temporal sequence of cell wall disassembly in rapidly ripening melon fruit. *Plant Physiology* 117, 345–361.

Seymour, G.B., Ryder, C.D., Cevik, V., Hammond, J.P., Popovich, A., King, G.J., Vrebalov, J., Giovannoni, J.J. and Manning, K. (2011) A *SEPALLATA* gene is involved in the development and ripening of strawberry (*Fragaria* × *ananassa* Duch.) fruit, a non-climacteric tissue. *Journal of Experimental Botany* 62, 1179–1188.

Stewart, I. and Wheaton, T.A. (1972) Carotenoids in citrus: their accumulation induced by ethylene. *Journal of Agricultural and Food Chemistry* 20, 448–449.

Sun, L., Sun, Y., Zhang, M., Wang, L., Ren, J., Cui, M., Wang, Y., Ji, K., Li, P., Li, Q., Chen, P., Dai, S., Duan, C., Wu, Y. and Leng, P. (2012a) Suppression of 9-*cis*-epoxycarotenoid dioxygenase, which encodes a key enzyme in abscisic acid biosynthesis, alters fruit texture in transgenic tomato. *Plant Physiology* 158, 283–298.

Sun, L., Yuan, B., Zhang, M., Wang, L., Cui, M., Wang, Q. and Leng, P. (2012b) Fruit-specific RNAi-mediated suppression of *SlNCED1* increases both lycopene and β-carotene contents in tomato fruit. *Journal of Experimental Botany* 63, 3097–3108.

Sunako, T., Sakuraba, W., Senda, M., Akada, S., Ishikawa, R., Niizeki, M. and Harada, T. (1999) An allele of the ripening-specific 1-aminocyclopropane-1-carboxylic acid synthase gene (*ACS1*) in apple fruit with a long storage life. *Plant Physiology* 119, 1297–1303.

Symons, G.M., Davies, C., Shavrukov, Y., Dry, I.B., Reid, J.B. and Thomas, M.R. (2006) Grapes on steroids. Brassinosteroids are involved in grape berry ripening. *Plant Physiology* 140, 150–158.

Tesniere, C., Pradal, M., El-Kereamy, A., Torregrosa, L., Chatelet, P., Roustan, J.P. and Chevin, C. (2004) Involvement of ethylene signalling in a non-climacteric fruit: new elements regarding the regulation of *ADH* expression in grapevine. *Journal of Experimental Botany* 55, 2235–2240.

Thompson, A.J., Tor, M., Barry, C.S., Vrebalov, J., Orfila, C., Jarvis, M.C., Giovannoni, J.J., Grierson, D. and Seymour, G.B. (1999) Molecular and genetic characterization of a novel pleiotropic tomato-ripening mutant. *Plant Physiology* 120, 383–389.

Tian, M.S., Prakash, S., Elgar, H.J., Young, H., Burmeister, D.M. and Ross, G.S. (2000) Responses of strawberry fruit to 1-methylcyclopropene (1-MCP) and ethylene. *Plant Growth Regulation* 32, 83–90.

Tingwa, P. and Young, R. (1975) The effect of indole-3-acetic acid and other growth regulators on the ripening of avocado fruits. *Plant Physiology* 55, 937–940.

Trainotti, L., Pavanello, A. and Casadoro, G. (2005) Different ethylene receptors show an increased expression during the ripening of strawberries: does such an increment imply a role for ethylene in the ripening of these non-climacteric fruits? *Journal of Experimental Botany* 56, 2037–2046.

Vardhini, B.V. and Rao, S.S.R. (2002) Acceleration of ripening of tomato pericarp discs by brassinosteroids. *Phytochemistry* 61, 843–847.

Vendrell, M. and Buesa, C. (1989) Relationship between abscisic acid content and ripening of apples. *Acta Horticulturae* 258, 389–396.

Villarreal, N.M., Rosli, H.G., Martínez, G.A. and Civello, P.M. (2008) Polygalacturonase activity and expression of related genes during ripening of strawberry cultivars with contrasting fruit firmness. *Postharvest Biology and Technology* 47, 141–150.

Vrebalov, J., Ruezinsky, D., Padmanabhan, V., White, R., Medrano, D., Drake, R., Schuch, W. and Giovannoni, J.J. (2002) A MADS-box gene necessary for fruit ripening at tomato *ripening-inhibitor* (*rin*) locus. *Science* 296, 343–346.

Vrebalov, J., Pan, I.L., Arroyo, A.J.M., McQuinn, R., Chung, M., Poole, M., Rose, J., Seymour, G., Grandillo, S., Giovannoni, J. and Irish, V.F. (2009) Fleshy fruit expansion and ripening are regulated by the tomato *SHATTERPROOF* gene *TAGL1*. *Plant Cell* 21, 3041–3062.

Wang, A., Yamakake, J., Kudo, H., Wakasa, Y., Hatsuyama, Y., Igarashi, M., Kasai, A., Li, T. and Harada, T. (2009) Null mutation of the *MdACS3* gene, coding for a ripening-specific 1-aminocyclopropane-1-carboxylate synthase, leads to long shelf life in apple fruit. *Plant Physiology* 151, 391–399.

Watkins, C.B. (2006) The use of 1-methylcyclopropene (1-MCP) on fruits and vegetables. *Biotechnology Advances* 24, 389–409.

Watkins, C.B. (2008) Overview of 1-methylcyclopropene trials and uses for edible horticultural crops. *HortScience* 43, 86–94.

Wheeler, S., Loveys, B., Ford, C. and Davies, C. (2009) The relationship between the expression of abscisic acid biosynthesis genes, accumulation of abscisic acid and the promotion of *Vitis vinifera* L. berry ripening by abscisic acid. *Australian Journal of Grape Wine Research* 15, 195–204.

Wilkinson, J.Q., Lanahan, M.B., Yen, H.C., Giovannoni, J.J. and Klee, H.J. (1995) An ethylene-inducible component of signal transduction encoded by *Never-ripe*. *Science* 270, 1807–1809.

Yamane, M., Abe, D., Yasui, S., Yokotani, N., Kimata, W., Ushijima, K., Nakano, R., Kubo, Y. and Inaba, A. (2007) Differential expression of ethylene biosynthetic genes in climacteric and non-climacteric Chinese pear fruit. *Postharvest Biology and Technology* 44, 220–227.

Yokotani, N., Nakano, R., Imanishi, S., Nagata, M., Inaba, A. and Kubo, Y. (2009) Ripening-associated ethylene biosynthesis in tomato fruit is autocatalytically and developmentally regulated. *Journal of Experimental Botany* 60, 3433–3442.

Zhang, M., Leng, P., Zhang, G. and Li, X. (2009a) Cloning and functional analysis of 9-*cis*-epoxycarotenoid dioxygenase (NCED) genes encoding a key enzyme during abscisic acid biosynthesis from peach and grape fruits. *Journal of Plant Physiology* 166, 1241–1252.

Zhang, M., Yuan, B. and Leng, P. (2009b) The role of ABA in triggering ethylene biosynthesis and ripening of tomato fruit. *Journal of Experimental Botany* 60, 1579–1588.

Zheng, X.Y. and Wolff, D.W. (2000) Ethylene production, shelf-life and evidence of RFLP polymorphisms linked to ethylene genes in melon (*Cucumis melo* L.). *Theoretical and Applied Genetics* 101, 613–624.

Zheng, X.Y., Wolff, D.W. and Crosby, K.M. (2002) Genetics of ethylene biosynthesis and restriction fragment length polymorphism (RFLPs) of ACC oxidase and synthase genes in melon (*Cucumis melo* L.). *Theoretical and Applied Genetics* 105, 397–403.

2 Fruit Ripening: Primary Metabolism

Sonia Osorio[1,2] and Alisdair R. Fernie[1]*
[1]*Max-Planck-Institut für Molekulare Pflanzenphysiologie, Potsdam-Golm, Germany;* [2]*IHSM-UMA-CSIC, Universidad de Málaga, Málaga, Spain*

2.1 Introduction

Fruit ripening is a complex and highly coordinated developmental process that yields succulent and flavourful tissues for organisms that consume and disperse the associated seed (Giovannoni, 2001). Ripening involves softening of the fruit tissues to facilitate seed dispersal. In addition to softening, fruits normally exhibit increased accumulation of sugars, acids, pigments and volatile compounds that increase interest and palatability to animals. Additionally, fruits are an important source of supplementary diet, providing minerals, vitamins, fibre and antioxidants for humans. From an agronomical point of view, nutritional value, flavour, processing qualities and shelf-life determine the quality of fruit. Additional fruit attributes, including early maturity, enhanced colour and increased size, constitute the selection of so-called domestication traits.

The main changes associated with ripening include colour (loss of green colour and increase in non-photosynthetic pigments that vary depending on species and cultivar), firmness (softening by cell-wall-degrading activities), taste (increase in sugar and decline in organic acids) and flavour (production of volatile compounds providing the characteristic aroma).

Developing tools that allow comprehensive phenotyping at the level of the transcriptome (Alba *et al.*, 2005; Vriezen *et al.*, 2008; Matas *et al.*, 2011; Rohrmann *et al.*, 2011), proteome (Lee *et al.*, 2004; Rose *et al.*, 2004; Saravanan and Rose 2004) and metabolome (Fait *et al.*, 2008; Lombardo *et al.*, 2011) enable us to get a detailed view of the metabolic network (Carrari *et al.*, 2006; Deluc *et al.*, 2007; Grimplet *et al.*, 2007; Enfissi *et al.*, 2010; Zamboni *et al.*, 2010; Osorio *et al.*, 2011; Rohrmann *et al.*, 2011), and the likely outcome of taking such an approach is a better understanding of the metabolic regulation underlying fruit development. This chapter will concentrate on central carbon metabolism, as this is the subject of the majority of the authors' own research.

2.2 Hormonal Control during Ripening

Based on the increase in respiration and concomitant increase in synthesis of ethylene during ripening, fleshy fruits are classified as either 'climacteric' or 'non-climacteric'. Ethylene synthesis in climacteric fruits such as tomato, apple and banana is essential for normal fruit ripening, and blocking either synthesis or perception prevents ripening. As expected,

* fernie@mpimp-golm.mpg.de

the synthesis and degradation of this hormone is highly regulated. Moreover, signal transduction is also a critical aspect and is regulated at multiple levels. The role of ethylene in ripening of climacteric fruits has been known for more than 50 years. Since its discovery, considerable effort has been focused on studies of ethylene biosynthesis (involving the enzymes *S*-adenosylmethionine (SAM) synthetase, 1-aminocyclopropane carboxylic acid (ACC) synthase (ACS) and ACC oxidase (ACO)), ethylene perception (by ethylene receptors), signal transduction (by ethylene response factors) and ethylene-regulated genes such as cell-wall-disassembling genes (endo-polygalacturonase, pectin methyl esterase and pectate lyase).

Tomato fruit has been used as a model for climacteric fruits because ethylene is an agriculturally important hormone. In ethylene biosynthesis, only the *ACS* and *ACO* genes are involved. SAM is converted to ACC by ACS, and this step is considered the limiting step. ACC is subsequently converted to ethylene by ACO. There are a few characterized *ACS* and *ACO* genes in tomato, but only *ACO1*, the most highly induced *ACO* during ripening, prevents ethylene synthesis and ripening (Hamilton *et al.*, 1990). On the other hand, only two *ACS* genes, *ACS2* and *ACS4*, are significantly increased during ripening (Rottmann *et al.*, 1991; Barry *et al.*, 1996), and both genes, as with *ACO1*, prevent ethylene synthesis and ripening (Oeller *et al.*, 1991). Recently, genomics approaches have provided insight into the primary ripening control upstream of ethylene. The tomato pleiotropic ripening mutations *ripening inhibitor* (*rin*), *non-ripening* (*nor*) and *Colourless non-ripening* (*Cnr*) have added much insight in this regard. The rin, nor and Cnr mutations are affected in all aspects of the tomato fruit-ripening process that are unable to respond to ripening-associated ethylene genes (Vrebalov *et al.*, 2002; Manning *et al.*, 2006). Furthermore, in fruits from these mutants, the ripening-associated ethylene genes are induced by exogenous ethylene, indicating that all three genes operate upstream of ethylene biosynthesis and are involved in a process controlled exclusively by ethylene. The three mutant loci encode putative transcription factors. The *rin* mutant encodes a partially deleted MADS-box protein of the SEPALATTA clade (Hileman *et al.*, 2006), whereas *Cnr* is an epigenetic change that alters the promoter methylation of SQUAMOSA promoter binding (SPB) proteins. It has been suggested that the *nor* loci encodes a transcription factor (J. Vrebalov and J. Giovannoni, unpublished results), although not a member of MADS-box family (Giovannoni, 2004). The observed ethylene-independent aspect of ripening suggests that the RIN, NOR and CNR proteins are candidates for conserved molecular mechanisms of fruit in both the climacteric and non-climacteric categories. Another study (Osorio *et al.*, 2011) in which transcriptome, proteome and targeted metabolite analysis were combined during development and ripening of *nor* and *rin* mutants has helped to refine the ethylene-regulated transcriptome and has added to our knowledge of the role of ethylene in both protein and metabolite regulation in tomato ripening. These data support the contention that *nor* and *rin* act together in a cascade to control ripening (Giovannoni *et al.*, 1995; Thompson *et al.*, 1999) and also suggest that *nor* has a more global effect on ethylene/ripening-related gene expression than *rin*, which indicates that *nor* probably operates upstream of *rin*.

Biochemical evidence suggests that ethylene production may be influenced or regulated by interactions between its biosynthesis and other metabolic pathways. One such example is provided by the fact that SAM is the substrate for both the polyamine pathway and nucleic acid methylation; competition for substrate has been demonstrated by the finding that overexpression of a SAM hydrolase is associated with inhibited ethylene production during ripening (Good *et al.*, 1994). On the other hand, the methionine cycle directly links ethylene biosynthesis to the central pathways of primary metabolism.

Non-climacteric fruits such as strawberry and grape do not increase respiration, although they produce a small amount of ethylene during ripening (Iannetta *et al.*, 2006). Different studies have tried to relate this hormone to the ripening process. However, in spite of many efforts, no results have been obtained that can demonstrate a clear relationship. In pepper fruit, some cultivars seem to be ethylene insensitive, whilst other cultivars, as well as grape berries treated with exogenous ethylene, are able to stimulate the expression of ripening-specific genes (Armitage 1989; Ferrarese *et al.*, 1995; Harpster *et al.*, 1997; El-Kereamy *et al.*, 2003). In strawberry, the situation is even more complex, because molecular data about the possible relations between ethylene and the expression of genes during ripening are not in agreement. The expression of two genes involved in softening of strawberries (expansin and cellulose) seems to be ethylene insensitive (Civello *et al.*, 1999). However, the expression of other ripening-related genes in strawberry (pectin methyl esterase and β-galactosidase) were modified by treatments with ethylene (Trainotti *et al.*, 2001; Castillejo *et al.*, 2004). Interestingly, this dual effect of ethylene has also been found in climacteric peach fruit where the role of this hormone can either be positive or negative according to the different genes (Trainotti *et al.*, 2003).

Currently, no single growth regulator appears to play a positive role analogous to the role played by ethylene in the ripening of climacteric fruits. However, it has been observed for some time that auxin can negatively control the ripening of some non-climacteric fruits. In strawberry, it has been shown that the expression of many ripening-specific genes can be down-regulated by treatments with an exogenous auxin. By contrast, the expression of ripening-specific genes is accelerated following the removal of the achenes, which are a source of endogenous auxin (Harpster *et al.*, 1998; Aharoni *et al.*, 2002). Also, in grape, auxin seems to play a negative role in the regulation of ripening. In fact, it has been shown that treatments with a synthetic auxin are able to delay the expression of a number of ripening-related genes (Davies *et al.*, 1997). However, detailed studies on the content, synthesis and signalling of this hormone in different fruit parts at different developmental stages are lacking. As a consequence of the prominent role of auxin in the development and ripening of some non-climateric fruits such as strawberry (*Fragaria ananassa*), little attention has been paid to possible roles of other plant hormones in these processes such as gibberellins (GAs). It has been reported that application of GA3 to ripening fruit caused a significant delay in the development of the red colour (Martinez *et al.*, 1996). In addition, external application of GA3 was able to modify the expression of fruit genes such as *FaGAST*, which encodes a protein involved in cell enlargement and final fruit size (de la Fuente *et al.*, 2006), and *FaXyl*, encoding a β-xylosidase (Bustamante *et al.*, 2009).

2.3 Central Carbon Metabolism

Metabolism in the fruit involves the conversion of high-molecular-weight precursors to smaller compounds that help to produce viable seeds and to attract seed-dispersing species. The flavour of fruit is generally determined by tens to hundreds of constituents, most generated during the ripening phase of the fruit growth and development process. Any study on the metabolic pathways leading to their synthesis must be considered in the context of this developmental process. Thus, it is known that, during the rapid growth phase, the fruits act as strong sinks that import massive amounts of photo-assimilates from photosynthesizing organs. Translocation occurs in the phloem, with sucrose being the main compound translocated, although in some species there are other predominant compounds such as polyalcohols (e.g. mannitol or sorbitol) and even oligosaccharides. These translocated compounds, which are products of primary metabolism, are the precursors of most of the metabolites that account for the fruit flavour, generally classified as secondary metabolites. Thus, the synthesis of these compounds is necessarily supported by the supply of the primary photoassimilate.

In general, there are three major classes of chemicals responsible for flavour (a combination of taste and smell): sugars, acids and volatiles. Some of these primary metabolites can be essential components of taste, as they may be, depending on the species, main components of the harvested fruit, being recognized by receptors for sweet taste.

2.3.1 Sugars

The sugar, or sugar alcohol, delivered to the fruit, is converted to starch (e.g. mango, banana, kiwifruit), stored as reducing sugar (e.g. tomato, strawberry) or stored as sucrose (e.g. wild tomato, watermelon, grape), or may even be converted to lipids (e.g. olive). Sucrose, glucose and fructose are the most abundant carbohydrates and are widely distributed food components derived from plants. The sweetness of fruits is the central characteristic determining fruit quality and is determined by the total sugar content and by the ratios among these sugars. Accumulation of sucrose, glucose and fructose in fruits such as melon, watermelon (Brown and Summers 1985), strawberry (Fait *et al.*, 2008) and peach (Lo Bianco and Rieger 2002) is evident during ripening. However, in domesticated tomato (*Solanum lycopersicum*), only a high accumulation of the two hexoses (glucose and fructose) is observed, whereas some wild tomato species (e.g. *Solanum chmielewskii*) accumulate mostly sucrose (Yelle *et al.*, 1991).

The variability in the content of sucrose and hexoses is the result of activities of the enzymes responsible for their degradation and synthesis, with invertase and sucrose synthase being the most studied. In tomato, the involvement of apoplastic invertase in the balance between sucrose and hexoses has been thoroughly studied by taking advantage of the fact that the wild species accumulates sucrose but the cultivated species accumulates hexoses (Klann *et al.*, 1996), allowing genetic and biochemical studies to be carried out that have provided evidence that the kinetic properties of the invertase from the domesticated cultivars accounts for the hexose accumulation in the fruit of these species (Fridman *et al.*, 2004). By contrast, there is little evidence of a role of sucrose synthase in fruit metabolism (Carrari and Fernie 2006). In tomato, utilizing a reverse genetics approach, Zanor *et al.* (2009a) reported that *LIN5* (a gene encoding a cell-wall invertase) antisense plants had decreased glucose and fructose in their fruit, indicating the importance of LIN5 *in planta* in the control of total soluble solids content. The transformants were characterized by an altered flower and fruit morphology, displaying increased numbers of petals and sepals per flower, an increased rate of fruit abortion and a reduction in fruit size.

Apoplastic invertase has been studied in the fruits of species other than tomato, such as strawberry, not only in terms of its critical role in determining the sucrose/hexose ratio but also because this ratio determines the sink strength of the fruit and, indirectly, fruit size. Thus, in strawberry, the levels of sucrose and hexoses (glucose and fructose) increased during fruit ripening, whilst other sugars such as xylose and galactose, and the polyol inositol, decreased (Fait *et al.*, 2008). The apoplastic invertase-deficient *miniature1* mutant of maize exhibits a dramatically decreased seed size as well as altered levels of phytohormones (Miller and Chourey 1992; Sonnewald *et al.*, 1997; LeClere *et al.*, 2008). This raises interesting questions regarding the regulation of carbon partitioning in fruit. Recently, a metabolic and transcriptional study using tomato introgression lines resulting from a cross between *S. lycopersicum* and *S. chmielewskii* revealed that the dramatic increase in amino acid content in the fruit is the result of upregulated transport of amino acids via the phloem, although the mechanism is still unknown (Thi Do *et al.*, 2010).

Starch is another carbohydrate that undergoes modifications during ripening and is metabolized to glucose and fructose, the two main sugars in ripe fruit (Ho *et al.*, 1983). Tomato introgression lines containing the exotic allele of LIN5 (IL 9-2-5) accumulated significantly more starch in both pericarp and columella tissues (Baxter *et al.*, 2005). This is in agreement with the finding that starch accumulation plays an important role in determining the soluble solids content, or Brix index, of mature fruit (Schaffer and Petreikov, 1997). Recently, in tomato fruit, it has been shown that malic acid influences starch content via redox control of the enzyme responsible for starch synthesis, ADP-glucose pyrophosphorylase (Centeno *et al.*, 2011). Given that starch synthesis is positively correlated with reducing sugar content in ripe fruit, the malic acid content of the immature fruit is predicted to be inversely correlated to the sugar content of the ripe fruit.

2.3.2 Organic acids

Organic acids are the other intermediate metabolites important as flavour components, either by themselves, because the organic acid:sugar ratio defines quality parameters at harvest time in fruits, or as precursors of other secondary metabolites. Therefore, organic acid manipulation is highly valuable in terms of metabolic engineering. The organic acid content is dependent on the activity of the main metabolic pathways such as glycolysis, the tricarboxylic acid (TCA) cycle and respiration. The main organic acids are the TCA intermediates citrate and malate, together with quinate, with other TCA intermediates like oxalate, succinate, isocitrate, fumarate and aconitate being present at lower levels. The pH of a ripe fruit is typically around pH 4. Interestingly, in an independent study that focused on early fruit development, the levels of both citrate and malate were also highly correlated to many important regulators of ripening (Mounet *et al.*, 2009).

The structure of the TCA cycle is well known in plants. However, until recently, its regulation was poorly characterized. In our laboratory, several studies have been done to determine the role of the mitochondrial TCA cycle in plants. Biochemical analysis of the *ACO1* mutant revealed a decreased flux through the TCA cycle, decreased levels of TCA cycle intermediates, enhanced carbon assimilation and dramatically increased fruit weight (Carrari *et al.*, 2003). Tomato plants with reduced mitochondrial malate dehydrogenase (mMDH) showed an increase in fruit weight that was probably due to enhanced photosynthetic activity and carbon assimilation in the leaves, which also led to increased accumulation of starch and sugars, as well as some organic acids (succinate, ascorbate and dehydroascorbate) (Nunes-Nesi *et al.*, 2005). In fruits, malic acid also influences starch content via redox control of the ADP-glucose pyrophosphorylase (Centeno *et al.*, 2011). Those plants also showed a small effect on the total fruit yield, as well as unanticipated changes in postharvest shelf-life and susceptibility to bacterial infection. Tomato plants with reduced fumarase activity showed the opposite effect to those with reduced mMDH (Nunes-Nesi *et al.*, 2007). Additionally, biochemical analyses of antisense tomato mitochondrial NAD-dependent isocitrate dehydrogenase plants showed a reduction in flux through the TCA cycle, decreased levels of TCA cycle intermediates and relatively few changes in photosynthetic parameters; however, fruit size and yield were reduced (Sienkiewicz-Porzucek *et al.*, 2010). Despite the fact that much research work is needed to understand the exact mechanism for the increase in the fruit dry weight, manipulation of central organic acids is clearly a promising approach to enhance fruit yield (Nunes-Nesi *et al.*, 2011).

2.4 Volatiles

Flavour is the sum of a large set of primary and secondary metabolites that are measured by the taste and olfactory systems. These metabolites comprise a diverse set of chemicals derived from essential amino acids (phenylalanine, leucine and isoleucine), essential fatty acids (principally linolenic acid) and carotenoids such as β-carotene, the precursor of retinol (vitamin A), which is the immediate precursor of one of the most important flavour volatiles, β-ionone (Goff and Klee, 2006). More than 400 volatiles compounds have been identified in tomato (Petro-Turza, 1987; Buttery, 1993; Buttery and Ling, 1993; Fulton *et al.*, 2002), although a smaller set of only 15–20 are made in sufficient quantities to have an impact on human perception (Baldwin *et al.*, 2000).

Any study of the metabolic pathways leading to the synthesis of its various metabolites must be considered in the context of this developmental process. During the rapid growth phase, fruits act as strong sinks, importing massive amounts of photoassimilates from photosynthesizing organs. The content of most, although not all, of the flavour volatiles in tomato increases at the onset of ripening and peaks either at or shortly before full ripening. This timing suggests that synthesis of flavour volatiles is highly regulated.

Recently, a metabolomic approach was used to describe the phenotypic variation of a broad range of primary and volatile metabolites across a series of tomato lines resulting from crosses between a cherry tomato and three independent large fruit cultivars ('Levovil', 'VilB' and 'VilD') (Zanor *et al.*, 2009b). The results of the most highly abundant primary metabolite analysis of cherry and large-fruited tomato lines were largely in accordance with those obtained from previous studies (Causse *et al.*, 2002). The low sugar and high malate content of the 'Levovil' parental and the corresponding very low sugar:acid ratio could explain the lower acceptance of the fruit by food panel tasters, especially as malate is perceived as tasting more sour than citrate (Marsh *et al.*, 2003). In addition to the changes observed in sugars and acids in cherry tomatoes, the glutamate level, known to be sensed as the fifth basic taste (umami), which evokes a savoury feeling, was found to be considerably higher in the cherry variety than in the large-fruited varieties. This finding is also in accordance with the fact that cherry tomatoes were found to be tastier than the other parental lines used in this study. Additionally, considerable correlations within the levels of primary metabolites or volatile compounds were also observed. However, there was relatively little association between the levels of primary metabolites and volatile compounds, implying that they are not tightly linked to one another, with the exception of sucrose, which showed a strong association with a number of volatile compounds (Zanor *et al.*, 2009b).

A broad profiling of tomato volatiles in a tomato introgression line population harbouring introgression of the wild species *Solanum pennellii* yielded over 100 quantitative trait loci (QTLs) that were reproducibly altered in one or more volatiles contributing to flavour (Tieman *et al.*, 2006b). These QTLs have been used as tools to identify the genes responsible for controlling the synthesis of many volatile compounds. Very few genes involved in the biosynthetic pathways of tomato flavour volatiles have been identified, although the detection of *malodorous*, a wild species allele that affects tomato aroma, allowed the identification of a QTL that is linked to a markedly undesirable flavour within the *S. pennellii* introgression line IL8-2 (Tadmor *et al.*, 2002). A complementary approach utilizing broad genetic crosses has been used to identify QTLs for organoleptic properties of tomatoes (Causse *et al.*, 2002). The lines identified as preferable by the consumer could now be comprehensively characterized with respect to volatile and non-volatile compounds alike. By using a combination of metabolic and flux profiling alongside reverse genetic studies on IL8-2, it was possible to confirm the biological pathway of a set of phenylalanine-derived volatiles, 2-phenylacetaldehyde and 2-phenylethanol, important aromatic compounds in tomato (Tieman *et al.*, 2006a). A combined metabolic, genomic and biochemical analysis of glandular trichomes from the wild tomato species

Solanum habrochaites identified a key enzyme in the biosynthesis of methylketones that serves this purpose (Fridman *et al.*, 2005). In recent years, there have been dramatic improvements in the knowledge of volatiles, and it is clear that synthesis of the flavour volatiles is associated with ripening. However, the regulatory mechanisms have not been established. It is likely that a large part of this regulation is mediated by a subset of the many ripening-associated transcription factors.

2.5 Cell-wall Metabolism

The primary cell wall is composed of numerous polymers. These vary in structure somewhat between species, but eight polymeric components (cellulose, three matrix glycans composed of neutral sugars, three pectins rich in d-galacturonic acid and structural proteins) are usually present. During ripening, cell-wall architecture and the polymers of which it is composed are progressively modified. The metabolic changes during ripening include alteration of cell structure involving changes in cell-wall thickness, permeability of the plasma membrane, hydration of the cell wall, decreases in structural integrity and increases in intracellular spaces (Redgwell *et al.*, 1997). In fruit such as strawberry and avocado, which develop a soft melting texture during ripening, swelling and softening of the cell wall is evident, but in fruit such as apple, which ripen to a crisp, fracturable texture, cell-wall swelling is not observed (Redgwell *et al.*, 1997). Ripening is also usually accompanied by a reduction in cell turgor, due to an increasing concentration of solutes in the cell-wall space and to wall loosening (Shackel *et al.*, 1991). Cell-wall disassembly rate and extent are crucial for the maintenance of fruit quality and integrity (Matas *et al.*, 2009). For this reason, maintenance of firmness has long been the target for breeders in many crops to minimize postharvest decay.

The conventional approach to elucidating fruit softening has typically been based on two strategies: (i) the identification of cell-wall components whose solubility increases and/or polymer size decreases in parallel with decreasing fruit firmness (Seymour *et al.*, 1987; Tucker and Grierson, 1987; Redgwell *et al.*, 1992); and (ii) the characterization of proteins that are expressed during ripening and whose biochemical activities can be mechanistically related to the observed cell-wall changes (Goulao and Oliveira, 2007; Vicente *et al.*, 2007). Among cell-wall hydrolases, pectin-degrading enzymes are mostly implicated in fruit softening. Increased solubilization of the pectin substances, progressive loss of tissue firmness and a rapid rise in the polygalacturonase (PG) activity accompany normal ripening in many fruits (Brady, 1987; Fisher and Bennett, 1991). A positive correlation between PG activity and initiation of softening is known in a number of fruits like guava (El-Zoghbi, 1994), papaya (Paull and Chen, 1983), mango (Roe and Bruemmer, 1981) and strawberry (Garcia-Gago *et al.*, 2009; Quesada *et al.*, 2009). However, experiments with transgenic tomatoes have shown that, even though PG is important for the degradation of pectins, it is not the sole determinant of tissue softening during ripening (Gray *et al.*, 1992). Pectin methyl esterase catalyses the de-esterification of pectin, and its activity together with that of PG increase remarkably during ripening in peach, tomato, pear and strawberry (Tucker and Grierson, 1987; Osorio *et al.*, 2010).

The regulation of texture and shelf-life is clearly far more complex than was envisaged previously and so new approaches are needed for a better understanding of the relationships among changes in the texture properties of specific fruit tissues, intact fruit 'firmness' and shelf-life.

Fruit softening is generally described as a textural change that is associated with cell-wall disassembly due to the activity of degrading enzymes. However, an alternative explanation is that polysaccharide degradation is not the sole determinant of fruit softening and that other structures and ripening-related physiological processes also play critical roles. The cuticle has a number of biological functions that could have an important impact on fruit quality and shelf-life including the ability to maintain fruit skin integrity (Hovav *et al.*, 2007), to restrict cuticular transpiration (Leide *et al.*, 2007) and to limit microbial infection. It has been proposed that differences in tomato fruit cuticle structure and composition may be associated with the substantial variations in tomato fruit shelf-life that have been reported in different tomato genotypes (Saladie *et al.*, 2007). Other reports have also highlighted other process that contribute to fruit softening such as turgor pressure (Saladie *et al.*, 2007; Thomas *et al.*, 2008; Wada *et al.*, 2008) and the possible associated developmental changes in apoplastic solute accumulation (Wada *et al.*, 2008).

2.6 Summary

Much effort has been made to gain an understanding of the hormonal regulators of ripening in climacteric and non-climacteric fruits that mediate the physiological changes such as fruit softening and the accumulation of metabolites such as pigments, sugars, acids and volatiles. Currently, no analytical techniques can provide detection of all metabolites in all samples. However, the shift from single-metabolite measurements to platforms that can provide information on hundreds of metabolites has led to the development of better models to describe the links both within the metabolites themselves and between the metabolism and other processes. It is clear, therefore, that, if we are to improve our ability to carry out rational manipulations of plant metabolism for engineering purposes, we will have to generate a more complete view of metabolism as a network. It is hoped that, in the future, this approach will allow a comprehensive understanding of genetic and metabolic networks that govern fruit metabolism and its compositional quality.

References

Aharoni, A., Keizer, L.C.P., Van den Broeck, H.C., Blanco-Portales, R., Munoz-Blanco, J., Bois, G., Smit, P., De Vos, R.C.H. and O'Connell, A.P. (2002) Novel insight into vascular, stress, and auxin-dependent and -independent gene expression programs in strawberry, a non-climacteric fruit. *Plant Physiology* 129, 1019–1031.

Alba, R., Payton, P., Fei, Z.J., McQuinn R., Debbie, P., Martin, G.B., Tanksley, S.D. and Giovannoni, J.J. (2005) Transcriptome and selected metabolite analyses reveal multiple points of ethylene control during tomato fruit development. *Plant Cell* 17, 2954–2965.

Armitage, A.M. (1989) Promotion of fruit ripening of ornamental peppers by ethephon. *Horticulture Science* 24, 962–964.

Baldwin, E.S., Scott, J.W., Shewmarker, C.K. and Schuch, W. (2000) Flavor trivia and tomato aroma: biochemistry and possible mechanims for control of important aroma components. *Horticulture Science* 35, 1013–1022.

Barry, C.S., Blume, B., Bouzayen, M., Cooper, W., Hamilton, A.J. and Grierson, D. (1996) Differential expression of the 1-aminocyclopropane-1-carboxylate oxidase gene family of tomato. *Plant Journal* 9, 525–535.

Baxter, C.J., Carrari, F., Bauke, A., Overy, S., Hill, S.A., Quick, P.W., Fernie, A.R. and Sweetlove, L.J. (2005) Fruit carbohydrate metabolism in an introgression line of tomato with increased fruit soluble solids. *Plant and Cell Physiology* 46, 425–437.

Brady, C.J. (1987) Fruit ripening. *Annual Review of Plant Physiology* 38, 155–178.

Brown, A.C. and Summers, W.L. (1985) Carbohydrate accumulation and color development in watermelon. *Journal of American Society of Horticulture Science* 110, 683–686.

Bustamante, C.A., Civello, P.M. and Martinez, G.A. (2009) Cloning of the promoter region of β-xylosidase (*FaXyl1*) gene and effect of plant growth regulators on the expression of FaXyl1 in strawberry fruit. *Plant Science* 177, 49–56.

Buttery, R.G. (1993) Quantitative and sensory aspects of flavour of tomato and other vegetables and fruit. In: Acree, T.E. and Teranishsi, R. (eds) *Flavor Science: Sensible Principles and Techniques*. American Chemical Society, Washington, DC, pp. 259–286.

Buttery, R.G. and Ling, L.C. (1993) Volatiles of tomato fruit and plant parts: relationship and biogenesis. In: Teranishsi, R., Buttery, R.G. and Sugisawa, H. (eds) *Bioactive Volatile Compounds from Plants*. ACS Books, Washington, DC, pp. 23–34.

Carrari, F. and Fernie, A.R. (2006) Metabolic regulation underlying tomato fruit development. *Journal of Experimental Botany* 57, 1883–1897.

Carrari, F., Nunes-Nesi A., Gibon, Y., Lytovchenko, A., Loureiro, M.E. and Fernie, A.R. (2003) Reduced expression of aconitase results in an enhanced rate of photosynthesis and marked shifts in carbon partitioning in illuminated leaves of wild species tomato. *Plant Physiology* 133, 1322–1335.

Carrari, F., Baxter, C., Usadel, B., Urbanczyk-Wochniak, E., Zanor, M.I., Nunes-Nesi, A., Nikiforova, V., Centero, D., Ratzka, A., Pauly, M., Sweetlove, L.J. and Fernie, A.R. (2006) Integrated analysis of metabolite and transcript levels reveals the metabolic shifts that underlie tomato fruit development and highlight regulatory aspects of metabolic network behavior. *Plant Physiology* 142, 1380–1396.

Castillejo, C., de la Fuente, J.I., Iannetta, P., Botella, M.A. and Valpuesta, V. (2004) Pectin esterase gene family in strawberry fruit: study of *FaPE1*, a ripening-specific isoform. *Journal of Experimental Botany* 55, 909–918.

Causse, M., Saliba-Colombani, V., Lecomte, L., Duffe, P., Rousselle, P. and Buret, M. (2002) QTL analysis of fruit quality in fresh market tomato: a few chromosome regions control the variation of sensory and instrumental traits. *Journal of Experimental Botany* 53, 2089–2098.

Centeno, D.C., Osorio, S., Nunes-Nesi, A., Bertolo, A.L.F., Carneiro, R.T., Araújo, W.L., Steinhauser, M.C., Michalska, J., Rohrmann, J., Geigenberger, P., Oliver, S.N., Stitt, M., Carrari, F., Rose, J.K.C. and Fernie, A.F. (2011) Malate plays a crucial role in starch metabolism, ripening, and soluble solid content of tomato fruit and affects postharvest softening. *Plant Cell* 23, 162–184.

Civello, P.M., Powell, A.L.T., Sabehat, A. and Bennett, A.B. (1999) An expansin gene expressed in ripening strawberry fruit. *Plant Physiology* 121, 1273–1279.

Davies, C., Boss, P.K. and Robinson, S.P. (1997) Treatment of grape berries, a nonclimacteric fruit with a synthetic auxin, retards ripening and alters the expression of developmentally regulated genes. *Plant Physiology* 115, 1155–1161.

de la Fuente, J.I., Amaya, I., Castillejo, C., Sanchez-Sevilla, J.F., Quesada, M.A., Botella, M.A. and Valpuest, V. (2006) The strawberry gene *FaGAST* affects plant growth through inhibition of cell elongation. *Journal of Experimental Botany* 57, 2401–2411.

Deluc, L.G., Grimplet, J., Wheatley, M.D., Tillett, R.L., Quilici, D.R., Osborne, C., Schooley, D.A., Schlauch, K.A., Cushman, J.C. and Cramer, G.R. (2007) Transcriptomic and metabolite analyses of Cabernet Sauvignon grape berry development. *BMC Genomics* 8, 429.

El-Kereamy, A., Chervin, C., Roustan, J.P., Cheynier, V., Souquet, J.M., Moutounet, M., Raynal, J., Ford, C., Latche, A., Pech, J.C. and Bouzayen, M. (2003) Exogenous ethylene stimulates the long-term expression of genes related to anthocyanin biosynthesis in grape berries. *Physiologia Plantarum* 119, 175–182.

El-Zoghbi, M. (1994) Biochemical changes in some tropical fruits during ripening. *Food Chemistry* 49, 33–37.

Enfissi, E.M., Barneche, F., Ahmed, I., Lichtle, C., Gerrish, C., McQuinn, R.P., Giovannoni, J.J., Lopez-Juez, E., Bowler, C., Bramley, P.M. and Fraser, P.D. (2010) Integrative transcript and metabolite analysis of nutritionally enhanced DE-ETIOLATED1 downregulated tomato fruit. *Plant Cell* 22, 1190–1215.

Fait, A., Hanhineva, K., Beleggia, R., Dai, N., Rogachev, I., Nikiforova, V.J., Fernie, A.R. and Aharoni, A. (2008) Reconfiguration of the achene and receptacle metabolic networks during strawberry fruit development. *Plant Physiology* 148, 730–750.

Ferrarese, L., Trainotti, L., Moretto, P., de Laureto, P.P., Rascio, N. and Casadoro, G. (1995) Differential ethylene-inducible expression of cellulase in pepper plants. *Plant Molecular Biology* 29, 735–747.

Fisher, R.L. and Bennett, A.B. (1991) Role of cell wall hydrolases in fruit ripening. *Annual Review of Plant Physiology and Plant Molecular Biology* 42, 675–703.

Fridman, E., Carrari, F., Liu, Y.S., Fernie, A.R. and Zamir, D. (2004) Zooming in on a quantitative trait for tomato yield using interspecific introgressions. *Science* 305, 1786–1789.

Fridman, E., Wang, J., Iijima, Y., Froehlich, J.E., Gang, D.R., Ohlrogge, J. and Pichersky, E. (2005) Metabolic, genomic, and biochemical analyses of glandular trichomes from the wild tomato species *Lycopersicon hirsutum* identify a key enzyme in the biosynthesis of methylketones. *Plant Cell* 17, 1252–1267.

Fulton, T.M., Buchell, P., Voirol, E., Lopez, J., Petiard, V. and Tanksley, S.D. (2002) Quantitative trait loci (QTL) affecting sugars, organic acids and other biochemical properties possibly contributing to flavour, identified in four advanced backcross populations of tomato. *Euphytica* 127, 163–177.

Garcia-Gago, J.A., Pose, S., Munoz-Blanco, J., Quesada, M.A. and Mercado, J.A. (2009) The polygalacturonase *FaPG1* gene plays a key role in strawberry fruit softening. *Plant Signaling and Behavior* 4, 766–768.

Giovannoni, J. (2001) Molecular biology of fruit maturation and ripening. *Annual Review of Plant Physiology and Plant Molecular Biology* 52, 725–749.

Giovannoni, J.J. (2004) Genetic regulation of fruit development and ripening. *Plant Cell* 16 (Suppl.), S170–S180.

Giovannoni, J.J., Noensie, E.N., Ruezinsky, D.M., Lu, X., Tracy, S.L., Ganal, M.W., Martin, G.B., Pillen, K., Alpert, K. and Tanksley, S.D. (1995) Molecular genetic analysis of the ripening-inhibitor and non-ripening loci of tomato: a first step in genetic map-based cloning of fruit ripening genes. *Molecular and General Genetics* 248, 195–206.

Goff, S.A. and Klee, H.J. (2006) Plant volatile compounds: sensory cues for health and nutritional value? *Science* 311, 815–819.

Good, X., Kellogg, J.A., Wagoner, W., Langhoff, D., Matsumura, W. and Bestwick, R.K. (1994) Reduced ethylene synthesis by transgenic tomatoes expressing *S*-adenosylmethionine hydrolase. *Plant Molecular Biology* 26, 781–790.

Goulao, L.F. and Oliveira, C.M. (2007) Cell wall modifications during fruit ripening: when a fruit is not a fruit. *Trends in Food Science and Technology* 19, 4–25.

Gray, J., Picton, S., Shabbeer, J., Schuch, W. and Grierson, D. (1992) Molecular biology of fruit ripening and its manipulation with antisense genes. *Plant Molecular Biology* 19, 69–87.

Grimplet, J., Deluc, L.G., Tillett, R.L., Wheatley, M.D., Schlauch, K.A., Cramer, G.R. and Cushman, J.C. (2007) Tissue-specific mRNA expression profiling in grape berry tissues. *BMC Genomics* 8, 187.

Hajirezaei, M.R., Kossmann, J., Heyer, A., Trethewey, R.N. and Willmitzer, L. (1997) Increased potato tuber size resulting from apoplastic expression of a yeast invertase. *Nature Biotechnology* 15, 794–797.

Hamilton, A.J., Lycett, G.W. and Grierson, D. (1990) Antisense gene that inhibits synthesis of the hormone ethylene in transgenic plants. *Nature* 346, 284–287.

Harpster, M.H., Lee, K.Y. and Dunsmuir, P. (1997) Isolation and characterization of a gene encoding endo-β-1,4-glucanase from pepper (*Capsicum annuum* L). *Plant Molecular Biology* 33, 47–59.

Harpster, M.H., Brummell, D.A. and Dunsmuir, P. (1998) Expression analysis of a ripening-specific, auxin-repressed endo-1,4-β-glucanase gene in strawberry. *Plant Physiology* 118, 1307–1316.

Hileman, L.C., Sundstrom, J.F., Litt, A., Chen, M., Shumba, T. and Irish, V.F. (2006) Molecular and phylogenetic analyses of the MADS-box gene family in tomato. *Molecular Biology and Evolution* 23, 2245–2258.

Ho, L.C., Sjut, V. and Hoad, G.V. (1983) The effect of assimilate supply on fruit-growth and hormone levels in tomato plants. *Plant Growth Regulation* 1, 155–171.

Hovav, R., Chehanovsky, N., Moy, M., Jetter, R. and Schaffer, A.A. (2007) The identification of a gene (*Cwp1*), silenced during *Solanum* evolution, which causes cuticle microfissuring and dehydration when expressed in tomato fruit. *Plant Journal* 52, 627–639.

Iannetta, P.P.M., Laarhoven, L.J., Medina-Escobar, N., James, E.K., McManus, M.T., Davies, H.V. and Harren, F.J.M. (2006) Ethylene and carbon dioxide production by developing strawberries show a correlative pattern that is indicative of ripening climacteric fruit. *Physiologia Plantarum* 127, 247–259.

Klann, E.M., Hall, B. and Bennett, A.B. (1996) Antisense acid invertase (*TIV1*) gene alters soluble sugar composition and size in transgenic tomato fruit. *Plant Physiology* 112, 1321–1330.

LeClere, S., Schmelz, E.A. and Chourey, P.S. (2008) Cell wall invertase-deficient *miniature1* kernels have altered phytohormone levels. *Phytochemistry* 69, 692–699.

Lee, S.J., Saravanan, R.S., Damasceno, C.M.B., Yamane, H., Kim, B.D. and Rose, J.K.C. (2004) Digging deeper into the plant cell wall proteome. *Plant Physiology Biochemistry* 42, 979–988.

Leide, J., Hildebrandt, U., Reussing, K., Riederer, M. and Vogg, G. (2007) The developmental pattern of tomato fruit wax accumulation and its impact on cuticular transpiration barrier properties: effects of a deficiency in a β-ketoacyl-coenzyme A synthase (LeCER6). *Plant Physiology* 144, 1667–1679.

Lo Bianco, R. and Rieger, M. (2002) Partitioning of sorbitol and sucrose catabolism within peach fruit. *Journal of American Society of Horticultural Science* 127, 115–121.

Lombardo, V.A., Osorio, S., Borsani, J., Lauxmann, M.A., Bustamante, C.A., Budde, C.O., Andreo, C.S., Lara, M.V., Fernie, A.R. and Drincovich, M.F. (2011) Metabolic profiling during peach fruit development and ripening reveals the metabolic networks that underpin each developmental stage. *Plant Physiology* 157, 1696–1710.

Manning, K., Tor, M., Poole, M., Hong, Y., Thompson, A.J., King, G.J., Giovannoni, J.J. and Seymour, G.B. (2006) A naturally occurring epigenetic mutation in a gene encoding an SBP-box transcription factor inhibits tomato fruit ripening. *Nature Genetics* 38, 948–952.

Marsh, K.B., Rossiter, K., Lau, K., Walker, S., Gunson, A. and MacRae, E.A. (2003) Using fruit pulps to explore flavour in kiwifruit. *Acta Hortuculturae* 610, 229–238.

Martinez, G.A., Chaves, A.R. and Añón, M.C. (1996) Effect of exogenous application of gibberellic acid on color change and phenylalanine ammonia-lyase, chlorophyllase, and peroxidase activities during ripening of strawberry fruit (*Fragaria × ananassa* Duch). *Journal of Plant Growth Regulation* 15, 139–146.

Matas, A.J., Gapper, N.E., Chung, M.Y., Giovannoni, J.J. and Rose, J.K. (2009) Biology and genetic engineering of fruit maturation for enhanced quality and shelf-life. *Current Opinion in Biotechnology* 20, 197–203.

Matas, A.J., Yeats, T.H., Buda, G.J., Zheng, Y., Chatterjee, S., Tohge, T., Ponnala, L., Adato, A., Aharoni, A., Stark, R., Fernie, A.R., Fei, Z.J., Giovannoni, J.J. and Rose, J.K.C. (2011) Tissue- and cell-type specific transcriptome profiling of expanding tomato fruit provides insights into metabolic and regulatory specialization and cuticle formation. *Plant Cell* 23, 3893–3910.

Miller, M.E. and Chourey, P.S. (1992) The maize invertase-deficient *miniature-1* seed mutation is associated with aberrant pedicel and endosperm development. *Plant Cell* 4, 297–305.

Mounet, F., Moing, A., Garcia, V., Petit, J., Maucourt, M., Deborde, C., Bernillon, S., Le Gall, G., Colquhoun, I., Defernez, M., Giraudel, J.L., Rolin, D., Rothan, C. and Lemaire-Chamley, M. (2009) Gene and metabolite regulatory network analysis of early developing fruit tissues highlights new candidate genes for the control of tomato fruit composition and development. *Plant Physiology* 149, 1505–1528.

Nunes-Nesi, A., Carrari, F., Lytovchenko, A., Smith, A.M., Loureiro, M.E., Ratcliffe, R.G., Sweetlove, L.J. and Fernie, A.R. (2005) Enhanced photosynthetic performance and growth as a consequence of decreasing mitochondrial malate dehydrogenase activity in transgenic tomato plants. *Plant Physiology* 137, 611–622.

Nunes-Nesi, A., Carrari, F., Gibon, Y., Sulpice, R., Lytovchenko, A., Fisahn, J., Graham, J., Ratcliffe, R.G., Sweetlove, L.J. and Fernie, A.R. (2007) Deficiency of mitochondrial fumarase activity in tomato plants impairs photosynthesis via an effect on stomatal function. *Plant Journal* 50, 1093–1106.

Nunes-Nesi, A., Araujo, W.L. and Fernie, A. (2011) Targeting mitochondrial metabolism and machinery as a means to enhance photosynthesis. *Plant Physiology* 155, 101–107.

Oeller, P.W., Wong, L.M., Taylor, L.P., Pike, D.A. and Theologis, A. (1991) Reversible inhibition of tomato fruit senescence by antisense RNA. *Science* 254, 437–439.

Osorio, S., Bombarely, A., Giavalisco, P., Usadel, B., Stephens, C., Aragüez, I., Medina-Escobar N., Botella, M.A., Fernie, A.R. and Valpuesta, V. (2010) Demethylation of oligogalacturonides by FaPE1 in the fruits of the wild strawberry *Fragaria vesca* triggers metabolic and transcriptional changes associated to defence and development of the fruit. *Journal of Experimental Botany* 62, 2855–2873.

Osorio, S., Alba, R., Damasceno, C.M., Lopez-Casado, G., Lohse, M., Zanor, M.I., Tohge, T., Usadel, B., Rose, J.K., Fei, Z., Giovannoni, J.J. and Fernie, A.R. (2011) Systems biology of tomato fruit development: combined transcript, protein, and metabolite analysis of tomato transcription factor (*nor*, *rin*) and ethylene receptor (*Nr*) mutants reveals novel regulatory interactions. *Plant Physiology* 157, 405–425.

Paull, R.E. and Chen, N.J. (1983) Postharvest variation in cell wall-degrading enzymes of papaya (*Carica papaya* L.) during fruit ripening. *Plant Physiology* 72, 382–385.

Petro-Turza, M. (1987) Flavor of tomato and tomato products. *Food Reviews International* 2, 309–351.

Quesada, M.A., Blanco-Portales, R., Pose, S., Garcia-Gago, J.A., Jimenez-Bermudez, S., Munoz-Serrano, A., Caballero, J.L., Pliego-Alfaro, F., Mercado, J.A. and Munoz-Blanco, J. (2009) Antisense down-regulation of the *FaPG1* gene reveals an unexpected central role for polygalacturonase in strawberry fruit softening. *Plant Physiology* 150, 1022–1032.

Redgwell, R.J., Melton, L.D. and Brasch, D.J. (1992) Cell wall dissolution in ripening kiwifruit (*Actinidia deliciosa*): solubilization of the pectic polymers. *Plant Physiology* 98, 71–81.

Redgwell, R.J., MacRae, E.A., Hallet, I., Fischer, M., Perry, J. and Harker, R. (1997) *In vivo* and *in vitro* swelling of cell walls during fruit ripening. *Planta* 203, 162–173.

Roe, B. and Bruemmer, J.H. (1981) Changes in pectin substances and enzymes during ripening and storage of "keitt" mangos. *Journal of Food Science* 46, 186–189.

Rohrmann, J., Tohge, T., Alba, R., Osorio, S., Caldana, C., McQuinn, R., Arvidsson, S., van der Merwe, M.J., Riano-Pachon, D.M., Mueller-Roeber, B., Fei, Z.J., Nesi, A.N., Giovannoni, J.J. and Fernie, A.R. (2011) Combined transcription factor profiling, microarray analysis and metabolite profiling reveals the transcriptional control of metabolic shifts occurring during tomato fruit development. *Plant Journal* 68, 999–1013.

Rose, J.K.C., Bashir, S., Giovannoni, J.J., Jahn, M.M. and Saravanan, R.S. (2004) Tackling the plant proteome: practical approaches, hurdles and experimental tools. *Plant Journal* 39, 715–733.

Rottmann, W.H., Peter, G.F., Oeller, P.W., Keller, J.A., Shen, N.F., Nagy, B.P., Taylor, L.P., Campbell, A.D. and Theologis, A. (1991) 1-Aminocyclopropane-1-carboxylate synthase in tomato is encoded by a multigene family whose transcription is induced during fruit and floral senescence. *Journal of Molecular Biology* 222, 937–961.

Saladie, M., Matas, A.J., Isaacson, T., Jenks, M.A., Goodwin, S.M., Niklas, K.J., Xiaolin, R., Labavitch, J.M., Shackel, K.A., Fernie, A.R., Lytovchenko, A., O'Neill, M.A., Watkins, C.B. and Rose, J.K. (2007) A reevaluation of the key factors that influence tomato fruit softening and integrity. *Plant Physiology* 144, 1012–1028.

Saravanan, R.S. and Rose, J.K.C. (2004) A critical evaluation of sample extraction techniques for enhanced proteomic analysis of recalcitrant plant tissues. *Proteomics* 4, 2522–2532.

Schaffer, A.A. and Petreikov, M. (1997) Sucrose-to-starch metabolism in tomato fruit undergoing transient starch accumulation. *Plant Physiology* 113, 739–746.

Seymour, G.B., Harding, S.E., Taylor, M.G., Hobson, G.E. and Tucker, G.A. (1987) Polyuronide solubilization during ripening of normal and mutant tomato fruit. *Phytochemistry* 26, 1871–1875.

Shackel, K.A., Greve, C., Labavitch, J.M. and Ahmadi, H. (1991) Cell turgor changes associated with ripening in tomato pericarp tissue. *Plant Physiology* 97, 814–816.

Sienkiewicz-Porzucek, A., Sulpice, R., Osorio, S., Krahnert, I., Leisse, A., Urbanczyk-Wochniak, E., Hodges, M., Fernie, A.R. and Nunes-Nesi, A. (2010) Mild reductions in mitochondrial NAD-dependent isocitrate dehydrogenase activity result in altered nitrate assimilation and pigmentation but do not impact growth. *Molecular Plant* 3, 156–173.

Tadmor, Y., Fridman, E., Gur, A., Larkov, O., Lastochkin, E., Ravid, U., Zamir, D. and Lewinsohn, E. (2002) Identification of *malodorous*, a wild species allele affecting tomato aroma that was selected against during domestication. *Journal of Agricultural and Food Chemistry* 50, 2005–2009.

Thi Do, P., Prudent, M., Sulpice, R., Causse, M. and Fernie, A.R. (2010) The influence of fruit load on the tomato pericarp metabolome in a *Solanum chmielewskii* introgression line population. *Plant Physiology* 154, 1128–1142.

Thomas, T.R., Shackel, K.A. and Matthews, M.A. (2008) Mesocarp cell turgor in *Vitis vinifera* L. berries throughout development and its relation to firmness, growth, and the onset of ripening. *Planta* 228, 1067–1076.

Thompson, A.J., Tor, M., Barry, C.S., Vrebalov, J., Orfila, C., Jarvis, M.C., Giovannoni, J.J., Grierson, D. and Seymour, G.B. (1999) Molecular and genetic characterization of a novel pleiotropic tomato-ripening mutant. *Plant Physiology* 120, 383–390.

Tieman, D., Taylor, M., Schauer, N., Fernie, A.R., Hanson, A.D. and Klee, H.J. (2006a) Tomato aromatic amino acid decarboxylases participate in synthesis of the flavor volatiles

2-phenylethanol and 2-phenylacetaldehyde. *Proceedings of the National Academy of Sciences USA* 103, 8287–8292.

Tieman, D.M., Zeigler, M., Schmelz, E.A., Taylor, M.G., Bliss, P., Kirst, M. and Klee, H.J. (2006b) Identification of loci affecting flavour volatile emissions in tomato fruits. *Journal of Experimental Botany* 57, 887–896.

Trainotti, L., Spinello, R., Piovan, A., Spolaore, S. and Casadoro, G. (2001) β-Galactosidases with a lectin-like domain are expressed in strawberry. *Journal of Experimental Botany* 52, 1635–1645.

Trainotti, L., Zanin, D. and Casadoro, G. (2003) A cell wall-oriented genomic approach reveals a new and unexpected complexity of the softening in peaches. *Journal of Experimental Botany* 54, 1821–1832.

Tucker, G.A. and Grierson, D. (1987) Fruit ripening. In: Davies, D. (ed.) *The Biochemistry of Plants.* Academic Press, New York, pp. 265–319.

Vicente, A.R., Saladié, M., Rose, J.K.C. and Labavitch, J.M. (2007) The linkage between cell wall metabolism and fruit softening: looking to the future. *Journal of the Science of Food and Agriculture* 87, 1435–1448.

Vrebalov, J., Ruezinsky, D., Padmanabhan, V., White, R., Medrano, D., Drake, R., Schuch, W. and Giovannoni, J. (2002) A MADS-box gene necessary for fruit ripening at the tomato ripening-inhibitor (*rin*) locus. *Science* 296, 343–346.

Vriezen, W.H., Feron, R., Maretto, F., Keijman, J. and Mariani, C. (2008) Changes in tomato ovary transcriptome demonstrate complex hormonal regulation of fruit set. *New Phytologist* 177, 60–76.

Wada, H., Shackel, K.A. and Matthews, M.A. (2008) Fruit ripening in *Vitis vinifera*: apoplastic solute accumulation accounts for preveraison turgor loss in berries. *Planta* 227, 1351–1361.

Yelle, S., Chetelat, R.T., Dorais, M., Deverna, J.W. and Bennett, A.B. (1991) Sink metabolism in tomato fruit: IV. Genetic and biochemical analysis of sucrose accumulation. *Plant Physiology* 95, 1026–1035.

Zamboni, A., Di Carli, M., Guzzo, F., Stocchero, M., Zenoni, S., Ferrarini, A., Tononi, P., Toffali, K., Desiderio, A., Lilley, K.S., Pe, M.E., Benvenuto, E., Delledonne, M. and Pezzotti, M. (2010) Identification of putative stage-specific grapevine berry biomarkers and omics data integration into networks. *Plant Physiology* 154, 1439–1459.

Zanor, M.I., Osorio, S., Nunes-Nesi A., Carrari, F., Lohse, M., Usadel, B., Kuhn, C., Bleiss, W., Giavalisco, P., Willmitzer, L., Sulpice, R., Zhou, Y.H. and Fernie, A.R. (2009a) RNA interference of LIN5 in tomato confirms its role in controlling Brix content, uncovers the influence of sugars on the levels of fruit hormones, and demonstrates the importance of sucrose cleavage for normal fruit development and fertility. *Plant Physiology* 150, 1204–1218.

Zanor, M.I., Rambla, J.L., Chaib, J., Steppa, A., Medina, A., Granell, A., Fernie, A.R. and Causse, M. (2009b) Metabolic characterization of loci affecting sensory attributes in tomato allows an assessment of the influence of the levels of primary metabolites and volatile organic contents. *Journal of Experimental Botany* 60, 2139–2154.

3 Cellular, Metabolic and Molecular Aspects of Chromoplast Differentiation in Ripening Fruit

Jean Claude Pech,[1,2]* Mondher Bouzayen[1,2] and Alain Latché[1,2]
[1]*Université de Toulouse, Castanet-Tolosan, France;* [2]*INRA, Castanet-Tolosan, France*

3.1 Introduction

Chromoplasts are non-green plastids that are responsible for the yellow, orange and red colours of many fruit. They evolve during fruit ripening by differentiation of other forms of plastids. In a number of fruit, such as tomatoes and peppers, coloured chromoplasts are derived from green chloroplasts with the disintegration of the thylakoid membranes and the formation of new carotenoid-bearing structures (Frey-Wyssling and Kreutzer, 1958; Rosso, 1968). In other fruit, such as the flesh of developing papayas, chromoplasts evolve from leuco- or proplastids, as no intermediate amyloplast or chloroplast structures are encountered (Schweiggert *et al.*, 2011). A very complex origin of chromoplasts has been found in mango where a dynamic interconversion of plastids occurs, although it was not possible to establish a sequential pattern (Vasquez-Caicedo *et al.*, 2006). In this fruit, a chloroplastic, pro-plastidial and amyloplastic origin is suspected due to the persistence of stroma thylakoid structures, of pro-plastid-like prolamellar bodies and of large starch grains, respectively. Whatever the origin of the chromoplast, the common feature is the accumulation of carotenoids due to the upregulation of carotenoid biosynthesis genes. In a few cases, such as the rind of cantaloupe melons, similar to senescent leaves, the yellowing is due to the unmasking of carotenoids as a consequence of chlorophyll degradation rather than to the new synthesis of yellow pigments (Flores *et al.*, 2001). In this situation, the yellow plastids generated from chloroplasts correspond to a senescence-specific form of plastids named gerontoplasts that cannot be considered as true chromoplasts. In contrast to all the other forms of plastids, the metabolism of gerontoplasts is solely catabolic (Matile *et al.*, 1999).

The cell biology of chromoplasts started to be elucidated with electron microscopy descriptions of their ultrastructure. Various types of chromoplast have been categorized, mainly as a function of the structure of pigment-containing bodies (Ljubesic *et al.*, 1991; Camara *et al.*, 1995). Because most of the pigments present in chromoplasts are carotenoids, the carotenoid biosynthesis

* jean-claude.pech@ensat.fr

  Fruit Ripening: Physiology, Signalling and Genomics (eds P. Nath *et al.*)

pathway and its regulation have been studied extensively (Ruiz-Sola and Rodriguez-Concepción, 2012). Nevertheless chromoplasts carry out many other functions such as synthesis of sugars, starch, lipids, aromatic compounds, vitamins (riboflavine, folate, tocopherols) and hormones (Neuhaus and Emes, 2000; Bouvier and Camara, 2007).

In recent years, transcriptomic and proteomic approaches have generated novel and extensive information on the specificity of the chromoplast proteome of the bell pepper (Siddique *et al.*, 2006), tomato (Barsan *et al.*, 2010) and sweet orange (Zeng *et al.*, 2011), and on the molecular and biochemical events occurring during the chloroplast-to-chromoplast transition in tomato (Kahlau and Bock, 2008; Egea *et al.*, 2010). In this chapter, we review the most recent findings on the metabolic shifts occurring during the biogenesis of chromoplasts and the accompanying structural and regulatory events. Most of the data comes from the study of the chloroplast-to-chromoplast transition in tomato, but reference will also be made to other fruit species.

3.2 Changes in Plastid Morphology and Structure during Chromoplast Differentiation

3.2.1 Changes in structure and morphology

Morphological changes in plastids during chromoplast differentiation have been investigated by confocal microscopy coupled with the plastid-located green fluorescent protein (GFP) (Köhler and Hanson, 2000; Waters *et al.*, 2004; Forth and Pyke, 2006). As fruit ripens, the red autofluorescence conferred by chlorophyll decreases concomitant with chlorophyll degradation, so fully ripe pericarp cells possess a large population of chromoplasts, appearing green due to the exclusive fluorescence of GFP (Forth and Pyke, 2006). Differences in plastid size and appearance among various tomato fruit tissues have been reported, with chromoplasts of the outer mesocarp having an oblong, needle-like appearance, and chromoplasts in the inner mesocarp being much larger and ovoid (Waters *et al.*, 2004). At the breaker stage, plastids show considerable intracellular variability in size and differentiation status, with the chromoplast being smaller than the chloroplasts. The chloroplast-to-chromoplast transition events are presumably not simultaneous throughout the fruit tissues, leading to a heterogeneous population of plastids within a whole fruit. However, an *in situ* real-time recording showed that the chloroplast-to-chromoplast transition was synchronous for all plastids of a single cell (Egea *et al.*, 2011). In addition, all chromoplasts are derived from pre-existing chloroplasts, thus confirming that plastid division ceases during chromoplast differentiation (Pyke and Howells, 2002; Waters *et al.*, 2004). This is further supported by the loss of several proteins involved in the plastid division machinery during the chloroplast-to-chromoplast transition in tomato fruit (Barsan *et al.*, 2012).

Fruit chromoplasts have been categorized according to the predominant structure of the carotenoid-bearing bodies, into globular (e.g. yellow and orange pepper, orange, yellow kiwi fruit), crystalline (e.g. tomato) and fibrillar (e.g. red pepper). However, it is rare that only one kind of substructure exists alone in a chromoplast. Different structures accumulating carotenoids have been described for a number of fruit by Jeffery *et al.* (2012). For instance, in mango chromoplasts, most of the carotenoids occur as globules, but different tubular membrane structures are also present (Vasquez-Caicedo *et al.*, 2006). In tomato, besides the dominant crystalloid bodies of lycopene, globuli and membranous structures are also encountered (Harris and Spurr, 1969). The structure of the carotenoid-containing bodies is of great importance for the bioavailability of carotenoids during human digestion. A decreasing order of bioavailability has been established from plastoglobules to crystals and membranes (Jeffery *et al.*, 2012). The nature of the carotenoid molecules seems to play a role in determining the type of structure. In red papayas, where lycopene is abundant, crystalloid structures are dominant, whilst in yellow papaya, rich in β-carotene and β-cryptoxanthin esters, tubulo-globular chromoplasts are predominant (Schweiggert *et al.*, 2011). Interestingly, the sequestration of carotenoids into crystals can be driven by the functional overexpression of phytoene synthase, a gene crucial for the biosynthesis of carotenoids. This has been observed in both non-green plastids of *Arabidopsis* seedlings not requiring a chromoplast developmental programme and in white carrots (Maas *et al.*, 2009). Therefore, the stimulation of the carotenoid biosynthesis pathway appears to play a major role in the development of carotenoid storage structures.

3.2.2 Plastoglobules and stromules

Plastoglobules (PGs) are oval or tubular lipid-rich subcompartments present in all plastid types, but their number and size is increased in ripening fruit in parallel with chromoplast differentiation (Harris and Spurr, 1969). PGs contain quinones, α-tocopherol and lipids and, in chromoplasts, carotenoids as well. They contain several enzymes involved in the synthesis of tocopherol and carotenoids (Austin *et al.*, 2006). Plastoglobules comprise proteins named plastoglubulins such as fibrillin, plastid lipid-associated proteins and carotenoid-associated proteins that form supramolecular lipoprotein structures with carotenoids and galacto- and phospholipids (Deruere *et al.*, 1994). PGs arise from the stroma-side layer of the thylakoid membrane. In chloroplasts, they form a functional metabolic link between the inner envelope and thylakoid membranes and play a role in the breakdown of carotenoids and oxidative stress defence (Ytterberg *et al.*, 2006). In chromoplasts, the metabolic link with thylakoids is lost and the production of plasto/phylloquinones disappears. Instead, as shown in sweet pepper chromoplasts, PGs acquire specific functions in the conversion of carotenoids through the presence of ζ-carotene desaturase (ZDS), lycopene β-cyclase (CYC-β) and two β-carotene β-hydroxylases (CrtR-β) operating in series in bicyclic carotenoid biosynthesis (Ytterberg *et al.*, 2006).

Stromules represent another structural element of plastids participating in the interconnection of plastids. They are motile protrusions of the plastid membrane into the cytoplasm that are involved in the trafficking of proteins between plastids (Köhler *et al.*, 1997; Kwok and Hanson, 2004). During chromoplast differentiation, they increase in length and number, at least in the inner mesocarp of tomato, suggesting increased interactions with the cytosol (Waters *et al.*, 2004).

3.2.3 Internal membrane remodelling

Early electron microscopy observations described a remodelling of the internal membrane system during chromoplast formation in pepper (Spurr and Harris, 1968) in which thylakoid grana and intergrana, characteristic of chloroplasts, undergo lysis while new membrane systems are formed. It has now been demonstrated in tomato fruit that the newly synthesized membranes are the site of formation of carotenoid crystals and that they are not related to the disassembly of thylakoids. Rather, they are derived from vesicles generated from the inner membrane of the plastid (Simkin *et al.*, 2007).

3.2.4 Loss of capacity for division

According to Cookson *et al.* (2003), the number of plastids remains constant during the tomato ripening process. This has been confirmed by confocal microscopy observations showing that all pre-existing chloroplasts differentiate into chromoplasts (Egea *et al.*, 2011). The cessation of plastid division in ripening tomato is accompanied by the disappearance after the mature green stage of proteins involved in plastid division such as Filamentous temperature-sensitive Z1 (FtsZ1), Accumulation and Replication of Chloroplasts (ARC6) and CRumpled Leaf (CRL) proteins (Barsan *et al.*, 2012).

3.3 Major Metabolic Shifts: Carotenoid Accumulation and Chlorophyll Degradation

3.3.1 Carotenoid biosynthesis and accumulation

The most visible change in chromoplast biogenesis is the accumulation of carotenoids. The carotenoid biosynthetic pathway (Plate 1) shows that different carotenoid derivatives accumulate in different fruits: lycopene in tomato, watermelon, guava, papaya and grapefruit; β-carotene in cantaloupe melon, mango, pumpkin and apricot; zeaxanthin in orange pepper, citrus fruit and persimmon; and capsanthin and capsorubin in red pepper. However, besides the dominant compounds, other carotenoids can also be present at reduced concentrations.

During the chloroplast-to-chromoplast transition, specific carotenoid biosynthesis genes are expressed. The first committed step in carotenoid biosynthesis corresponds to the condensation of two geranylgeranyl diphosphate molecules into phytoene, which is catalysed by phytoene synthase (PHY). In tomato fruit, two *PHY* genes are expressed. Phytoene synthase 1 (*PHY1*) is highly expressed in ripening fruit and is responsible for the formation of chromoplastic carotenoids, whilst phytoene synthase 2 (*PHY2*), which is responsible for the formation of chloroplastic carotenoids, is expressed exclusively in green tissues and therefore makes no contribution to carotenoid biosynthesis in ripening fruit (Fraser *et al.*, 1999).

The accumulation of lycopene in regular tomatoes is related to the low expression of the lycopene β-cyclase (*CYC-B*) gene involved in the conversion of lycopene into β-carotene (Ronen *et al.*, 1999). Proteomics studies dealing with the quantitative analysis of proteins involved in the carotenoid pathway during the transition from mature green to red tomatoes have shown that PHY1, ZDS and carotene isomerase (CRTISO or CIS) undergo a strong increase in abundance, whilst geranylgeranyl diphosphate synthase and phytoene desaturase remain equally abundant during the chloroplast-to-chromoplast transition (Fig. 3.1). Interestingly, proteins downstream of lycopene, such as CYC-B, were detected at low levels only and could not be quantified. The low level of the CYC-B protein reported in this proteomics study is consistent with the low expression of the corresponding gene, as mentioned above. When tomato fruit accumulate β-carotene, such as in the *Delta* mutants, the expression of the *CYC-B* gene is elevated (Ronen *et al.*, 1999), thus demonstrating the crucial role of CYC-B in controlling the accumulation of lycopene or β-carotene.

In Navel oranges (*Citrus sinensis*), two *CYC-B* genes are expressed, *CsB-LYC1* and *CsB-LYC2*, but only *CsB-LYC2* is expressed specifically in the chromoplast (Alquezar *et al.*, 2009). *CsB-LYC1* is expressed at low levels during fruit ripening, whilst *CsB-LYC2* shows a strong induction in the pulp and peel of Navel orange fruit. The accumulation of lycopene in the pulp of red grapefruit is associated with a low expression level of both *CsB-LYC2* and β-carotene hydroxylase (*CHX-B*) compared with Navel orange fruit. In addition, two alleles of *CsB-LYC2* (*a* and *b*) are present in Navel orange and in red grapefruit. The non-functional *b* allele is expressed in red grapefruit, whilst the functional allele *a* is expressed in Navel orange (Alquezar *et al.*, 2009). The yellow colour of the lemon fruit flavedo is due to the accumulation, in decreasing order of importance, of lutein, β-carotene and α-carotene. It is associated with a low expression of carotenoid biosynthesis genes, surprisingly including those leading to the β,ε-branch of carotenoids: ε-lycopene cyclase (*CYC-E*) and *CHX-B* (Kato *et al.*, 2004). In the flavedo of sweet orange, phytoene desaturase expression correlates well with the carotenoid content in developing fruit

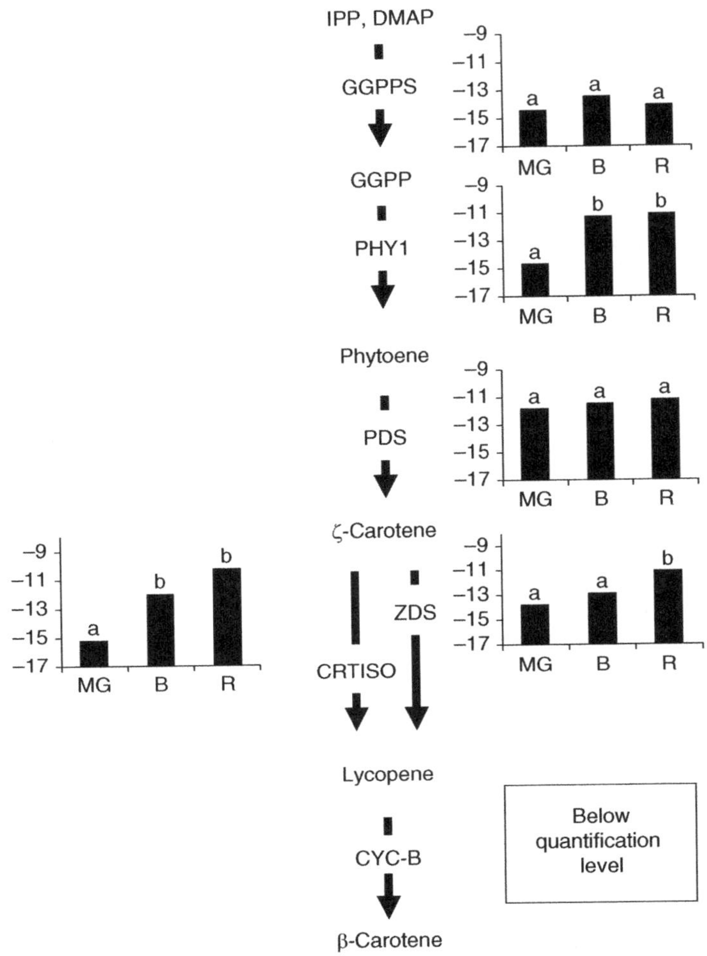

Fig. 3.1. Abundance of proteins of the carotenoid biosynthetic pathway leading to the accumulation of lycopene in tomato plastids during the chloroplast-to-chromoplast transition determined by quantitative proteomics. Proteins downstream of the pathway were absent or at very low abundance. Values of abundance ($\log_2$) are statistically different when indicated by different letters. MG, mature green; B, breaker; R, red; IPP, isopentenyl phosphate; DMAP, dimethylallyl diphosphate; GGPPS, geranylgeranyl diphosphate synthase; see Plate 1 for other abbreviations. Data from Barsan *et al.* (2012).

and the upregulation of *PHY* and *ZDS* genes at the onset of fruit coloration enhances the production of linear carotenes and the flux into the pathway. The shift from the β,ε-branch carotenoids (lutein and α-carotene) present in green fruit to the β,β-branch (e.g. β-carotene, zeaxanthin) present in red fruit occurs during the transition from chloroplast to chromoplast with the downregulation of *CYC-E* and upregulation of the *CYC-B* and *CHX-B* genes (Rodrigo *et al.*, 2004). The flavedo of orange fruit accumulates large amounts of zeaxanthin. This correlates well with a high expression of *CHX-B* converting β-carotene into zeaxanthin and of upstream genes providing the β-carotene substrate (Rodrigo *et al.*, 2004).

A chromoplast-specific *CYC-B* also appears to control the colour of papaya fruit. A mutation of the gene results in the accumulation of lycopene in red papayas, whilst in yellow papayas, the gene is thought to be functional allowing the accumulation of β-carotene (Blas *et al.*, 2010). The presence of β-carotene in red papayas is probably due to the expression of another gene, a chloroplast-specific *CYC-B* expressed at low levels in fruit (Blas *et al.*, 2010).

Zeaxanthin is converted into antheraxanthin and violaxanthin present in yellow peaches and orange flavedo by violaxanthin de-epoxidase and zeaxanthin epoxydase. In red peppers, the accumulation of red pigments, capsanthin and capsorubin, is associated with the expression of the capsanthin–capsorubin synthase (*CCS*) gene (Lefebvre *et al.*, 1998; Ha *et al.*, 2007). In yellow pepper, deletion of the *CCS* gene has been considered as responsible for the phenotype (Lefebvre *et al.*, 1998), although mutations of the gene rather than complete deletion can also be responsible for the yellow ripening phenotype (Ha *et al.*, 2007). The orange colour of some peppers can be due to either a deletion of the N terminus of CCS (Lang *et al.*, 2004) or a mutation of the *PHY* gene (Huh *et al.*, 2001). Other processes controlling the colour in some orange-pigmented peppers have also been described, such as post-transcriptional and post-translational regulations leading to the inhibition of expression of the *CCS* gene and the synthesis of the CCS protein (Rodriguez-Uribe *et al.*, 2012).

3.3.2 Loss of proteins involved in photosynthesis and in the machinery for the biogenesis of thylakoids

The loss of chlorophyll-synthesizing capacity during the chloroplast-to-chromoplast transition results in a strong decline in photosynthetic activity (Piechulla *et al.*, 1987; Carrara *et al.*, 2001). Chlorophyll-free chromoplasts isolated from daffodil flowers were unable to synthesize *in vitro* the isocyclic chlorophyll ring, whilst they were capable of generating all compounds of the upstream pathway of tetrapyrroles until the magnesium-protoporphyrin IX monomethyl ester (Lützow and Kleinig, 1990). A similar observation has been made in proteomic studies of tomato fruit plastids, indicating that, among proteins of the tetrapyrrole biosynthetic pathway, proteins involved in the chlorophyll biosynthesis branch downstream of protoporphyrin IX are absent in red chromoplasts, whilst all proteins upstream of the chlorophyll formation branch remain present in red chromoplasts (Barsan *et al.*, 2012). The photosynthetic apparatus is strongly impaired during chromoplast differentiation with a decrease in the amount of proteins and downregulation of many photosynthetic genes (Piechulla *et al.*, 1985). In bell pepper, for instance, nuclear genes encoding chloroplast proteins such as the major chlorophyll *a*/*b*-binding protein and the small subunit of ribulose-1,5-bisphosphate carboxylase disappeared in fruit chromoplasts (Kuntz *et al.*, 1989). Surprisingly, the expression of several plastid-encoded photosynthetic genes remains high in fruit chromoplasts of tomatoes (Piechulla *et al.*, 1985), peppers (Kuntz *et al.*, 1989) and squash (Obukosia *et al.*, 2003). In particular, transcripts of the plastid genes coding for the rbcL and psbA proteins were detected in fruit chromoplasts sometimes at higher levels than in chloroplasts (Kuntz *et al.*, 1989; Obukosia *et al.*, 2003). It was also demonstrated that the rbcL protein remained stable during the chloroplast-to-chromoplast transition in tomato (Barsan *et al.*, 2012). The function of proteins of the photosynthetic apparatus in chromoplasts, at a stage where chlorophylls have totally disappeared, is unknown. Early observations by electron microscopy showed that the differentiation of chromoplasts comprised a lysis of the grana and thylakoids (Rosso, 1968; Spurr and Harris, 1968), in agreement with the loss of photosynthetic activity. Proteomic studies have revealed that the abundance of a number of proteins participating in the build-up of thylakoids strongly decreases during the chloroplast-to-chromoplast transition in tomato and that they ultimately become absent in red chromoplasts (Table 3.1). They participate in processes such as the formation of thylakoid membranes, vesicular trafficking, protein import, provision of precursors, and assembly or repair of photosystems. The increase in Stay-GReen (SGR) protein is interesting. A mutation of the *SGR* gene prevents chlorophyll degradation (Thomas *et al.*, 1999). SGR interacts with five chlorophyll catabolic enzymes at the light-harvesting complex II to ensure chlorophyll degradation during leaf senescence in *Arabidopsis* (Sakuraba *et al.*, 2012).

3.4 Metabolic Re-orientations: Carbohydrate and Lipid Metabolisms

3.4.1 The Calvin cycle and oxidative pentose phosphate pathway (OxPPP)

During the chloroplast-to-chromoplast transition of sweet pepper fruit, the activity of transaldolase, which participates in the regenerative phase of the Calvin cycle, increases considerably (Thom *et al.*, 1998). In ripening tomato fruits, several enzymes of the Calvin cycle remain active (Obiadalla-Ali *et al.*, 2004). However, interference of the Calvin cycle with photosynthesis disappears due to degradation of the photosynthetic apparatus so that the Calvin cycle operates only for the recycling of carbon within the OxPPP. The OxPPP remains active during fruit ripening as indicated by a high activity of glucose-6-phosphate dehydrogenase (G6PDH), a key component of the OxPPP. G6PDH levels are even higher in fully ripe tomato fruit chromoplasts than in leaves or green fruits (Aoki *et al.*, 1998). A functional OxPPP has also been encountered in isolated buttercup chromoplasts (Tetlow *et al.*, 2003). The role of OxPPP in the chromoplasts is probably to support metabolic activities within the organelle.

3.4.2 Starch biosynthesis

In many fruits, including apples, bananas and kiwifruit, starch accumulates throughout development as granules in plastids and then undergoes complete degradation at maturity. In growing tomato and kiwi fruit, starch can reach up to 20 and 50% of the fruit dry weight, respectively. Starch pattern tests are sometimes used as maturity indices for some fruit such as apples. Starch accumulation results from an imbalance between synthesis and degradation. Indeed, tomato fruit can synthesize starch during the period of net starch breakdown, illustrating that these two mechanisms coexist (Luengwilai and Beckles, 2009). Proteins of both starch synthesis and starch degradation have been encountered in the plastids of ripening tomato fruit, including in chromoplasts (Bian *et al.*, 2011), supporting the persistence of intense starch turnover during chloroplast-to-chromoplast conversion. Also supporting starch turnover is the presence in chromoplasts of a glucose translocator for the export of sugars generated by starch degradation (Bian *et al.*, 2011). In olive fruit, active expression of a glucose transporter gene was observed at full maturity when the chromoplasts were devoid of starch (Butowt *et al.*, 2003). The absence of starch accumulation in plastids and the prevalence of degradation can be related to the action of proteins that facilitate the action of β-amylases, named starch-excess proteins (SEX1, corresponding to glucan water dikinase, and SEX4, corresponding to phosphoglucan phosphatase), which have been encountered in tomato fruit plastids, although they decrease in abundance during chromoplastogenesis (Barsan *et al.*, 2012). Plants harbouring mutants of these proteins accumulate large amounts of starch (Zeeman *et al.*, 2010). Interestingly, a β-amylase protein (Solyc01g067660) increases in abundance during the differentiation of chromoplasts (Barsan *et al.*, 2012). Another important regulatory mechanism is related to the presence in the tomato chromoplast proteome (Barsan *et al.*, 2010, 2012) of orthologues of the 14-3-3 proteins of the ε-subgroup family, which have been shown to be involved in the regulation of starch accumulation in *Arabidopsis* (Sehnke *et al.*, 2001). The 14-3-3 proteins participate

Table 3.1. List of proteins involved in thylakoid synthesis and assembly of photosystems that disappear during the chloroplast-to-chromoplast transition in tomato. The stay-green protein involved in the regulation of chlorophyll degradation shows increasing abundance. Data from Barsan *et al.* (2012).

Abbreviation	Full name	Functional description	Solyc or GI code	Reference
Disappear durng chloroplast-to-chromoplast transition				
MFP1	Matrix attachment region filament-binding protein 1	Accumulation of thylakoid membranes	Solyc03g120230	Jeong *et al.* (2003)
THF1	Thylakoid formation 1	Thylakoid formation through vesicular trafficking	Solyc07g054820	Huang *et al.* (2006)
FtsH2	Filamentous temperature-sensitive metalloprotease 2	Protease involved in the repair of PSII	Solyc07g055320	Liu *et al.* (2010a); Kato *et al.* (2012)
FtsH5	Filamentous temperature-sensitive metalloprotease 5	Protease involved in the repair of PSII	Solyc04g082250	Liu *et al.* (2010a); Kato *et al.* (2012)
FtsH6	Filamentous temperature-sensitive metalloprotease 6	Protease involved in the repair of PSII	Solyc02g081550	Liu *et al.* (2010a); Kato *et al.* (2012)
FtsH12	Filamentous temperature-sensitive metalloprotease 12	Protease involved in the repair of PSII	Solyc02g079000	Liu *et al.* (2010a); Kato *et al.* (2012)
EGY1	Ethylane-dependent gravitropism-deficient and yellow mutant 1	Thylakoid membrane biogenesis	Solyc10g081470	Chen *et al.* (2005)
SECA1	Sec translocase 1	Protein import	Solyc01g080840	Liu *et al.* (2010b)
SRP54	Signal recognition particle protein 54	Integration of LHCP into the thykaloid	Solyc09g009940	Li *et al.* (1995); Rutschow *et al.* (2008)
HDR	Hydroxymethylbutenyl diphosphate reductase	Production of MEP precursors for carotenoid synthesis	Solyc01g109300	Botella-Pavia *et al.* (2004)
LPA1	Low PSII accumulation protein 1	Assembly of PSII	Solyc09g074880	Peng *et al.* (2006)
LPA3	Low PSII accumulation protein 3	Assembly of PSII	Solyc06g068480	Peng *et al.* (2006); Cai *et al.* (2010)
HCF13	High chlorophyll fluorescence 136	Assembly of PSII	Solyc02g014150	Plücken *et al.* (2002)
YCF4	Hypothetical chloroplast reading frame family	Assembly of PSI	GI89241682	Krech *et al.* (2012)
Increase during chloroplast-to-chromoplast transition				
SGR	Stay-green protein	Regulates chlorophyll degradation	Solyc08g080090	Sakuraba *et al.* (2012)

in phosphorylation-mediated regulatory functions in plants and are involved in the post-translational regulation of enzymes of starch metabolism (Sehnke *et al.*, 2001; Tetlow *et al.*, 2004).

3.4.3 Lipid synthesis and metabolism

Lipids fulfil a specific storage/sequestering function in chromoplasts. The amount of phospholipids increases markedly during the chloroplast-to-chromoplast transition in relation to the *de novo* synthesis of new membranes required for the sequestration and growth of lycopene crystals (Lenucci *et al.*, 2012). Plastid lipids are located in both plastoglobules and membranes. In tomato, the total amount of diacylglycerides remains almost unchanged, whilst the triacyl glyceride content markedly decreases and the amount of phytosterol and phospholipids increases during plastid transition from chloroplasts to chromoplasts (Lenucci *et al.*, 2012). The lipid composition of chromoplast membranes has scarcely been evaluated specifically. However, some data are available for daffodil flowers, showing that phosphoglycerolipids constitute a minor proportion of the lipid content, whilst galactoglycerolipids are abundant (Liedvogel and Kleinig, 1977), as in chloroplasts (Andersson and Dörmann, 2009). In chloroplasts, the envelope contains very low amounts of sterols, and the thylakoid membrane is devoid of sterols (Hartmann, 1998). In daffodil chromoplasts, membrane-free sterols, steryl glycosides and acetylated steryl glycosides have been found associated with the activities of glycosylation and acylation of sterols (Liedvogel and Kleinig, 1977). During tomato fruit ripening, plastids undergo a four- to five-fold increase in phytosterols and phospholipids (Lenucci *et al.*, 2012), suggesting that the newly synthesized membranes of chromoplasts have a specific composition in terms of sterols. Chromoplasts possess the entire metabolic equipment for the synthesis of fatty acids. Isolated tomato chromoplasts are capable of synthesizing lipids upon uptake of glucose, pyruvate and malate precursors (Angaman *et al.*, 2012). The first step corresponding to the activation of acetate into acetyl CoA through the action of a pyruvate dehydrogenase has been characterized in daffodil chromoplasts (Kleinig and Lievogel, 1978). A chromoplastic pyruvate dehydrogenase has been encountered in tomato (Barsan *et al.*, 2010) and in sweet orange (Zeng *et al.*, 2011) chromoplast proteomes. Interestingly the amount of protein remains stable during chloroplast-to-chromoplast differentiation in tomato (Barsan *et al.*, 2012). All the subunits of acetyl CoA carboxylase involved in the formation of malonyl CoA from acetyl CoA, including the ACCD plastid-encoded protein, are present, generally at stable levels during chromoplast differentiation. The ACCD protein is the major product of the transcription and translation machinery of the chromoplast (Kahlau and Bock, 2008). Sustained biosynthesis of lipids therefore occurs in chromoplasts, probably for providing a lipid storage matrix for the accumulation of carotenoids. Among the proteins involved in lipid metabolism, of special interest is the presence in the tomato and sweet orange chromoplast proteomes of several proteins of the lipoxygenase (LOX) pathway leading to the generation of aroma volatiles (Barsan *et al.*, 2010; Zeng *et al.*, 2011). Fatty acids synthesized in the plastids can have several destinations: either they remain in the plastid or they are exported to the endoplasmic reticulum where they are glycosylated and phosphorylated (Wang and Benning, 2012). Analysis of the changes in the plastid proteome during tomato fruit ripening (Barsan *et al.*, 2012) indicates that proteins involved in the export pathway (acyl carrier proteins (ACPs)) remain abundant, whilst proteins of the endoplasmic reticulum pathway (trigalatosyldiacetylglycerol protein), which act as import proteins (Xu *et al.*, 2010), decrease in abundance. As the phospholipid content of plastids is increased significantly during fruit ripening (Lenucci *et al.*, 2012), it can be concluded that the increase in phospholipids observed in plastids during fruit ripening originates mostly from *in situ* biosynthesis through plastid phosphatidic acid phosphatase and lysophospatidic acid:acyl-ACP acyltransferase (Wang and Benning, 2012).

3.5 Development of an Active Antioxidant System

Reactive oxygen species (ROS) synthesized during fruit ripening contribute to the peroxidation of lipids and deterioration of membranes and are therefore deleterious for cell metabolism (Jimenez *et al.*, 2002). In order to counteract the action of ROS, fruit synthesize a number of molecules having antioxidant capacity. The antioxidant system includes reactive oxygen scavenging enzymes, superoxide dismutases, catalase and enzymes of the ascorbate glutathione cycle, which is a series of coupled redox reactions involving four enzymes: ascorbate peroxidase, monodehydroascorbate reductase, dehydroascorbate reductase and glutathione reductase, as well as $NADP^+$ dehydrogenases and the thiol-specific antioxidant proteins including peroxiredoxins (Bernier-Villamor *et al.*, 2004; Finkemeier *et al.*, 2005). The main non-enzymatic antioxidant molecules are ascorbate and glutathione, which are also components of the above-mentioned cycle, and α-tocopherol, α-carotene and flavonoids (Noctor and Foyer, 1998). The activity of antioxidant enzymes (superoxide dismutase and enzymes of the ascorbate–glutathione cycle) increases significantly in pepper plastids during the differentiation of chloroplasts into chromoplasts, as well as the content of ascorbate and glutathione (Marti *et al.*, 2009). Lipids, rather than proteins, seem to be an oxidation target in chromoplasts. The role of a high antioxidant system in the chromoplast could be to protect plastid components such as carotenoids against oxidation but also to mediate signalling between the chromoplast and the nucleus. ROS have been implicated in plastid-to-nucleus communication (Kleine *et al.*, 2009; Galvez-Valdivieso and Mullineaux, 2010). The role of ROS as second messengers has been demonstrated for the biosynthesis of carotenoids. Thus, ROS generated chemically in green pericarp discs of pepper fruits rapidly and simultaneously induced the expression of multiple carotenogenic genes responsible for the accumulation of capsanthin (Bouvier *et al.*, 1998). In addition, the application of ROS-generating compounds to pepper protoplasts transiently transformed with the CCS promoter coupled with the β-glucuronidase (*GUS*) reporter gene resulted in strong activation of the *GUS* gene (Bouvier *et al.*, 1998).

3.6 Import of Proteins and Metabolites and Provision of Energy Accompanying the Metabolic Changes

3.6.1 Import of proteins

A Toc/Tic (translocon at the outer/inner envelope membrane of chloroplast) import machinery performs the translocation of proteins carrying a transit peptide (Jarvis, 2008). In tomato (Barsan *et al.*, 2010) and sweet orange (Zeng *et al.*, 2011) fruit chromoplasts, Toc/Tic proteins have been encountered together with chaperonin-associated proteins, indicating that the system is probably still active in chromoplasts. In contrast, proteins involved in internal trafficking through the thylakoid to the lumen are absent as a result of the loss of thylakoid structure. However, intracellular vesicular transport as described in chloroplasts (Westphal *et al.*, 2001) seems to persist in chromoplasts with the presence of several proteins homologous to yeast vesicular trafficking components (Andersson and Sandelius, 2004; Barsan *et al.*, 2010).

3.6.2 Provision of energy, and import of precursors and metabolites from the cytosol

The synthesis of metabolites such as lipids and sugars depends greatly upon the provision of energy and import of precursors into the plastid. An adenylate (ATP/ADP) translocator has been characterized in *Narcissus* chromoplasts that is suggested to provide ATP for supporting biosynthetic activity in the plastid, as ATP external to chromoplasts stimulates fatty acid biosynthesis (Liedvogel and Kleinig, 1980). During the chromoplast differentiation process in tomato, ATP synthase subunits (nuclear and plastid encoded), as well as an ADP/ATP carrier and transporters of glucose-6-phosphate, phosphoenolpyruvate and triosephosphate, remain abundant (Barsan *et al.*, 2012). Isolated tomato fruit chromoplasts are capable of *de novo* ATP production through a respiratory pathway using NADPH as the electron donor. ATP synthesis involves an ATP synthase harbouring an atypical γ-subunit, which is induced during ripening and replaces the γ-subunit present in tomato leaf and green fruit chloroplasts (Pateraki *et al.*, 2013). These data indicate that the machinery for the provision of energy and precursors keeps very active to allow the synthesis of fatty acids and sucrose within the chromoplasts. It has been suggested that lipids or lipid precursors can be imported into the plastids via vesicles derived from the endoplasmic reticulum membrane and fused with Golgi membranes (Andersson and Dörmann, 2009; Benning, 2009). Interestingly, two proteins of the SEC translocase system, homologues of a yeast phosphatidylinositol transfer protein (Yakir-Tamang and Gerst, 2009), are highly expressed and increase continuously in abundance during the biogenesis of chromoplasts in tomato (Barsan *et al.*, 2012).

3.7 Regulatory Events Controlling Chromoplast Differentiation and Development of Carotenoid Storage Structures

Some genes have been described as potential players in regulating the process leading to the transition from choloroplast to chromoplast based mainly on the upregulation of their expression during the developmental shift corresponding to the onset of fruit ripening. Among these, the early light-inducible protein (*ELIP*) gene, showing homology with light-harvesting complex proteins, displays elevated expression during the breaker/turning stages of fruit ripening in tomato. Yet direct evidence for the role of *ELIP* in chromoplast differentiation is still lacking. New prospects of uncovering the mechanisms regulating chromoplast differentiation have been provided by the discovery that a mutated version of the cauliflower *Or* gene results in the accumulation of large amounts of β-carotene in tissues typically devoid of carotenoids (Li *et al.*, 2001; Lu *et al.*, 2006). The *Or* gene encodes a plastid-associated protein containing a cysteine-rich domain present in DnaJ-like chaperones and expression of the *Or* mutated form of the gene confers an orange pigmentation without significantly affecting the expression of carotenoid biosynthetic genes (Lu *et al.*, 2006; Lopez *et al.*, 2008). The *Or* mutation consists of the insertion of a retrotransposon in the *Or* gene that leads to the generation of multiple splicing variants. Strikingly, only ectopic expression of the *Or* mutated form is able to induce carotenoid accumulation in different plant species, whilst neither the wild-type gene nor the mutated version yielding the various splicing forms can reproduce the carotenoid-accumulating phenotype when expressed in the plant (Lu *et al.*, 2006). Moreover, downregulation of the *Or* gene also fails to produce the *Or*-associated phenotype, which, taken together with the absence of phenotypes in the overexpressing lines, suggest that carotenoid accumulation in the *Or* mutants results from a dominant-negative mutation (Lu *et al.*, 2006; Giuliano and Diretto, 2007). Even though the mechanisms by which the *Or* mutation works remain obscure, the functional role of the Or protein appears to be essential for the differentiation of uncoloured plastids into chromoplasts, which creates a deposition sink for carotenoid accumulation (Li and van Eck, 2007). Another remarkable feature displayed by the *Or* mutant is that chromoplast differentiation occurs mostly in the inflorescence tissues but not in the leaves, with the development of carotenoid membranous inclusions reminiscent of those found in carrot roots. Moreover, plastid divisions are arrested in the *Or* mutant, which may explain why the number of chromoplasts in affected cells is limited to one or two (Paolillo *et al.*, 2004). The *Or* mutant supports the idea that carotenoid sequestration capacity is an essential mechanism by which carotenoid stability is regulated and that creating a sink structure may confer the ability to accumulate provitamin A compounds in crops naturally devoid of carotenoids. In validating this hypothesis, tuber-specific expression of the *Or* gene in transgenic potato was found to result in enhanced accumulation of carotenoids without altering the expression of endogenous carotenoid biosynthetic genes, thus demonstrating that *Or* can provide an efficient genetic tool for improving the nutritional value of food crops (Lu *et al.*, 2006; Lopez *et al.*, 2008; Li *et al.*, 2012).

In addition to Or, other proteins may also participate in the differentiation of chromoplasts, such the plastidial heat-shock protein HSP21, which has been implicated in the promotion of colour changes during tomato fruit maturation (Neta-Sharir *et al.*, 2005). Some ATP-dependent casein lytic proteinases located in the stroma and known to participate in chloroplast development (Lee *et al.*, 2006) remain abundant during chloroplast-to-chromoplast transition, suggesting a sustained function during chromoplast differentiation (Barsan *et al.*, 2012).

3.8 Plastid-to-nucleus Communication

It is acknowledged that nuclear and plastid genes require tight coordination for expression. This involves an intracellular signalling network where plastids emit signals that regulate nuclear gene expression. Recent reviews on the topic (Waters *et al.*, 2009; Barajas-Lopez *et al.*, 2012) classify the plastid signals into five types originating from: (i) tetrapyrrole biosynthesis (Mg-protoporphyrin IX); (ii) carotenoid biosynthesis (methylerythritol cyclodiphosphate, MEcPP); (iii) ROS- and redox-related processes; (iv) metabolite pool changes; and (v) plastidial gene expression. As carotenoid-, ROS- and redox-related metabolisms are enhanced in ripening fruit, it may be assumed that signals originating from these metabolisms play a role in chromoplast formation. However, our knowledge of plastid-to-nucleus communication is restricted to the development of chloroplasts, studied by characterizing mutants impaired for chloroplast development such as the *gun* mutants. Several of the *gun* mutants are affected in tetrapyrrole biosynthesis, resulting in alteration of the expression of photosynthesis-associated genes (Woodson *et al.*, 2011).

The fact that several fruit ripening mutants are altered in terms of plastid development (see Chapter 15, this volume) supports the presence of plastid-to-nucleus signalling pathways in ripening fruit.

Compounds of the tetrapyrrole pathway have long been identified as possible messengers involved in retrograde signalling in chloroplasts. In chromoplasts, proteins of the upper part of the pathway up to the synthesis of Mg-protoporphyrin are present. Only the route to chlorophyll biosynthesis has disappeared (Barsan *et al.*, 2012). A possible role of tetrapyrrole signalling has been found from a study of the Golden 2-like 2 (*GLK2-like*) mutation in tomato. An allele of the *GLK2-like* gene encoding a truncated loss of function of the GLK protein is responsible for the uniformly light green colour and the absence of a green shoulder in many modern varieties of tomatoes (Powell *et al.*, 2012). In *Arabidopsis*, Golden 2-like 1 and 2 (GLK1/2) transcription factors are involved in the regulation of genes of the tetrapyrrole biosynthesis pathway and are required for chloroplast development (Waters *et al.*, 2009). In addition, the expression of GLKs responds to retrograde signals from the plastid. An alteration of tetrapyrrole biosynthesis in a tomato GLK2-like mutant can therefore be considered a plastid signal that will regulate nuclear gene expression, which will alter the fruit ripening phenotype. Evidence has been provided for the participation of an upstream metabolite of the tetrapyrrole pathway, 5-aminolevulinic acid (ALA), in retrograde signalling in *Arabidopsis* chloroplasts (Czarnecki *et al.*, 2012). Although the amount of glutamate 1-semialdehyde aminotransferase protein involved in the synthesis of ALA decreases during chloroplast-to-chromoplast transition (Barsan *et al.*, 2012), a role for ALA in retrograde signalling in chromoplasts remains possible.

The potential participation of components of the carotenoid biosynthetic pathway has been mentioned in *Arabidopsis* where a retrograde signalling mutant, *ceh1*, has been identified. This mutant is caused by a mutation in the 1-hydroxy-2-methyl-2-(E)-butenyl 4-diphosphate synthase *(HDS)* gene of the plastidial methylerythritol phosphate (MEP) pathway (Xiao *et al.*, 2012). The mutation results in the accumulation of high levels of the intermediate metabolite MEcPP, which, under stress conditions, stimulates the expression of nuclear genes, particularly those encoding the plastid-localized proteins hydroperoxide lyase involved in the production of aldehydes in the LOX pathway, and isochorismate synthase 1 involved in salicylic acid biosynthesis. The MEP pathway is active in ripening tomato fruits with high expression of the 1-deoxy-D-xylulose 5-phosphate synthase gene, the first step in the pathway (Lois *et al.*, 2000). The role of MEcPP signalling in chromoplast development deserves special attention inasmuch as hydroperoxide lyase has been encountered at substantial levels in chromoplasts (Barsan *et al.*, 2012) and has been involved in the synthesis of LOX-derived aroma volatiles.

Another argument for plastid-to-nucleus signalling can be found in the tomato fruit *lutescent1* (*l1*) and *lutescent2* (*l2*) mutants. These mutants are affected in terms of chloroplast development and possess chlorophyll-deficient phenotypes. The onset of fruit ripening is delayed by approximately 1 week, although, once ripening is initiated, the mutant fruit ripen at a normal rate and accumulation of carotenoids is not impaired (Barry *et al.*, 2012). The gene responsible for the *l2* mutation is the chloroplast-targeted zinc metalloprotease of the M50 family, which is homologous to the *Arabidopsis* gene *ETHYLENE-DEPENDENT GRAVITROPISM DEFICIENT AND YELLOW-GREEN1* (*EGY1*) (Chen *et al.*, 2005). The EGY1 protein has been detected at the mature green stage of tomato only (Barsan *et al.*, 2012). These data suggest a role for the chloroplast in mediating the onset of fruit ripening in tomato and indicate that chromoplast development in fruit does not depend on functional chloroplasts. A schematic representation of the possible signalling routes operating in plastids during fruit ripening is presented in Fig. 3.2. All these data strongly suggest the presence of plastid-to-nucleus signalling in ripening fruit, but we are still far from identifying the molecular events governing the interactions between plastid and nucleus and controlling the expression of ripening-related genes.

3.9 Conclusions and Perspectives

The processes leading to the formation of chromoplasts represent a major event in fruit ripening in many fruiting species. It corresponds to the conversion of chloroplast-like green plastids into the non-green plastids named chromoplasts associated with the accumulation of carotenoids. The process involves profound structural changes including the lysis of the thylakoid membrane system. In parallel, carotenoid-bearing bodies emerge surrounded by newly synthesized membranes derived from the inner membrane of the plastid (Fig. 3.3). These carotenoid-bearing bodies accumulate carotenoids in the form of fibrillar or crystalloid structures in combination with polar lipids and plastid- or carotenoid-associated proteins

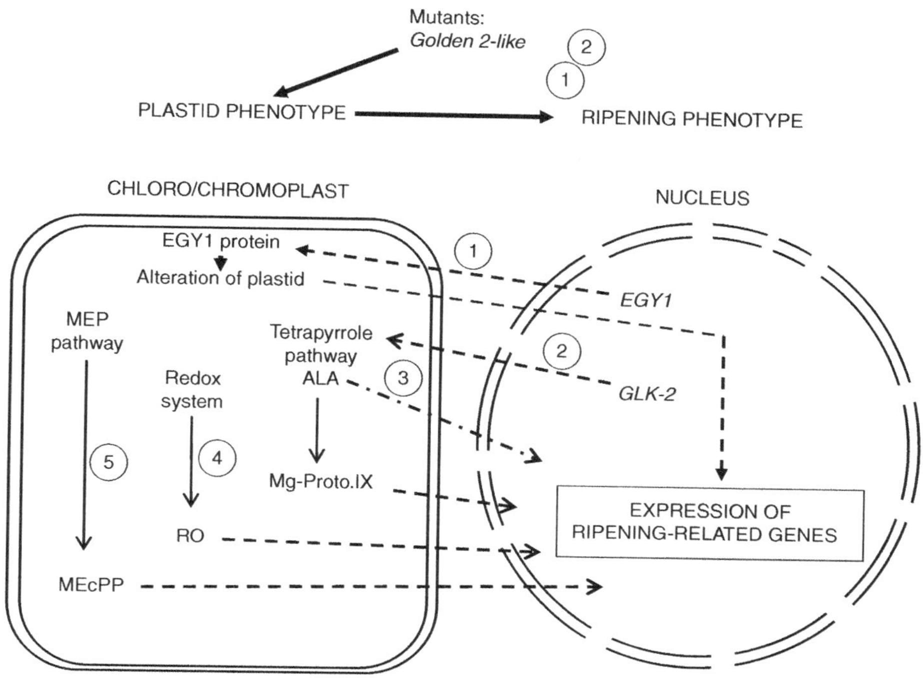

Fig. 3.2. Schematic representation of the possible signalling routes operating during the chloroplast-to-chromoplast transition between the plastids and the nucleus for regulating the expression of nuclear-encoded ripening-related genes. Experimental support for a role in plastid-to-nucleus signalling arises from the results of: (1) Barry *et al.* (2012) for the EGY1 protein of the *lutescent-2* mutant; (2) Powell *et al.* (2012) for the Golden 2-like protein (GLK-2) via the tetrapyrrole pathway; (3) Czarnecki *et al.* (2011) for 5-aminolevulinic acid (ALA); (4) Bouvier *et al.* (1998) and Marti *et al.* (2009) for reactive oxygen species (ROS); and (5) Xiao *et al.* (2012) for the methylerythritol phosphate (MEP) pathway via methylerythritol cyclodiphosphate (MEcPP).

that play a role in the sequestration of the carotenoids. The chromoplast acts as a metabolic sink requiring the presence of active systems for provision of energy (ATPases) for import of proteins and lipid precursors (Fig. 3.3). Trafficking between the plastid and the cytosol is also stimulated by the increase in size and length of stromules. The carotenoid biosynthesis pathway located in the plastid membrane is strongly upregulated. Part of the carotenoid biosynthesis pathway remains active in plastoglobules, which persist from green plastids and undergo enlargement during the ripening process.

Recent progress has been made into understanding the regulatory mechanisms governing the differentiation of chromoplasts. The discovery of the *Or* gene, which controls differentiation of non-coloured plastids into chromoplasts, has been an important step. Other regulatory genes, such as specific heat-shock proteins and ATP casein lytic proteinases, are also candidates for regulating chromoplast differentiation. Plastid-to-nucleus communication is another new field of investigation. Significant progress has been made and numerous reports published on the discovery of signalling events arising from chloroplasts to control the expression of nuclear-encoded photosynthesis genes. Research is much less advanced for chromoplasts. However, experimental data are starting to emerge that support the presence of plastid-to-nucleus signalling in ripening fruit. These discoveries open new perspectives

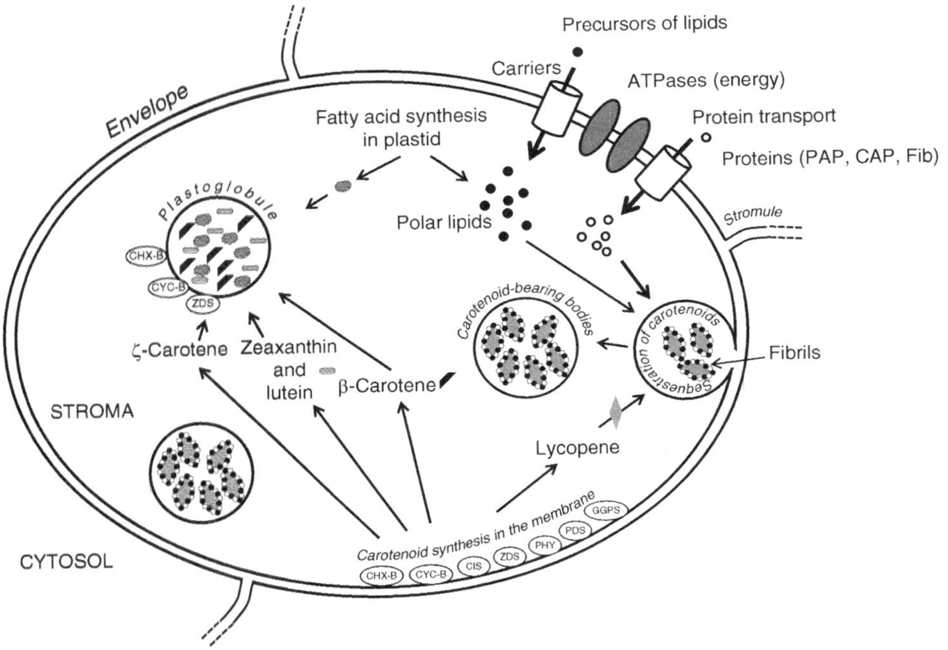

Fig. 3.3. Representation of a fully differentiated chromoplast acting as a metabolic sink with the accumulation of carotenoid-bearing structures. The following elements contribute to the metabolic sink: stimulation of carotenoid biosynthesis in the membrane, emergence of vesicles from the inner membrane that generate carotenoid-bearing bodies, import of proteins and lipid precursors that participate in the sequestration of carotenoids, active energy provision through ATPases, stimulation of the development of stromules, and accumulation of quinones, carotenoids and lipid material in plastoglobules. PAP, plastid lipid-associated protein; CAP, carotenoid-associated protein; Fib, fibrillin; see Plate 1 for abbreviations for the enzymes of the carotenoid biosynthesis pathway.

towards the discovery of the intimate molecular and signalling mechanisms involved in chromoplast differentiation and their participation in the fruit ripening process.

Acknowledgement

The personal work presented in this chapter was supported by the 'Laboratoire d'Excellence' (ANR-10-LABX-41).

References

Alquezar, B., Zacarías, L. and Rodrigo, M.J. (2009) Molecular and functional characterization of a novel chromoplast-specific lycopene β-cyclase from *Citrus* and its relation to lycopene accumulation. *Journal of Experimental Botany* 60, 1783–1797.

Andersson, M.X. and Dörmann, P. (2009) Chloroplast membrane lipid biosynthesis and transport. In: Sandelius, A.S. and Aronsson, H. (eds) *The Chloroplast. Interactions with the Environment.* Springer-Verlag, Berlin/Heidelberg, pp. 125–158.

Andersson, M.X. and Sandelius, A.S. (2004) A chloroplast-localized vesicular transport system: a bio-informatics approach. *BMC Genomics* 5, 40.

Angaman, D.M., Petrizzo, R., Hernadez-Gras, F., Romero-Segura, C., Pateraki, I., Busquets, M. and Boronat, A. (2012) Precursor uptake assays and metabolic analyses in isolated tomato fruit chromoplasts. *Plant Methods* 8, 1–10.

Aoki, K., Yamamoto, M. and Wada, K. (1998) Photosynthetic and heterotrophic ferredoxin isoproteins are colocalized in fruit plastids of tomato. *Plant Physiology* 118, 439–449.

Austin, J.R., Frost, E., Vidi, P., Kessler, F. and Staehelin, L.A. (2006) Plastoglobules are lipoproteins subcompartments of the chloroplast that are permanently coupled to thylakoid membranes and contain biosynthetic enzymes. *Plant Cell* 18, 1693–1703.

Barajas-Lopez, J.D., Blanco, N.E. and Strand, A. (2012) Plastid-to-nucleus communication, signals controlling the running of the plant cell. *Biochimica and Biophysica Acta* 1833, 425–437.

Barry, C.S., Aldridge, G.M., Herzog, G., Ma, Q., McQuinn, R.P., Hirschberg, J. and Giovannoni, J.J. (2012) Altered chloroplast development and delayed fruit ripening caused by mutations in a zinc metalloprotease at the *lutescent2* locus of tomato. *Plant Physiology* 159, 1086–1098.

Barsan, C., Sanchez-Bel, P., Rombaldi, C., Rossignol, M., Kuntz, M., Zouine, M., Latché, A., Bouzayen, M. and Pech, J.C. (2010) Characteristics of the tomato chromoplast proteome revealed by proteomic analysis. *Journal of Experimental Botany* 61, 2413–2431.

Barsan, C., Zouine, Z., Maza, E., Bian, W., Egea, I., Rossignol, M., Bouyssie, D., Pichereaux, C., Purgatto, E., Bouzayen, M., Latché, A. and Pech, J.C. (2012) Proteomic analysis of chloroplast-to-chromoplast transition in tomato reveals metabolic shifts coupled with disrupted thylakoid biogenesis machinery and elevated energy-production components. *Plant Physiology* 160, 708–725.

Benning, C. (2009) Mechanisms of lipid transport involved in organelle biogenesis in plant cells. *Annual Review of Cell and Developmental Biology* 25, 71–91.

Bernier-Villamor, L.N.E., Sevilla, F. and Lázaro, J.J. (2004) Cloning and characterization of a 2-Cys peroxiredoxin from *Pisum sativum*. *Journal of Experimental Botany* 55, 2191–2199.

Bian, W., Barsan, C., Egea, I., Purgatto, E., Chervin, C., Zouine, M., Latché, A., Bouzayen, M. and Pech, J.C. (2011) Metabolic and molecular events occurring during chromoplast biogenesis. *Journal of Botany* 2011, article ID 289859.

Blas, A.L., Ming, R., Liu, Z., Paull, R.E., Moore, P.H. and Yu, Q. (2010) Cloning of the papaya chromoplast-specific lycopene β-cyclase, *CpCYC-b*, controlling fruit flesh color reveals conserved microsynteny and a recombination hot spot. *Plant Physiology* 152, 2013–2022.

Botella-Pavia, P., Besumbes, O., Phillips, M.A., Carretero-Paulet, L., Boronat, A. and Rodriguez-Concepcion, M. (2004) Regulation of carotenoid biosynthesis in plants: evidence for a key role of hydroxymethylbutenyl diphosphate reductase in controlling the supply of plastidial isoprenoid precursors. *Plant Journal* 40, 188–199.

Bouvier, F. and Camara, B. (2007) The role of plastids in ripening fruits. In: Wise, R.R. and Hoober, J.K. (eds) *The Structure and Functions of Plastids*. Springer, Dordrecht, pp. 419–432.

Bouvier, F., Backhaus, R.A. and Camara, B. (1998) Induction and control of chromoplast-specific carotenoid genes by oxidative stress. *Journal of Biological Chemistry* 273, 30651–30659.

Butowt, R., Granot, D. and Rodriguez-Garcia, M.I. (2003) A putative glucose translocator is expressed in heterotrophic tissues that do not contain starch, during olive (*Olea europa* L.) fruit ripening. *Plant Cell Physiology* 44, 1152–1161.

Cai, W., Ma, J., Chi, W., Zou, M., Guo, J., Lu, C. and Zhang, L. (2010) Cooperation of LPA3 and LPA2 is essential for photosystem II assembly in Arabidopsis. *Plant Physiology* 154, 109–120.

Camara, B., Hugueney, P., Bouvier. F., Kuntz, M. and Monéger, R. (1995) Biochemistry and molecular biology of chromoplast development. *International Review of Cytology* 163, 175–247.

Carrara, S., Pardossi, A., Soldatini, G.F., Tognoni, F. and Guidi, L. (2001) Photosynthetic activity of ripening tomato fruit. *Photosynthetica* 39, 75–78.

Chen, G., Bi, Y.R. and Li, N. (2005) EGY1 encodes a membrane-associated and ATP-independent metalloprotease that is required for chloroplast development. *Plant Journal* 41, 364–375.

Cookson, P.J., Kiano, J., Fraser, P.D., Romer, S., Shipton, C.A., Schuch, W., Bramley, P.M. and Pyke, K.A. (2003) Increases in cell elongation, plastid compartment size and translational control of carotenoid gene expression underlie the phenotype of the *high pigment-1* mutant of tomato. *Planta* 217, 896–903.

Czarnecki, O., Gläßer, C., Chen, J.G., Mayer, K.F.X. and Grimm, B. (2012) Evidence for a contribution of ALA synthesis to plastid-to-nucleus signaling. *Frontiers in Plant Science* 3, 236.

Deruere, J., Romer, S., d'Harlingue, A., Backhaus, R.A., Kuntz, M. and Camara, B. (1994) Fibril assembly and carotenoid over-accumulation in chromoplasts: a model for supramolecular lipoprotein structures. *Plant Cell* 6, 119–133.

Egea, I., Barsan, C., Bian, W., Purgatto, E., Latché, A., Chervin, C., Bouzayen, M. and Pech, J.C. (2010) Chromoplast differentiation: current status and perspectives. *Plant and Cell Physiology* 51, 1601–1611.

Egea, I., Bian, W., Barsan, B., Jauneau, A., Pech, J.C., Latché, A., Li, Z.G. and Chervin, C. (2011) Chloroplast-to-chromoplast transition in tomato fruit: spectral confocal microscopy analyses of carotenoids and chlorophylls in isolated plastids and time-lapse recording on intact live tissue. *Annals of Botany* 108, 291–297.

Finkemeier, I., Goodman, M., Lamkemeyer, P., Kandlbinder, A., Sweetlove, L.J. and Dietz, K.J. (2005). The mitochondrial type II peroxiredoxin F is essential for redox homeostasis and root gowth of *Arabidopsis thaliana* under stress. *Journal of Biological Chemistry* 280, 12168–12180.

Flores, F.B., Martínez-Madrid, M.C., Sánchez-Hidalgo, F.J. and Romojaro, F. (2001) Differential rind and pulp ripening of transgenic antisense ACC oxidase melon. *Plant Physiology and Biochemistry* 39, 37–43.

Forth, D. and Pyke, K.A. (2006) The *suffulta* mutation in tomato reveals a novel method of plastid replication during fruit ripening. *Journal of Experimental Botany* 57, 1971–1979.

Fraser, P.D., Kiano, J.W., Truesdale, M.R., Schuch, W. and Bramley, P.M. (1999) Phytoene synthase-2 enzyme activity in tomato does not contribute to carotenoid synthesis in ripening fruit. *Plant Molecular Biology* 40, 687–698.

Frey-Wyssling, A. and Kreutzer, E. (1958) The submicroscopic development of chromoplasts in the fruit of *Capsicum annuum* L. *Journal of Ultrastructural Research* 1, 397–411.

Galvez-Valdivieso, G. and Mullineaux, P.M. (2010) The role of reactive oxygen species in signalling from chloroplasts to the nucleus. *Physiology Plantarum* 138, 430–439.

Giuliano, G. and Diretto, G. (2007) Of chromoplasts and chaperones. *Trends in Plant Science* 12, 529–531.

Ha, S.H., Kim, J.B., Park, J.S., Lee, S.W. and Cho, K.J. (2007) A comparison of the carotenoid accumulation in *Capsicum* varieties that show different ripening colours: deletion of the capsanthin-capsorubin gene is not a prerequisite for the formation of a yellow pepper. *Journal of Experimental Botany* 58, 3135–3144.

Harris, W. and Spurr, A. (1969) Chromoplasts of tomato fruits. II. The red tomato. *American Journal of Botany* 56, 380–389.

Hartmann, M.A. (1998). Plant sterols and the membrane environment. *Trends in Plant Science* 3, 170–175.

Huang, J., Taylor, J.P., Chen, J.G., Uhrig, J.F., Schnell, D.J., Nakagawa, T., Korth, K.L. and Jones, A.L. (2006) The plastid protein THYLAKOID FORMATION1 and the plasma membrane G-protein GPA1 interact in a novel sugar-signalling mechanism in Arabidopsis. *Plant Cell* 18, 1226–1238.

Huh, J., Kang, B., Nahm, S., Kim, S., Ha, K., Lee, M. and Kim, B. (2001) A candidate gene approach identified phytoene synthase as the locus for mature fruit colour in red pepper (*Capsicum* spp.). *Theoretical and Applied Genetics* 102, 524–530.

Jarvis, P. (2008). Targeting of nucleus-encoded proteins to chloroplasts in plants. *New Phytologist* 179, 257–285.

Jeffery, J., Holzenburg, A. and King, S. (2012) Physical barriers to carotenoid bioaccessibility. Ultrastructure survey of chromoplast and cell wall morphology in nine carotenoid-containing fruits and vegetables. *Journal of the Science of Food Agriculture* 92, 2594–2602.

Jeong, S.Y., Rose, A. and Meier, I. (2003) MPF1 is a thylakoid-associated, nucleoid-binding protein with a coiled-coil structure. *Nucleic Acids Research* 31, 5175–5185.

Jimenez, A., Creissen, G., Baldeep, K., Firmin, J., Robinson, S., Verhoeven, M. and Mullineaux, P. (2002) Changes in oxidative processes and components of the antioxidant system during tomato fruit ripening. *Planta* 214, 751–758.

Kahlau, S. and Bock, R. (2008) Plastid transcriptomics and translatomics of tomato fruit development and chloroplast-to-chromoplast differentiation: chromoplast gene expression largely serves the production of a single protein. *Plant Cell* 20, 856–874.

Kato, M., Ikoma, Y., Matsumoto, Y., Sugiura, M., Hyodo, H. and Yano, M. (2004) Accumulation of carotenoids and expression of carotenoid biosynthetic genes during maturation in citrus fruit. *Plant Physiology* 134, 824–837.

Kato, Y., Kouso, T. and Sakamoto, W. (2012) Variegated tobacco leaves generated by chloroplast FtsH suppression: implication of FtsH function in the maintenance of thylakoid membranes. *Plant Cell Physiology* 53, 391–404.

Kleine, T., Voigt, C. and Leister, D. (2009) Plastid signalling to the nucleus: messengers still lost in the mists? *Trends in Genetics* 25, 185–190.

Kleinig, H. and Lievogel, B. (1978) Fatty acid synthesis by isolated chromoplasts from daffodil. ^{14}C Acetate incorporation and distribution of labelled acids. *European Journal of Biochemistry* 83, 499–505.

Köhler, R.H. and Hanson, M.R. (2000) Plastid tubules of higher plants are tissue-specific and developmentally regulated. *Journal of Cell Science* 113, 81–89.

Köhler, R.H., Cao, J., Zipfel, W.R., Webb, W.W. and Hanson, M.R. (1997) Exchange of protein molecules through connections between higher plant plastids. *Science* 276, 2039–2042.

Krech, K., Ruf, S., Masduky, F.F., Thiele, W., Bednarczyk, D., Albus, C.A., Tiller, N., Hasse, C., Schöttler, M.A. and Bock, R. (2012) The plastid genome-encoded Ycf4 protein functions as a non-essential assembly factor for photosystem I in higher plants. *Plant Physiology* 159, 579–591.

Kuntz, M., Evrard, J.L., d'Harlingue, A., Weil, J.H. and Camara, B. (1989) Expression of plastid and nuclear genes during chromoplast differentiation in bell pepper (*Capsicum annuum*) and sunflower (*Helianthus annuus*). *Molecular and General Genetics* 216, 156–163.

Kwok, E.Y. and Hanson, M.R. (2004) Stromules and the dynamic nature of plastid morphology. *Journal of Microscopy* 214, 124–137.

Lang, Y., Yanagawa, S., Sasanuma, T. and Sasakuma, T. (2004) Orange fruit colour in Capsicum due to deletion of capsanthin-capsorubin synthesis gene. *Breeding Science* 54, 33–39.

Lee, U., Riofiorido, I., Hong, S.W., Larkindale, J., Waters, E.R. and Vierling, E. (2006) The Arabidopsis ClpB/Hsp100 family of proteins: chaperones for stress and chloroplast development. *Plant Journal* 49, 115–127.

Lefebvre, V., Kuntz, M., Camara, B. and Palloix, A. (1998) The capsanthin-capsorubin synthase gene: a candidate gene for the *y* locus controlling the red colour in pepper. *Plant Molecular Biology* 36, 785–789.

Lenucci, M.S., Serrone, L., De Caroli, M., Fraser, P.D., Bramley, P.M., Piro, G. and Dalessandro, G. (2012) Isoprenoid, lipid, and protein contents in intact plastids isolated from mesocarp cells of traditional and high-pigment tomato cultivars at different ripening stages. *Journal of Agricultural and Food Chemistry* 60, 1764–1775.

Li, L. and van Eck, J. (2007) Metabolic engineering of carotenoid accumulation by creating a metabolic sink. *Transgenic Research* 16, 581–585.

Li, L., Paolillo, D.J., Parthasarathy, M.V., DiMuzio, E.M. and Garvin, D.F. (2001) A novel gene mutation that confers abnormal patterns of β-carotene accumulation in cauliflower (*Brassica oleracea* var. *botrytis*). *Plant Journal* 26, 59–67.

Li, L., Yang, Y., Xu, Q., Owsiany, K., Welsch, R., Chitchumroonchokchai, C., Lu, S., Van Eck, J., Deng, X.X., Failla, M. and Thannhauser, T.W. (2012) The *Or* gene enhances carotenoid accumulation and stability during post-harvest storage of potato tubers. *Molecular Plant* 5, 339–352.

Li, X., Henry, R., Yuan, J., Cline, K. and Hoffman, N.E. (1995) A chloroplast homologue of the signal recognition particle subunit SRP54 is involved in the posttranslational integration of a protein into thylakoid membranes. *Proceedings of the National Academy of Sciences USA* 92, 3789–3793.

Liedvogel, B. and Kleinig, H. (1977) Lipid metabolism in chromoplast membranes from the daffodil: glycosylation and acylation. *Planta* 133, 249–253.

Liedvogel, B. and Kleinig, H. (1980) Phosphate translocator and adenylate translocator in chromoplast membranes. *Planta* 150, 170–173.

Liu, D., Gong, Q., Ma, Y., Li, P., Li, J., Yang, S., Yuan, L., Yu, Y., Pan, D., Xu, F. and Wang, N.N. (2010b) cpSecA, a thylakoid protein translocase subunit, is essential for photosynthetic development in *Arabidopsis*. *Journal of Experimental Botany* 61, 1655–1669.

Liu, X., Yu, F. and Rodermel, S. (2010a) Arabidopsis chloroplast FtsH, var2 and suppressors of var2 leaf variegation: a review. *Journal of Integrative Biology* 52, 750–761.

Ljubesic, N., Wrischer, M. and Devise, Z. (1991) Chromoplasts – the last stages in plastid development. *International Journal of Developmental Biology* 35, 251–258.

Lois, L.M., Rodriguez-Conception, M., Gallego, F., Campos, N. and Boronat, A. (2000) Carotenoid biosynthesis during tomato fruit development: regulatory role of 1-deoxy-d-xylulose 5-phosphate synthase. *Plant Journal* 22, 503–513.

Lopez, A.B., Van Eck, J., Conlin, B.J., Paolillo, D.J., O'Neill, J. and Li, L. (2008) Effect of the cauliflower *Or* transgene on carotenoid accumulation and chromoplast formation in transgenic potato tubers. *Journal of Experimental Botany* 59, 213–223.

Lu, S., Van Eck, J., Zhou, W., Lopez, A.B., O'Halloran, D.M., Cosman, K.M., Conlin, B.J., Paolillo, D.J., Garvin, D.F., Vrebalov, J., Kochian, L.V., Kupper, H., Earle, E.D., Cao, J. and Li, L. (2006). The cauliflower *Or* gene encodes a DnaJ cysteine-rich domain-containing protein that mediates high-levels of β-carotene accumulation. *Plant Cell* 18, 3594–3605.

Luengwilai, K. and Beckles, D.M. (2009) Starch granules in tomato fruit show a complex pattern of degradation, *Journal of Agricultural and Food Chemistry* 57, 8480–8487.

Lützow, M. and Kleinig, H. (1990) Chlorophyll-free chromoplasts from daffodil contain most of the enzymes for chlorophyll synthesis in a highly active form. *Archives of Biochemistry and Biophysics* 277, 94–100.

Maas, D., Arango, J., Wüst, F., Beyer, P. and Welsh, R. (2009) Carotenoid crystal formation in Arabidopsis and carrot roots caused increased phytoene synthase protein levels. *Plos ONE* 4, e6373.

Marti, M.C., Camejo, D., Olmos, E., Sandalio, L.M., Fernandez-Garcia, N., Jimenez, A. and Sevilla, F. (2009) Characterisation and changes in the antioxidant system of chloroplasts and chromoplasts isolated from green and mature pepper fruits. *Plant Biosystems* 11, 613–624.

Matile, P., Hortensteiner, S. and Thomas, H. (1999) Chlorophyll degradation. *Annual Review of Plant Physiology and Plant Molecular Biology* 50, 67–95.

Neta-Sharir, I., Isaacson, T., Lurie, S. and Weiss, D. (2005) Dual role for tomato heat shock protein 21: protecting photosystem II from oxidative stress and promoting color changes during fruit maturation. *Plant Cell* 17, 1829–1838.

Neuhaus, H.E. and Emes, M.J. (2000) Non-photosynthetic metabolism in plastids. *Annual Review of Plant Physiology and Plant Molecular Biology* 51, 111–140.

Noctor, G. and Foyer, C.H. (1998) Ascorbate and glutathione: keeping active oxygen under control. *Annual Review of Plant Physiology and Plant Molecular Biology* 49, 249–279.

Obiadalla-Ali, H., Fernie, A.R., Kossmann, J. and Lloyd, J.R. (2004) Developmental analysis of carbohydrate metabolism in tomato (*Lycopersicon esculentum* cv. Micro-Tom) fruits. *Physiology Plantarum* 120, 196–204.

Obukosia, S.D., Richards, C.M. and Boyer, C.D. (2003) Expression of plastid-encoded photosynthetic genes during chloroplast or chromoplast differentiation in *Cucurbita pepo* L. fruits. *Phytochemistry* 64, 1213–1221.

Paolillo, D.J.J., Garvin, D.F. and Parthasarathy, M.V. (2004). The chromoplasts of *Or* mutants of cauliflower (*Brassica oleracea* L. var. *botrytis*). *Protoplasma* 224, 245–253.

Pateraki, I., Renato, M., Joaquín Azcón-Bieto, J. and Boronat, A. (2013) An ATP synthase harboring an atypical γ-subunit is involved in ATP synthesis in tomato fruit chromoplasts. *Plant Journal* 74, 74–85.

Peng, L., Ma, J., Chi, W., Guo, J., Zhu, S., Lu, Q., Lu, C. and Zhang, L. (2006) LOW PSII ACCUMULATION1 is involved in efficient assembly of photosystem II in *Arabidopsis thaliana*. *Plant Cell* 18, 955–969.

Piechulla, B., Cholones Imlay, K.R. and Gruissem, W. (1985) Plastid gene expression during fruit ripening in tomato. *Plant Molecular Biology* 5, 373–384.

Piechulla, B., Glick, R.E., Bahl, H., Melis, A. and Gruissem, W. (1987) Changes in photosynthetic capacity and photosynthetic protein pattern during tomato fruit ripening. *Plant Physiology* 84, 911–917.

Plücken, H., Müller, B., Grohmann, D., Westhoff, P. and Eichacker, L.A. (2002) The HCF136 protein is essential for assembly of the photosystem II reaction center in *Arabidopsis thaliana*. *FEBS Letters* 532, 85–90.

Powell, A.L.T., Nguyen, C.V., Hill, T., Cheng, K.L., Figueroa-Balderas, R., Aktas, H., Ashrafi, H., Pons, C., Fernandez-Munos, R., Vicente, A., Lopez-Baltazar, J., Barry, C.S., Liu, Y., Chetelat, R., Granell, A., Deynze, A.V., Giovannoni, J.J. and Bennett, A.B. (2012) Uniform ripening encodes a Golden 2-like transcription factor regulating tomato fruit chloroplast development. *Science* 336, 1711–1715.

Pyke, K.A. and Howells, C.A. (2002) Plastid and stromule morphogenesis in tomato. *Annals of Botany* 90, 559–566.

Rodrigo, M.J., Marcos, J.F. and Zacarias, L. (2004) Biochemical and molecular analysis of carotenoid biosynthesis in flavedo of orange (*Citrus sinensis* L.) during fruit development and maturation. *Journal of Agricultural and Food Chemistry* 52, 6724–6731.

Rodriguez-Uribe, L., Guzman, I., Rajapakse, W., Richins, R.D. and O'Connell, M.A. (2012) Carotenoid accumulation in orange-pigmented *Capsicum annuum* fruit, regulated at multiple levels. *Journal of Experimental Botany* 63, 517–526.

Ronen, G., Cohen, M., Zamir, D. and Hirschberg, J. (1999) Regulation of carotenoid biosynthesis during tomato fruit development: expression of the gene for lycopene epsilon-cyclase is down regulated during ripening and is elevated in the mutant *Delta*. *Plant Journal* 17, 341–351.

Rosso, S.W. (1968) The ultrastructure of chromoplast development in red tomatoes. *Journal of Ultrastructural Research* 25, 307–322.

Ruiz-Sola, M.A. and Rodriguez-Concepción, M. (2012). Carotenoid biosynthesis in Arabidopsis: a colorful pathway. *Arabidopsis Book* 10, e0158.

Rutschow, H., Ytterberg, J., Friso, G., Nilsson, R. and van Wijk, K.J. (2008) Quantitative proteomics of a chloroplast *SRP54* sorting mutant and its genetic interaction with *CLP1* in Arabidopsis. *Plant Physiology* 148, 156–175.

Sakuraba, Y., Schelbert, S., Park, S.Y., Han, S.H., Lee, B.D., Andrès, C.B., Kessler, F., Hörtensteiner, S. and Paek, N.C. (2012) STAY-GREEN and chlorophyll catabolic enzymes interact at light-harvesting complex II for chlorophyll detoxification during leaf senescence in *Arabidopsis*. *Plant Cell* 24, 507–518.

Schweiggert, R.M., Steingass, C.B., Heller, A., Esquivel, P. and Carle, R. (2011). Characterization of chromoplasts and carotenoids of red- and yellow-fleshed papaya (*Carica papaya* L.). *Planta* 234, 1031–1044.

Sehnke, P.C., Chung, H.J., Wu, K. and Ferl, R.J. (2001) Regulation of starch accumulation by granule-associated plant 14-3-3 proteins. *Proceedings of the National Academy of Sciences USA* 98, 765–770.

Siddique, M.A., Grossmann, J., Gruissem, W. and Baginsky, S. (2006) Proteome analysis of bell pepper (*Capsicum annuum* L.) chromoplasts. *Plant and Cell Physiology* 47, 1663–1673.

Simkin, A.J., Gaffé, J., Alcaraz, J.P., Carde, J.P., Bramley, P.M., Fraser, P.D. and Kuntz, M. (2007) Fibrillin influence on plastid ultrastructure and pigment content in tomato fruit. *Phytochemistry* 68, 1545–1556.

Spurr, A.R. and Harris, W.M. (1968) Ultrastructure of chloroplasts and chromoplasts in *Capsicum annuum*. I. Thylakoid membrane changes during fruit ripening. *American Journal of Botany* 55, 1210–1224.

Tetlow, I.J., Bowsher, C.G. and Emes, M.J. (2003) Biochemical properties and enzymic capacities of chromoplasts isolated from wild buttercup (*Ranunculus acris* L.). *Plant Science* 165, 383–394.

Tetlow, I.J., Wait, R., Lu, Z., Akkasaeng, R., Bowsher, C.G., Esposito, S., Kosar-Hashemi, B., Morell, M.K. and Emes, M.J. (2004) Protein phosphorylation in amyloplasts regulates starch branching enzyme activity and protein–protein interactions. *Plant Cell* 16, 694–708.

Thom, E., Mohlmann, T., Quick, P.W., Camara, B. and Neuhaus, H.E. (1998) Sweet pepper plastids: enzymic equipment, characterisation of the plastidic pentose-phosphate pathway, and transport of phosphorylated intermediates across the envelope membrane. *Planta* 204, 226–233.

Thomas, H., Morgan, W.G., Thomas, A.M. and Ougham, H.J. (1999) Expression of the stay-green character introgressed into *Lolium temulentum* Ceres from a senescence mutant of *Festuca pratensis*. *Theoretical and Applied Genetics* 99, 92–99.

Vasquez-Caicedo, A.L., Heller, A., Neidhart, S. and Carle, R. (2006) Chromoplast morphology and β-carotene accumulation during postharvest ripening of mango Cv 'Tommy Atkins'. *Journal of Agricultural and Food Chemistry* 54, 5769–5776.

Wang, Z. and Benning, C. (2012) Chloroplast lipid synthesis and lipid trafficking through ER–plastid membrane contact sites. *Biochemical Society Transactions* 40, 457–463.

Waters, M.T., Fray, R.G. and Pyke, K.A. (2004) Stromule formation is dependent upon plastid size, plastid differentiation status and the density of plastids within the cell. *Plant Journal* 39, 655–667.

Waters, M.T, Wang, P., Korkaric, M., Capper, R.G., Saunders, N.J. and Langdale, J.A. (2009) GLK transcription factors coordinate expression of the photosynthetic apparatus in Arabidopsis. *The Plant Cell* 21, 1109–1128.

Westphal, S., Soll, J. and Vothknecht, U.C. (2001) A vesicle transport system inside chloroplasts. *FEBS Letters* 506, 257–261.

Woodson, J.D., Pérez-Ruiz, J.M. and Chory, J. (2011). Heme synthesis by plastid ferrochelatase I regulates nuclear gene expression in plants. *Current Biology* 21, 897–903.

Xiao, Y., Savchenko, T., Baidoo, E.E.K., Chehab, W.E., Hayden, D.M., Tolstikov, V., Corwin, J.A., Kliebenstein, D.J., Keasling, J.D. and Dehesh, K. (2012) Retrograde signaling by the plastidial metabolite MEcPP regulates expression of nuclear stress-response genes. *Cell* 149, 1525–1535.

Xu, C., Moellering, E.R., Muthan, B., Fan, J. and Benning, C. (2010) Lipid transport mediated by Arabidopsis TGD proteins is unidirectional from the endoplasmic reticulum to the plastid. *Plant and Cell Physiology* 51, 1019–1028.

Yakir-Tamang, L. and Gerst, J.E. (2009) A phosphatidylinositol-transfer protein and phosphatidylinositol-4-phosphate 5-kinase control Cdc42 to regulate the actin cytoskeleton and secretory pathway in yeast. *Molecular Biology of the Cell* 20, 3583–3597.

Ytterberg, A.J., Peltier, J.B. and van Wijk, K.J. (2006). Protein profiling of plastoglobules in chloroplasts and chromoplasts. A surprising site for differential accumulation of metabolic enzymes. *Plant Physiology* 140, 984–997.

Zeeman, S.C., Kossmann, J. and Smith, A.M. (2010) Starch: its metabolism evolution and biotechnological modification in plants. *Annual Review of Plant Biology* 61, 209–234.

Zeng, Y., Pan, Z., Ding, Y., Zhu, A., Cao, H., Xu, Q. and Deng, X. (2011) A proteomic analysis of the chromoplasts isolated from sweet orange fruits [*Citrus sinensis* (L.) Osbeck]. *Journal Experimental Botany* 62, 5297–5309.

4 Cell-wall Metabolism and Softening during Ripening

Mark L. Tucker*
US Department of Agriculture, Beltsville, MD, USA

4.1 Introduction

The final stage of fruit development is ripening, which typically includes transformation from an inedible hard organ into a palatable softer version. Fruit softening is a combination of changes in firmness and texture. Firmness can be defined as compressibility or the force required to deform the surface of the fruit (Brookfield *et al.*, 2011). Texture is defined as a sensory attribute and is more difficult to measure with instrumentation (Mohamed *et al.*, 1982; Garcia-Ramos *et al.*, 2005; Brookfield *et al.*, 2011). Texture includes crispness, viscosity and juiciness (Brookfield *et al.*, 2011). Texture is often best measured by human tasters (Brookfield *et al.*, 2011). Before moving on to the specific details of what happens during ripening to contribute to softening, let us identify a few basic principles to help visualize what softening really is. The edible parts of fruit are not woody (lignified), not even before they ripen. In other words, the cell walls in the edible parts of fruit are not rigid. They can flex. The flexibility of the cell wall is more obvious in thinner structures like leaves. Leaves are not as 'stiff as a board'. The primary force that maintains structure in a leaf and firmness in fruit is turgor pressure. Although many factors can affect turgor pressure, understanding the consequences of water loss is not too complicated. A partially dehydrated leaf will be more flexible. Similarly, a dehydrated fruit will be softer than a fully hydrated fruit. It also seems fairly obvious that if you loosen up the cell wall to make it more flexible and extensible, the fruit will be softer. However, the cell wall is a complex structure and how specific changes in the cell wall affect softening is not always obvious.

Let us continue with some additional basic concepts that should help us understand the role of the cell wall in a large multicellular organ like a fruit. Let us begin by building a mental picture of the basic function of the cell wall by comparing it to two types of balloons, a latex balloon and a foil balloon. The latex balloon starts out small and continues to expand as you fill it with gas until at some point it explodes. The foil balloon is flat, flimsy and flexible until it is completely full of gas (turgid) and does not expand much more once completely filled but rather explodes if too high a pressure is reached. The fruit cell walls act more like the foil balloon. Once the cell is turgid, it does not readily expand. The key here is

* mark.tucker@ars.usda.gov

that the cell wall is flexible, as is the foil balloon, but it does not stretch easily unless modified. If the cell loses some of its water through evaporation, it is softer and more flexible, like the foil balloon would be if it lost some of its air. To take this analogy a bit further, the foil balloon can be made to rupture at lower or higher pressures depending on the thickness and strength of the material used. Similarly, the pressure required to rupture a fruit cell depends on the strength and composition of the cell wall. None the less, a fruit is a bit more complicated than the single balloon because the fruit is not just one cell but many cells that can be made to slide past one another, and slippage can also contribute towards softening.

Imagine a bag full of marbles. If the marbles are glued tight to each other, they act as one big incompressible rigid block, but loosen the glue between the marbles, and the marbles readily move about inside the bag. The middle lamella between plant cells acts like a glue that adheres adjacent cells. Completely dissolving the middle lamella causes the fruit to become softer. However, if softening were simply dissolution of the middle lamella to allow cells to slide past each other, eating a fruit might be more like eating a bag of sand. The cells must also rupture to allow the ripened contents to escape.

Thus it can be seen that fruit softening is a combination of changes in turgor pressure, primary cell-wall construction that affects the pressure needed to rupture the cell, and cell-to-cell adhesion that affects how readily the cells slide past one another. However, there are limits to how much these parameters can change whilst still maintaining a living organ and palatable fruit. At this point, it is worth mentioning a basic principle necessary to maintain a living cell deep inside a fairly large organ like a fruit. All plant cells, including fruit cells, require efficient gas exchange (Ho *et al.*, 2011). Oxygen is taken up for respiration, and carbon dioxide is given off. If the environment surrounding the cell is anaerobic (i.e. oxygen depleted) for too long, the cell will die. Animals facilitate gas exchange by using a circulatory system, but plants typically rely on passive diffusion of oxygen to and from the cell surface. The primary cell wall is filled with liquid, but some portion of the intercellular space between cells must be gaseous (Ho *et al.*, 2011). There must be a contiguous air channel from the surface of the fruit to its centre to facilitate gas diffusion. If the primary cell-wall structure and middle lamella were to be degraded to the point that the intercellular gas space was lost, either from collapse of the fruit under its own weight or swelling of cells to the point that gas exchange was greatly compromised, the fruit would spoil and become rotten. We all know what a rotten fruit looks like. Rotten fruit can be caused by pathogens, mechanical damage, or over-ripening and collapse of the fruit structure.

4.1.1 Turgor: waxing and packaging

Water relationships are not the focus of this chapter, and factors that affect turgor pressure and water loss are complex. None the less, because water relationships are so important to softening and maintaining fruit quality, we will briefly discuss what has been done to understand and regulate this important parameter. You might be surprised how many fruits and vegetables are waxed: apples, pears, tomatoes, cucumbers and many more (Hardenburg, 1967). The commercial practice of waxing fruits and vegetables to reduce water loss and extend shelf-life has been used for many years (Hardenburg, 1967). In addition to wax, the coating often incorporates fungicides to reduce postharvest infections (Hall, 1989). In some cases, the wax may reduce not only water loss but also gas exchange enough that it may limit oxidative respiration without the fruit becoming anaerobic (Jeong *et al.*, 2003). Limiting oxidative respiration goes hand in hand with reducing metabolic rate and slowing the ripening process, which includes softening (Tucker and Laties, 1985; Asif *et al.*, 2006; Zabalza *et al.*,

2009). In recent years, considerable research has gone into developing modified atmosphere packaging that reduces mechanical damage and water loss, and creates a gaseous environment that limits oxygen exchange but not so much as to create anaerobic conditions within the fruit (Rojas-Graü *et al.*, 2009; Sandhya, 2010; Singh, 2011).

Of particular interest in regard to water relationships is recent research on a tomato variety that has a long shelf-life (remains firm and palatable for an extended period of time) (Saladie *et al.*, 2007). The authors of this study discovered that the long shelf-life was due to a mutation that altered the cuticle that surrounds and protects the fruit. This is of interest because it suggests an avenue for engineering or breeding that can have profound effects on water loss and maintenance of firm fruit.

None the less, the cell wall must ripen along with the rest of the fruit. What is cell-wall ripening? It is changes in the cell wall that make the fruit desirable – the perfect firmness and texture. But desirability depends on which fruit you are eating. An apple or pear is not the same as a grape, blackberry, banana or tomato. However, there are some general concepts about cell structure and cell-wall metabolism that can be applied to many fruits.

4.2 Cell-wall Structure

Before discussing how the cell wall changes during ripening, we must know something about how it formed. When cells divide to increase cell numbers, a cell plate forms between the dividing cells during a process called cytokinesis (Allen, 1901; Carpita and McCann, 2000). The cell plate will become the middle lamella – the glue that binds cells together. The middle lamella consists primarily of pectins. Very little or no cellulose is found in the middle lamella (Allen, 1901; Carpita and McCann, 2000). Soon after the formation of the cell plate, the primary cell wall is synthesized on either side of the plate. The primary cell wall provides structure and tensile strength to the cell (Carpita and Gibeaut, 1993). However, as mentioned above, there must be a contiguous gas space between cells to sustain adequate exchange of gases. The gas space can, however, vary widely depending on the fruit; for example, in pear it is ~5% (Verboven *et al.*, 2008), in apple ~23% (Drazeta *et al.*, 2004; Verboven *et al.*, 2008) and in tomato ~6% (Calbo and Nery, 1995). It is at this early step in cell division and primary cell-wall synthesis that the air spaces form. The air space forms at the corners where three or four cells make contact with each other. When the middle lamella joins with the middle lamella from an existing cell, this region (corner) is modified. Carbohydrate-specific antibodies have been used to demonstrate that this region of the middle lamella contains more unesterified homogalacturonans (pectin polymers) than the middle lamella that binds adjacent cells together (Ordaz-Ortiz *et al.*, 2009). Although the airspace at these corners may change in volume during fruit development, a contiguous gas space must be maintained during cell expansion and ripening (Ho *et al.*, 2011).

A typical primary cell wall of a dicot cell consists of 50% cellulose/xyloglucan, 30% pectic polysaccharides and 20% structural proteins (Carpita and Gibeaut, 1993). Polymers of cellulose (1,4-β-glucans) are twisted together to form a cellulose microfibril (Carpita and Gibeaut, 1993). A single microfibril can include several dozen glucan polymers twisted together to form a very strong thread that is not easily stretched or broken, yet is flexible (Carpita and Gibeaut, 1993). The direction in which the microfibrils are synthesized and laid down in the cell wall largely determines the direction in which the cell can expand. For example, if the microfibrils are laid down in a helical fashion around the cell, the cell expands primarily in the longitudinal direction, much like stretching a spring (Carpita and Gibeaut, 1993). A helical orientation of microfibrils would be the structure found in elongating cells immediately behind the stem or root meristem. If the microfibrils are laid down in a more random pattern, the cell may expand in all directions, called isodiametric expansion. The cellulose microfibrils give the cell wall its tensile strength (Carpita and Gibeaut, 1993). However, to provide structure to the cell, the microfibrils must be tethered to one another. The polysaccharides that provide this function are often called hemicelluloses and consist primarily of xyloglucans (Brummell, 2006). The cellulose microfibrils and tethering hemicelluloses are both embedded in a pectin matrix. The pectin fraction consists primarily of polygalacturonic acid (homogalacturonans) and rhamnogalacturans (Carpita and Gibeaut, 1993; Brummell, 2006). Pectins add strength to the cell wall and determine pore size in the cell wall, which may limit the access of enzymes to target sites on the xyloglucan matrix and other polymers (Carpita and Gibeaut, 1993). The unesterified charged ends of the individual subunits in polygalacturonic acid are often cross-linked by calcium ions to form a stable egg-box-like structure (Carpita and Gibeaut, 1993). The availability of calcium and esterification of polygalacturonic acid with methyl or acetyl groups limits cross-linking by calcium, which can in turn affect the structural properties of the cell wall. Proteins are another major fraction of cell walls (Carpita and Gibeaut, 1993). There are several types of cell-wall proteins, but the bulk of the protein is made up of extensins – hydroxyproline-rich glycoproteins (Carpita and Gibeaut, 1993). These proteins coalesce into rod-shaped structures that add strength to the cell wall and may play a role in cell-wall assembly, cell shape and formation of the intercellular spaces (Carpita and Gibeaut, 1993; Carpita and McCann, 2000). The rod-shaped protein structures may also anchor the plasma membrane to the cell wall (Knox, 1995).

4.3 Cell-wall Metabolism during Ripening

The tools available to a scientist determine the experiments they can perform. Thus, research on fruit ripening in the early part of the 20th century focused on respiration. In 1925, Kidd and West reported that apples underwent an increase in respiration during ripening, which they called the climacteric (Laties, 1995). At approximately the same time, it was discovered that a gaseous substance emitted by a ripe apple could stimulate the ripening of an unripe apple, and that exposure to ethylene could induce a similar response (Laties, 1995). In 1934, Gane published a paper demonstrating that some fruit produce ethylene. Assaying for changes in enzyme activity during ripening followed as these tools became available. This included assays for enzymes known to affect the cell wall. In 1979, several articles were published describing an increase in cellulase and polygalacturonase (PG) activity in apples and pears (Ben-Arie and Kislev, 1979), avocado (Awad and Young, 1979) and tomato (Poovaiah and Nukaya, 1979). In apples and pears, it was also noted that, at advanced stages of ripening, the middle lamella was clearly degraded and the orderly arrangement of microfibrils was lost (Ben-Arie and Kislev, 1979). Changes in cell-wall morphology appeared to fit with the rise in cellulase and PG activity. With the advent of tools for cloning mRNA, it was not too long before cellulase and PG were cloned from avocado and tomato and their expression correlated with changes in enzyme activity (Christoffersen *et al.*, 1984; Grierson *et al.*, 1986; Kutsunai *et al.*, 1993; Lashbrook *et al.*, 1994). Clones for PG, cellulases, pectin esterases and xyloglucanases in many different fruit soon followed (Marin-Rodriguez *et al.*, 2002; Rose *et al.*, 2002, 2004; Brummell, 2006; Vicente *et al.*, 2007; Bapat *et al.*, 2010). With improved techniques (Bräutigam *et al.*, 2011; Osorio *et al.*, 2011), we now have the tools to sequence and identify expression patterns for literally hundreds of different cell-wall proteins in practically any fruit we want to examine.

However, expression data alone is simply correlative and does not really tell us how the individual proteins function during fruit softening. The fact that the gene transcript for a protein increases during ripening suggests it plays a role in ripening, but the true function of the protein is not always obvious. To ascertain a function for the cell-wall proteins, researchers have relied on natural occurring variants (mutants) of a particular fruit or artificially created mutants made by genetic transformation.

4.3.1 Pectin

Let us look at a few examples where the expression of a protein has been altered to see how the loss or overexpression of the protein changed firmness, texture and cell-wall degradation products. PGs are of interest because they are expressed abundantly in many fruits during the softening phase (Crookes and Grierson, 1983). PG can degrade both the middle lamella between cells and the pectin matrix within the primary cell wall (Crookes and Grierson, 1983). Also important is the fact that it is the pectin matrix that determines the porosity of the cell wall and accessibility of enzymes and proteins to the cellulose microfibrils and hemicelluloses embedded in it (Carpita and Gibeaut, 1993). Thus, it seems reasonable to expect that the pectin fraction must first be modified in order to provide protein access to the hemicellulose and cellulose components of the cell wall. Tomato is an important model for genetic manipulation because protocols for genetic transformation of tomato were established fairly early and, unlike tree fruits like apple, pear and avocado, tomato plants bear fruit within a month or two of germination. Two separate research groups used antisense constructs in tomato to suppress accumulation of PG transcripts, one in the USA (Sheehy *et al.*, 1988) and another in the UK (Schuch *et al.*, 1991). These early transformation events are of particular interest because these transgenic plants were the first genetically modified plant products to be commercialized. Calgene in the USA commercialized the transgenic fresh fruit using the trade name Flavr Savr (Kramer and Redenbaugh, 1994), and Zeneca Seeds in the UK commercialized their product as a tomato purée paste (Khachatourians, 2002). Both groups reported improved shelf-life for the fruit (Sheehy *et al.*, 1988; Schuch *et al.*, 1991). The firmness, as measured by compressibility, of the freshly picked fruit was not much different from the non-transgenic fruit; however, after several days of storage, the transgenic fruit were significantly firmer (Kramer *et al.*, 1992; Brummell and Labavitch, 1997). The fruit were also more resistant to some common fruit pathogens, which improved storage quality. Both groups measured a significant increase in the viscosity of the processed fruit, e.g. purées (Schuch *et al.*, 1991; Kramer *et al.*, 1992). A later study of the Flavr Savr fruit confirmed and extended the earlier work by quantifying the depolymerization of polyuronides in the transgenic and non-transgenic (wild-type) fruit (Brummell and Labavitch, 1997). Depolymerization of polyuronides, as measured by size fractionation of *trans*-1,2-cyclohexanediaminetetraacetic acid (CDTA, a divalent cation chelator)-extractable cell-wall carbohydrate, changed markedly during ripening in both PG-suppressed and wild-type fruit (Brummell and Labavitch, 1997). However, they observed a small difference in depolymerization of polyuronides in PG-suppressed fruit compared with wild-type fruit, which correlated with a small shift in fruit firmness (Brummell and Labavitch, 1997). At least two reasons can be envisaged for only a slight difference in depolymerization of polyuronides at harvest when the polyuronides were clearly fragmented in both lines. It is possible that PG2A, the abundant transcript inhibited in these plants, is not responsible for fragmentation of polyuronides during ripening or that there are additional enzymes that can depolymerize pectins, such as pectate lyases. Giovannoni *et al.* (1989) did an interesting experiment that sheds light on this problem. They transformed a ripening inhibited mutant of tomato, *rin*, with a construct (gene) that could express active PG2A at a developmental stage that corresponded to when the mutant should normally ripen. They demonstrated that PG2A does cause depolymerization of polyuronides, but the fruit did not significantly soften. This indicates that PG2A can fragment pectins but that in the transgenic plants where PG2A is suppressed there are other enzymes that can also fragment pectin.

The results with overexpression of PG2A in *rin* and suppression in Flavr Savr further suggest that PG2A by itself may not be enough to cause significant alterations in tomato fruit firmness; but before moving on to a different topic, let us look at another fruit. In peach, there are naturally occurring varieties that produce fruit called melting flesh (MF) and non-melting flesh (NMF). Examination of NMF fruit found that PGs were not expressed or were not secreted into the cell wall (Callahan *et al.*, 2004; Ghiani *et al.*, 2011). The NMF varieties all had deletions of one form or another in a genetic locus for a family of PGs (Callahan *et al.*, 2004). In this particular case, the correlation between MF texture and the absence of PG was very compelling. Moreover, what is also important was that they found that fruit firmness (compressibility) was not greatly different between the MF and NMF varieties (Ghiani *et al.*, 2011). These authors concluded that cell turgor plays a more important role in changes in firmness but that PG activity is responsible for the marked difference in fruit texture. To restate, texture is defined as a sensory attribute and is more difficult to measure with instrumentation than firmness (Brookfield *et al.*, 2011).

β-Galactans are branch polymers on rhamnogalacturonans, which are a major component, in addition to polygalacturonic acid, in the pectin matrix of the primary cell wall (Carpita and Gibeaut, 1993). Antisense suppression of β-galactosidase 4 reduced the amount of free galactose in ripening fruit and increased fruit firmness (Smith *et al.*, 2002). Another enzyme that can affect pectin structure and fragmentation are pectin esterases (Rose *et al.*, 2004). It is thought that polyuronides are synthesized as a methyl or acetyl ester that blocks the carboxyl acid group (Carpita and Gibeaut, 1993; Rose *et al.*, 2004). After the polyuronides are laid down in the cell wall; the esterified polyuronides are de-esterified by pectin esterases. The charged carboxyl group can then interact with calcium to form a stable egg-box-like structure (Carpita and Gibeaut, 1993). It is thought that de-esterification of polyuronide is necessary before PGs can act on this substrate (Tieman *et al.*, 1992). Suppression of a pectin methyl esterase (PME) in tomato fruit reduced the de-esterification of polyuronides and its depolymerization during ripening, but, although an increase in soluble solids was observed in the transgenic fruit, no change in firmness or texture were noted (Tieman *et al.*, 1992). A separate group suppressed a different pectin esterase in tomato fruit and also observed a significant increase in esterified pectin but did not identify any significant differences in softening; however, unlike the earlier report on the suppression of a PME, they did not observe any difference in soluble solids (Hall *et al.*, 1993).

4.3.2 Cellulose

Cellulose is another major component of fruit cell walls, and it is the cellulose microfibrils that provide the tensile strength to the cell wall. Intuitively, it makes sense that modification of this fraction of the cell wall would have marked effects on softening. However, in most fruits examined, the cellulose content of the cell wall does not change much during ripening (Maclachlan and Brady, 1994; Newman and Redgwell, 2002; Brummell, 2006). Before discussing the results for mutants with suppressed levels of cellulases, it is worth noting that analysis is complicated by the fact that we do not know the true substrate for the plant enzymes called cellulase. Cellulases are defined by their ability to degrade and reduce the viscosity of carboxymethylcellulose (CMC) in an *in vitro* assay. CMC is an artificial water-soluble cellulose. For example, purified avocado cellulase readily degrades CMC but when added to microfibrils or crystalline cellulose, the enzyme is incapable of significantly degrading these substrates (O'Donoghue *et al.*, 1994). A similar result has been obtained with other plant cellulases (Urbanowicz *et al.*, 2007; Vicente *et al.*, 2007). It has been suggested that the native substrate for plant cellulases might be xyloglucans (Hatfield and Nevins, 1986); however, subsequent studies with purified avocado cellulase demonstrated that this cellulase could not degrade purified xyloglucans (O'Donoghue *et al.*, 1994). What is the substrate for cellulases? We still do not know for certain; however, incubation of avocado cellulase with cell walls purified from unripe avocado fruit caused a loss in the cohesiveness of the microfibrils (O'Donoghue *et al.*, 1994). These authors concluded that avocado cellulase does not cleave the xyloglucans that tether the cellulose microfibrils but rather cleaves cellulose at accessible sites on the periphery of the microfibrils, which causes a loss of integrity within the fibril structure and alters the binding of associated cell-wall matrix polysaccharides, such as xyloglucans. In avocado, which produces an inordinate amount of cellulase, it was concluded that, although cellulase might not depolymerize crystalline cellulose microfibrils, cellulase might modify the cellulose polymers exposed to the surface of the microfibril, which would then affect the tethering and organization of the microfibrils in the wall (O'Donoghue *et al.*, 1994). Unfortunately, transformation of avocado to suppress cellulase is not currently possible.

Tomato fruit includes two cellulases that are expressed during fruit ripening, Cel1 and Cel2 (Lashbrook *et al.*, 1994). When each gene – Cel1 (Lashbrook *et al.*, 1998) and Cel2 (Brummell *et al.*, 1999a) – was suppressed individually, there was no significant effect on fruit ripening or softening. This result left open the possibility that both genes needed to be suppressed in order to have an effect on softening. However, in a subsequent study where both Cel1 and Cel2 were suppressed simultaneously, no obvious change in softening was reported, but there was a clear effect on infection by *Botrytis cinerea* (a common fungal pathogen of fruit), which suggested to the authors that these cellulases altered the cell wall in a manner that made the fruit more susceptible to infection (Flors *et al.*, 2007). Interestingly, expression of Cel2 in the non-ripening *rin* mutant, which does not normally express Cel2, did cause the fruit to partially soften but not ripen (Flors *et al.*, 2007). What this suggests is that Cel2 might affect tethering of the cellulose microfibrils but that, in a normal ripening process, there are other enzymes or processes that have a similar effect on tethering of microfibrils.

Before moving on to look at other components of the cell wall, it is of interest to examine research with strawberry fruit. Suppression of PGs had a clear effect on fruit firmness when measured by extrusion of the pulp through an orifice during fruit compression (Quesada *et al.*, 2009). The measurement of fruit firmness for strawberry was done differently from that for tomato fruit and might be better compared with the decrease in viscosity of the tomato paste prepared from transgenic tomato with reduced PG activity and also the MF characteristic of peach. Also of comparative interest is that suppression of a cellulase (Cel1) in strawberry, like tomato, had no significant effect on fruit firmness (Woolley *et al.*, 2001).

Whilst on the topic of untethering of microfibrils, it is necessary to discuss work on a protein called expansin. Expansins were first discovered in the stems of elongating seedlings (McQueen-Mason *et al.*, 1992; McQueen-Mason and Cosgrove, 1994) and have since been identified in many plant tissues (Rose *et al.*, 1997; Rose *et al.*, 2000; Cosgrove *et al.*, 2002). Expansins are an interesting class of protein because although they have no known *in vitro* enzyme activity, they clearly play a role in loosening of the cell wall during cell elongation (Cosgrove,

1999; Cosgrove *et al.*, 2002). However, based on the effect of expansins on plant tissues and cellulose paper, it was concluded that expansins act on hydrogen bonding at the interface of cellulose microfibrils (McQueen-Mason and Cosgrove, 1994). If the tethering of microfibrils to each other is loosened during ripening, a marked effect on both firmness and texture might be expected. Expansins are a family of genes in tomato and several are expressed in fruit, but one in particular (EXP1) increases specifically during ripening (Rose *et al.*, 1997, 2000; Brummell *et al.*, 1999c). To examine a possible role for EXP1 in fruit softening its expression was both suppressed and enhanced in tomato fruit (Brummell *et al.*, 1999b). Suppression of EXP1 caused a significant increase in fruit firmness measured at all stages of ripening fruit including over-ripe fruit. Also of interest is that when EXP1 was overexpressed throughout fruit development, the fruit were less firm, even at the mature green stage. Fruit where EXP1 was overexpressed were also smaller and had what was described as a 'rubbery texture'. Thus, results with EXP1 are consistent with expansins affecting the tethering of cellulose microfibrils to each other.

4.3.3 Xyloglucans

Expansins may affect the tethering of cellulose microfibrils, but it is proposed that xyloglucans are the polysaccharide substrate that cross-links and ties the microfibrils together (Carpita and Gibeaut, 1993). Modification or depolymerization of the cross-linking xyloglucan network might also be expected to affect fruit softening. Xyloglucanases, like cellulase and expansins, are another interesting family of enzymes. In addition to hydrolysing xyloglucan polymers, most of these enzymes also appear to be able to join together two xyloglucan polymers. This family of enzymes have been called xyloglucan endotransglucosylase/hydrolases (XTHs) (Rose *et al.*, 2002). Thus, because xyloglucan polymers can be broken and rejoined, it is possible for cellulose microfibrils to slide past one another during cell expansion without evidence of xyloglucan depolymerization. Nevertheless, it has been demonstrated that xyloglucans are fragmented (depolymerized) during ripening (Brummell, 2006). In tomato, 25 XTHs have been identified (Ohba *et al.*, 2011) and at least ten are expressed in ripe fruit, but only two appeared to increase with ripening whilst most of the others decreased during ripening (Miedes and Lorences, 2009). Overexpression of XTH1 during fruit development actually caused a slight decrease in depolymerization of xyloglucans during ripening (Miedes *et al.*, 2010). These researchers concluded that the increased endotransglucosylase activity over the hydrolase activity in the XTHs was responsible for the lower xyloglucan depolymerization in fruit and suggested that the role of XTHs during fruit growth and ripening might be to maintain the structural integrity of the cell wall. They suggested that a decrease in XTH activity, rather than an increase, during ripening might contribute to fruit softening. A separate group both overexpressed and inhibited expression of XTH1 in tomato and found that the level of expression correlated directly with fruit size (Ohba *et al.*, 2011). These researchers concluded that XTH1 clearly plays a role in cell expansion in fruit development, but they did not identify a clear link to softening; nevertheless, they postulated that a spike in XTH activity at the turning stage early in the ripening process might be linked to xyloglucan depolymerization and a decrease in firmness. Of interest in this regard is recent research using carbohydrate-specific antibodies, which showed that hemicellulose-like polymers are also associated with the cell adhesion layer (middle lamella) and that these polymers tend to disappear in ripe tomato fruit (Ordaz-Ortiz *et al.*, 2009). Currently, it is not clear how important the loss of hemicelluloses in the adhesion layer is to fruit softening.

4.3.4 Protein

The protein component of cell walls is probably the least well understood part of the cell wall. It seems logical that something that accounts for as much as 20% of the mass of the cell wall (Carpita and Gibeaut, 1993) might be important in softening. Mutants of cell-wall proteins and manipulation of the glycosylation of these proteins clearly affects cell shape and development (Knox, 1995), but how do changes in this component of the cell wall affect fruit softening? Not much is known about this. An early study of tomato demonstrated that total nitrogen content from cell-wall protein changed very little during fruit ripening, but the amount of salt soluble protein that could be extracted from the cell wall increased approximately twofold from mature green to red-ripe fruit (Hobson *et al.*, 1983). However, the increase in extractable protein may have been due to changes in the carbohydrate fraction rather than the proteins themselves. In a recent paper, the change in accumulation of several thousand transcripts and a few hundred proteins was examined during tomato fruit development (Osorio *et al.*, 2011). The results suggested that neither transcription nor the protein content of structural cell-wall proteins changed much during ripening. None the less, it is possible that structural proteins in the cell wall are simply modified, which might affect softening, but this remains to be determined.

4.3.5 Other factors (pH, ionic composition, synthesis and non-uniformity)

The pH of most cell walls is typically between pH 6 and 7 (Almeida and Huber, 1999). When fruit ripen, the pH of the tomato cell-wall fluid (apoplast) drops from 6.5 in mature green fruit to pH 4.5 in ripe fruit (Almeida and Huber, 1999). A drop in the apoplastic pH may be common during fruit ripening. Changes in pH can affect enzyme activity (Chun and Huber, 1998) and possibly ionic interactions with calcium (Virk and Cleland, 1988). In this regard, an early theory proposed that auxin induced cell growth by acidification of the cell wall, known as the acid growth theory (Rayle and Cleland, 1970; Rayle and Cleland, 1980; Rayle and Cleland, 1992). The acid growth theory was originally explained as an effect on charge-related interactions between polymers or changes in enzyme activity (Rayle and Cleland, 1970). Subsequent to this early work, it was demonstrated that expansins were activated by cell-wall acidification and played a major role in loosening the cell walls during auxin-induced growth (McQueen-Mason and Cosgrove, 1994). As discussed above, expansins probably also play a role in fruit softening, but the role of pH in this process has not been examined (Brummell *et al.*, 1999b). Nevertheless, although the optimum pH for tomato fruit expansins was not reported, the activity of fruit expansin was assayed at pH 4.5 (Rose *et al.*, 2000), presumably because the pH optimum for the tomato fruit expansin is low, as it is for other expansins examined (McQueen-Mason and Cosgrove, 1994). Also of interest in regard to pH is that the activity maximum for the tomato fruit PG is approximately pH 5 and remains relatively high at pH 4.5 (Chun and Huber, 1998), which is the pH of the apoplast in ripe fruit (Almeida and Huber, 1999). At pH 6 and above, the pH of the apoplast in mature green fruit, PG has very low activity (Chun and Huber, 1998). Thus, the pH of the cell wall may be important for the activation of expansin, PG and other enzymes.

A change in the concentration of calcium can also affect cell-wall extensibility (Virk and Cleland, 1988), but the concentration of calcium in the cell wall of tomato does not change much during ripening (Almeida and Huber, 1999). None the less, the importance of calcium in fruit firmness and storage is exemplified by postharvest studies with apple (Mason *et al.*, 1975; Sams and Conway, 1984). More than simply a research interest, fruits and fruit slices are often commercially dipped in a solution containing $CaCl_2$ to improve firmness and storage life. Fruits dipped in $CaCl_2$ include but are not limited to apples (Saftner *et al.*, 1998), pears (Rosen and Kader, 1989), tomatoes (Floros *et al.*, 1992), blueberries (Camire *et al.*, 1994) and strawberries (García *et al.*, 1996).

Up to this point, we have focused on degradative processes in the cell wall, but biosynthesis of cell-wall polysaccharides continues throughout tomato fruit ripening (Mitcham *et al.*, 1989; Greve and Labavitch, 1991). In one study, the researchers injected radioactive sucrose into the pedicel of tomato fruit and then collected pericarp tissue at several stages of fruit development (Mitcham *et al.*, 1989). Radioactivity increased slightly in the pectin and hemicellulose fractions extracted from the fruit pericarp early in tomato fruit ripening and then declined at the red-ripe stage. The greatest increase in label per gram of pericarp was in the pectin fraction. It is not clear, however, whether the polymers synthesized during ripening were different from those synthesized earlier in development and how the newly synthesized polymers affected fruit softening.

Of possible importance to fruit firmness and more importantly texture, which includes juiciness, is that the cell wall is not the same all around the cell. We have already mentioned that the middle lamella is different where intercellular air spaces form, but the primary cell wall is also non-uniform. Pitfields are one example. Pitfields are regions of the cell wall where plasmodesmata transverse the walls between adjacent cells (Burch-Smith *et al.*, 2011). The plasmodesmata create channels between cells that play a role in cell-to-cell communication (Burch-Smith *et al.*, 2011). Immunoanalysis of the cell wall shows that regions near the plasmodesmata have a different carbohydrate composition from other parts of the wall (Orfila and Knox, 2000). There was an absence of 1,4-β-galactan near the plasmodesmata and the pectin material in this region was not easily extracted with calcium chelators (e.g. CDTA). It was concluded that the pectin structure (porosity) around the plasmodesmata might exclude some enzymes from reaching the plasmodesmata. Ultrastructural studies with apple and pear have indicated that the microdomain around plasmodesmata is indeed resistant to decomposition during ripening (Ben-Arie and Kislev, 1979; Roy *et al.*, 1997).

4.4 Conclusions

Ultrastructural (electron microscopy) studies indicate decomposition and spreading of the middle lamella during ripening of apple and pear (Ben-Arie and Kislev, 1979) and tomato (Crookes and Grierson, 1983). Disassembly of the middle lamella is probably a common theme in fruit ripening. The same ultrastructural studies also showed a disintegration of the fibrillar arrangement in the primary cell wall late in ripening. The disruption of fibrillar organization is even more pronounced in ripening avocado fruit, which, when ripe, has a MF texture (O'Donoghue *et al.*, 1994). The disruption of fibrillar organization can be accomplished simply by untethering the cellulose microfibrils with expansin (Cosgrove, 1999), depolymerization of the xyloglucan matrix that cross-links the microfibrils (Miedes and Lorences, 2009) or possibly by modification of peripheral cellulose polymers with cellulase (O'Donoghue *et al.*, 1994). However, the porosity of the cell wall is determined largely by the pectin matrix that the cellulose microfibrils and hemicelluloses are embedded in, and partial disassembly of the pectin matrix may be necessary to give enzymes access to the cellulose and hemicellulose polysaccharides (Carpita and Gibeaut, 1993). Extraction of cell-wall fractions at different stages of ripening supports extensive depolymerization of pectin but does not distinguish between the middle lamella and the pectin matrix in the primary cell wall (Brummell, 2006). Depolymerization of the hemicellulose (xyloglucan-rich) fraction does occur during ripening but is not nearly as extensive as the fragmentation of pectins (Brummell, 2006).

The recent immunological observation of hemicelluloses in the middle lamella, which disappear during ripening (Ordaz-Ortiz *et al.*, 2009), may contribute to the slight depolymerization of xyloglucans observed by others (Brummell, 2006).

Much of the research discussed above was performed simply to discover the fundamentals and basic biology of fruit ripening; none the less, the practical goal of using this knowledge to develop shipping and storage practices and to engineer high-quality fruit with a long shelf-life is never forgotten. In considering this practical goal, we must also keep in mind that, although we have selectively bred fruit for human consumption, the fruit has evolved for millions of years for reproduction of the species and not for commercial practice. Not very many fruits have evolved for optimal transcontinental shipping. Seed dispersal by consumption by an animal is one reason for ripening, but sometimes the fruit simply drops to the ground where the seed will then germinate. Over-ripening may be important to create a nursery environment for efficient seed germination. For example, many fruits, including tomato, will soften to the point that the fruit will collapse under its own weight. However, for human consumption, we typically want firm fruit that can easily be transported and a delay in the over-ripening of the fruit. We currently do this commercially by picking fruit before it is completely ripe, dipping fruit in calcium, waxing fruit or storing the fruit in reduced oxygen atmospheres, but we have also tried to do this through genetic engineering, such as the Flavr Savr tomato. Reducing the expression of a single gene or even two genes may not always produce the desired phenotype. Cell-wall ripening is complex. It may be necessary to suppress multiple genes or identify regulatory genes (proteins) that modulate the expression of several genes that affect the cell wall. None the less, we now have the tools to identify all the changes in gene expression and protein accumulation that occur during ripening and to use this information to engineer fruit with very special characteristics that include delayed softening to improve transport and delay the over-ripening response to extend shelf-life.

References

Allen, C.E. (1901) On the origin and nature of the middle lamella. *Botanical Gazette* 32, 1–34.

Almeida, D.P.F. and Huber, D.J. (1999) Apoplastic pH and inorganic ion levels in tomato fruit: a potential means for regulation of cell wall metabolism during ripening. *Physiologia Plantarum* 105, 506–512.

Asif, M., Trivedi, P. and Solomos, T. (2006) Effects of low oxygen and MCP, applied singly or together, on apple fruit ripening. *Acta Horticulturae* 712, 253–260.

Awad, M. and Young, R.E. (1979) Postharvest variation in cellulase, polygalacturonase, and pectinmethylesterase in avocado (*Persea americana* Mill, cv. Fuerte) fruits in relation to respiration and ethylene production. *Plant Physiology* 64, 306–308.

Bapat, V.A., Trivedi, P.K., Ghosh, A., Sane, V.A., Ganapathi, T.R. and Nath, P. (2010) Ripening of fleshy fruit: molecular insight and the role of ethylene. *Biotechnology Advances* 28, 94–107.

Ben-Arie, R. and Kislev, N. (1979) Ultrastructural changes in the cell walls of ripening apple and pear fruit. *Plant Physiology* 64, 197–202.

Bräutigam, A., Mullick, T., Schliesky, S. and Weber, A.P.M. (2011) Critical assessment of assembly strategies for non-model species mRNA-Seq data and application of next-generation sequencing to the comparison of C3 and C4 species. *Journal of Experimental Botany* 62, 3093–3102.

Brookfield, P.L., Nicoll, S., Gunson, F.A., Harker, F.R. and Wohlers, M. (2011) Sensory evaluation by small postharvest teams and the relationship with instrumental measurements of apple texture. *Postharvest Biology and Technology* 59, 179–186.

Brummell, D.A. (2006) Cell wall disassembly in ripening fruit. *Functional Plant Biology* 33, 103–119.

Brummell, D.A. and Labavitch, J.M. (1997) Effect of antisense suppression of endopolygalacturonase activity on polyuronide molecular weight in ripening tomato fruit and in fruit homogenates. *Plant Physiology* 115, 717–725.

Brummell, D.A., Hall, B.D. and Bennett, A.B. (1999a) Antisense suppression of tomato endo-1,4-β-glucanase Cel2 mRNA accumulation increases the force required to break fruit abscission zones but does not affect fruit softening. *Plant Molecular Biology* 40, 615–622.

Brummell, D.A., Harpster, M.H., Civello, P.M., Palys, J.M., Bennett, A.B. and Dunsmuir, P. (1999b) Modification of expansin protein abundance in tomato fruit alters softening and cell wall polymer metabolism during ripening. *Plant Cell* 11, 2203–2216.

Brummell, D.A., Harpster, M.H. and Dunsmuir, P. (1999c) Differential expression of expansin gene family members during growth and ripening of tomato fruit. *Plant Molecular Biology* 39, 161–169.

Burch-Smith, T.M., Stonebloom, S., Xu, M. and Zambryski, P.C. (2011) Plasmodesmata during development: re-examination of the importance of primary, secondary, and branched plasmodesmata structure versus function. *Protoplasma* 248, 61–74.

Calbo, A.G. and Nery, A.A. (1995) Methods for measurement of gas volume of fruits and vegetables. *Journal of the American Society for Horticultural Science* 120, 217–221.

Callahan, A.M., Scorza, R., Bassett, C., Nickerson, M. and Abeles, F.B. (2004) Deletions in an endopolygalacturonase gene cluster correlate with non-melting flesh texture in peach. *Functional Plant Biology* 31, 159–168.

Camire, M.E., Ismail, S., Work, T.M., Bushway, A.A. and Halteman, W.A. (1994) Improvements in canned lowbush blueberry quality. *Journal of Food Science* 59, 394–398.

Carpita, N. and McCann, M. (2000) The cell wall. In: Buchanan, B.B., Gruissem, W. and Jones, R.L. (eds) *Biochemistry and Molecular Biology of Plants.* American Society of Plant Biologists, Rockville, MD, pp. 52–108.

Carpita, N.C. and Gibeaut, D.M. (1993) Structural models of primary cell walls in flowering plants: consistency of molecular structure with the physical properties of the walls during growth. *Plant Journal* 3, 1–30.

Christoffersen, R.E., Tucker, M.L. and Laties, G.G. (1984) Cellulase gene expression in ripening avocado fruit: the accumulation of cellulase mRNA and protein as demonstrated by cDNA hybridization and immunodetection. *Plant Molecular Biology* 3, 385–391.

Chun, J.P. and Huber, D.J. (1998) Polygalacturonase-mediated solubilization and depolymerization of pectic polymers in tomato fruit cell walls. Regulation by pH and ionic conditions. *Plant Physiology* 117, 1293–1299.

Cosgrove, D.J. (1999) Enzymes and other agents that enhance cell wall extensibility. *Annual Review of Plant Physiology and Plant Molecular Biology* 50, 391–417.

Cosgrove, D.J., Li, L.C., Cho, H.T., Hoffmann-Benning, S., Moore, R.C. and Blecker, D. (2002) The growing world of expansins. *Plant and Cell Physiology* 43, 1436–1444.

Crookes, P.R. and Grierson, D. (1983) Ultrastructure of tomato fruit ripening and the role of polygalacturonase isoenzymes in cell wall degradation. *Plant Physiology* 72, 1088–1093.

Drazeta, L., Lang, A., Hall, A.J., Volz, R.K. and Jameson, P.E. (2004) Air volume measurement of 'Braeburn' apple fruit. *Journal of Experimental Botany* 55, 1061–1069.

Floros, J.D., Ekanayake, A., Abide, G.P. and Nelson, P.E. (1992) Optimization of a diced tomato calcification process. *Journal of Food Science* 57, 1144–1148.

Flors, V., Leyva Mde, L., Vicedo, B., Finiti, I., Real, M.D., Garcia-Agustin, P., Bennett, A.B. and Gonzalez-Bosch, C. (2007) Absence of the endo-β-1,4-glucanases Cel1 and Cel2 reduces susceptibility to *Botrytis cinerea* in tomato. *Plant Journal* 52, 1027–1040.

Gane, R. (1934) Production of ethylene by some ripening fruits. *Nature* 134, 1008.

García, J.M., Herrera, S. and Morilla, A. (1996) Effects of postharvest dips in calcium chloride on strawberry. *Journal of Agricultural and Food Chemistry* 44, 30–33.

Garcia-Ramos, F.J., Valero, C., Homer, I., Ortiz-Canavate, J. and Ruiz-Altisent, M. (2005) Non-destructive fruit firmness sensors: a review. *Spanish Journal of Agricultural Research* 3, 61–73.

Ghiani, A., Onelli, E., Aina, R., Cocucci, M. and Citterio, S. (2011) A comparative study of melting and non-melting flesh peach cultivars reveals that during fruit ripening endo-polygalacturonase (endo-PG) is mainly involved in pericarp textural changes, not in firmness reduction. *Journal of Experimental Botany* 62, 4043–4054.

Giovannoni, J.J., Dellapenna, D., Bennett, A.B. and Fischer, R.L. (1989) Expression of a chimeric polygalacturonase gene in transgenic *rin* (ripening inhibitor) tomato fruit results in polyuronide degradation but not fruit softening. *Plant Cell* 1, 53–63.

Greve, L.C. and Labavitch, J.M. (1991) Cell wall metabolism in ripening fruit: V. Analysis of cell wall synthesis in ripening tomato pericarp tissue using a D-[U-^{13}C]glucose tracer and gas chromatography-mass spectrometry. *Plant Physiology* 97, 1456–1461.

Grierson, D., Tucker, G.A., Keen, J., Ray, J., Bird, C.R. and Schuch, W. (1986) Sequencing and identification of a cDNA clone for tomato polygalacturonase. *Nucleic Acids Research* 14, 8595–8602.

Hall, D.J. (1989) Postharvest treatment of Florida fresh market tomatoes with fungicidal wax to reduce decay. *Proceedings of the Florida State Horticultural Society* 102, 365–367.

Hall, L.N., Tucker, G.A., Smith, C.J.S., Watson, C.F., Seymour, G.B., Bundick, Y., Boniwell, J.M., Fletcher, J.D., Ray, J.A., Schuch, W., Bird, C.R. and Grierson, D. (1993) Antisense inhibition of pectin esterase gene expression in transgenic tomatoes. *Plant Journal* 3, 121–129.

Hardenburg, R.E. (1967) *Wax and Related Coatings for Horticultural Products; a Bibliography.* US Department of Agriculture, Agricultural Research Service, Washington, DC.

Hatfield, R. and Nevins, D.J. (1986) Characterization of the hydrolytic activity of avocado cellulase. *Plant and Cell Physiology* 27, 541–552.

Ho, Q.T., Verboven, P., Verlinden, B.E., Herremans, E., Wevers, M., Carmeliet, J. and Nicolai, B.M. (2011) A three-dimensional multiscale model for gas exchange in fruit. *Plant Physiology* 155, 1158–1168.

Hobson, G.E., Richardson, C. and Gillham, D.J. (1983) Release of protein from normal and mutant tomato cell walls. *Plant Physiology* 71, 635–638.

Jeong, J., Huber, D.J. and Sargent, S.A. (2003) Delay of avocado (*Persea americana*) fruit ripening by 1-methylcyclopropene and wax treatments. *Postharvest Biology and Technology* 28, 247–257.

Khachatourians, G.G. (2002) *Transgenic Plants and Crops*. Marcel Dekker, New York.

Knox, J.P. (1995) Developmentally regulated proteoglycans and glycoproteins of the plant cell surface. *FASEB Journal* 9, 1004–1012.

Kramer, M., Sanders, R., Bolkan, H., Waters, C., Sheeny, R.E. and Hiatt, W.R. (1992) Postharvest evaluation of transgenic tomatoes with reduced levels of polygalacturonase: processing, firmness and disease resistance. *Postharvest Biology and Technology* 1, 241–255.

Kramer, M.G. and Redenbaugh, K. (1994) Commercialization of a tomato with an antisense polygalacturonase gene: the FLAVR SAVR™ tomato story. *Euphytica* 79, 293–297.

Kutsunai, S.Y., Lin, A.C., Percival, F.W., Laties, G.G. and Christoffersen, R.E. (1993) Ripening-related polygalacturonase cDNA from avocado. *Plant Physiology* 103, 289–290.

Lashbrook, C.C., Gonzalez-Bosch, C. and Bennett, A.B. (1994) Two divergent endo-β-1,4-glucanase genes exhibit overlapping expression in ripening fruit and abscising flowers. *Plant Cell* 6, 1485–1493.

Lashbrook, C.C., Giovannoni, J.J., Hall, B.D., Fischer, R.L. and Bennett, A.B. (1998) Transgenic analysis of tomato endo-β-1,4-glucanase gene function. Role of cel1 in floral abscission. *Plant Journal* 13, 303–310.

Laties, G.G. (1995) Franklin Kidd, Charles West and F.F. Blackman: the start of modern postharvest physiology. *Postharvest Biology and Technology* 5, 1–10.

Maclachlan, G. and Brady, C. (1994) Endo-1,4-β-glucanase, xyloglucanase, and xyloglucan endo-transglycosylase activities versus potential substrates in ripening tomatoes. *Plant Physiology* 105, 965–974.

Marin-Rodriguez, M.C., Orchard, J. and Seymour, G.B. (2002) Pectate lyases, cell wall degradation and fruit softening. *Journal of Experimental Botany* 53, 2115–2119.

Mason, J.L., Jasmin, J.J. and Granger, R.L. (1975) Softening of 'McIntosh' apples reduced by postharvest dip in calcium chloride solution plus thickeners. *HortScience* 10, 524–525.

McQueen-Mason, S. and Cosgrove, D.J. (1994) Disruption of hydrogen bonding between plant cell wall polymers by proteins that induce wall extension. *Proceedings of the National Academy of Sciences USA* 91, 6574–6578.

McQueen-Mason, S., Durachko, D.M. and Cosgrove, D.J. (1992) Two endogenous proteins that induce cell wall extension in plants. *Plant Cell* 4, 1425–1433.

Miedes, E. and Lorences, E.P. (2009) Xyloglucan endotransglucosylase/hydrolases (XTHs) during tomato fruit growth and ripening. *Journal of Plant Physiology* 166, 489–498.

Miedes, E., Herbers, K., Sonnewald, U. and Lorences, E.P. (2010) Overexpression of a cell wall enzyme reduces xyloglucan depolymerization and softening of transgenic tomato fruits. *Journal of Agricultural and Food Chemistry* 58, 5708–5713.

Mitcham, E.J., Gross, K.C. and Ng, T.J. (1989) Tomato fruit cell wall synthesis during development and senescence: *in vivo* radiolabeling of wall fractions using [^{14}C]sucrose. *Plant Physiology* 89, 477–481.

Mohamed, A.A.A., Jowitt, R. and Brennan, J.G. (1982) Sensory and instrumental measurement of food crispness: II – In a high moisture food. *Journal of Food Engineering* 1, 123–147.

Newman, R.H. and Redgwell, R.J. (2002) Cell wall changes in ripening kiwifruit: ^{13}C solid state NMR characterisation of relatively rigid cell wall polymers. *Carbohydrate Polymers* 49, 121–129.

O'Donoghue, E.M., Huber, D.J., Timpa, J.D., Erdos, G.W. and Brecht, J.K. (1994) Influence of avocado (*Persea americana*) Cx-cellulase on the structural features of avocado cellulose. *Planta* 194, 573–584.

Ohba, T., Takahashi, S. and Asada, K. (2011) Alteration of fruit characteristics in transgenic tomatoes with modified expression of a xyloglucan endotransglucosylase/hydrolase gene. *Plant Biotechnology* 28, 25–32.

Ordaz-Ortiz, J.J., Marcus, S.E. and Knox, J.P. (2009) Cell wall microstructure analysis implicates hemicellulose polysaccharides in cell adhesion in tomato fruit pericarp parenchyma. *Molecular Plant* 2, 910–921.

Orfila, C. and Knox, J.P. (2000) Spatial regulation of pectic polysaccharides in relation to pit fields in cell walls of tomato fruit pericarp. *Plant Physiology* 122, 775–781.

Osorio, S., Alba, R., Damasceno, C.M., Lopez-Casado, G., Lohse, M., Zanor, M.I., Tohge, T., Usadel, B., Rose, J.K., Fei, Z., Giovannoni, J.J. and Fernie, A.R. (2011) Systems biology of tomato fruit development: combined transcript, protein, and metabolite analysis of tomato transcription factor (nor, rin) and ethylene receptor (Nr) mutants reveals novel regulatory interactions. *Plant Physiology* 157, 405–425.

Poovaiah, B.W. and Nukaya, A. (1979) Polygalacturonase and cellulase enzymes in the normal Rutgers and mutant *rin* tomato fruits and their relationship to the respiratory climacteric. *Plant Physiology* 64, 534–537.

Quesada, M.A., Blanco-Portales, R., Pose, S., Garcia-Gago, J.A., Jimenez-Bermudez, S., Munoz-Serrano, A., Caballero, J.L., Pliego-Alfaro, F., Mercado, J.A. and Munoz-Blanco, J. (2009) Antisense down-regulation of the *FaPG1* gene reveals an unexpected central role for polygalacturonase in strawberry fruit softening. *Plant Physiology* 150, 1022–1032.

Rayle, D.L. and Cleland, R. (1970) Enhancement of wall loosening and elongation by acid solutions. *Plant Physiology* 46, 250–253.

Rayle, D.L. and Cleland, R.E. (1980) Evidence that auxin-induced growth of soybean hypocotyls involves proton excretion. *Plant Physiology* 66, 433–437.

Rayle, D.L. and Cleland, R.E. (1992) The Acid Growth Theory of auxin-induced cell elongation is alive and well. *Plant Physiology* 99, 1271–1274.

Rojas-Graü, M.A., Oms-Oliu, G., Soliva-Fortuny, R. and Martín-Belloso, O. (2009) The use of packaging techniques to maintain freshness in fresh-cut fruits and vegetables: A review. *International Journal of Food Science and Technology* 44, 875–889.

Rose, J.K., Lee, H.H. and Bennett, A.B. (1997) Expression of a divergent expansin gene is fruit-specific and ripening-regulated. *Proceedings of the National Academy of Sciences of the United States of America* 94, 5955–5960.

Rose, J.K., Cosgrove, D.J., Albersheim, P., Darvill, A.G. and Bennett, A.B. (2000) Detection of expansin proteins and activity during tomato fruit ontogeny. *Plant Physiology* 123, 1583–1592.

Rose, J.K., Braam, J., Fry, S.C. and Nishitani, K. (2002) The XTH family of enzymes involved in xyloglucan endotransglucosylation and endohydrolysis: current perspectives and a new unifying nomenclature. *Plant Cell Physiology* 43, 1421–1435.

Rose, J.K., Saladie, M. and Catala, C. (2004) The plot thickens: new perspectives of primary cell wall modification. *Current Opinion in Plant Biology* 7, 296–301.

Rosen, J.C. and Kader, A.A. (1989) Postharvest physiology and quality maintenance of sliced pear and strawberry fruits. *Journal of Food Science* 54, 656–659.

Roy, S., Watada, A.E. and Wergin, W.P. (1997) Characterization of the cell wall microdomain surrounding plasmodesmata in apple fruit. *Plant Physiology* 114, 539–547.

Saftner, R.A., Conway, W.S. and Sams, C.E. (1998) Effects of postharvest calcium and fruit coating treatments on postharvest life, quality maintenance, and fruit-surface injury in 'Golden Delicious' apples. *Journal of the American Society for Horticultural Science* 123, 294–298.

Saladie, M., Matas, A.J., Isaacson, T., Jenks, M.A., Goodwin, S.M., Niklas, K.J., Xiaolin, R., Labavitch, J.M., Shackel, K.A., Fernie, A.R., Lytovchenko, A., O'Neill, M.A., Watkins, C.B. and Rose, J.K. (2007) A reevaluation of the key factors that influence tomato fruit softening and integrity. *Plant Physiology* 144, 1012–1028.

Sams, C.E. and Conway, W.S. (1984) Effect of calcium infiltration on ethylene production, respiration rate, soluble polyuronide content, and quality of 'Golden Delicious' apples. *Journal of the American Society for Horticultural Science* 109, 53–57.

Sandhya (2010) Modified atmosphere packaging of fresh produce: current status and future needs. *LWT – Food Science and Technology* 43, 381–392.

Schuch, W., Kanczler, J., Robertson, D., Hobson, G., Tucker, G., Grierson, D., Bright, S. and Bird, C. (1991) Fruit quality characteristics of transgenic tomato fruit with altered polygalacturonase activity. *Horticulture Science* 26, 1517–1520.

Sheehy, R.E., Kramer, M. and Hiatt, W.R. (1988) Reduction of polygalacturonase activity in tomato fruit by antisense RNA. *Proceedings of the National Academy of Sciences USA* 85, 8805–8809.

Singh, S. (2011) Preservation technologies for fresh fruits and vegetables. *Stewart Postharvest Review* 7, 1–7.

Smith, D.L., Abbott, J.A. and Gross, K.C. (2002) Down-regulation of tomato β-galactosidase 4 results in decreased fruit softening. *Plant Physiology* 129, 1755–1762.

Tieman, D.M., Harriman, R.W., Ramamohan, G. and Handa, A.K. (1992) An antisense pectin methylesterase gene alters pectin chemistry and soluble solids in tomato fruit. *Plant Cell* 4, 667–679.

Tucker, M.L. and Laties, G.G. (1985) The dual role of oxygen in avocado fruit respiration: kinetic analysis and computer modelling of diffusion-affected respiratory oxygen isotherms. *Plant, Cell and Environment* 8, 117–127.

Urbanowicz, B.R., Bennett, A.B., Del Campillo, E., Catalá, C., Hayashi, T., Henrissat, B., Höfte, H., McQueen-Mason, S.J., Patterson, S.E., Shoseyov, O., Teeri, T.T. and Rose, J.K.C. (2007) Structural organization and a standardized nomenclature for plant endo-1,4-β-glucanases (cellulases) of glycosyl hydrolase family 9. *Plant Physiology* 144, 1693–1696.

Verboven, P., Kerckhofs, G., Mebatsion, H.K., Ho, Q.T., Temst, K., Wevers, M., Cloetens, P. and Nicolai, B.M. (2008) Three-dimensional gas exchange pathways in pome fruit characterized by synchrotron X-ray computed tomography. *Plant Physiology* 147, 518–527.

Vicente, A.R., Saladié, M., Rose, J.K.C. and Labavitch, J.M. (2007) The linkage between cell wall metabolism and fruit softening: looking to the future. *Journal of the Science of Food and Agriculture* 87, 1435–1448.

Virk, S.S. and Cleland, R.E. (1988) Calcium and the mechanical properties of soybean hypocotyl cell walls: possible role of calcium and protons in cell-wall loosening. *Planta* 176, 60–67.

Woolley, L.C., James, D.J. and Manning, K. (2001) Purification and properties of an endo-β-1,4-glucanase from strawberry and down-regulation of the corresponding gene, cel1. *Planta* 214, 11–21.

Zabalza, A., Van Dongen, J.T., Froehlich, A., Oliver, S.N., Faix, B., Gupta, K.J., Schmalzlin, E., Igal, M., Orcaray, L., Royuela, M. and Geigenberger, P. (2009) Regulation of respiration and fermentation to control the plant internal oxygen concentration. *Plant Physiology* 149, 1087–1098.

5 Aroma Volatiles

Bo Zhang and Kun-Song Chen*
Laboratory of Fruit Quality Biology, Zhejiang University, Hangzhou, PR China

5.1 Introduction

Plants are multifaceted chemical factories that produce at least 1000 volatile compounds, and the molecules involved in the biosynthesis and release of these volatiles comprise more than 1% of plant secondary metabolites (Qualley and Dudareva, 2009; Dicke and Loreto, 2010). Some plants allocate up to 10% of their carbon to the production of volatile secondary metabolites (Firn and Jones, 2006). Several years ago, an argument was raised focusing on the role of volatile compounds released by plants. This argument was presented in an article entitled, 'Plant volatiles: a lack of function or a lack of knowledge?' (Pichersky *et al.*, 2006). In recent years, thanks to the development of integrative biological approaches, more and more papers regarding volatiles have been published, giving rise to a molecular understanding of the function of volatiles. In March 2010, a special issue of *Trends in Plant Science*, entitled 'Induced biogenic volatile organic compounds from plants', discussed the role of volatiles. These studies have formidably enhanced our knowledge of the volatiles that are involved in plant responses to biotic and abiotic factors and in plant communication with other organisms.

As an important factor for quality, together with textural and visual cues and with sugars and acids, the aroma volatiles of fruits have a major impact on consumer preference (see Plate 2). Fruit aroma is determined by a complex mixture of volatiles, including aldehydes, alcohols, esters, ketones, lactones and terpenoids. Although the production of these compounds accounts for only 10^{-7}–10^{-4}% of the fresh fruit weight, these compounds can be detected by the human olfactory system and are regarded as the characteristic flavours of the fruit (Jiang and Song, 2010). Identification of the character-impacting aroma volatiles of major fruit species has been studied intensively, and volatile compounds are known to be influenced by various factors such as cultivar, maturity stage, postharvest treatment and analysis technique. Generally, the content of most volatile compounds increases at the onset of fruit ripening and peaks either at or shortly before full ripening (Goff and Klee, 2006).

Fruit aroma volatiles are represented by fatty acid derivatives, terpenoids and amino acid derivatives, and the diversity of their origins within the metabolome has been summarized by Schwab *et al.* (2008). Improving fruit aroma quality is becoming an important goal of breeding and

* akun@zju.edu.cn

biotechnological programmes and is of keen public interest. Metabolic engineering of volatile compounds using transgenic techniques has been widely applied in plants (Dudareva and Pichersky, 2008) and depends on the discovery of genes involved in the biosynthesis of volatiles. Advances in gene cloning associated with the formation of aroma volatiles during fruit ripening have been reviewed recently (Defilippi *et al.*, 2009a), and more reviews in terms of gene evaluation and functions involved in the biosynthesis of volatile compounds are also available (Chen *et al.*, 2011; Strommer, 2011).

A decline in fruit flavour quality has occurred over the last two decades (Klee, 2010). Recently, a growing number of consumers are willing to pay a premium for better aroma quality, and the number of scientific papers investigating the characterization and regulation of aroma volatiles has increased greatly. In addition, with the development and application of state-of-the-art techniques, certain metabolic pathways involved in the biosynthesis of aroma volatiles have been established recently. Even for defined metabolic pathways, the regulatory mechanisms during fruit ripening are not fully understood. The formation and regulation of aroma volatiles is becoming an important research topic. To further our understanding of the mechanism of aroma quality formation in ripening fruit, the present chapter should serve as a guide to review the latest progress in the identification of characteristic aroma volatiles of ripening fruit, summarizing the regulation of the biosynthesis of volatiles that are clearly associated with ripening.

5.2 Identification of Aroma Volatiles

5.2.1 Technologies used for analysis of aroma volatiles

A key prerequisite to understanding the function and biosynthesis of aroma volatiles released from ripening fruit is the separation and identification of volatile compounds within the complex mixtures. Over the past few decades, a variety of technologies have been developed to capture and identify the aroma volatiles of ripening fruit, including solid-phase microextraction (SPME), gas chromatography (GC), mass spectrometry (MS) and data analysis.

Distillation, employed since the middle ages, is the early separation process to isolate volatile compounds from fruit. In 1925, over 2000 l of Valencia orange juice was used as starting material to concentrate volatiles to levels at which they could be measured (Rouseff *et al.*, 2009). With the application of GC and MS, the initial required volume of grapefruit juice was reduced to 100 l to identify sulfur volatiles. The volume was further reduced to 10 ml with improvements in instrumentation such as headspace SPME, which is a rapid and simple analytical technique. The principle of SPME is the partitioning process of the analyte between the fibre coating and the sample. The development and application of SPME in aroma volatiles have been reviewed (Augusto *et al.*, 2000; Wardencki *et al.*, 2004; Qualley and Dudareva, 2009).

GC is a common chemical analysis technique for separating and analysing compounds in a complex sample. A modern GC instrument was invented in 1952 by James and Martin and was applied to the separation of volatile compounds from citrus juices in the 1950s and 1960s (Rouseff *et al.*, 2009). However, the identification of volatiles was insufficient until GC was coupled to MS in the 1960s. Although more and more forms of MS have been developed, the fragmentation pattern based on electron-impact quadrupole MS is probably the most-used MS form. Currently, the available NIST Mass Spectral library (http://www.sisweb.com/software/ms/nist.htm) is used to guide the identification of compounds, which has greatly facilitated separating and identifying fruit aroma volatiles.

SPME-GC-MS has been widely used to profile volatiles released from fruit. Conventional statistical analysis is not applicable to the large and multivariate data derived from SPME-GC-MS analysis, which can be regarded as a high-throughput metabolomic tool. Principal component analysis (PCA) and partial least squares projection to latent structure (PLS), together with PLS-discriminant analysis (PLS-DA), are popular and efficient statistical methods to explore and extract data sets from volatiles in ripening fruit (Aprea *et al.*, 2011). In addition, an increased demand for spectral data annotation and mining promoted the occurrence of a web-based database. The Golm Metabolome Database (http://gmd.mpimp-golm.mpg.de) has been established to allow researchers to compare spectral data and generate hypotheses about the changes in volatiles for mutants of genes of unknown function.

Fruit aroma volatiles are traditionally analysed using GC-MS. However, GC analysis is not a particularly practical technology because of the complexity of the volatile compounds.

Because of this fact and the need to link aroma analysis with human perception, electronic nose (E-nose) technology has been developed over the past decade to investigate the sensory properties of many fruits (Brezmes *et al.*, 2005). Such studies have included the characterization of different fruit cultivars and maturity stages with fruits such as mangoes (Lebrun *et al.*, 2008), mandarins (Gómez *et al.*, 2006) and peaches (Zhang *et al.*, 2011). In addition, the E-nose has also been used to determine fruit ripeness during the postharvest storage of fruits such as apples (Saevels *et al.*, 2004), apricots (Defilippi *et al.*, 2009b), bananas (Llobet *et al.*, 1999) and peaches (Infante *et al.*, 2008). Four peach fruit cultivars could be separated into corresponding clusters using an E-nose on the day of harvest, and the fruit from each cultivar exhibited a clear distribution according to ripening stage (Benedetti *et al.*, 2008). In another study, an E-nose was used to evaluate the aroma quality of peach fruit during cold storage, and fruits with chilling injury (CI) symptoms after 21 days of storage at a chilling-inducing temperature (5°C) were separated from those without the disorders at 0 and 8°C (Zhang *et al.*, 2011).

5.2.2 Fruits with different genetic backgrounds exhibit differences in aroma volatiles

As mentioned above, fruit aroma volatiles are represented by complex mixtures of compounds, and the volatile profiles are highly dependent on fruit species and varieties. Table 5.1 outlines the volatile compounds that are responsible for some fruit aroma qualities.

Analysis of fruit aroma volatiles has been used as a tool to characterize and classify varieties. The production of volatile compounds by five ripening tomato fruit mutants (*rin*, *nor*, *Nr*, *Cnr* and *hp1*) and the isogenic wild-type ‘Ailsa Craig’ were compared. Distinct modifications of the volatile profiles were found in ripening mutants, and the differences were most dramatically caused by fatty acid-derived volatiles such as hexanal and hexenal (Kovács *et al.*, 2009). In apple fruit, the developed PLS-DA model and the calculated variable importance values were successfully applied to identify characteristic volatiles of the four varieties (Aprea *et al.*, 2011). Among these volatile compounds, 3-methylpentanol, propyl 2-methylbutanoate and isobutyl 2-methylbutanoate are characteristic peaks for the ‘Golden Delicious’ variety, and *N*-phenylaniline is a marker for ‘Granny Smith’ apples. ‘Pinova’ apples show higher levels of 2-ethylphenol and hexyl 2-methylbutanoate, whilst ‘Stark Delicious’ fruit has higher emissions of ethyl butanoate, 5-hexenyl acetate and butyl hexanoate. According to volatile profiles, 50 peach and nectarine fruits with different genotypic backgrounds could be separated into four groups, in which Chinese wild peaches and ‘Wutao’ had high levels of terpenoids and esters, cultivars of American and European origin showed high linalool, ‘Ruipan 14’ and ‘Babygold 7’ had high lactones, and the remaining cultivars were without characteristic

Table 5.1. Volatile compounds responsible for fruit aroma. From Jiang and Song (2010).

Fruit type	Representative fruits	Aroma classification	Main characteristic aroma compounds
Stone fruit	Peach	Lactone	γ-Octalactone
		Lactone	γ-Decalactone
		Lactone	δ-Octalactone
	Nectarine	Terpenes	Linalool
		Terpenes	α-Terpinene
		Terpenes	γ-Terpinene
	Plum	Terpenes	Linalool
		Aldehydes	Benzaldehyde
		Terpenes	α-Terpineol
	Cherry	Aldehydes	Phenylacetaldehyde
		Aldehydes	(*E*)-2-hexenal
	Apricot	Aldehydes	(*E*,*Z*)-2,6-nonadienal
		Ketones	(*Z*)-1,5-octadien-3-one
Pome fruit	Apple	Esters	Butyl acetate
		Esters	Ethyl butyrate
	Pear	Esters	Hexyl acetate
		Esters	Ethyl butyrate
Berry fruit	Grape	Terpenes	Linalool
		Terpenes	Geraniol
	Strawberry	Ketones	2,5-Dimethyl-4-hydroxy- 3(2H)furanone
	Kiwifruit	Aldehydes	(*E*)-2-hexenal
		Esters	Ethyl butyrate
	Tomato	Aldehydes	(*Z*)-3-hexenal
		Ketones	β-ionone
Citrus	Citrus	Terpenes	D-Limonene
	Sour orange	Terpenes	Linalool
	Sweet orange	Esters	Ethyl butanoate
Others	Banana	Phenols	Eugenol
		Ethers	Eugenol methyl ether
	Coconut	Lactone	δ-Decalactone
		Lactone	γ-Dodecalactone
	Pineapple	Esters	Ethyl 2-methylthion propionate
		Esters	Methyl 2-methylthion propionate
	Lychee	Ketones	β-damascenone

volatile compounds (Wang *et al.*, 2009). Differences in the aroma volatiles of fruit cultivars have also been reported in plum (Chai *et al.*, 2012), mango (Pino *et al.*, 2005) and kiwifruit (Garcia *et al.*, 2012).

5.3 Pathways for the Biosynthesis of Aroma Volatiles

The volatile compounds that are responsible for the aroma quality of fruit are usually small molecules with low boiling points and high vapour pressures at ambient temperature. These compounds contain different chemical functional groups such as aldehydes, alcohols, esters, lactones, alkenes and ethers, which are derived from many different biosynthesis pathways. Generally, fatty acids, carbohydrates and amino acids are the major natural carbon pools for the biosynthesis of aroma volatiles (Fig. 5.1).

5.3.1 Carbohydrate pathway

Terpenoids consist of hemiterpenes (C5), monoterpenes (C10), sesquiterpenes (C15) and diterpenes (C20) and are the largest class of plant secondary metabolites. Three phases are involved in the basic pathway of volatile terpenoid biosynthesis: (i) formation of the C5 units; (ii) condensation of C5 units into C10, C15 or C20 prenyl diphosphates; and (iii) conversion of the resulting prenyl diphosphates to end products (reviewed by Dudareva *et al.*, 2004). In plants, the five-carbon compound isopentenyl pyrophosphate (IPP) and its isomer dimethylallyl pyrophosphate (DMAPP) are produced through either the plastidic methylerythritol phosphate pathways or cytosolic mevalonate. The methylerythritol phosphate pathway is responsible for the formation of hemiterpene, monoterpene and diterpene by providing IPP and DMAPP. In the cytosol, one DMAPP and two IPP molecules are condensed, resulting in farnesyl pyrophosphate (FPP), which is further converted into sesquiterpene. Recently, a novel pathway for sesquiterpene biosynthesis from *Z,Z*-FPP in the wild tomato has been discovered, suggesting that sesquiterpene could be generated using IPP and DMAPP from the plastidic 1-deoxy-D-xylulose 5-phosphate (DXP) pathway (Sallaud *et al.*, 2009).

Terpene synthases (TPSs) catalyse the formation of hemi-, mono-, sesqui- and diterpenes from DMAPP, geranyl

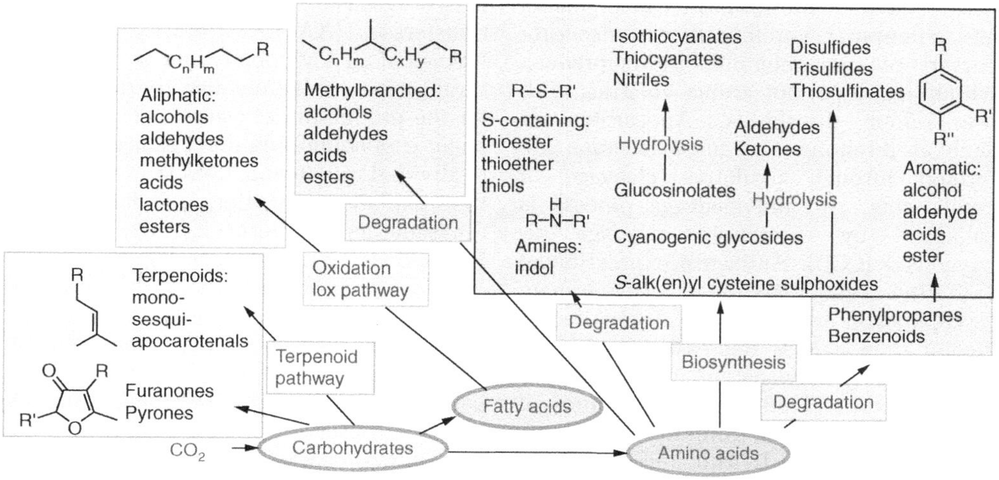

Fig. 5.1. Biosynthesis pathways of volatile compounds in plants. From Schwab *et al.* (2008).

pyrophosphate, farnesyl pyrophosphate and geranylgeranyl pyrophosphate, respectively. The TPS family can be separated into clusters according to the substrate used, and the current knowledge of this mid-sized gene family in plants has been reviewed by Chen *et al.* (2011). In tomato, 44 TPS genes located on eight chromosomes have been cloned, including 29 genes that have open reading frames (Falara *et al.*, 2011). In contrast, the remaining 15 TPS genes appear to have mutations and deletions. Among these 29 functional or potentially functional genes, 12 tomato TPS genes encode presumed cytosolic synthases and belong to the TPS-a clade, which is responsible for sesquiterpene biosynthesis, eight genes belong to the TPS-b clade encoding monoterpene synthases, two genes belong to the TPS-c clade, five genes belong to the TPS-e/f clade and two genes belong to the TPS-g clade. d-Limonene is the most abundant volatile compound in orange fruit, and a gene named *CitMTSE1* was confirmed to be involved in the biosynthesis of this compound (Rodríguez *et al.*, 2011). In addition, *Cstps1* has been identified as a key gene in the production of the sesquiterpene valencene in citrus fruit (Sharon-Asa *et al.*, 2003).

Carotenoids are widely distributed compounds in plants. In addition to their involvement in the formation of pigments and colourants, carotenoids are also precursors of apocarotenoids (norisoprenes), which are important aroma volatiles with low odour thresholds. Apocarotenoids such as β-ionone and geranylacetone are formed through oxidative cleavage of carotenoids, and the cleavage process is catalysed by carotenoid cleavage dioxygenase (CCD). Antisense expression of *LeCCD1A* and *LeCCD1B* in tomato (*Lycopersicon esculentum*) fruit resulted in greater than a 50% decrease in β-ionone and 60% in geranylacetone, indicating that *LeCCD1* is involved in the formation of aroma apocarotenoids (Simkin *et al.*, 2004). Yellow-fleshed peach fruit and its white-fleshed mutant were used to study the role of CCD in the metabolism of carotenoids and apocarotenoids. Significantly up-regulated expression of CCD4 at later ripening stages was observed in the white-fleshed peach fruit, which is concomitant with significantly higher levels of any identified apocarotenoid volatiles throughout peach fruit ripening (Brandi *et al.*, 2011). In melon (*Cucumis melo*) fruit, *CmCCD1* was suggested to be involved in the generation of important apocarotenoid aroma compounds (Ibdah *et al.*, 2006).

5.3.2 Fatty acid pathway

Saturated and unsaturated fatty acids are the most important precursors for the majority of fruit aroma volatiles, including straight-chain aldehydes, alcohols, esters, lactones and ketones. These compounds are biosynthesized mainly through the lipoxygenase (LOX) pathway and β-oxidation. The current knowledge of the fatty acid pathway involved in the biosynthesis of volatiles has been reviewed by Schwab *et al.* (2008).

In the LOX pathway, linoleic (18:2) and linolenic acid (18:3) are catalysed into hydroperoxide isomers, which are further cleaved by hydroperoxide lyase (HPL) to form hexanal and hexenal, respectively. The aldehydes are subsequently reduced to the corresponding C6 alcohols by alcohol dehydrogenase (ADH). Alcohol acyltransferase (AAT) catalyses the final linkage of an acyl moiety and an alcohol to form esters and is thus directly responsible for the production of esters. LOX is a non-haem iron-containing dioxygenase and can be divided into 9- and 13-LOX according to the oxidation position of fatty acids (Feussner and Wasternack, 2002; Porta and Rocha-Sosa, 2002). Generally, there is a close association of both LOX genes and enzyme activity with fruit ripening and associated aroma quality development (Defilippi *et al.*, 2009a). The relationship between LOX genes and aroma volatiles has been observed in ripening fruits such as apple (Schaffer *et al.*, 2007), apricot (González-Agüero *et al.*, 2009), banana (Yang *et al.*, 2011), kiwifruit (Zhang *et al.*,

2009), olive (Padilla *et al.*, 2009), peach (Zhang *et al.*, 2010) and strawberry (Leone *et al.*, 2006). Particular evidence for the involvement of LOX in the generation of fruit aroma volatiles came from transgenic work in tomato, in which the specific downregulation of *TomLOXC* resulted in a significant reduction of grassy C6 aldehydes (Chen *et al.*, 2004). A recent report showed that the expression of a plastid-located *TomLOXC* plays a key role in determining the volatile composition of tomato ripening mutations (*hp1*, *rin*, *Cnr*, *nor* and *Nr*) and wild-type controls ('Ailsa Craig') (Kovács *et al.*, 2009).

HPL belongs to a novel family of cytochrome P450s (CYP74) and can be divided into four subfamilies based on sequence identity or grouped into three clades according to preferred substrates (Padilla *et al.*, 2010). Defilippi *et al.* (2009b) outlined the progress in cloning HPL genes from tomato, guava, cucumber and melon fruit. Characterization of HPL related to the biosynthesis of aroma aldehydes has recently been reported in olive and peach fruit. The expression of peach (*Prunus persica*) *PpHPL* was at high levels on the harvest day and decreased progressively with ripening, which is consistent with changes in *n*-hexanal and (*E*)-2-hexenal, which are major contributors to peach fruit aroma at harvest (Zhang *et al.*, 2010). Recombinant olive (*Olea europaea*) *OepHPL1* is specific for 13-hydroperoxide derivatives of linolenic acid, producing (*E*)-2-hexenal and (*Z*)-3-hexenal (Padilla *et al.*, 2010). Moreover, the expression profile of *OepHPL1* has been shown to be spatially and temporally regulated during olive fruit development and ripening.

ADH is a Zn-binding enzyme that acts as a dimer and is involved in the reversible conversion of aldehydes to their corresponding alcohols. The structure, biochemistry, function and evaluation of the ADH gene family in plants have been reviewed recently by Strommer (2011). Multiple ADH genes have been characterized in a number of fruits, including apricot (González-Agüero *et al.*, 2009), banana (Yang *et al.*, 2011), melon (Manríquez *et al.*, 2006), peach (Zhang *et al.*, 2010) and tomato (Longhurst *et al.*, 1994). In apple and peach fruit, ADH expression and alcohol levels are the highest on the initial ripening day, followed by a progressive decline with further ripening and senescence (Schaffer *et al.*, 2007; Zhang *et al.*, 2010). Ripening-dependent changes in ADH activity and alcohols have also been observed in tomato (Longhurst *et al.*, 1994) and melon (Manríquez *et al.*, 2006) fruit. These results indicate that the expression of ADH is under tight developmental regulation in ripening fruit. In tomato fruit, over-expression of ADH2 resulted in significantly increased hexanol and (*Z*)-3-hexenol, producing more of a 'ripe fruit' flavour (Speirs *et al.*, 1998).

Identification of AAT genes and enzymes responsible for ester biosynthesis has been reported in a number of ripening fruits, including apple (Souleyre *et al.*, 2005; Li *et al.*, 2006), apricot (González-Agüero *et al.*, 2009), banana (Yang *et al.*, 2011), melon (Yahyaoui *et al.*, 2002; El-Sharkawy *et al.*, 2005), strawberry (Aharoni *et al.*, 2000) and peach (Zhang *et al.*, 2010). Acyltransferases are a large family of proteins that are widely distributed in plants and consists of five major clades according to preferred substrate or to the condition under which genes and enzymes are active (D'Auria, 2006). There is considerable divergence among AATs, not only among species but also within species. The maximum sequence identity between *Prunus armeniaca* (*PaAAT1*), *Pyrus communis* (*PAAT1*) and *Malus domestica* (*MdAAT2*) is only 58%, although all belong to the *Rosaceae* family. In *Cucumis melo*, the sequence identity between *CmAAT1* and *CmAAT4* is only 22%, whereas *CmAAT1*, *CmAAT2* and *CmAAT3* are more closely identical (58–84%) (El-Sharkawy *et al.*, 2005). In the case of the substrate utilized and the product generated, melon *CmAAT4* has a strong preference for producing cinnamoyl acetate, strawberry SAAT prefers to yield methyl hexanoate and hexyl butyrate, and apple MdAAT2 has a preference for the formation of pentyl acetate and hexyl acetate (Defilippi *et al.*, 2009a). Despite strong sequence identity, SAAT and VAAT (wild strawberry) proteins differ in their substrate preference (Beekwilder *et al.*, 2004). Site-directed mutagenesis has demonstrated that a threonine residue plays a crucial role in AAT enzyme activity (El-Sharkawy *et al.*, 2005). AAT enzyme activity and gene expression increase during fruit ripening (Defilippi *et al.*, 2009a).

The breakdown of fatty acids through oxidation at the β-carbon and the subsequent removal of two carbon units was first discovered in 1904, and the detailed mechanisms of β-oxidation have been reviewed (Baker *et al.*, 2006). Analysis of mutants has revealed essential roles for β-oxidation in plant development and in response to stresses (Goepfert and Poirier, 2007). Involvement of β-oxidation in the biosynthesis of volatile aroma lactones has been suggested (Schwab *et al.*, 2008). Aliphatic long-chain acyl-CoA is first converted to 2-*trans*-enoyl-CoA by acyl-CoA oxidase (ACX) and finally yields an acyl-CoA molecule. During the β-oxidation of fatty acids, the breakdown of acetyl-CoA can be stopped between β-oxidation cycles or inside the reaction sequence as a result of many factors, resulting in the liberation of volatile lactones (Husan, 2010). Although lactones play an important sensory role in fruit aroma quality, there is a lack of information on the characterization of both the enzymes and the genes associated with their biosynthesis. ACX is the first enzyme involved in fatty acid β-oxidation and is regarded as a key step controlling flux through the pathway (Arent *et al.*, 2008). ACX is widely involved in embryo development, seed germination, seedling establishment, natural senescence and the biosynthesis of jasmonic acid in response to stresses (Baker *et al.*, 2006; Yang and Ohlrogge, 2009). Because model plants such as *Arabidopsis*, rice and tomato lack lactone compounds, an association of ACX with fruity-note lactone formation is unclear. In peach fruit, there are at least four ACX gene members, of which *PpACX1* has a distinct expression profile. Transcript levels of *PpACX1* increase with accumulated lactones during peach fruit ripening, and a positive correlation between long-chain ACX activity and lactones has been reported (Xi *et al.*, 2012).

5.3.3 Amino acid pathway

Branched-chain volatile compounds, including alcohols, aldehydes, esters, lactones, acids and sulfur-containing aroma compounds, are important for fruit aroma. The important amino acids responsible for the biosynthesis of aromatic volatile compounds are leucine, isoleucine, valine, alanine, phenylalanine, tyrosine and tryptophan (Defilippi *et al.*, 2009a; Tzin and Galili, 2010). Volatile compounds derived from leucine, such as 3-methyl butanoic acid and 2-methyl butanoic acid, and from phenylalanine, such as benzaldehyde, phenylacetaldehyde, benzyl alcohol, 2-phenylethanol, eugenol and chavicol, have been identified from peach, grape, tomato and strawberry fruit (Schwab *et al.*, 2008; Eduardo *et al.*, 2010). Branched-chain esters, including 2-methylpropyl acetate, 2-methylbutyl acetate, 3-methylbutyl acetate and 3-methylbutyl butanoate, were the most abundant volatiles in ripe banana fruit (Yang *et al.*, 2011). The ester 2-methylbutyl acetate has a strong apple scent and is associated with apple fruit, and 2-methyl butanoate determines the characteristic aroma of prickly pear (reviewed by Schwab *et al.*, 2008).

The biosynthesis pathways of aromatic amino acids and their regulation have been explored extensively in bacteria because of their utility in the food and drug industry. In plants, the aromatic amino acids are synthesized through the shikimate pathway, with chorismate serving as a major intermediate branch point metabolite (Fig. 5.2). The progress in terms of the characterization of enzymes and genes and the regulation of aromatic amino acid biosynthesis has recently been reviewed by Tzin and Galili (2010), whose evidence is mainly based on results from the model plant *Arabidopsis*.

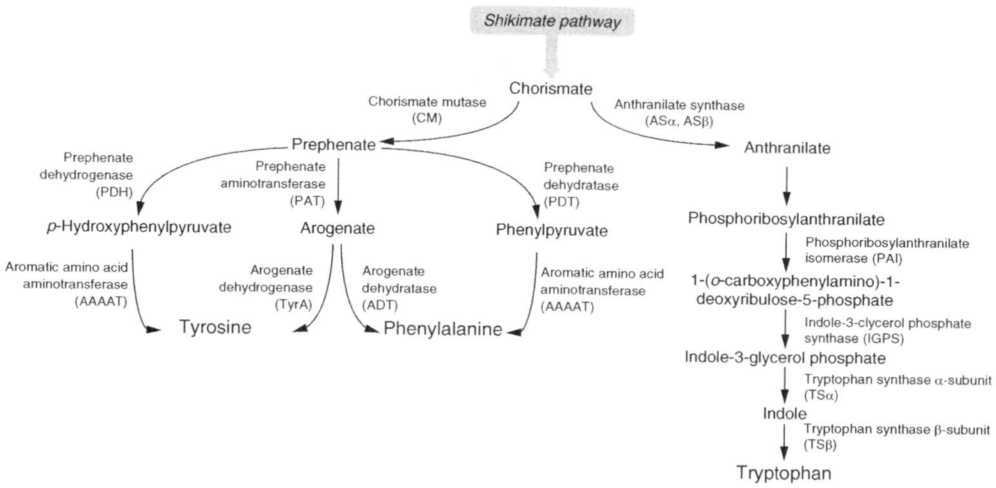

Fig. 5.2. Biosynthesis pathways of the aromatic amino acids in plants . From Tzin and Galili (2010).

Although amino acids are known as important precursors in the generation of aroma volatiles, the initial steps in the catabolism of amino acids into volatiles remains unclear. The results from tomato fruit showed that conversion of L-phenylalanine into aroma volatiles is initially catalysed by decarboxylation followed by deamination (Tieman *et al.*, 2006). Three gene family members encoding aromatic L-amino acid decarboxylase (AADC) were cloned from tomato fruit, in which overexpression of *LeAADC1A* and *LeAADC2* resulted in tenfold increases in the emission of aroma volatiles derived from phenethylamine, whilst antisense reduction of *LeAAD2* produced significantly lower levels (Tieman *et al.*, 2006). In contrast, studies from melon fruit showed that transamination of amino acids through aromatic amino acid transaminase (ArAT) or branched-chain amino acid transaminase (BCAT) was the initial step involved in the formation of branched-chain volatiles (Gonda *et al.*, 2010). Functional expression of *CmArAT1* and *CmBCAT1* in *Escherichia coli* had aromatic and branched-chain amino acid activities and converted amino acids into corresponding aroma volatiles. Accumulation of aroma volatiles is concomitant with increased expression of *CmArAT1* and *CmBCAT1* during melon fruit ripening. In addition, ripe melon fruit of climacteric aromatic cultivars showed high expression of *CmArAT1* and *CmBCAT1* during ripening in contrast to non-climacteric non-aromatic fruit (Gonda *et al.*, 2010). In banana fruit, the production of branched-chain esters such as 3-methylbutyl acetate, 3-methylbutyl butanoate and 3-methylbutyl 2-methylbutanoate was significantly increased by ethylene treatment during postharvest ripening (Yang *et al.*, 2011). The expression of banana *BanBCAT* was induced during fruit ripening and showed significantly higher levels in ethylene-treated fruit than in the controls (Yang *et al.*, 2011).

5.4 Regulation of Fruit Aroma Volatiles

5.4.1 Cultivation practices

The formation of aroma volatiles is a dynamic process, and changes in the volatile profile are both qualitative and quantitative during fruit growth and ripening. Grassy-note aroma volatiles such as C6 aldehydes and alcohols showed a decreasing trend during peach and nectarine fruit growth, whilst fruity-note lactones such as decalactone and dodecalactone accumulated greatly upon further maturity (Visai and Vanoli, 1997). The sesquiterpene valencene plays an important role in the overall aroma quality of orange fruit and accumulates during fruit development. A minor valencene peak was found in young Valencia orange fruit, followed by significantly higher levels at approximately 1–2 months after fruit colour break and continued accumulation until the fruits were fully mature (Sharon-Asa *et al.*, 2003). In strawberry fruit, the character aroma furanones and esters increased significantly and were closely correlated with skin colour development (Ménager *et al.*, 2004).

The generation of volatile compounds is greatly influenced by multiple environmental factors such as light and temperature, and the effects of these factors on the emission of plant volatiles have been reviewed by Holopaninen and Gershenzon (2010). During peach fruit development, the fruits were bagged under different levels of sunlight transmission (15, 50 and 80%) at approximately 50 days after full bloom, and non-bagged fruits were exposed to direct sunlight (100%) (Jia *et al.*, 2005). The results showed that bagging did not influence the total aroma volatiles released by whole fruit and the flesh tissue, but significant differences were observed in the skin tissue. Significantly higher levels of γ- and δ-decalactone and total aroma volatiles were found in peach fruit treated in bags with 15% transmission of sunlight, indicating that bagging can improve aroma quality during fruit growth. Differences in the volatile profiles of bagged fruit may be caused by accelerated ripening by bagging (Jia *et al.*, 2005; Wang *et al.*, 2010). In tomato fruit, a significantly higher content of hexanal was observed in traditional open-field conditions compared with protected cultivation in a screenhouse (a system usually used to prevent virus transmission by thrips and whiteflies) (Cebolla *et al.*, 2011).

In addition, preharvest calcium sprays enhanced the accumulation of the volatiles contributing to the overall aroma quality, indicating a suitable procedure for the improvement of fruit aroma at harvest. 'Fuji' apple fruits were sprayed weekly with $CaCl_2$ (1.6%, w/v) from 81 days after full bloom until the commercial harvest date, resulting in significant increases in ester production, both quantitatively and qualitatively (Ortiz *et al.*, 2011). The enhanced emission of eight straight-chain esters and four branched-chain esters was suggested to be caused by increased activity of pyruvate decarboxylase (PDC) and ADH, leading to a better supply of alcohols and acyl-CoAs for ester formation in calcium-treated apple fruit (Ortiz *et al.*, 2011). A recent study by Salas *et al.* (2011) showed that application of aminoethoxyvinylglycine to apple trees 4 weeks prior to harvest resulted in a reduction of volatile compounds during postharvest storage.

5.4.2 Ethylene

Ethylene is the major regulator of fruit ripening in many fleshy fruits. The relationship between ethylene and the biosynthesis of volatiles in ripening fruit has been reported in apricot (González-Agüero *et al.*, 2009), citrus (Mayuoni *et al.*, 2011), kiwifruit (Zhang *et al.*, 2009), peach (Zhang *et al.*, 2010), tomato (Qin *et al.*, 2012) and other fruits (reviewed by Defilippi *et al.*, 2009a). The evidence for a regulatory role of ethylene in the development of aroma quality is based mainly on the accumulation of aroma volatiles during fruit ripening and the effects of ethylene and its inhibitors on changes in volatile compounds.

The production of aroma volatiles was compared in 15 Charentais melon cultivars with different ethylene emission and ripening dates, and a large reduction (49–87%) of volatiles was found in long-shelf-life cultivars compared with the original type and mid-shelf-life cultivars (Lucchetta *et al.*, 2009). Differences in the aroma volatiles of climacteric and non-climacteric melon fruit have also been analysed

(Obando-Ulloa *et al.*, 2008). A transgenic line of 'Royal Gala' apple fruit produced no detectable ethylene using antisense 1-aminocyclopropane-1-carboxylic acid (ACC) oxidase (ACO), resulting in a significant reduction in aroma volatiles (Schaffer *et al.*, 2007). Furthermore, the transgenic hybrids, produced by crossing Charentais melons with an antisense ACO line, showed a 60–80% lower ester content with a low odour threshold (potent odourants). These results showed that enzymes located in the final step of ester biosynthesis, AATs, are regulated by ethylene during fruit ripening (Schaffer *et al.*, 2007; Lucchetta *et al.*, 2009).

In the case of non-climacteric fruit, a significant accumulation of volatile esters was found during strawberry (*Fragaria chiloensis*) fruit ripening, which was correlated with increased AAT activity and *FcAAT1* expression (González *et al.*, 2009). In citrus fruit, there was induced expression of the TPS-encoding gene *Cstps1* accompanying a considerable increase in the sesquiterpene valencene in response to exogenous ethylene treatment during fruit growth (Sharon-Asa *et al.*, 2003). Ripening-related increases in volatiles have also been observed in other non-climacteric fruits such as raspberry and koubo (Defilippi *et al.*, 2009a).

Loss of function of ripening inhibitor (RIN) protein caused a delay in ripening and extended the shelf-life by regulating the expression of ACC synthase genes in tomato fruit (Fujisawa *et al.*, 2011). A recent report by Qin *et al.* (2012) demonstrated the role of transcription factor RIN in the regulation of the biosynthesis of ethylene and aroma volatiles during tomato fruit ripening. Five target gene promoters that could be bound by RIN were identified using high-resolution two-dimensional electrophoresis coupled with chromatin immunoprecipitation techniques. Furthermore, RIN was found to be involved in the modulation of aroma formation derived from the LOX pathway through the direct and rigorous regulation of genes such as *TomLOXC*, HPL and ADH2 (Qin *et al.*, 2012). Increased biosynthesis of volatile compounds during tomato fruit ripening depends on ethylene, as there were no ripening-associated increases in the ethylene-insensitive *Nr* mutant (Klee and Giovannoni, 2011). The above results taken together suggest that ethylene preferentially controls the generation of aroma volatiles.

5.4.3 Temperature

Fruits ripen and senesce rapidly at ambient temperature after harvest, and low-temperature storage is the primary technology used to delay quality deterioration. Tomato fruits were stored at 21°C (control) and 6°C (chilled) for 3 days followed by transfer to ambient temperature for up to 3 days to mimic storage in the retail chain. The content of major aroma volatiles, such as the C6 compounds isobutylthiazole and methylbutanol, were significantly reduced after chilling storage and a subsequent shelf-life (Boukobza and Taylor, 2002). In the case of fruits that are susceptible to CI, the overall levels of volatile compounds and aroma acceptability showed a gradual decrease during storage and transfer to shelf-life at ambient temperature. A distinct separation between peach fruit with CI and without CI was observed based on E-nose analysis, and the differences were caused mainly by decreases in volatile C6 compounds, esters and lactones during cold storage and shelf-life after transfer (Zhang *et al.*, 2011). Gradual decreases in fruity-note esters were observed in durian fruit with longer chilling temperature storage (Voon *et al.*, 2007).

Low-temperature-induced changes in aroma volatile profiles could be caused by a decrease in enzyme activity or a decline in substrate abundance. In tomato fruit, a significant decrease in ADH activity was correlated with changes in volatile alcohols during cold storage (León-Sánchez *et al.*, 2009). Reduced levels of fruity-note esters in peach fruit with CI were the consequence of reduced expression of *PpLOX1*, *PpLOX3* and *PpAAT1* (Zhang *et al.*, 2011). The content of aroma-characteristic lactones was the lowest at CI-inducing temperatures, whilst low-temperature conditioning alleviated the development of CI and maintained higher levels of lactones. Furthermore, changes in the lactones of peach fruit during cold storage are suggested to be a consequence of the altered expression of *PpACX1* and long-chain enzyme activity (Xi *et al.*, 2012). The addition of exogenous linoleic acid to chilled tomato fruit resulted in an increase in the corresponding volatile hexanal (Boukobza and Taylor, 2002). In a recent study in tomato fruit, increased linoleic and linolenic acid content caused by the overexpression of fatty acid desaturase (FAD) resulted in a significant accumulation of C6 compounds (Domínguez *et al.*, 2010).

5.4.4 Atmosphere

An enormous volume of research has been reported on the controlled atmosphere (CA) storage of fruit to retain quality for a longer period. The results from 'Rich Lady' peach fruit showed that CA storage (3% O_2 + 10% CO_2 at 2°C) was characterized by higher production of characteristic δ-decalactone and γ-dodecalactone compared with air-stored fruit, resulting in improved flavour perception and thus consumer acceptability after CA storage (Ortiz *et al.*, 2009). However, emission of straight-chain esters in 'Tardibelle' peach fruit was significantly reduced after 7 days of shelf-life after CA storage (Ortiz *et al.*, 2010). The blockage of ester-formation capacity was more severe and even unrecoverable in apple (Lara *et al.*, 2007) and pear fruit (Lara *et al.*, 2003) after extended storage under CA, leading to decreased overall aroma quality and fruit acceptability.

The observed differences in AAT activity alone were not enough to explain the deduced straight-chain esters caused by CA storage, indicating that an altered supply of alcohol and acyl-CoA precursors has an important role in the modulation of volatile esters (Lara *et al.*, 2003, 2006). For short-term storage, the inhibition of LOX activity in CA-stored 'Mondial Gala' apple fruit resulted in decreased emission of esters despite substantial ester-forming capacity that allowed for the recovery of ester production during shelf-life, whilst in the case of long-term storage (6 months), strong inhibition of AAT activity in combination with LOX caused unrecoverable diminution of the generation of esters (Lara *et al.*, 2007). A recent study suggested that the deduced respiration rate might influence the modulation of the different energy-carrying compounds and the overall activity of the enzymes such as LOX, ADH, PDC and AAT (Ortiz *et al.*, 2010), resulting in lower fruit aroma quality and consumer acceptability after CA storage.

5.4.5 Metabolic engineering

Notable success has been reported in enhancing and improving fruit aroma quality using transgenic techniques. The first attempt to add a new volatile compound to fruit was performed in tomato by introducing a *Clarkia breweri* linalool synthase gene under the control of the fruit-specific *E8* promoter, leading to significant accumulation of (*S*)-linalool and 8-hydroxylinalool (Lewinsohn *et al.*, 2001). The first successful enrichment of aroma and flavour in fruit through metabolic engineering was achieved by expressing lemon basil (*Ocimum basilicum*) geraniol synthase in tomato (see Plate 5). The genetically engineered tomato fruit had an increased content of monoterpenes such as nerol, citronellol, citronellal, citronellic acid, citronellyl acetate and rose oxide (Davidovich-Rikanati *et al.*, 2007), giving the fruit lemon and rose aroma properties. A panel of 82 people tested the genetically modified tomato fruit, which was preferred by 49 members of the panel, whilst four expressed no preference. The effects of modified expression of fruit TPS, CCD, ADH, LOX and FAD on volatile profiles have been reviewed by Dudareva and Pichersky (2008).

Apart from the functions involved in aroma quality, recent reports have indicated that changes in the content of volatiles also play important roles in the fruit response to biotic and abiotic stresses. Downregulation of the citrus limonene synthase gene (*CitMTSE1*) in the antisense orientation produced approximately 85 and 50 times less (+)-limonene and β-myrcene, respectively, in the peels of transgenic lines compared with the peels of wild-type fruit (Rodríguez *et al.*, 2011). Transgenic *CitMTSE1* citrus fruit showed increased resistance to economically important fungal and bacterial citrus pathogens and resulted in the repulsion of a major insect pest (Rodríguez *et al.*, 2011), suggesting a promising method for developing broad-spectrum resistance or tolerance in fruit and other crops. Transgenic tomato fruit generated by overexpressing the ω-3 fatty acid desaturases FAD3 and FAD7 had a significant increase in linolenic acid, a considerable accumulation of (*Z*)-hex-3-enal and enhanced resistance to CI (Domínguez *et al.*, 2010).

5.5 Conclusions

Volatiles are important determinants in the overall aroma quality and taste of fruit. In nature, volatiles serve as signals involved in protecting fruit against various stresses and contributing to seed dispersion by increasing fruit attractiveness. However, in the past few decades, breeders have selected new cultivars based mainly on yield, visual characteristics, sugar content and postharvest storability, whilst less attention is devoted to enhancing or even maintaining aroma quality. Recently, consumers have complained about a loss of flavour quality in fruit and are thus willing to pay higher prices for a product with a better flavour quality. The introduction of new long-shelf-life cultivars using traditional genetics or the inhibition of ethylene production using biotechnology has generally been accompanied by a loss of aroma quality and low consumer acceptance. The reintroduction of fruit aroma volatiles can be obtained using classical crossing with relatives that are rich in volatile compounds, as performed in tomato fruit by Kamal *et al.* (2001). It therefore would be important and necessary to analyse and compare volatile profiles among fruit species or cultivars with different genetic backgrounds, with the aim of finding 'the lost aroma volatiles' caused by evolution or breeding. However, although the biosynthesis of aroma volatiles is clearly ripening dependent and is associated with ethylene, the regulatory mechanisms are still not clear. A major limitation to such an understanding is the identification of genes and enzymes responsible for the biosynthesis of aroma volatiles. Once the biosynthetic pathways for fruit aroma volatiles have been established, the isolation of genes and transcription factors can be performed by exploiting the extensive genome and expressed sequence tag (EST) databases.

References

Aharoni, A., Keizer, L.C.P., Bouwmeester, H.J., Sun Z., Alvarez-Huerta, M., Verhoeven, H.A., Blaas, J., van Houwelingen A.M.M.L., De Vos, R.C.H., van der Voet, H., Jansen, R.C., Guis, M., Mol, J., Davis, R.W., Schena, M., van Tunen, A.J. and O'Conell, A.P. (2000) Identification of the SAAT gene involved in strawberry flavor biogenesis by use of DNA microarrays. *Plant Cell* 12, 647–661.

Aprea, E., Gika, H., Carlin, S., Theodoridis, G., Vrhovsek, U. and Mattivi, F. (2011) Metabolite profiling on apple volatile content based on solid phase micro extraction and gas-chromatography time of flight mass-spectrometry. *Journal of Chromatography A* 1218, 4517–4524.

Arent, S., Pye, V.E. and Henriksen, A. (2008) Structure and function of plant acyl-CoA oxidases. *Plant Physiology and Biochemistry* 46, 292–301.

Augusto, F., Valente, A.L.P., Tada, E.S. and Rivellino, S.R. (2000) Screening of Brazilian fruit aromas using solid-phase micro extraction-gas chromatography-mass spectrometry. *Journal of Chromatography A* 873, 117–127.

Baker, A., Graham, I.A., Holdsworth, M., Smith, S.M. and Theodoulou, F.L. (2006) Chewing the fat: β-oxidation in signalling and development. *Trends in Plant Science* 11, 125–132.

Beekwilder, J., Alavarez-Huerta, M., Neef, E., Verstappen, F.W.A., Bouwmeester, H.J. and Aharoni, A. (2004) Functional characterization of enzymes forming volatile esters from strawberry and banana. *Plant Physiology* 135, 1865–1878.

Benedetti, S., Buratti, S., Spinardi, A., Mannono, S. and Mignani, I. (2008) Electronic nose as a non-destructive tool to characterise peach cultivars and to monitor their ripening stage during shelf-life. *Postharvest Biology and Technology* 47, 181–188.

Boukobza, F. and Taylor, A.J. (2002) Effect of postharvest treatment on flavour volatiles of tomatoes. *Postharvest Biology and Technology* 25, 321–331.

Brandi, F., Bar, E., Mourgues, F., Horváth, G., Turcsi, E., Giuliano, G., Liverani, A., Tartarini, S., Lewinsohn, E. and Rosati, C. (2011) Study of 'Redhaven' peach and its white-fleshed mutant suggests a key role of CCD4 carotenoid dioxygenase in carotenoid and norisoprenoid volatile metabolism. *BMC Plant Biology* 11, 24.

Brezmes, J., Fructuoso, M.L.L., Llobet, E., Vilanova, X., Recasens, I., Orts, J., Saiz, G. and Correig, X. (2005) Evaluation of an electronic nose to assess fruit ripeness. *IEEE Sensors Journal* 5, 97–108.

Cebolla, J.C., Roselló, S., Valcárcel, M., Serrano, E., Beltrán, J. and Nuez, F. (2011) Evaluation of genotype and environment effects on taste and aroma flavor components of Spanish fresh tomato varieties. *Journal of Agricultural and Food Chemistry* 59, 2440–2450.

Chai, Q.Q., Wu, B.H., Liu, W.S., Wang, L.J., Yang, C.X., Wang, Y.J., Fang, J.B., Liu, Y.C. and Li, S.H. (2012) Volatiles of plums evaluated by HS-SPME with GC-MS at the germplasm levels. *Food Chemistry* 130, 432–440.

Chen, F., Tholl, D., Bohlmann, J. and Pichersky, E. (2011) The family of terpene synthases in plants: a mid-size family of genes for specialized metabolism that is highly diversified throughout the kingdom. *Plant Journal* 66, 212–229.

Chen, G.P., Hackett, R., Walker, D., Taylor, A., Lin, Z.F. and Grierson, D. (2004) Identification of a specific isoform of tomato lipoxygenase (TomloxC) involved in the generation of fatty acid-derived flavor compounds. *Plant Physiology* 136, 2641–2651.

Conzález, M., Gaete-Eastman, C., Valdenegro, M., Figueroa, C.R., Fuentes, L., Herrera, R., Alejandra, M. and Moya-León, M.A. (2009) Aroma development during ripening of *Fragaria chiloensis* fruit and participation of an alcohol acyltransferase (*FcAAT1*) gene. *Journal of Agricultural and Food Chemistry* 57, 9123–9132.

D'Auria, J.C. (2006) Acyltransferases in plants: a good time to be BAHD. *Current Opinion in Plant Biology* 9, 331–340.

Davidovich-Rikanati, R., Siytrit, Y., Tadmor, Y., Iijima, Y., Bilenko, N., Bar, E., Carmona, B., Fallik, E., Dudai, N., Simon, J.E., Pichersky, E. and Lewinsohn, E. (2007) Enrichment of tomato flavor by diversion of the early palstidial terpenoid pathway. *Nature Biotechnology* 25, 899–901.

Defilippi, B.G., Manriquez, D., Leungwilai, K. and Gonzalez-Aguero, M. (2009a) Aroma volatiles: biosynthesis and mechanisms of modulation during fruit ripening. In: Kader, J.C. and Delseny, M. (eds) *Advances in Botanical Research* 50, 1–38.

Defilippi, B.G., Juan, W.S., Valdes, H., Moya-Leon, M.A., Infante, R. and Campos-Vargas, R. (2009b) The aroma development during storage of Castlebrite apricots as evaluated by gas chromatography, electronic nose, and sensory analysis. *Postharvest Biology and Technology* 51, 212–219.

Dicke, M. and Loreto, F. (2010) Induced plant volatiles: from genes to climate change. *Trends in Plant Science* 15, 115–117.

Domínguez, T., Hernández, M.L., Pennycooke, J.C., Jiménez, P., Martinez-Rivas, J.M., Sanz, C., Stockinger, E.J., Sánchez-Serrano, J.J. and Sanmartín, M. (2010) Increasing ω-3 desaturase expression in tomato results in altered aroma profile and enhanced resistance to cold stress. *Plant Physiology* 153, 655–665.

Dudareva, N. and Pichersky, E. (2008) Metabolic engineering of plant volatiles. *Current Opinion in Biotechnology* 19, 1–9.

Dudareva, N., Pichersky, E. and Gershenzon, J. (2004) Biochemistry of plant volatiles. *Plant Physiology*, 135, 1893–1902.

Eduardo, I., Chietera, G., Bassi, D., Rossini, L. and Vecchietti, A. (2010) Identification of key odor volatile compounds in the essential oil of nine peach accessions. *Journal of the Science of Food and Agriculture* 90, 1146–1154.

El-Sharkawy, I., Manríquez, D., Flores, F.B., Regad, F., Bouzayen, M., Latché, A. and Pech, J.C. (2005) Functional characterization of a melon alcohol acyl-transferase gene family involved in the biosynthesis of ester volatiles. Identification of the crucial role of a threonine residue for enzyme activity. *Plant Molecular Biology* 59, 345–362.

Falara, V., Akhtar, T.A., Nguyen, T.T.H., Spyropoulou, E.A., Bleeker, P.M., Schauvinlhold, I., Matsuba, Y., Bonini, M.E., Schilmiller, A.L., Last, R.L., Schuurink, R.C. and Pichesky, E. (2011) The tomato terpene synthase gene family. *Plant Physiology* 157, 770–789.

Feussner, I. and Wasternack, C. (2002) The lipoxygenase pathway. *Annual Review of Plant Biology* 53, 275–297.

Firn, R.D. and Jones, C.G. (2006) Do we need a new hypothesis to explain plant VOC emission? *Trends in Plant Science* 11, 112–113.

Fujisawa, M., Nakano, T. and Ito, Y. (2011) Identification of potential target genes for the tomato fruit-ripening regulator RIN by chromatin immunoprecipitation. *BMC Plant Biology* 11, 26.

Garcia, C.V., Quek, S.Y., Stevenson, R.J. and Winz, R.A. (2012) Kiwifruit flavour: a review. *Trends in Food Science and Technology* 24, 82–91.

Goepfert, S. and Poirier, Y. (2007) β-Oxidation in fatty acid degradation and beyond. *Current Opinion in Plant Biology* 10, 245–251.

Goff, S.A. and Klee, H.J. (2006) Plant volatile compounds: sensory cues for health and nutritional value. *Science* 311, 815–819.

Gómez, A.H., Wang, J., Hu, G.X. and Pereira, A.G. (2006) Electronic nose technique potential monitoring mandarin maturity. *Sensors and Actuators B* 113, 347–353.

Gonda, I., Bar, E., Portnoy, V., Lev, S., Burger, J., Schaffer, A.A., Tadmor, Y., Gepstein, S., Giovannoni, J.J., Katzir, N. and Lewinsohn, E. (2010) Branched-chain and aromatic acid catabolism into aroma volatiles in *Cucumis melo* L. fruit. *Journal of Experimental Botany* 61, 1111–1123.

González-Agüero, M., Troncoso, S., Gudenschwager, O., Gampos-Vargas, R., Moya-León, M.A. and Defilippi, B.G. (2009) Differential expression levels of aroma-related genes during ripening of apricot (*Prunus armeniaca* L.). *Plant Physiology and Biochemistry* 47, 435–440.

Holopaninen, J.K. and Gershenzon, J. (2010) Multiple stress factors and the emission of plant VOCs. *Trends in Plant Science* 15, 176–184.

Husan, Q. (2010) Chemistry and biochemistry of some vegetable flavors. In: Hui, Y.H (ed.) *Handbook of Fruit and Vegetable Flavors*. John Wiley & Sons, Hoboken, NJ, pp. 575–625.

Ibdah, M., Azulay, Y., Portnoy, V., Wasserman, B., Bar, E., Meir, A., Burger, Y., Hirschberg, J., Schaffer, A.A., Katzir, N., Tadmor, Y. and Lewinsohn, E. (2006). Functional characterization of *CmCCD1*, a carotenoid cleavage dioxygenase from melon. *Phytochemistry* 67, 1579–1589.

Infante, R., Farcuh, M. and Meneses, C. (2008) Monitoring the sensorial quality and aroma through an electronic nose in peach during cold storage. *Journal of the Science of Food and Agriculture* 88, 2073–2078.

Jia, H.J., Araki, A. and Okamoto, G. (2005) Influence of fruit bagging on aroma volatiles and skin coloration of 'Hakuho' peach (*Prunus persica* Batsch). *Postharvest Biology and Technology* 35, 61–68.

Jiang, Y.M. and Song, J. (2010) Fruits and fruit flavor: classification and biological characterization. In: Hui, Y.H. (ed.) *Handbook of Fruit and Vegetable Flavors*. John Wiley & Sons, Hoboken, NJ, pp. 3–23.

Kamal, A.H.M., Takashina, T., Egashira, H., Satoh, H. and Imanishi, S. (2001) Introduction of aromatic fragrance into cultivated tomato from the '*peruvianum* complex'. *Plant Breeding* 120, 179–181.

Klee, H.J. (2010) Improving the flavor of fresh fruits: genomics, biochemistry, and biotechnology. *New Phytologist* 187, 44–56.

Klee, H.J. and Giovannoni, J.J. (2011) Genetics and control of tomato fruit ripening and quality attributes. *Annual Review of Genetics* 45, 41–49.

Kovács, K., Fray, R.G., Tikunov, Y., Graham, N., Bradley, G., Seymour, G.B., Bovy, A.G. and Grierson, D. (2009) Effect of tomato pleiotropic ripening mutations on flavour volatile biosynthesis. *Phytochemistry* 70, 1003–1008.

Lara, I., Miró, R.M., Fuentes, T., Sayez, G., Graell, J. and López, M.L. (2003) Biosynthesis of volatile aroma compounds in pear fruit stored under long-term controlled-atmosphere conditions. *Postharvest Biology and Technology* 29, 29–39.

Lara, I., Graell, J., López, M.L. and Echeverría, G. (2006) Multivariate analysis of modification in biosynthesis of volatile compounds after CA storage of 'Fuji' apples. *Postharvest Biology and Technology* 39, 19–28.

Lara, I., Echevrria, G., Graell, J. and Lopez, M.L. (2007) Volatile emission after controlled atmosphere storage of Mondial Gala apple (*Malus domestica*): relationship to some involved enzyme activities. *Journal of Agricultural and Food Chemistry* 55, 6087–6095.

Lebrun, M., Plotto, A., Goodner, K., Ducamp, M.N. and Baldwin, E. (2008) Discrimination of mango fruit maturity by volatiles using the electronic nose and gas chromatography. *Postharvest Biology and Technology* 48, 122–131.

León-Sánchez, F.D., Pelayo-Zaldívar, C., Rivera-Cabrera, F., Ponce-Valadez, M., Ávila-Alejandre, X., Fernández, F.J., Escalona-Buendía, H.B. and Pérez-Flores, L.J. (2009) Effect of refrigerated storage on aroma and alcohol dehydrogenase activity in tomato fruit. *Postharvest Biology and Technology* 54, 93–100.

Leone, A., Bleve-Zacheo, T., Gerardi, C., Melillo, M.T., Leo, L. and Zacheo, G. (2006) Lipoxygenase involvement in ripening strawberry. *Journal of Agricultural and Food Chemistry* 54, 6835–6844.

Lewinsohn, E., Schalechet, F., Wilkinson, J., Matsui, K., Tadmor, Y., Nam, K.H., Amar, O., Lastochkin, E., Larkov, O., Ravid, U., Hiatt, W., Gepstein, S. and Pickersky, E. (2001) Enhanced levels of the aroma and flavor compound *S*-linalool by metabolic engineering of the terpenoid pathway in tomato fruit. *Plant Physiology* 127, 1256–1265.

Li, D.P., Xu, Y.F., Xu, G.M., Gu, L.K., Li, D.Q. and Shu, H.R. (2006) Molecular cloning and expression of a gene encoding alcohol acyltransferase (*MdAAT2*) from apple (cv. Golden Delicious). *Phytochemistry* 67, 658–667.

Llobet, E., Hines, E.L., Gardner, J.W. and Franco, S. (1999) Non-destructive banana ripeness determination using a neural network-based electronic nose. *Measurement Science and Technology* 10, 538–548.

Longhurst, T., Lee, E., Hinde, R., Brady, C. and Speirs, J. (1994) Structure of the tomato *Adh2* gene and *Adh2* pseudogenes, and a study of *Adh2* gene expression in fruit. *Plant Molecular Biology* 26, 1073–1084.

Lucchetta, L., Manríquez, D., El-Sharkawy, I., Flores, F.B., Latché, A. and Pech, J.C. (2009) The role of ethylene in the expression of genes involved in the biosynthesis of aroma volatiles in melon. In: Ramina, A., Chang, C., Giovannoni, J., Klee, H., Perera, P. and Woltering, E. (eds) *Advances in Plant Ethylene Research: Proceedings of the 7th International Symposium on the Plant Hormone Ethylene*, Springer, Dordrecht, the Netherlands, pp. 189–195.

Manríquez, D., El-Sharkawy, I., Flores, F.B., El-Yahyaoui, F., Regad, F., Bouzayen, M., Latché, A. and Pech, J.C. (2006) Two highly divergent alcohol dehydrogenases of melon exhibit fruit ripening-specific expression and distinct biochemical characteristics. *Plant Molecular Biology* 61, 675–685.

Mayuoni, L., Tietel, Z., Patil, B.S. and Porat, R. (2011) Does ethylene degreening affect internal quality of citrus fruit. *Postharvest Biology and Technology* 62, 50–58.

MenagerMénager, I., Jost, M. and Aubert, C. (2004) Changes in physicochemical characteristics and volatile constituents of strawberry (cv. Cigaline) during maturation. *Journal of Agriculture and Food Chemistry* 52, 1248–1254.

Obando-Ulloa, J.M., Moreno, E., García-Mas, J., Nicolai, B., Lammertyn, J., Monforte, A.J. and Fernández-Trujillo, J.P. (2008) Climacteric or non-climacteric behavior in melon fruit. *Postharvest Biology and Technology* 49, 27–37.

Ortiz, A., Echeverría, G., Graell, J. and Lara, I. (2009) Overall quality of 'Rich Lady' peach fruit after air- or CA storage. The importance of volatile emission. *LWT – Food Science and Technology* 42, 1520–1529.

Ortiz, A., Graell, J., López, M.L., Echeverría, G. and Lara, I. (2010) Volatile ester-synthesising capacity in 'Tardibelle' peach fruit in response to controlled atmosphere and 1-MCP treatment. *Food Chemistry* 123, 698–704.

Ortiz, A., Graell, J. and Lara, I. (2011) Preharvest calcium sprays improve volatile emission at commercial harvest of 'Fuji Kiku-8' apples. *Journal of Agricultural and Food Chemistry* 59, 335–341.

Padilla, M.N., Hernández, M.L, Pérez, A.G., Sanz, C. and Martínez-Rivas, J.M. (2009) Functional characterization of two 13-lipoxygenase genes from olive fruit in relation to the biosynthesis of volatile compounds of virgin olive oil. *Journal of Agricultural and Food Chemistry* 57, 9097–9107.

Padilla, M.N., Hernández, M.L., Pérez, A.G., Sanz, C. and Martínez-Rivas, J.M. (2010) Isolation, expression, and characterization of a 13-hydroperoxide lyase gene from olive fruit related to the biosynthesis of the main virgin olive oil aroma compounds. *Journal of Agricultural and Food Chemistry* 58, 5649–5657.

Pichersky, E., Sharkey, T.D. and Gershernzon, J. (2006) Plant volatiles: a lack of function or a lack of knowledge? *Trends in Plant Sciences* 11, 421.

Pino, J.A., Mesa, J., Munoz, Y., Marti, M.P. and Marbot, R. (2005) Volatile components from Mango (*Mangifera indica* L.) cultivars. *Journal of Agricultural and Food Chemistry* 53, 2213–2223.

Porta, H. and Rocha-Sosa, M. (2002) Plant lipoxygenases. Physiological and molecular features. *Plant Physiology* 130, 15–21.

Qin, G.Z., Wang, Y.Y., Cao, B.H., Wang, W.H. and Tian, S.P. (2012) Unraveling the regulatory network of the MADS box transcription factor RIN in fruit ripening. *Plant Journal* 70, 243–255.

Qualley, A.V. and Dudareva, N. (2009) Metabolomics of plant volatiles. In: Belostotsky, D.A. (ed.) *Plant Systems Biology* 553. Humana Press, a part of Springer Science+Business Media, LLC, Clifton, NJ, pp. 329–343.

Rodríguez, A., Andres, V.S., Cervera, M., Redondo, A., Alquezar, B., Schimade, T., Gadea, J., Rodrigo, M.J., Zacarias, L.Z., Paou, L., Lopez, M.M., Castanera, P. and Pene, L. (2011) Terpene down-regulation in orange reveals the role of fruit aromas in mediating interactions with insect herbivores and pathogens. *Plant Physiology* 156, 793–802.

Rouseff, R.L., Perez-Cacho, P.R. and Jabalpurwala, F. (2009) Historical review of citrus flavor research during the past 100 years. *Journal of Agricultural and Food Chemistry* 57, 8115–8124.

Saevels, S., Lammertyn, J., Berna, A.Z., Veraverbeke, E.A., Natale, C.D. and Nicolai, B.M. (2004) An electronic nose and a mass spectrometry-based electronic nose for assessing apple quality during shelf life. *Postharvest Biology and Technology* 31, 9–19.

Salas, N.A., Molina-Corral, F.J., Gonzalez-Aguilar, G.A., Otero, A., Sepulveda, D.R. and Olivas, G.I. (2011) Volatile production by 'Golden Delicious' apple is affected by preharvest application of aminoethoxyvinylglycine. *Scientia Horticulturae* 130, 436–444.

Sallaud, C., Rontein, D., Onillon, S., Jabes, F., Duffé, P., Giacalone, C., Thoraval, S., Escoffier, C., Herbette, G., Leonhardt, N., Causse, M. and Tissier, A. (2009) A novel pathway for sesquiterpene biosynthesis from *Z,Z*-farnesyl pyrophosphate in the world tomato *Solanum habrochaites. Plant Cell* 21, 301–317.

Schaffer, R.J., Friel, E.N., Souleyre, E.J.F., Bolitho, K., Thodey, K., Ledger, S., Bowen, J.H., Ma, J.H., Nain, B., Cohen, D., Gleave, A.P., Crowhurst, R.N., Janssen, B.J., Yao, J.L. and Newcomb, R.D. (2007) A genomics approach reveals that aroma production in apple is controlled by ethylene predominantly at the final step in each biosynthetic pathway. *Plant Physiology* 144, 1899–1912.

Schwab, W., Davidovich-Rikanati, R. and Lewinsohn, E. (2008) Biosynthesis of plant-derived flavor compounds. *Plant Journal* 544, 712–732.

Sharon-Asa, L., Shalit, M., Frydman, A., Bar, E., Holland, D., Or, E., Lavi, U., Lewinsohn, E. and Eyal, Y. (2003) Citrus fruit flavor and aroma biosynthesis: isolation, functional characterization, and developmental regulation of *Cstps1*, a key gene in the production of the sesquiterpene aroma compound valencene. *Plant Journal* 36, 664–674.

Simkin, A.J., Schwartz, S.H., Auldridge, M., Taylor, M.G. and Klee, H.J. (2004) The tomato carotenoid cleavage dioxygenase 1 genes contribute to the formation of the flavor volatiles beta-ionone, pseudoionone, and geranylacetone. *Plant Journal* 40, 882–892.

Souleyre, E.J.F., Greenwood, D.R., Friel, E.N., Karunairetnam, S. and Newcomb, R.D. (2005) An alcohol acyl transferase from apple (cv. Royal Gala), MpAAT1, produces esters involved in apple fruit flavor. *FEBS Journal* 272, 3132–3144.

Speirs, J., Lee, E., Holt, K., Yong-Duk, K., Scott, N.S., Loveys, B. and Schuch, W. (1998) Genetic manipulation of alcohol dehydrogenase levels in ripening tomato fruit affects the balance of some flavor aldehydes and alcohols. *Plant Physiology* 117, 1047–1058.

Strommer, J. (2011) The plant ADH gene family. *Plant Journal* 66, 128–142.

Tieman, D., Taylor, M., Schauer, N., Fernie, A.R., Hanson, A.D. and Klee, H.J. (2006) Tomato aromatic amino acid decarboxylases participate in synthesis of the flavor volatiles 2-phenylethanol and 2-phyenylacetaldehyde. *Proceedings of the National Academy of Sciences USA* 103, 8287–8292.

Tzin, V. and Galili, G. (2010) New insights into the shikimate and aromatic amino acids biosynthesis pathways in plants. *Molecular Plant* 3, 956–972.

Visai, C. and Vanoli, M. (1997) Volatile compound production during growth and ripening of peaches and nectarines. *Scientia Horticulturae* 70, 15–24.

Voon, Y.Y., Hamid, N.S.A., Rusul, G., Osman, A. and Quek, S.Y. (2007) Volatile flavour compounds and sensory properties of minimally processed durian (*Durio zibethinus* cv. D24) fruit during storage at 4°C. *Postharvest Biology and Technology* 46, 76–85.

Wang, Y.J., Yang, C.X., Li, S.H., Yang, L., Wang, Y.N., Zhao, J.B. and Jiang, Q. (2009) Volatile characteristics of 50 peaches and nectarines evaluated by HP-SPME with GC-MS. *Food Chemistry* 116, 356–364.

Wang, Y.J., Yang, C.X., Liu, C.Y., Xu, M., Li, S.H., Yang, L. and Wang, Y.N. (2010) Effects of bagging on volatiles and polyphenols in 'Wanmi' peaches during endocarp hardening and final fruit rapid growth stages. *Journal of Food Science* 75, S455–S460.

Wardencki, W., Michulec, M. and Curylo, J. (2004) A review of theoretical and practical aspects of solid-phase microextraction in food analysis. *International Journal of Food Science and Technology* 39, 703–717.

Xi, W.P., Zhang, B., Liang, L., Shen, J.Y., Wei, W.W., Xu, C.J., Allan, A.C., Ferguson, I.B. and Chen, K.S. (2012) Postharvest temperature influences volatile lactone production via regulation of acyl-CoA oxidases in peach fruit. *Plant, Cell & Environment* 35, 534–545.

Yahyaoui, F.E.L., Wong-Aree, C., Latché, A., Hackett, R., Grierson, D. and Pech, J.C. (2002) Molecular and biochemical characteristics of a gene encoding an alcohol acyl-transferase involved in the generation of aroma volatile esters during melon ripening. *European Journal of Biochemistry* 269, 2359–2366.

Yang, X.T., Song, J., Fillmore, S., Pang, X.Q. and Zhang, Z.Q. (2011) Effect of high temperature on color, chlorophyll fluorescence and volatile biosynthesis in green-ripe banana fruit. *Postharvest Biology and Technology* 62, 246–257.

Yang, Z. and Ohlrogge, J.B. (2009) Turnover of fatty acids during natural senescence of Arabidopsis, *Brachypodium*, and switchgrass and in Arabidopsis β-oxidation mutants. *Plant Physiology* 150, 1981–1989.

Zhang, B., Yin, X.R., Li, X., Yang, S.L., Ferguson, I. and Chen, K.S. (2009) Lipoxygenase gene expression in ripening kiwifruit in relation to ethylene and aroma production. *Journal of Agricultural and Food Chemistry* 57, 2875–2881.

Zhang, B., Shen, J.Y., Wei, W.W., Xi, W.P., Xu, C.J., Ferguson, I. and Chen, K.S. (2010) Expression of genes associated with aroma formation derived from the fatty acid pathway during peach fruit ripening. *Journal of Agricultural and Food Chemistry* 58, 6157–6165.

Zhang, B., Xi, W.P., Wei, W.W., Shen, J.Y., Ferguson, I. and Chen, K.S. (2011) Changes in aroma-related volatiles and gene expression during low temperature storage and subsequent shelf-life of peach fruit. *Postharvest Biology and Technology* 60, 7–16.

6 Making the Surface of Fleshy Fruit: Biosynthesis, Assembly and Role of the Cuticular Layer

Justin Lashbrooke,[1,2,3] Fabrizio Costa[2] and Asaph Aharoni[1]*

[1]Department of Plant Sciences, Weizmann Institute of Science, Rehovot, Israel; [2]Research and Innovation Centre, Fondazione Edmund Mach, TN, Italy; [3]Institute for Wine Biotechnology, Stellenbosch University, Stellenbosch, South Africa

6.1 Introduction to the Plant Cuticle

The primary barrier between the atmosphere and the aerial parts of higher plants is the cuticular membrane or the cuticle. The constituents of this hydrophobic extracellular membrane, typically comprised of soluble waxes and polymerized lipids, are produced and secreted by the plant's epidermal cells (Kunst and Samuels, 2003; Pollard *et al.*, 2008). Lipids consisting mostly of C16 to C20 fatty acids are polymerized to form a matrix known as cutin (Schreiber, 2010). The cutin matrix is both embedded with waxes (intracuticular) and covered with a thin layer of surface (epicuticular) waxes (Kunst and Samuels, 2003; Samuels *et al.*, 2008) consisting of very-long-chain, saturated, non-polar hydrocarbons and their derivatives including alcohols, aldehydes and alkanes (Kunst and Samuels, 2003; Bargel *et al.*, 2006). The innermost part of the cuticular layer forms an interface with the cell wall of the underlying epidermal cells and contains branched polysaccharides (López-Casado *et al.*, 2007; Pollard *et al.*, 2008). This general structure of the cutin matrix and embedded waxes is conserved throughout many plants and organs; yet, a massive variety exists in both cuticle structure and function throughout the plant kingdom. This can be attributed to differences in monomer composition of the cutin and waxes as well as the inclusion (or omission) of a number of secondary metabolites (e.g. triterpenoids and aromatic compounds).

The primary role of the cuticle is to act as an interface between the plant and its environment. Due to the varied nature of the aerial organs found in higher plants, including stems, leaves, flowers, fruits and seeds, the specific function of the cuticle (as well as the structure) is relatively diverse. Typically, the cuticle provides a waterproof barrier between the epidermal cells and the relatively dry environment and regulates gaseous exchange. It also provides resistance against biotic and abiotic stresses such as mechanical damage

* asaph.aharoni@weizmann.ac.il

caused by microorganisms and damaging UV light (Bargel *et al.*, 2006). Additionally, the cuticle provides mechanical support to the plant organ and acts as a division to prevent, or in some cases promote, the fusion of plant organs during development. Being the outermost part of the plant, the cuticle can be regarded as a detector for environmental changes and plays a role in transmitting signals to the interior of the plant (Curvers *et al.*, 2010).

The cuticles of the various organs of higher plants have been adapted to allow optimal functionality of the organ. This is reflected in the diversity of compounds found in plant cuticle membranes. Analysis of a variety of cutin polymers has found the following compounds to be most abundant: unsubstituted fatty acids (1–25% of the cutin matrix), ω-hydroxy fatty acids (1–32%), dicarboxylic acids (typically less than 5%), mid-chain polyhydroxy fatty acids (16–92%), fatty alcohols (0–8%), glycerol (1–14%) and phenolics (0–1%) (Pollard *et al.*, 2008). As in many fields of plant research, much work on cuticle biology has been performed in *Arabidopsis*. In retrospect, this may not have been the most representative model, as the cutin in green tissue of *Arabidopsis* comprises in excess of 50% dicarboxylic acids, whilst the majority of other plants contain less than 5% dicarboxylic acids in the cutin. However, it has been shown that *Arabidopsis* petals have a very similar cutin composition when compared with the cutin found in fleshy fruits. Specifically, both fleshy fruits and *Arabidopsis* flowers contain high levels of 10,16-dihydroxypalmitic acid (10,16-DHPA) (Li-Beisson *et al.*, 2009). Other models commonly used include maize, which has been the model of choice in the study of wax composition during phase transition (Sturaro *et al.*, 2005), and more recently tomato, which is used as a model for the cuticle of fleshy fruit (Mintz-Oron *et al.*, 2008; Isaacson *et al.*, 2009; Matas *et al.*, 2011).

The leaf, being the primary photosynthetic organ of the plant, has asymmetric cuticle deposition, with more deposition occurring on the adaxial than the abaxial side (Jetter and Schäffer, 2001). This extra protection for the sun-exposed side of the leaf probably protects the underlying cells from excessive radiation damage. The cuticle covers not only the leaf's external epidermal cells but also the epidermal cells lining the substomatal cavity (Roth-Nebelsick *et al.*, 2013). This suggests that the inner epidermal cells have significantly reduced transpiration. In seeds, the outer coat comprises a cuticular membrane that isolates the seed from the plant with the exception of a vascular bundle, which provides nutrients (Van Dongen *et al.*, 2003). Once the seed has reached maturity, this cutin membrane extends to surround the entire seed. This protective barrier limits the rate of water uptake and is therefore the primary control of germination. The cuticle of the flower is adapted to serve the role of this organ, which is the attraction of pollinators. This is illustrated when examining *Arabidopsis thaliana* mutants that are deficient in cutin biosynthesis (Li-Beisson *et al.*, 2009). These mutants display an absence of petal nanoridges (cuticular ridges), which are distinctive to flower organs and have been suggested to attract pollinators in a number of ways, including trapping reflective dew drops, reflecting light in a specific pattern, creating a surface that provides perceptible stimulation and providing a desirable surface on which to walk (Li-Beisson *et al.*, 2009).

Whilst in the past the majority of studies have focused on the cuticle of vegetative tissues, in recent years there has been a growing interest in understanding the biology of the specialized fruit cuticle (Leide *et al.*, 2007; Isaacson *et al.*, 2009; Shi *et al.*, 2011). During fruit development, the surface area of the fruit may increase rapidly in size, whilst the biosynthesis of the cuticle is seldom able to match the fruit expansion (Mintz-Oron *et al.*, 2008; Domínguez *et al.*, 2012). The cuticle must therefore be elastic enough to tolerate the high tensions created by the expanding fruit. As the primary means of protection of the fruit, the cuticle has significant economic importance. It plays a major role in postharvest shelf-life as well as conferring a number of quality parameters including colour and texture. Although research has been performed in other fleshy fruit species, including apple (Brendolise *et al.*, 2011), grape (Mahjoub *et al.*, 2009) and cherry (Peschel *et al.*, 2007; Alkio *et al.*, 2012), the model of choice for the characterization of genes involved in fleshy fruit cuticle is tomato. Tomato is a valuable model for studying the cuticle for a number of reasons: (i) it has a relatively short life cycle; (ii) genetic transformation of this plant species is relatively simple; (iii) the fruit possesses a relatively thick fruit cuticle that is comparatively easy to isolate and study; (iv) the ripe fruit peel lacks stomata; and (v) the genome has been sequenced and is well annotated (The Tomato Genome Consortium, 2012). An interesting phenomenon of the tomato fruit cuticle that is not observed in the cuticle of leaves is that epidermal cells are often entirely surrounded by the cuticle (Fig. 6.1) (Mintz-Oron *et al.*, 2008).

This chapter will describe the structure and assembly of the cuticle, focusing on the genes and enzymes implicated in constructing the cuticle of fleshy fruit and its role in this organ. It is worth noting that, in the majority of studies, the 'peel' or 'skin' of the fruit is discussed. This is not a scientific term and describes in general the fruit outer layer including the pericarp and epidermal cells as well as the cuticle.

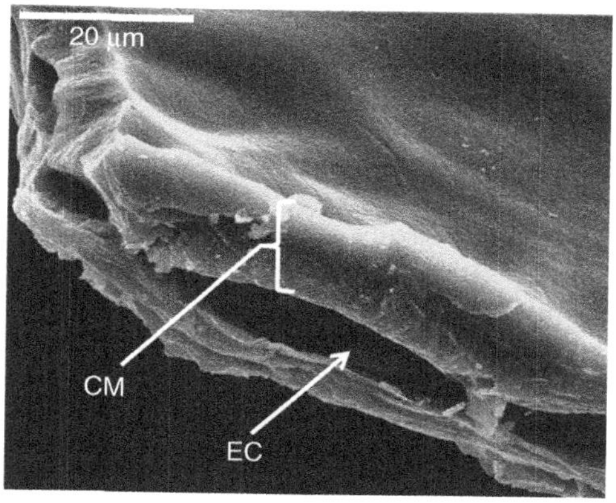

Fig. 6.1. Scanning electron micrograph of tomato fruit surface. A cross-section of a ripe tomato peel (cv. *Ailsa Craig*) showing the thick cuticular membrane (CM) surrounding the entire epidermal cell (EC).

6.2 Composition, Structure and Function of the Cuticle

6.2.1 General: composition and structure of the cuticle

The structure and composition of the plant cuticle can vary widely among plant species, and among organs and developmental stages within a species. This is illustrated in the typical range of thickness (1–10 µm) and quantity (100–1000 µg cm^{-2}) of the deposited cuticle (Riederer and Muller, 2006). The major components of the plant cuticle are the cutin matrix and epicuticular and intracuticular waxes. In some species, a non-hydrolysable polymer matrix named cutan may also be present (Pollard *et al.*, 2008). The cutin is attached to the underlying epidermal cells, whilst a thin layer of wax crystals covers the outermost surface of the cutinized layer. This hydrophobic epicuticular wax allows the plant to repel water and is thus important as a transpiration barrier. A cutinized cell wall is formed at the inner surface of the cutin where it is interconnected with the polysaccharides of the epidermal cell wall (López-Casado *et al.*, 2007). The typical fatty acid monomers found in cutin are C16 and C18 ω-hydroxy fatty acids and glycerol. The cuticular waxes are comprised of very-long-chain saturated non-polar hydrocarbons (typically C20–C60) and their derivatives (alcohols, aldehydes and alkanes), as well as secondary metabolites such as triterpenoids and phenylpropanoids (e.g. flavonoids) (Kunst and Samuels, 2003). The polysaccharides present in the cutin matrix are understood to be the major factor contributing to the elasticity of the matrix, whilst the cutin provides the strength (López-Casado *et al.*, 2007, 2010).

Transmission electron microscopy (TEM) has identified six different types of cuticle fine structures, but how these relate to the molecular structure of the cutin matrix is not yet known (Riederer and Muller, 2006). This lack of knowledge of the polymer structure makes the deciphering of cutin biosynthesis an interesting but complex task. The specific monomer composition of numerous species and organs is, however, well described. The fatty acid monomers of cutin are thought to be partially polymerized into di- or tri-acylglycerols, which are then transported to the exterior of the cell where further polymerization may occur (Panikashvili and Aharoni, 2008). Polymerization of the ω-hydroxy fatty acids then results in a linear polymer. However, depending on species and organ, a large percentage (up to 90% in tomato) of the ω-hydroxy fatty acid monomers may contain mid-chain hydroxyls (Mintz-Oron *et al.*, 2008), which allow branching of the polymer via esterification of the hydroxyls to ω-hydroxy fatty acids. Whilst the involvement of glycerol in the biosynthesis of the initial cutin oligomers has been reported (Li *et al.*, 2007; Chen *et al.*, 2011), its contribution to cutin polymer assembly is poorly understood. It has been suggested that it may lead to further branch points in the polymer via increased cross-linking and subsequently a larger cutin matrix. Additionally, phenolics (such as ferulates, which are found in low quantities in the cutin matrix) may be able to act as branching points for fatty acids, or to allow cross-linking within the cutin polymer or to the polysaccharide and lignin components of the cell wall (Li *et al.*, 2007; Pollard *et al.*, 2008).

6.2.2 The cuticle of developing fleshy fruit

The cuticle of developing fleshy fruit is adapted to maintain its integrity while the fruit rapidly expands, as well as to attract agents of seed dispersal. The structure of the fruit cuticle has therefore evolved to perform these roles and to be able to withstand both biotic and abiotic stresses (Domínguez *et al.*, 2012). The function played by the cuticle to reduce water loss and maintain structural support for the fruit is illustrated in the tomato *delayed fruit deterioration* (*dfd*) mutant (Saladie *et al.*, 2005). Cutin deposition in this mutant continues throughout fruit development, in contrast to the typical deposition profile, which shows significant reduction at the 'breaker' stage of fruit development. Fruit of *dfd* mutants therefore have significantly more cutin than normal fruit at the ripening stage. These fruit have a dramatically longer shelf-life than wild-type fruit and show a reduction in fruit softening. Saladie *et al.* (2005) showed that transpirational water loss was reduced and that there was elevated cellular turgor in the fruit of *dfd* mutants. A combination of increased physical support provided by the thicker cuticle and reduced water loss probably contributed to the reduction in fruit softening (Saladie *et al.*, 2005).

Not only is the cuticle able to maintain its integrity and cope with the increase in turgor pressure during fruit development and ripening, it also plays a central role in protection of the fruit from external factors. Both biotic stresses (e.g. insects and fungi) and abiotic stresses (e.g. UV radiation) must be coped with. There are a number of specialized (i.e. secondary) metabolites found in the cuticle that contribute to protection from environmental conditions and deter potential pathogens (Mintz-Oron *et al.*, 2008). High light levels and temperature cause an increase in the concentration of reactive oxygen species in exposed tissues and these may result in oxidative damage to the fruit. Antioxidants are therefore important metabolites in the outer layer of the fruit to moderate the potential damage. The yellow flavonoid naringenin chalcone accumulates in the epidermal cells as well as in the cuticular membrane of tomato fruit (Adato *et al.*, 2009; Domínguez *et al.*, 2009), where it probably protects against excessive UV radiation whilst also serving to attract agents of seed dispersal due to their intense pigmentation. Another group of specialized metabolites found in the tomato cuticle are pentacyclic triterpenoids. Although suggested to perform a structural role in the cuticle membrane, triterpenoids also possess antimicrobial properties and contribute to defence against fungal pathogens (Brendolise *et al.*, 2011; Wang *et al.*, 2011).

6.2.3 The inner fruit epidermis contains a cuticular layer

An inner epidermal layer (or endoderm) containing a cuticle-like structure was first suggested in the mid-20th century (Kraus, 1949) but was only recently described in tomato by Mintz-Oron *et al.* (2008) and further characterized by Matas *et al.* (2011). This internal cuticle covers the cells of the endocarp, separating the fleshy tissue pericarp from the locular region (i.e. the placental tissue, locular tissue and seeds). The cuticle is impermeable to water during the early stages of development, but as the fruit matures, the permeability increases (Matas *et al.*, 2011). The inner epidermal cuticle possesses similar properties with regard to the thicker external cuticle and, according to expression analysis, also shares common routes of biosynthesis. Whilst a number of known cuticle biosynthesis genes are found to be expressed in the endodermal cell layer, there are some noteworthy exceptions such as *CYP77A*, which encodes a protein catalysing the mid-chain hydroxylation of fatty acids (Li-Beisson *et al.*, 2009) and has 40-fold lower expression in the endodermal layer (Matas *et al.*, 2011). The differences observed in expression of biosynthesis genes are reflected in the structure of the inner epidermal cuticle, which, whilst being structurally similar to the outer cuticle, does show marked differences. Chemical analysis of the cutin monomer composition of the inner and outer layer of tomato found the same two dominant monomers (16-hydroxypalmitic acid and 10,16-DHPA). However, although in the outer cuticle 10,16-DHPA (a mid-chain fatty acid) was the most abundant monomer, the most abundant monomer in the inner cuticle was 16-hydroxypalmitic acid (lacking mid-chain hydroxylation).

6.3 Assembly of the Cutin Polymer

The precise mechanism of cutin polymer assembly and polymerization still remains largely undescribed, whilst the biosynthesis of the monomers/oligomers that form the building blocks of the polymer is far better described. Cutin polymer assembly can be divided into three parts: (i) monomer/oligomer biosynthesis; (ii) extracellular transport; and (iii) polymerization (Fig. 6.2). The order of these reactions is not completely clear, and it is likely that there is some degree of overlap between the processes.

Biosynthesis of the fatty acid monomers occurs in the plastids. Free fatty acids (FFAs) (mostly C16 and C18) undergo three major reactions on their way to incorporation into ω-hydroxy acylglycerols, namely, oxidation, acyl activation and acyl transfer (Tang *et al.*, 2007; Li-Beisson *et al.*, 2009; Chen *et al.*, 2011). The sequence of these reactions is yet to be determined; however, individual knockouts in *Arabidopsis* of the genes involved in these processes results in significantly reduced biosynthesis of acylglycerols. A possible candidate for the first reaction is acyl activation of the FFAs through a member of the long-chain acyl-CoA synthetase (LACS) family (Schnurr *et al.*, 2004; Bessire *et al.*, 2007). Metabolism of acyl-CoAs may then lead to a variety of pathways such as membrane lipid biosynthesis, storage lipid biosynthesis, and cuticular wax and cutin biosynthesis. Acyl-CoAs destined for cutin biosynthesis will undergo oxidation by members of the cytochrome P450 (CYP450) family (the CYP86A and CYP77A subfamilies) and transfer of the acyl chains to a glycerol-based acceptor mediated by glycerol 3-phosphate:acyl-CoA *sn*-1 acyltransferase (GPAT) (Li *et al.*, 2007; Li-Beisson *et al.*, 2009; Shi *et al.*, 2011). CYP86 family members and CYP77A6 have been shown

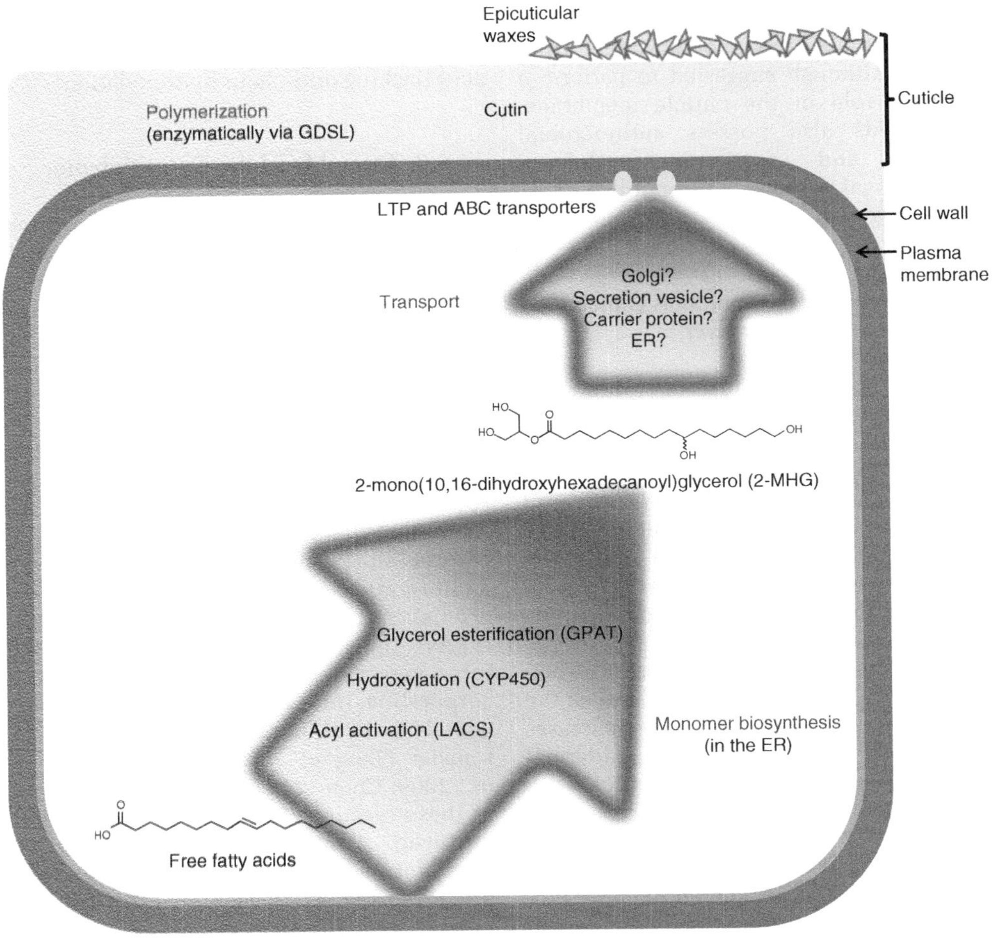

Fig. 6.2. Schematic of proposed routes of cutin monomer and polymer biosynthesis in an epidermal plant cell. Enzymatic modifications of free fatty acids occur in the endoplasmic reticulum (ER) resulting in a variety of acylglycerols (here represented by 2-mono(10,16-dihydroxyhexadecanoyl)glycerol (2-MHG)). Important genes in this process include long-chain acyl-CoA synthetases (LACS), cytochrome P450s (CYP450) and glycerol-3-phosphate-acyltransferases (GPAT). Acylglycerol oligomers are then transported to the exterior of the cell via ATP-binding cassette (ABC) and lipid-transfer protein (LTP) transporters. It is possible that secretion vesicles and carrier proteins play additional roles in the extracellular transport of cutin. Cutin oligomers are further polymerized outside the cell by enzymes including GDSL-motif lipases.

to be responsible for ω-hydroxylase and mid-chain hydroxylase activity, respectively. It is not known if the substrate for the P450 ω-hydroxylases are acyl-CoAs or acylglycerol moieties or even the non-acyl activated FFAs. Whilst P450s have been shown to act on FFAs *in vitro*, this does not exclude their action on acyl-CoAs and/or acylglycerols. P450s are not the only enzymes to be able to act on FFAs, as in the case of LACSs, which have demonstrated activity on both normal and ω-hydroxy FFAs. It is too early to assume that we fully understand the metabolic pathway(s) leading to ω-oxidized acylglycerols.

Regardless of the order of the reactions, the resultant ω-hydroxy acylglycerol molecules are considered the putative structural element of lipid polymers. However, the extracellular transportation of these molecules and their subsequent polymerization raises further questions. Analysis of *Arabidopsis* mutants has revealed the major role of a BAHD acyl transferase in cutin formation (Panikashvili *et al.*, 2009). DCR (DEFECTIVE IN CUTICULAR RIDGES) was suggested to be required for the incorporation of 10,16-DHPA into cutin. This mid-chain ω-hydroxy fatty acid is the major monomer of cutin in both *Arabidopsis* flowers and fruit of species such as tomato (Mintz-Oron *et al.*, 2008), cherry (Peschel *et al.*, 2007) and gooseberry (Kolattukudy, 2001). DCR mutants in *Arabidopsis* displayed almost undetectable levels of 10,16-DHPA in their flowers and leaves, and were susceptible to salinity, osmotic and water-deprivation stress (Panikashvili *et al.*, 2009). DCR seems likely to carry out partial polymerization of the cutin monomers. Some possible mechanisms of action for DCR have been suggested – it may catalyse: (i) the acylation of aromatics or hydroxy fatty acids with the CoA of 10,16-DHPA; or (ii) the acylation of cutin dimers or trimers using aromatic or aliphatic CoAs or the CoA of 10,16-DHPA. These reactions could result in linear and branched chains of 10,16-DHPA (Panikashvili *et al.*, 2009; Rani *et al.*, 2010). Panikashvili *et al.* (2009) localized the DCR protein to the cytosol, suggesting that polymerization of cutin commences inside the cell.

Whilst partial polymerization seems to occur in the cell, the extent of this polymerization is yet to be determined. The degree of intracellular polymerization will have a direct effect on extracellular transport and the mechanisms capable of this process. The waxes embedded in the cutin matrix have been shown to require transport to the apoplast by ATP-dependant ATP-binding cassette (ABC) transporters (Bird *et al.*, 2007; Panikashvili and Aharoni, 2008). However, the strict substrates of these transporters are yet to be discovered. Of particular interest is the level of cutin polymerization that occurs before transport and therefore the degree of oligomerized cutin that is accepted by ABC transporters. If monomers are exclusively transported out of the cell, the entire cutin matrix must be polymerized extracellularly. It is likely that there are multiple routes for extracellular transport depending on the specific monomer or the degree of intracellular polymerization. Highly polymerized molecules would probably require transport via vesicles. Studies in *Arabidopsis* have shown that ABC transporters are at least partially responsible for cutin and wax monomer transport (Pighin *et al.*, 2004; Bird *et al.*, 2007; Panikashvili *et al.*, 2007, 2011; Panikashvili and Aharoni, 2008; Bessire *et al.*, 2011).

One such ABC transporter that was characterized by Panikashvili *et al.* (2011) is ABCG13. *Arabidopsis* mutants for *abcg13* showed flower-specific phenotypes including fusion of flower organs and abnormal epidermal cell development (Panikashvili *et al.*, 2011). These phenotypes are likely to be the result of the significant reduction in cutin seen in *abcg13* flowers. The first ABC transporter characterized for wax transport was CER5 (Pighin *et al.*, 2004). *Arabidopsis cer5* mutants showed reduced deposition of stem cuticular wax. Electron microscopy analysis found that wax inclusions had formed in the cytoplasm of the epidermal cells indicating that wax biosynthesis was normal, but the extracellular transport was defective. The *CER5* gene was found to encode an ABC transporter (ABCG12) specific for wax monomers (Pighin *et al.*, 2004).

Arabidopsis ABCG11 and ABCG32 mutants also show reduced deposition of cutin (Panikashvili *et al.*, 2007; Bessire *et al.*, 2011). Analysis of the *permeable cuticle 1* (*pec1*) mutant of *Arabidopsis* revealed the role of ABCG32 in cuticular lipid transport (Bessire *et al.*, 2011). The *pec1* mutation was mapped to the *ABCG32* gene, a member of the *PLEIOTROPIC DRUG RESISTANCE* gene family. The *pec1* mutants displayed phenotypes associated with cuticle permeability, and chemical analysis of the flowers revealed a 40% reduction in both ω-hydroxylated fatty acids and 10,16DHPA. As expected for an extracellular transport protein, ABCG32 is localized to the plasma membrane of the epidermal cells, but, interestingly, localization occurs in a polar manner so that the proteins are found on the surface side of the epidermal cells (Bessire *et al.*, 2011).

Another protein implicated in the extracellular transport of cuticular lipids is the *Arabidopsis* glycosylphosphatidylinositol-anchored lipid-transfer protein (LTPG) (DeBono *et al.*, 2009). LTPG is able to bind lipids *in vitro* and was localized to the plasma membrane of epidermal cells in growth regions of *Arabidopsis*, whilst mutants for LTPG showed a reduction in wax load on the stem surface (DeBono *et al.*, 2009).

Once the monomers/oligomers are outside the cell, polymerization taken place through the formation of ester bonds between the monomers or oligomers. Lipases are likely candidates for driving these reactions and may be found outside the cell in the cuticular membrane. The most characterized enzyme is BODYGUARD (Kurdyukov *et al.*, 2006), an extracellular epidermal protein with lipase domains, whilst other lipase candidates are the monoacylglycerol or glycine–aspartic acid–serine-leucine (GDSL) motif lipases. *Arabidopsis* mutants for BODYGUARD in fact show an increase in cutin and wax monomers, but this is coupled with phenotypes characteristic of a plant with a deficient cuticle. It is suggested that the increase in cutin monomers is a response of the plant after sensing a disrupted cuticle, but the monomers remain unpolymerized (Kurdyukov *et al.*, 2006). Recently, a GDSL from tomato (*Solanum lycopersicum*) (SlGDSL1) has been shown to be localized to the exterior of the cell in the cuticular membrane, where it is able to polymerize monoacylglycerol cutin monomers (Girard *et al.*, 2012; Yeats *et al.*, 2012).

6.4 Genes Involved in Fleshy Fruit Cuticle Biosynthesis and Assembly

6.4.1 Building the cutin polymer

Whilst the majority of the fundamental work to describe cuticle biosynthesis has been performed in *Arabidopsis*, a dry fruit-bearing species, several genes have also been characterized in various fleshy fruit species, and these will be discussed in this section (see Table 6.1). Mapping of the *cutin deficient 1* (*cd1*) mutant gene led to the identification of *SlGDSL1*, a candidate for the extracellular polymerization of cutin monomers (Isaacson *et al.*, 2009). The *cd1* mutants in tomato possessed a significantly thinner cuticle than that of wild-type tomato as well as a reduction in cutin monomers and in the ester bonds cross-linking the cutin matrix (Isaacson *et al.*, 2009; Yeats *et al.*, 2012). This was coupled with an increase in cuticle permeability and postharvest water loss. SlGDSL1 was characterized as an acyltransferase that is able to perform the polymerization of cutin monomers, specifically 2-mono(10,16-dihydroxyhexadecanoyl)glycerol (Yeats *et al.*, 2012). Importantly, this protein has been localized to the exterior of the cell in tomato fruit, providing evidence for polymerization occurring after the extracellular transport of monomers and/or oligomers.

A gene involved in the biosynthesis of fruit cutin monomers was discovered through the analysis of the tomato *cuticle deficient 3* (*cd3*) mutant (Isaacson *et al.*, 2009). The mutation was mapped to SlCYP86A69, a cytochrome P450 oxidase (Shi *et al.*, 2013). SlCYP86A69 shares an amino acid sequence similarity of 96% with petunia CYP86A22, a fatty acyl-CoA ω-hydroxylase shown to be required for the production of ω-hydroxy fatty acids and the biosynthesis of cutin polymer in the petunia stigma (Han *et al.*, 2010). Whilst *cd3* mutants display a strong reduction in cutin, there is an insignificant

Table 6.1. Genes involved in cuticle biosynthesis that have been functionally characterized in fleshy fruit species.

Gene	Species	Function	Reference(s)
GDSL1 (CD1)	*Solanum lycopersicum*	Extracellular polymerization of cutin monomers	Isaacson *et al.* (2009); Girard *et al.* (2012); Yeats *et al.* (2012)
CYP86A89	*Solanum lycopersicum*	Cutin monomer biosynthesis	Shi *et al.* (2013)
CER6 (KCS6)	*Solanum lycopersicum*	Very-long-chain fatty acid elongation	Vogg *et al.* (2004); Mintz-Oron *et al.* (2008)
PS	*Solanum lycopersicum*	Decarbonylation of waxes	Leide *et al.* (2011)
TTS2	*Solanum lycopersicum*	Amyrin synthase	Wang *et al.* (2011)
OSC1 and *OSC3*	*Malus domestica*	Amyrin synthase	Brendolise *et al.* (2011)
F3H	*Solanum lycopersicum*	Flavonoid biosynthesis	Verhoeyen *et al.* (2002)
FLS	*Solanum lycopersicum*	Flavonoid biosynthesis	Verhoeyen *et al.* (2002)
CHS	*Solanum lycopersicum*	Flavonoid biosynthesis	Schijlen *et al.* (2007); Adato *et al.* (2009); Ballester *et al.* (2010)
ANL2 (CD2)	*Solanum lycopersicum*	Homeodomain transcription factor	Isaacson *et al.* (2009)
SHINE3	*Solanum lycopersicum*	Apatella transcription factor	Shi *et al.* (2013)
MYB12	*Solanum lycopersicum*	MYB domain transcription factor	Adato *et al.* (2009); Ballester *et al.* (2010)
MYB5b	*Vitis vinifera*	MYB domain transcription factor	Mahjoub *et al.* (2009)

change in cuticular wax composition. Chemical analyses of *slcyp86a69* mutants showed a significant reduction in all cutin monomers in tomato fruit cuticle (Isaacson *et al.*, 2009; Shi *et al.*, 2013). The plants had an increased susceptibility to microbial infection as well as an increase in susceptibility to dehydration stress (see Plate 4). Enzymatic assays performed with SlCYP86A89 found that the enzyme preferentially catalysed the hydroxylation of C18:1 fatty acid to C18-hydroxyoleic acid, but it was also able to hydroxylate C14 and C16 fatty acids (at a significantly lower activity) (Shi *et al.*, 2013). This acyl hydroxylation is a key step in the biosynthesis of the cutin monomers, as the hydroxyl groups allow increased branching during the subsequent polymerization reactions.

6.4.2 Wax-associated genes in the fleshy fruit

Genes involved in wax biosynthesis in fleshy fruit cuticles have been studied in a number of species, although tomato is the species most investigated to date. Mutations in, or RNA interference knockdowns of, genes involved in wax biosynthesis typically lead to a greater increase in water permeability than observed in cutin mutants. The composition of the cuticular waxes of tomato is made up primarily of *n*-alkanes followed by triterpenoids and sterol derivatives and alkanoic acids (Schreiber, 2010).

An orthologue of the *Arabidopsis CER6* has been studied in tomato fruit, where the gene was shown to encode a *β*-ketoacyl-CoA synthase involved in very-long-chain fatty acid elongation (Vogg *et al.*, 2004;

Mintz-Oron *et al.*, 2008). Expression of a reporter gene driven by the upstream promoter region of *SlCER6* was localized to both the exocarp and endocarp of tomato fruit (see Plate 3) (Mintz-Oron *et al.*, 2008). Analysis of *slcer6* tomato lines revealed that, starting from the mature green stage of fruit development, there is a reduction of *n*-alkanes with a chain length greater than C28 when compared with wild-type tomato (Vogg *et al.*, 2004). Substrates for SlCER6 are therefore likely to include *n*-alkanes with chain lengths longer than C28. This decrease in *n*-alkanes (*n*>C28) is coupled with an increase in cyclic triterpenoids and sterols, as well as an increase in the weight of the cuticle, considered to be a compensating mechanism. The triterpenoids, however, do not sufficiently compensate for the permeability of the cuticle as the mutants display a three- to eightfold increase in water loss when compared with wild-type tomato. The effect of the absence of β-ketoacyl-CoA synthase is not limited to *n*-alkanes, as *slcer6* lines also show a complete lack of cuticular alkene (C33, C35), aldehyde (C24, C26, C32), alkenol (C24, C26) and alkadienol (C22, C24, C26) formation, probably caused by the absence of a common precursor (Vogg *et al.*, 2004).

Another tomato mutant that has shed light on the biosynthesis of fruit cuticular wax is the recessive *positional sterile* (*ps*) mutant (Leide *et al.*, 2011). The *ps* phenotype is characterized by floral organ fusions (a commonly observed phenotype in plants with depleted cuticles), positional sterility and the formation of wrinkled ripe and over-ripe fruit. The surface area of the *ps* fruit at the fully ripe stage was approximately 33% of that seen in the wild type (Leide *et al.*, 2011). Chemical analysis of the cuticular wax composition of the *ps* mutants revealed a severe depletion of alkanes and aldehydes. As with the *slcer6* mutant, an increase in triterpenoids and sterol derivatives was observed as well as a five- to eightfold increase in water permeability. There was, however, almost no effect on the cutin composition of the *ps* mutant cuticle. The observed modification of the cuticular waxes coupled with the lack of change seen in the cutin composition indicate that the *ps* mutation causes disabling of the decarbonylation pathway of wax biosynthesis in the fruit epidermal cells (Leide *et al.*, 2011).

6.4.3 Triterpenoid-related genes in the fleshy fruit

In a number of fruits, including tomato, triterpenoids are found in the cuticle layer and in particular as constituents of the intracuticular waxes (Leide *et al.*, 2007; Mintz-Oron *et al.*, 2008). Typically, they are more abundant in fruit than leaf cuticles and are suggested to be a contributing factor to the increased permeability of fruit cuticles, as they have been shown to provide a less effective barrier to water transport than long-chain hydrocarbons (Leide *et al.*, 2007). A number of genes that code for triterpenoid synthases (TTSs) have been characterized in tomato (Wang *et al.*, 2011) and apple (Brendolise *et al.*, 2011) fruit. Characterization of these enzymes has focused on the oxidosqualene cyclases (OSCs), which catalyse the first committed step of triterpenoid biosynthesis, the cyclization of epoxysqualene into various triterpene alcohol isomers. Whilst the expression of *OSC* genes contributes significantly to the distribution of triterpenoids in plants, there are additional points of control. This can be observed by the only partial correlation between expression levels and cuticular triterpenoids across tomato cultivars. Within a cultivar, however, *OSC* expression between organs correlated well with triterpenoid accumulation (Wang *et al.*, 2011).

The product specificity (or lack thereof) categorizes OSCs and determines the triterpenoid profile found in fruit waxes. Tomato SlTTS1 is a monofunctional β-amyrin synthase, whilst SlTTS2 is a multifunctional amyrin synthase producing predominantly δ-amyrin and six other terpenoid products (Wang *et al.*, 2011). The major role of the genes as the producers of cuticular triterpenoids in the fruit is illustrated in their exclusively fruit epidermal expression pattern and the fact that the combined enzyme activities correlate with all the triterpenoids present in the cuticular wax of tomato fruit (Wang *et al.*, 2011). The characterization apple (*Malus domestica*) *OSC* genes led to the discovery of the first TTS (MdOSC1), which is primarily an α-amyrin synthase (Brendolise *et al.*, 2011). MdOSC1 produces both α-amyrin and β-amyrin in a 5:1 ratio. The fact that α-amyrin is the precursor to ursolic acid, the major terpenoic acid found in the cuticular wax of apple fruit, indicates the importance of MdOSC1 in wax terpenoid biosynthesis in apple. Additional apple TTS-encoding genes identified include *MdOSC2* and *MdOSC3*. MdOSC3 shares a greater than 99% amino acid identity with MdOSC1, exhibits a similar fruit peel-specific expression pattern and is thought to have the same activity (Brendolise *et al.*, 2011). *MdOSC2*, however, exhibits a significantly reduced expression level and has no observed activity when assayed with epoxysqualene (Brendolise *et al.*, 2011). Although the activity of MdOSC3 has not been tested, it is likely to have the same activity as MdOSC1 due to its high homology, whilst *MdOSC3* is hypothesizsed to be a pseudogene.

6.4.4 Genes involved in flavonoid accumulation in the fleshy fruit

Flavonoids are typically found embedded in the cuticle of tomato fruit but have not been found in many other fleshy fruit cuticles (Mintz-Oron *et al.*, 2008). They comprise a diverse group of polyphenolic compounds that can be divided based on their core structure – the aglycone – into chalcones, flavanones, aurones, flavonols and anthocyanins. The diversity of flavonoids is largely the result of the activity of a number of modifying enzymes such as glycosyl-, malonyl-, acyl- and methyltransferases. They fulfil a number of diverse roles in plant growth and development that include pigments for the attraction of pollinators and agents of seed dispersal, pathogen resistance and protection against damage from UV light (Harborne and Williams, 2000; Adato *et al.*, 2009; Domínguez *et al.*, 2009). Transcript and metabolite analysis during tomato fruit development indicates that the synthesis of cuticular lipids precedes phenylpropanoid and flavonoid biosynthesis. Expression levels of *CHALCONE SYNTHASE* (*CHS*), *FLAVANONE- 3-HYDROLASE* (*F3H*) and *FLAVONOL SYNTHASE* (*FLS*) have all been shown to increase during ripening of tomato fruit, peaking at the breaker stage (Verhoeyen *et al.*, 2002; Schijlen *et al.*, 2007; Mintz-Oron *et al.*, 2008).

Whilst anthocyanins contribute to red, purple and blue pigments in many flowers and fruit, other flavonoids are responsible for imparting yellow pigmentation, such as chalcones and aurones. Analysis of the tomato *y* mutant and the wild tomato species *Solanum chmielewskii*, which both remain pink when fully ripe, led to the characterization of SlCHS and a MYB transcription factor (SlMYB12) that affect chalcone accumulation in the cuticle and consequently the fruit colour (Adato *et al.*, 2009; Ballester *et al.*, 2010). The major contributor to tomato fruit colour is the red carotenoid lycopene, found throughout the flesh of ripe tomatoes. Pink tomatoes are caused by a colourless epidermis, which is devoid of the yellow-coloured flavonoid naringenin chalcone (Adato *et al.*, 2009). The combination of a yellow epidermal layer and the pink flesh results in the typical red–orange colour of wild-type tomatoes. RNA interference studies performed on *SlCHS* demonstrated that SlCHS was required for the accumulation of naringenin chalcone (Schijlen *et al.*, 2007). A mutation in *SlMYB12* was identified as the genetic basis for the lack of accumulation of naringenin chalcone in the pink mutants (Adato *et al.*, 2009). Virus-induced gene silencing of *SlMYB12* resulted in a similar phenotype and a decrease in naringenin chalcone accumulation in the peel (Ballester *et al.*, 2010). Therefore, SlMYB12 is responsible for regulation of the flavonoid biosynthetic pathway in tomato fruit epidermal cells.

6.4.5 Transcriptome and proteome analyses for the identification of genes associated with fleshy fruit cuticle assembly

Large-scale transcriptomic studies in tomato have identified a number of genes that display peel-enriched expression profiles. Mintz-Oron *et al.* (2008) performed genome-wide transcriptional profiling and metabolomic analysis of tomato peel and flesh at five stages of development. Functional annotation of the profiled genes found that 17% of the 574 genes that were upregulated in peel (by at least twofold) were involved in cutin, wax, flavonoid or fatty acid metabolism. Many of these genes probably encode proteins whose orthologues have been demonstrated to be involved in *Arabidopsis* cuticle biosynthesis. Peel-enriched genes putatively encoding cutin biosynthesis enzymes included *HOTHEAD* (*HTH*), *LACS* and *GDSL*, whilst genes encoding wax biosynthesis enzymes included *CER6* and *FIDDLEHEAD* (*FDH*). Matas *et al.* (2011) profiled gene expression in various tomato fruit tissues after laser-capture microdissection. They were able to isolate five distinct cell types: (i) outer and (ii) inner epidermal layers; (iii) collenchyma; (iv) parenchyma; and (v) vascular tissue. Gene expression analysis found a number of genes that have been proposed as related to cuticle metabolism to be expressed specifically in either the outer epidermis or in both the outer and inner epidermis. These genes included putative lipid-transfer protein, flavonoid biosynthetic genes, GDSL family genes, cytochrome P450 and ABC transporter genes (Matas *et al.*, 2011).

These large-scale transcriptome studies provide an important overview of the genes involved in cuticle biosynthesis. Even more importantly, they provide novel genes for more detailed functional characterization. Proteome analysis has also led to a number of novel genes that are putatively involved in cuticle biosynthesis being identified. Yeats *et al.* (2010) extracted and identified 200 surface proteins from tomato fruit. A number of the proteins identified are likely to be involved in the transport, deposition or modification of the cuticle and include lipid-transfer proteins, GDSL and an MD-2-related lipid recognition domain-containing protein (Yeats *et al.*, 2010).

6.5 Regulation and Hormonal Control of Fruit Cuticle Biosynthesis

6.5.1 Phytohormone regulation

Plant cuticle metabolism is strongly regulated in response to developmental and environmental cues such as osmotic stress. Biosynthesis of waxes and cutin appears to be under independent control, as their synthesis and deposition is not necessarily correlated. As discussed above, intracuticular waxes play a vital role in the waterproofing of the plant. Wax deposition is typically increased when the plant is under osmotic stress or in organs that require greater protection from osmotic stress, such as fleshy fruit, whilst rapidly expanding tissues often show an increase in cutin deposition that is not always accompanied by a concurrent deposition of cuticular waxes (Bargel *et al.*, 2006).

The phytohormone abscisic acid (ABA) and gibberellin have both been linked to the regulation of cuticle biosynthesis (Knoche and Peschel, 2007; Curvers *et al.*, 2010). Application of gibberellins to developing tomato fruit causes an increase in the mass of the cuticle layer per unit surface area. Knoche and Peschel (2007) demonstrated that this effect is more dramatic when gibberellins are applied to young fruit that are actively synthesizing cuticle components. They did not, however, describe whether wax, cutin or both monomers were increasing in the tomato cuticle, but studies on pea stem sections have shown that gibberellins increase the incorporation of palmitic acid into the cutin matrix (Bowena and Walton, 1988). Gibberellins are routinely applied to apples as a measure to prevent russeting, which arises due to cuticle failure resulting in the deposition of suberin (Knoche *et al.*, 2011).

The gibberellins probably increase cuticle deposition in the apple peel, thus preventing mechanical failure of the cuticle. Further analysis of epidermal gene expression after gibberellin treatment will provide a better understanding of the role played by gibberellins in cuticle biosynthesis.

Whilst a relationship between ABA biosynthesis and cuticle biosynthesis is expected due to the involvement of both elements in the response to osmotic stress, there is relatively little understanding of the extent of this relationship. Analysis of *Arabidopsis* mutants that fail to express ABA biosynthetic genes in response to osmotic stress have highlighted this connection, as mutant lines were shown to be deficient for an allele of the cutin polymerase gene, *BODYGUARD* (discussed earlier), rather than for ABA biosynthesis genes (Kurdyukov *et al.*, 2006). How the cuticle, and in particular cuticle biosynthesis, regulates ABA biosynthesis is not known, but analysis of additional cuticle biosynthesis mutants has shown that a functioning cuticle is required for proper induction of ABA biosynthetic genes in response to osmotic stress (Cominelli *et al.*, 2008; Curvers *et al.*, 2010). It would appear that a properly functioning cuticle mediates a stress signal that seems to influence ABA regulation in response to osmotic stress.

6.5.2 Transcriptional regulation of biosynthesis pathways

Only a handful of genes have been implicated in the regulation of fruit cuticle biosynthesis. These genes typically encode transcription factors. Regulation of the biosynthesis of cuticular lipids by these transcription factors typically leads to wide-ranging effects on the permeability of the cuticle and epidermal morphology.

A cuticle-related homeodomain transcription factor was identified in tomato by characterization and mapping of the *cuticle deficient 2* (*cd2*) mutant (Isaacson *et al.*, 2009). The protein is a homologue of the *Arabidopsis* ANTHOCYANINLESS2 (ANL2), which has been associated with anthocyanin distribution in epidermal cells (Kubo *et al.*, 1999). The CD2 protein contains a DNA-binding homeodomain and a steroidogenic acute regulatory lipid-transfer (START) domain. The function of the START domain in plants has not been shown but is thought to facilitate the binding of regulatory lipids (Schrick *et al.*, 2004). The *cd2* mutant shows a dramatic reduction in cutin content in the tomato fruit peel but an insignificant change in cuticular waxes (Isaacson *et al.*, 2009). The fact that *cd2* mutants are not severely affected in terms of water retention illustrates that the cuticular waxes and the remaining cutin are sufficient for waterproofing. Fruit peels of the *cd2* mutants did, however, display an increase in stiffness and susceptibility to microbial infection (Isaacson *et al.*, 2009). This suggests that a correctly formed cutin matrix is necessary for the cuticle to maintain its elasticity and to protect the tissues from microbial infection.

A grape (*Vitis vinifera*) R2R3-MYB transcription factor (VvMYB5b) has been implicated in the regulation of cuticular wax accumulation (Mahjoub *et al.*, 2009). When *VvMYB5b* was overexpressed in tomato plants, a variety of pleiotropic phenotypes were observed. These included modified leaf structure, alterations of floral morphology and glossy fruit appearance. Chemical analysis of the fruit cuticular waxes revealed a decrease in total amyrin content. Oleanolic acid is the major component of grape waxes and its precursor, *β*-amyrin, showed the strongest decrease in the transgenic tomato lines (Mahjoub *et al.*, 2009). This modification of cuticular wax was confirmed by scanning electron microscopy. One must be cautious when inferring the action of transcription factors from overexpression in a non-native species, but, whilst the native targets of VvMYB5b are not yet determined, it is possible that VvMYB5b controls similar pathways in grapevine. More research is, however, required to determine the grapevine VvMYB5b downstream target genes.

Expression correlation analysis in large-scale transcriptome studies has resulted in the identification of a number of candidate transcription factors that may control cuticle assembly. Experiments in tomato identified orthologues of the *Arabidopsis* SHINE and MIXTA-LIKE transcription factors (Mintz-Oron *et al.*, 2008). The *Arabidopsis AtSHINE* genes encode APETALA2-domain proteins and primarily regulate cutin biosynthesis genes (Shi *et al.*, 2011). Functional characterization of the tomato *SHINE3* gene implicates this transcription factor as a major regulator of cuticle biosynthesis in tomato (Shi *et al.*, 2013). Transgenic tomato lines reduced in *SHINE3* expression exhibit a severe phenotype in the tomato peel, including increased susceptibility to fungal parthenogenesis and an increased rate of dehydration. The cuticles of transgenic lines were significantly thinner than those of the wild type and showed a reduction of up to 40% in cutin monomers (Shi *et al.*, 2013). A number of putative cutin biosynthesis gene targets of SHINE3 are downregulated in the transgenic lines and include: *LACS2*, *GPAT4*, *CYP86A* and *GDSL1*. Shi *et al.* (2013) suggested that SHINE3 is required for cutin biosynthesis and possibly cutin polymerization.

6.6 The Applied Side of Fleshy Fruit Cuticle Research and Future Prospects

The cuticle membrane of fleshy fruit is a highly specialized and complex layer. Whilst we have a relatively good understanding of the biosynthesis and assembly of the cuticle, we still have much to learn about the regulation of cuticle biosynthesis. It is likely that regulation of cuticle biosynthesis together with solving the final problem of cutin oligomer polymerization and transport will be a focus of future work. The economic importance of understanding these processes is clear when one considers the scale of the fresh fruit market and the impact that the cuticle has on fruit quality. Mechanical failure of the fruit cuticle can lead to considerable economic loss for farmers through splitting or cracking of the cuticle. Additionally, the fruit surface is the major feature by which the consumer is able to judge fruit quality. Traits such as colour, glossiness, texture and uniformity are all determined by the cuticle. As our understanding of the relationship between cuticle structure and fruit surface characteristics advances, we will increasingly be able to manipulate these important traits. The range of biological roles to which the fruit cuticle contributes is extensive, and includes sensitivity and response to biotic and abiotic stresses as well as the development of the mature fruit. Although characterization of a number of genes and their involvement in fruit cuticle biosynthesis has begun, the genetics of fruit cuticle biosynthesis still remains largely uncharacterized. Despite the limitations in our understanding, current knowledge does allow the molecular breeding and engineering of many important traits into fruit. Traits ranging from fruit quality and aesthetics, such as shininess and shelf-life, to agriculturally important traits, including disease resistance and sensitivity to water stress, are all associated with the cuticle. Identification of the key genes involved in conferring these traits to fruit will provide breeders with the ability to screen large breeding populations for these important genes. The traits can then be imported into fruit lines through classical breeding or via biotechnological methods. This has the possibility of leading to improvements in fruit production and quality.

References

Adato, A., Mandel, T., Mintz-Oron, S., Venger, I., Levy, D., Yativ, M., Domínguez, E., Wang, Z., De Vos, R.C., Jetter, R., Schreiber, L., Heredia, A., Rogachev, I. and Aharoni, A. (2009) Fruit-surface flavonoid accumulation in tomato is controlled by a SlMYB12-regulated transcriptional network. *PLoS Genetics* 5, 823–851.

Alkio, M., Jonas, U., Sprink, T., van Nocker, S. and Knoche, M. (2012) Identification of putative candidate genes involved in cuticle formation in *Prunus avium* (sweet cherry) fruit. *Annals of Botany* 110, 101–112.

Ballester, A.R., Molthoff, J., de Vos, R., Hekkert, B., Orzaez, D., Fernández-Moreno, J.P., Tripodi, P., Grandillo, S., Martin, C., Heldens, J., Ykema, M., Granell, A. and Bovy, A. (2010) Biochemical and molecular analysis of pink tomatoes: deregulated expression of the gene encoding transcription factor SlMYB12 leads to pink tomato fruit color. *Plant Physiology* 152, 71–84.

Bargel, H., Koch, K., Cerman, Z. and Neinhuis, C. (2006) Structure–function relationships of the plant cuticle and cuticular waxes – a smart material? *Functional Plant Biology* 33, 893–910.

Bessire, M., Chassot, C., Jacquat, A.C., Humphry, M., Borel, S., Petétot, J.M., Métraux, J.P. and Nawrath, C. (2007) A permeable cuticle in *Arabidopsis* leads to a strong resistance to *Botrytis cinerea*. *EMBO Journal* 26, 2158–2168.

Bessire, M., Borel, S., Fabre, G., Carraça, L., Efremova, N., Yephremov, A., Cao, Y., Jetter, R., Jacquat, A.C., Métraux, J.P. and Nawrath, C. (2011) A member of the PLEIOTROPIC DRUG RESISTANCE family of ATP binding cassette transporters is required for the formation of a functional cuticle in Arabidopsis. *Plant Cell* 23, 1958–1970.

Bird, D., Beisson, F., Brigham, A., Shin, J., Greer, S., Jetter, R., Kunst, L., Wu, X., Yephremov, A. and Samuels, L. (2007) Characterization of Arabidopsis ABCG11/WBC11, an ATP binding cassette (ABC) transporter that is required for cuticular lipid secretion. *Plant Journal* 52, 485–498.

Bowena, D.J. and Walton, T.J. (1988) Cutin composition and biosynthesis during gibberellic acid-induced stem extension of *Pisum sativum* var. Meteor. *Plant Science* 55, 115–127.

Brendolise, C., Yauk, Y.K., Eberhard, E.D., Wang, M., Chagne, D., Andre, C., Greenwood, D.R. and Beuning, L.L. (2011) An unusual plant triterpene synthase with predominant α-amyrin-producing activity identified by characterizing oxidosqualene cyclases from Malus × domestica. *FEBS Journal* 278, 2485–2499.

Chen, X., Truksa, M., Snyder, C.L., El-Mezawy A., Shah, S. and Weselake, R.J. (2011) Three homologous genes encoding *sn*-glycerol-3-phosphate acyltransferase 4 exhibit different expression patterns and functional divergence in *Brassica napus*. *Plant Physiology* 155, 851–865.

Cominelli, E., Sala, T., Calvi, D., Gusmaroli, G. and Tonelli, C. (2008) Over-expression of the Arabidopsis *AtMYB41* gene alters cell expansion and leaf surface permeability. *Plant Journal* 53, 53–64.

Curvers, K., Seifi, H., Mouille, G., de Rycke, R., Asselbergh, B., Van Hecke, A., Vanderschaeghe, D., Höfte, H., Callewaert, N., Van Breusegem, F. and Höfte, M. (2010) Abscisic acid deficiency causes changes in cuticle permeability and pectin composition that influence tomato resistance to *Botrytis cinerea*. *Plant Physiology* 154, 847–860.

Debono, A., Yeats, T.H., Rose, J.K., Bird, D., Jetter, R., Kunst, L. and Samuels, L. (2009) Arabidopsis LTPG is a glycosylphosphatidylinositol-anchored lipid transfer protein required for export of lipids to the plant surface. *Plant Cell* 21, 1230–1238.

Domínguez, E., Luque, P. and Heredia, A. (2009) Sorption and interaction of the flavonoid naringenin on tomato fruit cuticles. *Journal of Agriculture and Food Chemistry* 57, 7560–7564.

Domínguez, E., Fernández, M.D., Hernández, J.C., Parra, J.P., España, L., Heredia, A. and Cuartero, J. (2012) Tomato fruit continues growing while ripening, affecting cuticle properties and cracking. *Physiologia Plantarum* 146, 473–486.

Girard, A.L., Mounet, F., Lemaire-Chamley, M., Gaillard, C., Elmorjani, K., Vivancos, J., Runavot, J.L., Quemener, B., Petit, J., Germain, V., Rothan, C., Marion, D. and Bakan, B. (2012) Tomato GDSL1 is required for cutin deposition in the fruit cuticle. *Plant Cell* 24, 3119–3134.

Han, J., Clement, J.M., Li, J., King, A., Ng, S. and Jaworski, J.G. (2010) The cytochrome P450 CYP86A22 is a fatty acyl-CoA ω-hydroxylase essential for Estolide synthesis in the stigma of *Petunia hybrida*. *Journal of Biological Chemistry* 285, 3986–3996.

Harborne, J.B. and Williams, C.A. (2000) Advances in flavonoid research since 1992. *Phytochemistry* 55, 481–504.

Isaacson, T., Kosma, D.K., Matas, A.J., Buda, G.J., He, Y., Yu B., Pravitasari, A., Batteas, J.D., Stark, R.E., Jenks, M.A. and Rose, J.K. (2009) Cutin deficiency in the tomato fruit cuticle consistently affects resistance to microbial infection and biomechanical properties, but not transpirational water loss. *Plant Journal* 60, 363–377.

Jetter, R. and Schäffer, S. (2001) Chemical composition of the *Prunus laurocerasus* leaf surface. Dynamic changes of the epicuticular wax film during leaf development. *Plant Physiology* 126, 1725–1737.

Knoche, M. and Peschel, S. (2007) Gibberellins increase cuticle deposition in developing tomato fruit. *Plant Growth Regulation* 51, 1–10.

Knoche, M., Khanal, B.P. and Stopar, M. (2011) Russeting and microcracking of 'Golden delicious' apple fruit concomitantly decline due to gibberellin A4+7 application. *Journal of American Society of Horticultural Science* 136, 159–164.

Kolattukudy, P.E. (2001) Polyesters in higher plants. *Advances in Biochemical Engineering and Biotechnology* 71, 1–49.

Kraus, G. (1949) Morphologic–anatomical studies on development-conditioned changes in axis, leaf and ovary in some berry fruits. *Oesterreichische botanische Zeitschrift* 96, 325–360.

Kubo, H., Peeters, A.J., Aarts, M.G., Pereira, A. and Koornneef, M. (1999) *ANTHOCYANINLESS2*, a homeobox gene affecting anthocyanin distribution and root development in Arabidopsis. *Plant Cell* 11, 1217–1226.

Kunst, L. and Samuels, A.L. (2003) Biosynthesis and secretion of plant cuticular wax. *Progress in Lipid Research* 42, 51–80.

Kurdyukov, S., Faust, A., Nawrath, C., Bär, S., Voisin, D., Efremova, N., Franke, R., Schreiber, L., Saedler, H., Métraux, J.P. and Yephremov, A. (2006) The epidermis-specific extracellular BODYGUARD controls cuticle development and morphogenesis in Arabidopsis. *Plant Cell* 18, 321–339.

Leide, J., Hildebrandt, U., Reussing, K., Riederer, M. and Vogg, G. (2007) The developmental pattern of tomato fruit wax accumulation and its impact on cuticular transpiration barrier properties: effects of a deficiency in a β-ketoacyl-coenzyme A synthase (LeCER6). *Plant Physiology* 144, 1667–1679.

Leide, J., Hildebrandt, U., Vogg, G. and Riederer, M. (2011) The *positional sterile* (*ps*) mutation affects cuticular transpiration and wax biosynthesis of tomato fruits. *Journal of Plant Physiology* 168, 871–877.

Li, Y., Beisson, F., Koo, A.J., Molina, I., Pollard, M. and Ohlrogge, J. (2007) Identification of acyltransferases required for cutin biosynthesis and production of cutin with suberin-like monomers. *Proceedings of National Academy of Sciences USA* 104, 18339–18344.

Li-Beisson, Y., Pollard, M., Sauveplane, V., Pinot, F., Ohlrogge, J. and Beisson, F. (2009) Nanoridges that characterize the surface morphology of flowers require the synthesis of cutin polyester. *Proceedings of National Academy of Sciences USA* 106, 22008–22013.

López-Casado, G., Matas, A.J., Domínguez, E., Cuartero, J. and Heredia, A. (2007) Biomechanics of isolated tomato (*Solanum lycopersicum* L.) fruit cuticles: the role of the cutin matrix and polysaccharides. *Journal of Experimental Botany* 58, 3875–3883.

López-Casado, G., Salamanca, A. and Heredia, A. (2010) Viscoelastic nature of isolated tomato (*Solanum lycopersicum*) fruit cuticles: a mathematical model. *Physiologia Plantarum* 140, 79–88.

Mahjoub, A., Hernould, M., Joubès, J., Decendit, A., Mars, M., Barrieu, F., Hamdi, S. and Delrot, S. (2009) Overexpression of a grapevine R2R3-MYB factor in tomato affects vegetative development, flower morphology and flavonoid and terpenoid metabolism. *Plant Physiology and Biochemistry* 47, 551–561.

Matas, A.J., Yeats, T.H., Buda, G.J., Zheng, Y., Chatterjee, S., Tohge, T., Ponnala, L., Adato, A., Aharoni, A., Stark, R., Fernie, A.R., Fei, Z., Giovannoni, J.J. and Rose, J.K. (2011) Tissue- and cell-type specific transcriptome profiling of expanding tomato fruit provides insights into metabolic and regulatory specialization and cuticle formation. *Plant Cell* 23, 3893–3910.

Mintz-Oron, S., Mandel, T., Rogachev, I., Feldberg, L., Lotan, O., Yativ, M., Wang, Z., Jetter, R., Venger, I., Adato, A. and Aharoni, A. (2008) Gene expression and metabolism in tomato fruit surface tissues. *Plant Physiology* 147, 823–851.

Panikashvili, D and Aharoni, A. (2008) ABC-type transporters and cuticle assembly: linking function to polarity in epidermis cells. *Plant Signaling and Behavior* 3, 806–809.

Panikashvili, D., Savaldi-Goldstein, S., Mandel, T., Yifhar, T., Franke, R.B., Höfer, R., Schreiber, L., Chory, J. and Aharoni, A. (2007) The Arabidopsis DESPERADO/AtWBC11 transporter is required for cutin and wax secretion. *Plant Physiology* 145, 1345–1360.

Panikashvili, D., Shi, J.X., Schreiber, L. and Aharoni, A. (2009) The Arabidopsis *DCR* encoding a soluble BAHD acyltransferase is required for cutin polyester formation and seed hydration properties. *Plant Physiology* 151, 1773–1789.

Panikashvili, D., Shi, J.X., Schreiber, L. and Aharoni, A. (2011) The Arabidopsis ABCG13 transporter is required for flower cuticle secretion and patterning of the petal epidermis. *New Phytology*, 190, 113–124.

Peschel, S., Franke, R., Schreiber, L. and Knoche, M. (2007) Composition of the cuticle of developing sweet cherry fruit. *Phytochemistry* 68, 1017–1025.

Pighin, J.A., Zheng, H., Balakshin, L.J., Goodman, I.P., Western, T.L., Jetter, R., Kunst, L. and Samuels, A.L. (2004) Plant cuticular lipid export requires an ABC transporter. *Science* 306, 702–704.

Pollard, M., Beisson, F., Li, Y. and Ohlrogge, J.B. (2008) Building lipid barriers: biosynthesis of cutin and suberin. *Trends in Plant Sciences* 13, 236–246.

Rani, S.H., Krishna, T.H., Saha, S., Negi, A.S. and Rajasekharan, R. (2010) Defective in cuticular ridges (DCR) of *Arabidopsis thaliana*, a gene associated with surface cutin formation, encodes a soluble diacylglycerol acyltransferase. *Journal of Biological Chemistry* 285, 38337–38347.

Riederer, M. and Muller, C. (eds) (2006) *Biology of the Plant Cuticle. Annual Plant Reviews*, Vol. 23. Blackwell, Oxford, UK.

Roth-Nebelsick, A., Fernandez, V., Peguero-Pina, J.J., Sancho-Knapik, D. and Gil-Pelegrin, E. (2013) Stomatal encryption by epicuticular waxes as a plastic trait modifying gas exchange in a Mediterranean evergreen species (*Quercus coccifera* L.). *Plant, Cell and Environment* 36, 579–589.

Saladie, M., Rose, J.K.C. and Watkins, C.B. (2005) Characterization of *DFD* (*delayed fruit deterioration*): a new tomato mutant. *Acta Horticulturae* 682, 79–84.

Samuels, L., Kunst, L. and Jetter, R. (2008) Sealing plant surfaces: cuticular wax formation by epidermal cells. *Annual Review of Plant Biology* 59, 683–707.

Schijlen, E.G., de Vos, C.H., Martens, S., Jonker, H.H., Rosin, F.M., Molthoff, J.W., Tikunov, Y.M., Angenent, G.C., van Tunen, A.J. and Bovy, A.G. (2007) RNA interference silencing of chalcone synthase, the first step in the flavonoid biosynthesis pathway, leads to parthenocarpic tomato fruits. *Plant Physiology* 144, 1520–1530.

Schnurr, J., Shockey, J. and Browse, J. (2004) The acyl-CoA synthetase encoded by *LACS2* is essential for normal cuticle development in Arabidopsis. *Plant Cell* 16, 629–642.

Schreiber, L. (2010) Transport barriers made of cutin, suberin and associated waxes. *Trends in Plant Science* 15, 546–553.

Schrick, K., Nguyen, D., Karlowski, W.M. and Mayer, K.F. (2004) START lipid/sterol-binding domains are amplified in plants and are predominantly associated with homeodomain transcription factors. *Genome Biology* 5, R41.

Shi, J.X., Malitsky, S., De Oliveira, S., Branigan, C., Franke, R.B., Schreiber, L. and Aharoni, A. (2011) SHINE transcription factors act redundantly to pattern the archetypal surface of Arabidopsis flower organs. *PLoS Genetics* 7, e1001388.

Shi, J.X., Adato, A., Alkan, N., He, Y., Lashbrooke, J., Matas, A.J., Meir, S., Malitsky, S., Isaacson, T., Prusky, D., Leshkowitz, D., Schreiber, L., Granell, A.R., Widemann, E., Grausem, B., Pinot, F., Rose, J.K., Rogachev, I., Rothan, C. and Aharoni, A. (2013) The tomato SlSHINE3 transcription factor regulates fruit cuticle formation and epidermal patterning. *New Phytology* 197, 468–480.

Sturaro, M., Hartings, H., Schmelzer, E., Velasco, R., Salamini, F. and Motto, M. (2005) Cloning and characterization of *GLOSSY1*, a maize gene involved in cuticle membrane and wax production. *Plant Physiology* 138, 478–489.

Tang, D., Simonich, M.T. and Innes, R.W. (2007) Mutations in *LACS2*, a long-chain acyl-coenzyme A synthetase, enhance susceptibility to avirulent *Pseudomonas syringae* but confer resistance to *Botrytis cinerea* in Arabidopsis. *Plant Physiology* 144, 1093–1103.

The Tomato Genome Consortium (2012) The tomato genome sequence provides insights into fleshy fruit evolution. *Nature* 485, 635–641.

Van Dongen, J.T., Ammerlaan, A.M., Wouterlood, M., Van Aelst, A.C. and Borstlap, A.C. (2003) Structure of the developing pea seed coat and the post-phloem transport pathway of nutrients. *Annals of Botany* 91, 729–737.

Verhoeyen, M.E., Bovy, A., Collins, G., Muir, S., Robinson, S., de Vos, C.H. and Colliver, S. (2002) Increasing antioxidant levels in tomatoes through modification of the flavonoid biosynthetic pathway. *Journal of Experimental Botany* 53, 2099–2106.

Vogg, G., Fischer, S., Leide, J., Emmanuel, E., Jetter, R., Levy, A.A. and Riederer, M. (2004) Tomato fruit cuticular waxes and their effects on transpiration barrier properties: functional characterization of a mutant deficient in a very-long-chain fatty acid β-ketoacyl-CoA synthase. *Journal of Experimental Botany* 55, 1401–1410.

Wang, Z., Guhling, O., Yao, R., Li, F., Yeats, T.H., Rose, J.K. and Jetter, R. (2011) Two oxidosqualene cyclases responsible for biosynthesis of tomato fruit cuticular triterpenoids. *Plant Physiology* 155, 540–552.

Yeats, T.H., Howe, K.J., Matas, A.J., Buda, G.J., Thannhauser, T.W. and Rose, J.K. (2010) Mining the surface proteome of tomato (*Solanum lycopersicum*) fruit for proteins associated with cuticle biogenesis. *Journal of Experimental Botany* 61, 3759–3771.

Yeats, T.H., Martin, L.B.B., Viart, H.M.-F., Isaacson, T., He, Y., Zhao, L., Matas, A.J., Buda, G.J., Domozych, D.S., Clausen, M.H. and Rose, J.K.C. (2012) The identification of cutin synthase: formation of the plant polyester cutin. *Nature Chemical Biology* 8, 609–611.

7 Antioxidants and Bioactive Compounds in Fruits

Angelos K. Kanellis[1]* and George A. Manganaris[2]

[1]Department of Pharmaceutical Sciences, Aristotle University of Thessaloniki, Thessaloniki, Greece; [2]Department of Agricultural Sciences, Biotechnology and Food Science, Cyprus University of Technology, Lemesos, Cyprus

7.1 Introduction

Quality is determined based mainly on external appearance, background colour and fruit size, whilst flavour, taste and aroma perception occur during or following consumption (Struik *et al.*, 2005). Fruits are also considered to be beneficial sources of antioxidant potency due to their phytochemical properties. Numerous bioactive compounds are theorized to have a role in preventing or ameliorating various chronic human diseases such as cancer, coronary vascular disease, Alzheimer's disease and diabetes. Metabolic pathways are not completely understood and as-yet-undefined non-antioxidant mechanisms may be responsible. Nevertheless, the consumer cannot perceive such attributes. Indisputably, however, plants have formed the basis of the human diet since the existence of human kind, a fact that led Hippocrates, the ancient Greek physician, to say 'let food be your medicine and medicine be your food'. Today, consumers' belief in the health benefits of selected foods and their components appears to be increasing at an unprecedented pace. However, one should keep in mind that dietary factors act in synergy and thus the complex mixture of nutrients and chemopreventive agents present in horticultural crops renders it impossible to identify a single 'miracle' nutrient for protection from diseases. Current evidence suggests that this 'natural' mix is more important than any one single nutrient that can be added to the diet in the form of a supplement. This dietary complexity has been linked with certain dietary habits such as the Mediterranean diet.

Specific phytochemicals may also affect fruit colour and taste, both desirable quality attributes. Health-conscious consumers now demand fruits with improved nutritional value, making development of rapid methods for the determination of phytochemical profiles a necessary tool for producers. Fruit ripening is a result of a complex network of biological processes yet to be fully elucidated, and hence quality traits are not usually subjected to modelling (Struik *et al.*, 2005; Genard *et al.*, 2007). Fruit ripening, among other events, results in enhanced accumulation of reactive oxygen species (ROS) (Giovannoni, 2004),

* kanellis@pharm.auth.gr

yet the relationship between common quality attributes and phytochemical profiles has not been examined in great detail.

Quality evaluation is an important issue in the fruit industry and in breeding programmes, whilst studies of phytochemical attributes (e.g. carotenoids, tocopherols and phenolics) related to antioxidant properties have recently commenced in breeding populations. The exploitation of natural populations is an integral part of modern marker-assisted breeding, with special reference to nutritional quality and bioactive compounds. Recently, phytochemicals present in green, yellow and red wild relatives of tomato have been analysed during fruit development and ripening, and noticeable differences in terms of antioxidant properties have been monitored (Mélendez-Martinez *et al.*, 2010).

Over the last decade, numerous studies have quantified the phytochemical content and/or the total antioxidant capacity with an array of *in vitro* assays in cultivars/ genotypes of the same species. Such analyses are usually conducted with samples from a single maturity stage (physiological or commercial), and only later studies have measured quantities of such compounds during fruit ripening both on and off the vine. As a result, information regarding phytochemical profiles of fruit during maturation and ripening under various postharvest conditions represents an ongoing research emphasis.

Testing methods employed to monitor the phytochemical composition and antioxidant properties of fruits are another critical factor. Most studies determine the total content of phenolic compounds, carotenoids and flavonoids using relatively simple spectrophotometric methods. However, it should be noted that, due to the complex nature of bioactive compounds in fruit, one-step characterization of such compounds is difficult. The selection of the solvent, drying and extraction method is of critical importance in phytochemical analysis (Goulas and Manganaris, 2012). Hyphenated techniques, such as high-performance liquid chromatography with UV detection (HPLC-UV) and liquid chromatography–mass spectrometry (LC-MS) or spectroscopic techniques such as nuclear magnetic resonance (NMR) and Fourier transform infrared spectroscopy (FT-IR) offer insights in to the composition of specific bioactive compounds in fruit. Emerging research is employing metabolomic platforms to monitor fruit antioxidant/phytochemical profiles (Lombardo *et al.*, 2011; Rohrmann *et al.*, 2011).

The antioxidant levels of fruit crops may be attributed to on- and off-the-vine factors including environmental conditions during growth and development, as well as those in the postharvest environment. In addition to fruit manipulation and postharvest environmental conditions, the accumulation and degradation of antioxidant compounds of fruits during the ripening process may also be attributed to both genetic and environmental factors, such as cultivar, cultural practices, environmental factors and developmental stage (Vicente *et al.*, 2009). Duration of holding at room temperature, in cold storage and other postharvest treatments including controlled storage temperature and modified atmospheres and/or other physical treatments all have the potential to change the phytochemical profile of the fruit under study. A recent review highlighted changes that occurred in phytochemical concentration during on-tree ripening or after the employment of several postharvest technologies and processing methods (Serrano *et al.*, 2011). Technologies such as heat and calcium treatments, modified atmosphere packaging and ethylene action inhibition through application of 1-methylcyclopropene (1-MCP) and polyamine can be applied before or along with cold storage. Intriguingly, these specific treatments increased the bioactive content of fruit commodities compared with conventional cold storage alone (Serrano *et al.*, 2011).

An overview of the health-promoting properties of horticultural commodities on a species basis as well as on general aspects of fruit nutritional quality has been published recently (Vicente *et al.*, 2009;

Terry, 2011). This chapter provides an initial overview of the principal antioxidants and bioactive compounds found in ripening fruit, as well as the methods employed for determination. A thorough description of the diversity of phytochemical compounds found in fruits and factors affecting their content is provided. Lastly, recent advances in the developmental regulation of bioactive compound accumulation are discussed.

7.2 The Main Antioxidants and Bioactive Compounds in Ripening Fruits

Typical nutritional components of fleshy fruits are water, organic acids, proteins, lipids and fatty acids, metabolizable carbohydrates and dietary fibre (reviewed by Vicente *et al.*, 2009). Apart from the so-called 'traditional' components, fruits are rich sources of phytochemical compounds, also referred to as bioactive compounds. Antioxidant (polyphenols and carotenoids) and non-antioxidant (phytosterols) bioactive compounds are strongly linked to high consumption of fruit having a significant health-promoting role (Saura-Calixto and Goni, 2009). Many phytochemicals are present in a wide range of fruits, but some are distributed only among specific species. In addition, they are considered non-nutrient constituents with significant biological activity, with each fruit crop having a distinct profile (Schreiner and Huyskens-Keil, 2006). Fruit phytochemicals have variable chemical structures and functions and are categorized into phenolic compounds, carotenoids and vitamins (C and E). An overview of the principal phytochemical compounds is provided in Fig. 7.1. Details about the structure of the antioxidant compounds are specified in a recent review by Vicente *et al.* (2009).

7.2.1 Phenolic compounds or polyphenols

This group encompasses a wide diversity of compounds derived from the aromatic amino acids phenylalanine and tyrosine, having variable degrees of hydroxylation, methoxylation and glycosidation. They usually act as sunscreens, deterrents of potential predators and antimicrobials, and also contribute to fruit pigmentation, astringency and the bitter taste of some products (Mattila *et al.*, 2006). In general, phenolic compounds accumulate more in the peel than in the pulp of the fruits.

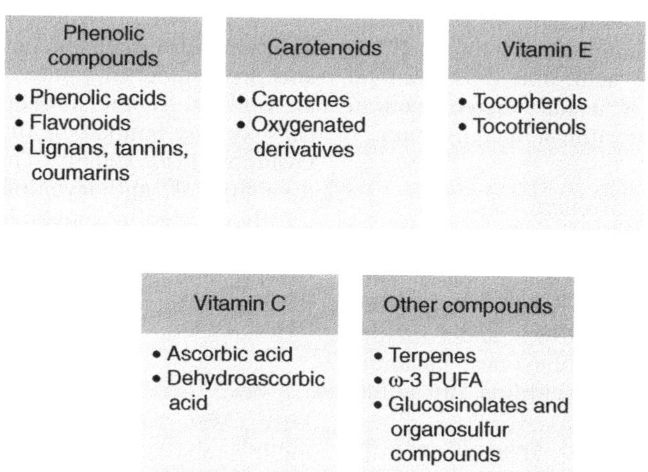

Fig. 7.1. The main antioxidants in ripening fruits. PUFA, polyunsaturated fatty acids.

Although *in vivo* data are rather limited, there is accumulative *in vitro* evidence showing that phenolic compounds affect many cellular processes. The antioxidant activity of phenolics is attributable to the electron delocalization over the aromatic ring and their high redox potential, which allows them to act as reducing agents, hydrogen donors and singlet oxygen quenchers (Serrano *et al.*, 2011). The main categories of phenolic compounds are phenolic acids and flavonoids. This classification is made based on the structure of their basic skeleton. Other compounds that belong to this category are lignans, stilbenes, tannins, coumarins and lignins.

Phenolic acids

Phenolic acids (C6–C1) are categorized as: (i) benzoic acid derivatives (syringic acid, gallic acid, vanillic acid, *p*-hidroxybenzoic); (ii) cinnamic acid derivatives (*p*-coumaric acid, caffeic acid, ferulic acid, sinapic acid, chlorogenic acid); and (iii) hydroxyphenylacetic acids (4-hydroxyphenylacetic acid, dihydroxyphenylacetic acid). Such derivatives differ in the degree of hydroxylation, methoxylation and alkylation of the aromatic ring. A structure–antioxidant activity relationship has been found for phenolic acids (Exarchou and Gerothanasis, 2006). The most common phenolic acids in fruits are caffeic and gallic acid, and thus the total phenol content is usually determined in terms of caffeic or gallic acid equivalents.

Flavonoids

Flavonoids have a structure of two aromatic rings, which are associated together by a three-carbon oxygenated heterocycle (C6–C3–C6), and include flavonols, flavones, isoflavones, flavanols, flavanones, proanthocyanidins and anthocyanidins.

Rutin, luteolin and apigenin are the most common flavones, quercetin and kampferol are typical flavonols, and catechin and epicatechin are flavanols (Manach *et al.*, 2004). The genus *Citrus* is characterized by the accumulation of flavanone glycosides. Orange juice is a source of the flavanone glycoside hesperidin (Tripoli *et al.*, 2007), whilst the flavanols catechin and epicatechin are common in the peel and seeds of grape and apple (Rice-Evans *et al.*, 1997). Isoflavones are not commonly found in fruits.

Proanthocyanidins are oligomeric flavonoids, usually dimers or oligomers of the flavanols catechin and epicatechin. Anthocyanidins are pigments giving several fruits their characteristic red, blue or purple colours (e.g. berries, grape, cherry, pomegranate, plum, apple), although under some conditions they remain uncoloured. In addition, anthocyanidins have great antioxidant potency, with the distribution of hydroxyls in the molecule affecting the antioxidant capacity of the different anthocyanidins; the differences between them result from the OH, H and OCH_3 substitutions on the phenolic ring. Among the 23 anthocyanidins that have been described, the most common ones found in fruits are cyanidin, delphynidin, pelargonidin, peonidin, petunidin and malvidin.

Anthocyanins are glycosides containing a sugar moiety and an anthocyanidin unit. More than 635 different anthocyanins exist and their structural differences are related to the number of hydroxyl groups in the anthocyanidin skeleton, and the position and the number of bonded sugar residues, as well as by the aliphatic or aromatic carboxylates bonded to them (Castañeda-Ovando *et al.*, 2009). The content and diversity of anthocyanins in fruits is greatly affected by genetic factors, environmental conditions, agricultural practices, harvest maturity, storage conditions, postharvest treatments and processing.

Other phenolic compounds

Fruit also contain polyphenolic compounds, derived from the polymerization of simple phenolics. The most well-known group of these compounds is tannins.

Tannins are segregated into hydrolysable tannins, with a base unit of a gallic acid moiety, and condensed tannins, with a flavone base unit. Tannins are found mainly in pomegranate, persimmon, berries and nuts, and are major contributors to the bitter, astringent taste component of fruits. Lignans are polyphenolic substances derived from phenylalanine via dimerization of substituted cinnamic alcohols, such as cinnamic acid. Apricots, strawberries and acai fruits are rich in lignans. An analytical description of the properties of major fruit polyphenols, as well as biosynthetic and regulatory processes that affect their composition, are presented by Ageorges *et al.*, Chapter 9, this volume.

7.2.2 Carotenoids

Carotenoids are liposoluble pigments, responsible for the yellow, orange and red colour of several fruits, divided into carotenes (e.g. α-carotene, β-carotene and lycopene) and oxygenated derivatives known as xanthophylls (e.g. lutein, cryptoxanthin, violaxanthin and zeaxanthin). Some carotenoids (like α-carotene, β-carotene and cryptoxanthin) have provitamin A activity (Kopsell and Kopsell, 2006; Meléndez-Martínez *et al.*, 2007). Carotenoids are usually present at low concentrations and their levels vary significantly among species. Fruit are generally not such good sources of carotenoids as vegetable crops, although there are a few notable exceptions such as apricot, mango, citrus, papaya and watermelon (Vicente *et al.*, 2009). To date, more than 600 different carotenoids have been identified, but few are common; for example, 25 carotenoids have been identified in loquat cultivars (de Faria *et al.*, 2009). β-Carotene is the most studied carotenoid, and lycopene is common in tomato and watermelon. As lycopene is regarded as a powerful natural antioxidant, high-lycopene tomato cultivars are being promoted in the market (Ilahy *et al.*, 2011).

7.2.3 Vitamin C

Ascorbic acid (AsA) is usually mentioned as being synonymous with vitamin C; however, vitamin C is considered the sum of AsA and its first oxidation product, dehydroascorbic acid (DHA). Ascorbic acid can even be oxidized during eating while the food is being chewed. However, it is important to consider that the first breakdown product of AsA, DHA, still has vitamin C activity, and all activity is lost if oxidation proceeds beyond this stage (Salunkhe *et al.*, 1991). Ascorbic acid is a water-soluble carbohydrate-derived compound showing antioxidant and acidic properties due to the presence of a 2,3-enediol moiety. The efficacy of AsA in disease prevention has been associated with its capacity to neutralize ROS. AsA is highly susceptible to oxidation, either directly or through the enzyme ascorbate oxidase, catalysing the oxidation of AsA to DHA, with the concomitant reduction of molecular oxygen to water (Sanmartin *et al.*, 2007). Ascorbic acid concentration is highly dependent on the individual commodity considered (Lee and Kader, 2000). Persimmon, strawberry, kiwifruit, and citrus fruit can be considered excellent sources of AsA (reviewed by Davey *et al.*, 2000; Vicente *et al.*, 2009). Wide variations in its content within cultivars of the same species also exist. For instance, AsA content in fuzzy kiwifruit (*Actinidia deliciosa*) varies from 29 to 80 mg 100 g^{-1} of fresh weight (Nishiyama *et al.*, 2004) and from 14 to 103 mg 100 g^{-1} of fresh weight in berry fruits (Pantelidis *et al.*, 2007). The retention of AsA is also markedly affected by storage and processing, with high temperatures having the highest impact (Lee and Kader, 2000).

Complex control networks affected by multiple factors regulate ascorbate biosynthesis and metabolism in higher plants. Perturbation of the ascorbate redox state by either increasing or decreasing the reduced ascorbate status alters the signalling pathway and consequently the plant responses to various stimuli and stresses. The ascorbate redox state seems to be a critical factor regulating various processes, stress responses and plant–pathogen interactions (Davey *et al.*, 2000; Sanmartin *et al.*, 2003; Fotopoulos *et al.*, 2006, 2008; Ioannidi *et al.*, 2009; Fotopoulos and Kanellis, 2013).

The main operational ascorbate biosynthetic pathway in tomato (*Solanum lycopersicum*) fruit tissues was recently shown to be the mannose/L-galactose one (Mellidou *et al.*, 2012). A combination of metabolite analyses, non-labelled and radio-labelled substrate feeding experiments, enzyme activity measurements and gene expression studies showed that, in the low-AsA cultivar ('Ailsa Craig'), alternative routes of AsA biosynthesis may supplement biosynthesis via L-galactose, whilst in the high-AsA cultivar ('Santorini'), enhanced AsA recycling activities appeared to be responsible for AsA accumulation in the later stages of ripening. Furthermore, it was shown that GDP–L-galactose phosphorylase 1 (*SlGGP1*) and two orthologues of NAD(P)-dependent monodehydroascorbate reductase (*SlMDHAR*) were closely correlated with AsA+DHA concentrations during ripening and are potentially good candidates for breeding purposes and marker selection (Mellidou *et al.*, 2012). It was suggested that the AsA pool is always regulated by the combination of synthesis, recycling and breakdown, without excluding a possible feedback inhibition of biosynthesis. It should be noted that no accumulation of a high ratio of AsA:total AsA early in fruit development and ripening has been recorded, even though AsA biosynthetic capacity is enhanced. However, AsA can accumulate later, with the onset of ripening (breaker stage), as a result of increased biosynthesis (Ioannidi *et al.*, 2009), recycling and decreased breakdown (Mellidou *et al.*, 2012).

Based on previous data (Ioannidi *et al.*, 2009; Mellidou *et al.*, 2012), overexpression of *GDP–L-galactose phosphorylase* (*GGP* or *VTC2*) and *L-galactose-1-phosphate phosphatase* (*GPP* or *VTC4*) under the control of three different promoters (*35S*, a constitutive promoter; *phosphoenolpyruvate carboxylase* (*PPC2*), a green fruit-specific promoter; and *polygalacturonase* (*PG*), a ripening fruit-specific promoter) resulted in T3 transgenic red tomato fruit with threefold increased levels of AsA (Kanellis *et al.*, unpublished data). Lastly, suppression of ascorbate oxidase in melon resulted in significant increase in AsA, but at the same time, it caused a reduction in fruit size in agreement with suggested participation of ascorbate oxidase in the fast growth of Curcubitaceae fruit (Kanellis *et al.*, unpublished data). An analytical description of vitamin C biosynthesis and regulation in fleshy fruits is provided by Baldet *et al.*, Chapter 8, this volume.

7.2.4 Vitamin E

Vitamin E includes tocopherols and tocotrienols. These occur in eight different forms (four tocopherols and four tocotrienols). All the isomers have aromatic rings with a hydroxyl group that can donate hydrogen atoms to reduce ROS. The different isomers are named α, β, γ and δ related to the number and position of methyl groups in the ring. Each of the forms has its own vitamin E activity, with α-tocopherol being the more active. In general, vitamin E levels are more abundant in oily seeds, olives, nuts, peanuts, avocados and almonds. Vitamin E is highly susceptible to oxidation during storage and processing. An analytical description of vitamin E biosynthesis and regulation in fleshy fruits is provided by Baldet *et al.*, Chapter 8, this volume.

7.2.5 Terpenes

Terpenes are a large group of phytochemicals and are derived from isoprene units (C5H8) that can be linked to form carbon skeletons such as C5, C10, C15, C20 up to C40. According to Nobelist Leopold Ruzicka, the isoprene units may be linked together 'head to tail' to form linear chains or they may be arranged to form rings. The number of isoprene units (C5H8)*n* is used for the classification of terpenes, as shown in Table 7.1 (Graßmann, 2005).

Terpenes of low molecular weight are components of essential oils such as limonene, pinene, linalool, etc. (Bakkali, *et al.*, 2008). The most interesting triterpenes are oleanolic and maslinic acid with high potent biological activities (Kontogianni, *et al.*, 2009). In addition, limonoids such as limonin, nomilin and nomilinic acid are found in citrus fruits and are the most well-known tetraterpenes (Manners, 2007).

7.2.6 Other bioactive compounds

Arginine

Arginine is an important amino acid in human nutrition. It is an essential amino acid for the fetus and neonate and a conditionally essential amino acid for adults (Wu *et al.*, 2000). An increasing number of medical surveys support the beneficial effect of a diet rich in arginine (Flynn *et al.*, 2002, and references therein). This amino acid appears to be a powerful mediator of multiple biological processes, including the release of several hormones, collagen synthesis during wound healing, antitumor activity, immune cell responses and the prevention of cardiovascular diseases (Lewis and Langkamp-Henken, 2000; de Nigris *et al.*, 2003; Grillo and Colombatto, 2004; Hayashi *et al.*, 2005; Arnold and Barbul, 2006). Arginine is also the precursor in nitric oxide synthesis in vascular cells (Palmer *et al.*, 1988). Nitric oxide is a ubiquitous signalling molecule, a cytotoxic free radical and a neurotransmitter, and is involved in many physiological and pathological processes including immune responses, cardiovascular disease and tumorigenesis (Radomski *et al.*, 1990).

In higher plants, besides a role in protein synthesis, arginine is utilized as nitrogen storage in seeds immobilized during seedling growth (Chen *et al.*, 2004) and as a precursor for biosynthesis of polyamines via arginine decarboxylase, proline, nitric oxide and plant alkaloids (Delauney and Verma, 1993; Crawford, 2006). Beyond its apparent role in metabolism, arginine is involved in many physiological processes including abiotic stress responses. Arginine participates in the detoxification of plant tissues from overaccumulation of ammonia during stress (Rabe and Lovatt, 1986). Other metabolites further up the arginine biosynthetic pathway such as citrulline and ornithine have also been associated with stress resistance. Citrulline is involved in resistance to hyperosmotic stress acting as a hydroxyl radical ($^{\cdot}OH$) scavenger (Takahara *et al.*, 2005). Ornithine is also a precursor of polyamine biosynthesis via ornithine decarboxylase, which is also involved in stress resistance by stabilizing cellular structures (Tiburcio *et al.*, 1994; Handa and Mattoo, 2010), acting as a scavenger of ROS and as an osmolyte (Nambeesan *et al.*, 2010), and having important biological functions in human physiology and the prevention of certain diseases (Kalac and Krausova, 2005; Larqué *et al.*, 2007). Polyamines occur in a variety of fruits and vegetables, with putrescine scoring the highest values among the other polyamines, namely spermidine and spermine (Mattoo *et al.*, 2010). Different citrus fruit (oranges,

Table 7.1. Classification of terpenes.

Number of carbons	Isoprene units	Subgroup
10	2	Monoterpenes
15	3	Sesquiterpenes
20	4	Diterpenes
25	5	Sesterterpenes
30	6	Triterpenes
40	8	Tetraterpenes

mandarins and grapefruit) and juices thereof and processed foods such as sauerkraut, ketchup, frozen green peas and fermented soy products are rich in putrescine, whereas spermidine content was recorded as highest in legumes, especially soybean, pear, cauliflower and broccoli (Kalac and Krausova, 2005). Spermine is usually found in legumes.

We are interested in understanding arginine biosynthesis and its regulation in tomato fruit. A single copy of the *N-acetyl-L-glutamate synthase* gene (*SlNAGS1*), the first gene in arginine biosynthesis, has been isolated from tomato and overexpressed in *Arabidopsis thaliana* under the control of the *35S* promoter. The transgenic plants accumulated high levels of ornithine and citrulline and exhibited a higher tolerance to salt and drought stress compared with wild-type plants (Kalamaki *et al.*, 2009). Recently, all genes in arginine biosynthesis in tomato fruit have been cloned, and the *SlNAGS1* gene was specifically overexpressed in tomato fruit under the *PG* promoter, which resulted in high levels of arginine in pink and red fruit (Kanellis *et al.*, unpublished data).

Selenium

Among the mineral elements, selenium (Se) can be regarded as an essential micronutrient for human health. It exhibits an antioxidant function, and as such can be considered as a chemopreventive agent in cancer and as an important nutrient in the immune system (Pedrero and Madrid, 2009). This function was attributed to its participation in the 25 selenoproteins in the human proteome including gluthathione peroxidase and thioredoxin reductase (White and Broadley, 2009; Hasanuzzaman *et al.*, 2010).

Generally, Se occurs in low amounts in plants and fruits, and thus it can affect their antioxidant properties and chemical composition, whereas at higher amounts it can act as a pro-oxidant, resulting in reduced plant growth (Pezzarossa *et al.*, 2012). Selenium can delay tomato fruit ripening and senescence through the inhibition of ethylene biosynthesis (Pezzarossa *et al.*, 1999). In lettuce and chicory, Se improved the keeping quality through the inhibition of ethylene production and phenylalanine ammonia lyase activity (Malorgio *et al.*, 2009). Se concentrations were increased in peach and pear fruit by spraying leaves and fruit with sodium selenate, resulting in an extended shelf-life of the fruit and delayed softening (Pezzarossa *et al.*, 2012).

7.3 Determination of Phytochemical Content and Antioxidant Potency

The extraction of phytochemicals from fruit tissues is a prerequisite step for the quantification of phytochemical content in fruits. A plethora of extraction methods such as solid–liquid extraction, Soxhlet, ultrasound-assisted extraction, pressurized liquid extraction, supercritical extraction and microwave-assisted extraction have been described for recovery of phytochemicals from diverse fruit materials (Ignat *et al.*, 2011). The extraction of phytochemicals is performed using polar or apolar solvents, depending on the polarity of the phytochemicals being examined. The temperature of extraction is also of critical importance, as it usually increases the efficiency of extraction, but some phytochemicals are degraded at high temperatures (Galanakis *et al.*, 2013).

The quantification of phytochemicals is usually carried out by instrumental analytical methods. The most common group of analyses are simple spectophotometric methods that allow measurement of the total concentration of compounds such as total phenolics, total anthocyanins or total carotenoids. These methods are usually based on a reagent that reacts with phenols to form chromogens that can be detected spectrophotometrically, as in the Folin–Ciocalteu method or the determination of total AsA using the dye 2,6-dichloroindophenol. Other spectophotometric protocols exploit the characteristic absorbance wavelengths of a group of compounds, as Obied *et al.*

(2005) suggested for the determination of phenolic acids, hydroxycinnamic acid derivatives, flavonols and anthocyanins. The quantification of total carotenoids using absorbance-based equations was proposed by Nagata and Yamashita (1992). In addition, there are protocols based on derivatization of the studied compounds: for example, the evaluation of flavanol content is performed by the reaction of flavanol with vanillin to produce a red chromophore moiety (Tabart *et al.*, 2010). Overall, these methods are rapid, low cost and easy, and inexpensive instrument set-up is required; however, other compounds often interfere in these measurements. The Folin–Ciocalteu reagent may react with non-phenolic reducing substances like AsA, citric acid and sugars, which are found at high concentrations in fruit (Apak *et al.*, 2007), and produce erroneous readings.

Traditionally, fruit extracts are subjected to fractionation and isolation procedures in order to obtain pure compounds for their identification. This laborious and time-consuming step is also a prerequisite for performing bioassays such as the evaluation of antioxidant activity. Nowadays, high-resolution screening techniques have been developed that combine a separation technique with fast post-column (bio) chemical detection to allow the rapid pinpointing of antioxidant compounds in complex fruit extracts, without previous isolation procedures (Niederlander *et al.*, 2008). This on-line technique has been applied successfully in large-scale routine analysis of complex plant extracts. Bandoniene and Murkovic (2002) demonstrated the on-line 2,2-diphenyl-1-picrylhydrazyl (DPPH) method for evolution of antioxidant activity of individual phenolic compounds from different apple cultivars.

HPLC coupled with different detectors has been thoroughly exploited for the qualitative and quantitative determinations of individual phytochemicals in fruits. Reversed-phase chromatography is used for the phytochemical study of polar compounds such as phenolics, terpenes and AsA, whilst less-polar compounds such as carotenoids and tocopherols are analysed by normal-phase chromatography. Mass and UV-Vis detectors are the most commonly used in phytochemical analysis of fruits. Techniques such as HPLC-MS with diode array (HPLC-DAD-MS) and MS with electrospray ionization (MS-ESI) have been used widely for the identification and quantification of phenols, carotenoids and vitamin C in fruits (Sancho *et al.*, 2011). Fluorescence detectors have also been used for these purposes for the determination of fluorescent secoiridoid compounds of olive fruits or the determination of triterpenic acids with on-line derivatization (Li *et al.*, 2011). Fluorescence detection is also preferred for the quantification of phytochemicals at lower concentrations. The combination of LC with electrochemical detection is often used for the quantification of catechins (Trojanowicz, 2011). On the other hand, the application of GC is quite limited because: (i) an extra step of derivatization is usually required; and (ii) many phytochemicals are degraded at high temperatures. However, GC is often used for the analysis of terpene composition. The derivatization of flavonoids and phenolic acids for GC-MS analysis has also been described (Proestos *et al.*, 2006).

NMR spectroscopy is one of the most powerful instrumental analysis techniques for phytochemical characterization. One- and two-dimensional methodologies provide an array of experiments for the structure elucidation of phytochemicals. NMR spectroscopy allows the simultaneous detection and quantification of individual phytochemicals of all groups in a complex fruit extract, thus avoiding time-consuming and tedious chromatographic techniques for their separation. Recently, Charisiadis *et al.* (2011) demonstrated the utility of ultrahigh-resolution hydroxyl group ^{1}H-NMR analysis, creating new directions for phytochemical analysis. Nowadays, NMR fingerprints are also used to compare the phytochemical composition of different fruits (Ali *et al.*, 2011; Kim *et al.*, 2011).

The antioxidant activity/antioxidant capacity per se is a well-established biomarker and is determined using an array of protocols. Niederlander *et al.* (2008) divided the antioxidant assays into three main categories:

- assays involving actual ROS–oxidizable substrate interactions;
- assays involving a relatively stable single oxidizing reagent; and
- assays relating antioxidant activity to electrochemical behaviour.

In the first category of assays, the oxidizing agents are those that are also active in biological systems and play a role in consumer product deterioration. These are applied to oxidize a substrate of which the concentration can easily be determined. The antioxidant activity of a phytochemical introduced into the system is related to the decrease in substrate conversion due to competition between the substrate and the phytochemical. In this group, assays use hydrogen peroxide and/or superoxide anion as the active ROS.

In the second category of assays, the oxidizing agent is a relatively stable reagent that represents the oxidizing pressure exercised on a potential reductant, e.g. a substrate or an antioxidant. Upon reaction with a phytochemical introduced into the system, conversion of the reagent can be monitored through a change in a specific characteristic (e.g. its UV/Vis absorbance). In the absence of phytochemicals with antioxidant activity, no reagent conversion will take place. Antioxidant activity is related to the rate and extent of reagent conversion. The most common assays of this group involve DPPH radical scavenging activity, 2,2-azinobis(3-ethylbenzothiazoline 6-sulfonic acid) ($ABTS^+$), ferric reducing/antioxidant power (FRAP), trolox equivalent antioxidant capacity (TEAC), phosphomolybdenum and β-carotene bleaching inhibition. Measurement of the antioxidant activity of each assay is based on different mechanisms. The activity of antioxidants depends mainly on substrates, solvents, pH, conditions and stages of oxidation, and localization of antioxidants in different phases (Frankel and Meyer, 2000). For example, the evaluation of extract antioxidant activity using the β-carotene bleaching inhibition method permits the determination of antioxidants in an emulsion, whilst the FRAP method measures the hydrophilic antioxidants and the DPPH assay detects antioxidants soluble in organic solvents. The antioxidant activity of citrus extracts, determined with four *in vitro* assays, showed a higher correlation coefficient between the phosphomolybdenum assay and the β-carotene bleaching inhibition assay. On the other hand, correlations between the DPPH method and the phosphomolybdenum and β-carotene bleaching inhibition assays were substantially lower (Goulas and Manganaris, 2011a). Thus, an approach with multiple assays for determination of antioxidant activity is highly recommended (Sacchetti *et al.*, 2005; Goulas and Manganaris, 2011b).

Further to the commonly used methods for assessing antioxidant activity, an antioxidant effect can be determined over time using bulk oil and emulsions (Goulas and Manganaris, 2011b). Such assays have a series of advantages: they do not require expensive instrumentation, they measure both hydrophilic and lipophilic antioxidants, and they determine the antioxidant effect on substrates with different physicochemical properties of interest to the food industry. However, the disadvantages include the overall duration of the assays, the oxidation conditions that influence the estimated activity and the effect on the antioxidant activity by interfacial and phase distribution properties, not to mention the complexity of interpreting the results.

It should be noted that assays of antioxidant capacity are based on discrete mechanisms using different means to generate radicals or oxidants used to test compounds suspected of having antioxidant potential. As a result, each individual assay generates a unique value that belies direct comparison. There is no direct evidence that the beneficial effects of polyphenol-rich foods can be attributed to the measured antioxidant properties of these foods. It is impossible to extrapolate the data generated by laboratory methods to *in vivo* (human) effects, and clinical trials to test benefits of dietary antioxidants have produced mixed results. Antioxidant molecules in food have a wide range of functions, many of which are unrelated to the ability to absorb free radicals.

7.4 Fruits as Reservoirs of Antioxidants

Over the last decade, numerous screening studies have monitored the antioxidant profile and phenolic, carotenoid and vitamin C contents in an array of fruits at their commercial maturity stage. Such studies have focused on the genotype variation among cultivars of the same species grown under similar environmental conditions and subjected to common agricultural practices. Epidemiological studies have already documented an inverse association between fruit consumption and chronic diseases, such as different types of cancer and cardiovascular disease. In promotion of the consumption of fresh horticultural products, the five-a-day campaign substantially helped to deliver the message of a healthy diet based on fruit consumption (Havas *et al.*, 1994; Sorensen *et al.*, 1999).

Modern crop breeding has developed many tomato cultivars with increased yield performance, disease tolerance and extended shelf-life, paying little attention to qualitative traits such as flavour, nutrition and human health (Dorais *et al.*, 2008). A recent study compared the bioactive content and antioxidant potency of tomato cultivars with high lycopene content during different maturity stages (green, green-orange, orange-red and red-ripe) with 'Rio Grande', a common cultivar (Ilahy *et al.*, 2011).

Determination of phytochemical profiles has led to the reconsideration of some overlooked commodities such as small fruits (blackberries, blueberries, gooseberries, red currants) based on their size and also possibly due to their relatively minor economic importance. Recently, these fruits have received significant attention due to their superior phytochemical profiles and antioxidant potency (Pantelidis *et al.*, 2007; reviewed by Szajdek and Borowska, 2008). Another review work described the principal bioactive compounds of berry fruits and subsequent influence of growth conditions. This work also offered the possibility of new genotypes selected for enhanced phytochemical content (Battino *et al.*, 2009). The beneficial health properties of berry fruits are partly associated with the presence of relatively high levels of phenolic compounds (Seeram *et al.*, 2006). Other 'forgotten' berries, such as sea buckthorn (*Hippophae rhamnoides* L.) are currently of particular interest for their high healthy phytochemical content of, for example, tocopherols and tocotrienols. It should be noted that the concentration of such compounds is cultivar specific and highly dependent on harvest date, while seasonal year-to-year differences have been monitored (Andersson *et al.*, 2008).

Pome and stone fruits are the most extensively studied temperate fruit crops because of their value in the marketplace. Presently, fruit quality retention after harvest is also perceived in terms of bioactive compound content. This is particularly important when new peach and apple cultivars are launched into the marketplace. The main polyphenols in apples are catechin, epicatechin, chlorogenic acid, phloridzin, quercetin-3-glucoside, quercetin-3-rhamnoside, kaempferol-3-glucoside and procyanidins B1 and B2. Measurement studies have shown a relatively high content of such compounds in indigenous or 'forgotten' cultivars. Whilst apple has been studied adequately in terms of its phytochemical profile, a scarcity of information exists regarding other pome fruits, primarily pear. As European and Japanese pears are also characterized by different ripening patterns, there is an emerging need to clarify their phytochemical content fluctuations during storage.

The main bioactive compounds found in *Prunus* species are carotenoids, AsA, vitamin E and phenolic compounds (reviewed by Vicente *et al.*, 2009). Initial studies that monitored differences in the phytochemical profile of peach, nectarine and plum fruits, either white- or yellow-fleshed, have been carried out within the last 10 years (Tomas-Barberan *et al.*, 2001; Gil *et al.*, 2002). Hydroxycinnamates, procyanidins, flavonols and anthocyanins were quantified in stone fruits using state-of-the-art techniques (HPLC-DAD-MS-ESI), which allowed the identification of procyanidin trimers in plums (Tomas-Barberan *et al.*, 2001). A close correlation between total phenolics and antioxidant activity was monitored; this was not always the case in studies with citrus fruits (Goulas and Manganaris, 2011a). Apparently, different phytochemicals extant in distinct fruit types differentially affect correlation with *in vitro* antioxidant assays. It must be noted that unmarketable over-ripe fruits can be utilized as sources of bioactive compounds (e.g. dietary supplements, functional foods) (Puerta-Gomez and Cisneros-Zevallos, 2011).

Extensive studies in terms of their content have also been carried out in citrus fruits, mainly in orange, mandarin, lemon, mandarin, grapefruit and other citrus species of relatively minor commercial importance (reviewed by Simonne and Ritenour, 2011). Carotenoids are the main pigments reported in citrus fruits, evident both in the peel and in the juice. Citrus crops are also characterized by the presence of naringin and hesperidin flavanones and highly oxygenated triterpenoid acids (limonoids) such as limonin, limonin glucoside and neoriocitrin (Yu *et al.*, 2005). Besides their involvement in the bitter taste, such compounds have been proved to possess significant antioxidant capacity.

Tropical fruit seem to possess strong antioxidant activity (Terry and Thompson, 2011). Within this context, 'Maradol' papaya fruit exocarp was found to be rich in phenolic compounds such as ferulic acid, caffeic acid and rutin, whereas lycopene, β-cryptoxanthin and β-carotene were identified in the mesocarp as the major carotenoids (Rivera-Pastrana *et al.*, 2010). Ferulic acid, caffeic acid and rutin tend to decrease whilst lycopene and β-cryptoxanthin tend to increase during papaya postharvest ripening (Rivera-Pastrana *et al.*, 2010). Kiwifruit is another exotic fruit crop with well-advertised health-promoting properties, with special reference to its high AsA content. The bioactive profile of the most common kiwifruit cultivar ('Hayward') indicates the presence of caffeic acid glucosyl derivatives, coumarin glucosydes, β-sitosterol, stigmasterol, campesterol and chlorogenic acid, as well as flavone and flavanol molecules (Fiorentino *et al.*, 2009). As there can be wide differences among cultivars of the same species, such studies are particularly important where a relatively low number of fruit cultivars account for the greatest amount of production worldwide.

Pomegranate is receiving considerable attention as a 'new' crop owing to its high bioactive compound content, primarily anthocyanins (delphinidin, cyanidin and pelargonidin) (Faria and Calhau, 2011). Evidence exists describing the anti-carcinogenic properties of pomegranate extracts with potential for human cancer prevention (Lansky and Newman, 2007). Research is lacking to determine how the phytochemical profile of pomegranate is affected by cultivar variation, environmental conditions or postharvest treatments.

In contrast to other fleshy fruits, avocado and olive are characterized by high oil content. The fat content comprises 30–70% of avocado and olive dry mass, whilst other fleshy fruits have less than 1%. Avocado fruit cv. 'Hass' showed higher antioxidant capacity in lipophilic than in hydrophilic extracts (Villa-Rodriguez *et al.*, 2011). Avocado is a rich source of lipophilic phytochemicals such as monounsaturated fatty acids, carotenoids, vitamin E and phytosterols; however, relatively few studies exist regarding the changes in phytochemical content of avocado during maturation or postharvest ripening (Meyer *et al.*, 2011).

7.5 Aspects Controlling the Accumulation of Bioactive Compounds in Fruits

The antioxidant levels of a fruit crop may be attributed to pre- and postharvest factors. It has been established that the accumulation of natural substances with bioactive and antioxidant activities in fruit and vegetables is affected by genetic control (Bramley, 2002; George *et al.*, 2004; Leskovar *et al.*, 2004; Milella *et al.*, 2011; Tlili *et al.*, 2011; Fitzpatrick *et al.*, 2012), environmental stimuli (Dumas *et al.*, 2003; Atkinson *et al.*, 2011), and developmental and ripening influences (Perkins-Veazie *et al.*, 2001; Bramley, 2002; Serrano *et al.*, 2005; Fitzpatrick *et al.*, 2006, 2012; Ioannidi *et al.*, 2009; Martí *et al.*, 2011; Tlili *et al.*, 2011), as well as preharvest and postharvest conditions (Lee and Kader, 2000; Ioannidi *et al.*, 2009; Vicente *et al.*, 2009; Atkinson *et al.*, 2011; Vallverdú-Queralt *et al.*, 2012). Furthermore, as the ripening process is thought to involve oxidative phenomena, a corresponding antioxidant system is needed to balance out the subsequent build-up of ROS (Jimenez *et al.*, 2002).

7.5.1 Genetic influence

Genetic variability, as depicted by the plethora of different fruit crop cultivars in use today, is an important factor influencing the accumulation of all bioactive compounds studied to date, including vitamins C and E and other compounds with antioxidant activity such as carotenoids (β-carotene, lycopene and xanthophylls), phenylpropanoids, glucosinolates, polyamines and specific amino acids (arginine and citrulline).

This has been documented in tomatoes (Abushita *et al.*, 2000; George *et al.*, 2004; Ilahy *et al.*, 2011), watermelon (Perkins-Veazie *et al.*, 2001, 2006; Tlili *et al.*, 2011), cherries (Gao and Mazza, 1995; Ferretti *et al.*, 2010), strawberries (Olsson *et al.*, 2004; Tulipani *et al.*, 2008) and peppers (Brand *et al.*, 2012). Interestingly, these compounds do not all segregate in the same fashion; for example, naringenin chalcone content segregates as a monogenic trait independently from carotenoids and chlorophylls in melon (Tadmor *et al.*, 2010). Traits linked to fruit nutritional quality are complex and have polygenic control and quantitative inheritance. However, a considerable number of quantitative trait loci linked to fruit nutritional quality have recently been characterized in fruits (Davey *et al.*, 2006; Fernie *et al.*, 2006; Stevens *et al.*, 2008; Almeida *et al.*, 2011; Farre *et al.*, 2011).

7.5.2 Developmental and ripening regulation

The accumulation of phytochemicals with bioactive properties and the corresponding total antioxidant activity during fruit ripening seem to be under developmental and ethylene control and represent part of the overall ripening changes consisting of chlorophyll loss, the appearance of carotenoids including lycopene and xanthophylls and the accumulation of phenylpropanoids, ascorbate, α- and β-tocopherol and others (Hancock *et al.*, 2007; Giorio *et al.*, 2008; Fraser *et al.*, 2009; Ioannidi *et al.*, 2009; Martí *et al.*, 2011; Fitzpatrick *et al.*, 2012). These changes occur in all types of fruit, regardless of the characteristic ripening pattern (climacteric versus non-climacteric). However, control of phytochemical accumulation varies with ripening classification. Although critical assessment of the developmental control of fruit ripening and of the main bioactive compounds can be found in other chapters, it is thought appropriate to briefly mention here the most recent findings with respect to the aforementioned compounds. Carotenoid increase during ripening is a result of the coordinated gene expression of the biosynthetic genes phytoene synthase (PSY), especially *PSY1* (Bramley, 2002; Giorio *et al.*, 2008), and genes encoding desaturases and isomerases, whereas both lycopene β- and ε-cyclases cease expression resulting in the rise of lycopene (Galpaz *et al.*, 2006;

Fraser *et al.*, 2009; Águila Ruiz-Sola and Rodríguez-Concepción, 2012). Additional evidence has indicated that carotenoid accumulation in ripening tomato fruit is under the control of the negative-regulator *APETALA2/ERF* gene *SlAP2a*, opening new directions for manipulating the nutritional quality of tomato (Chung *et al.*, 2010).

The biosynthetic pathway of ascorbate has recently been fully characterized (Smirnoff, 2011). It has also been shown that AsA accumulation is developmentally regulated during ripening in tomato (Ioannidi *et al.*, 2009), melon (Pateraki *et al.*, 2004; Sanmartin *et al.*, 2007), apple (Davey *et al.*, 2004; Li *et al.*, 2009), blackcurrant (Hancock *et al.*, 2007), peach (Imai *et al.*, 2009), grape (Cruz-Rus *et al.*, 2010), strawberry (Cruz-Rus *et al.*, 2011) and kiwifruit (Li *et al.*, 2010). However, although AsA is a co-factor of ACC oxidase, the last enzyme of ethylene biosynthesis, it is not yet clear whether its biosynthesis is regulated by ethylene during fruit ripening. On the other hand, Ioannidi *et al.* (2009) found that the only ascorbate biosynthetic gene that is ethylene regulated is GPP.

Phenylpropanoid accumulation is developmentally regulated during ripening in a variety of fruit (Singh *et al.*, 2010). In climacteric fruit, ethylene plays the main role in inducing the biosynthetic genes, starting from phenylalanine ammonia lyase and ending with more specialized enzymes catalysing the formation of various phenols, flavonoids, proanthocyanidins and anthocyanins. However, in non-climacteric fruit, abscissic acid seems to play a critical role in regulating the ripening process and the accumulation of the products of the phenylpropanoid pathway (Zifkin *et al.*, 2012). It should be noted that this pathway is controlled by a number of transcription factors belonging to the MYB family in combination with basic helix–loop–helix (bHLH) and WD-40 partners.

Other bioactive compounds, such as, among others, folates (Waller *et al.*, 2010; Hanson and Gregory, 2011) and tocopherols (Mélendez-Martinez *et al.*, 2010), are also developmentally regulated.

7.5.3 Environmental factors

Cultivation systems and environmental signals may control the composition of bioactive compounds, with the most important environmental factors being light and temperature. In addition, other atmospheric factors can be considered: air carbon dioxide concentration, air humidity and air pollutants, including ozone (Dorais *et al.*, 2008; Vigneault *et al.*, 2012).

It is known that light quality and intensity, including UV radiation, can regulate the accumulation of vitamin C (Davey *et al.*, 2000; Davuluri *et al.*, 2004, 2005; Pateraki *et al.*, 2004; Smirnoff, 2011; Alimohammadi *et al.*, 2012), carotenoids (Davuluri *et al.*, 2004, 2005; Fraser *et al.*, 2009; Azari *et al.*, 2010; Águila Ruiz-Sola and Rodríguez-Concepción, 2012) and phenylpropanoid compounds (Davuluri *et al.*, 2004, 2005; Wang *et al.*, 2009b; Azari *et al.*, 2010). Studies on the interplay of light and shade have suggested that light influences ascosbate recycling and biosynthesis processes of apples, primarily in the peel and leaf but not in the flesh (Davey *et al.*, 2004; Li *et al.*, 2009). In tomatoes, ascorbate accumulation was greatly affected by high light intensity (Gautier *et al.*, 2009), which seems to be regulated by the expression of GGP and GPP (Massot *et al.*, 2012). Interestingly, a recent study suggested that a different light composition controls the accumulation of distinctive phenypropanoid compounds; that is, proanthocyanidin biosynthesis and composition were stimulated primarily by visible light, whereas flavonols were specifically triggered by UV light (Koyama *et al.*, 2012).

In terms of carotenoids, it has been shown that light is an important factor affecting the expression of biosynthetic genes, but also that light regulates their accumulation by controlling the light-signalling apparatus (Azari *et al.*, 2010; Águila Ruiz-Sola and Rodríguez-Concepción, 2012). Within this context, four different types of light-responsive promoter elements were detected in all ascorbate-related genes in *Arabidopsis* (Ioannidi *et al.*, 2009). One light-responsive element was detected in 17 out of 21 of these gene promoters. However, no such elements were detected in the promoters of GPP, superoxide dismutase (SOD), chloroplastic iron SOD and ascorbate peroxidase. ABA-, gibberellin-, heat shock-, wounding-, fungal elicitor- and endosperm-responsive elements were also detected in the promoters (Ioannidi *et al.*, 2009).

Thus, manipulation of the light-signalling apparatus in plants might be useful for manipulation of fruit phytochemicals (Azari *et al.*, 2010). More information on progress using this valuable approach, accentuating the outcome of genetic and transgenic modulation of light-signalling elements on the functional properties of fruit, can be found in other chapters in this volume covering the individual phytonutrients.

Temperature regulates bioactive compound pools in different fruits through the general effect on metabolism (Dorais *et al.*, 2008). However, a recent study suggested that low temperature induces the expression of genes implicated in anthocyanin biosynthesis and regulation in a number of tree fruits (Xie *et al.*, 2012). This was done by moderation of a molecular mechanism in which the central role was played by the cold-induced bHLH transcription factor gene *MdbHLH3*, and the resultant activation of genes participating in anthocyanin accumulation and fruit coloration in apples (*Malus domestica*) (Xie *et al.*, 2012). In terms of AsA, among all of the genes participating in its biosynthesis, oxidation and recycling, only the expression of GPP was induced by low temperature (Ioannidi *et al.*, 2009). The *GPP* gene is believed to play an important role during tomato ripening, and apparently the same gene was also upregulated by wounding and ethylene but suppressed by high temperature. Generally, AsA accumulation is negatively regulated by high and positively by low temperature, a phenomenon coinciding with the expression profile of GPP. On the other hand, anoxia for 48 h provoked an increase in AsA content in tomatoes upon transfer to air, an observation corresponding with the induction of all AsA biosynthetic genes (Ioannidi *et al.*, 2009). This response could be tightly connected with the generation of ROS under the above stress conditions (Pucciariello *et al.*, 2012).

7.5.4 Cold storage

Cold storage is a common practice to extend market life and maintain the perceived quality of horticultural commodities. However, the effect of this 'stress condition' on the phytochemical profile of flesh fruits and vegetables is largely unexplored. Preliminary data indicate that the pronounced loss of fruit-keeping quality during apple ripening at room temperature after prolonged low-temperature storage coincided with decreased content of bioactive compounds (V. Goulas and G.A. Manganaris, unpublished data). This was not the case for plum fruit, where no significant loss of bioactive compounds and antioxidant capacity of cold-stored plums for an extended period was noticed (Diaz-Mula *et al.*, 2009).

In papaya, chilling temperatures (1°C) negatively affected the content of major carotenoids, except for β-carotene, but preserved or increased the ferulic and caffeic acid levels compared with fruit held at room temperature (25°C) (Rivera-Pastrana *et al.*, 2010).

The phytochemical dynamics of each fruit commodity should also be considered based on the effect of extended cold storage with special reference to the incidence of cold-storage disorders. Such symptoms vary by commodity and there are few research reports regarding the relationship, if any, between chilling injury symptoms and fruit phytochemical profile.

7.5.5 Postharvest application of signalling molecules

Application of methyl jasmonate and benzothiadiazole, a synthetic functional analogue of salicyclic acid, was found to lead to a substantial increase in the anthocyanin, flavonol and proanthocyanidin content of grapes through the induction of enzymes implicated in the phenylpropanoid biosynthetic pathway; such compounds may gain interest in the wine industry by improving grape quality (Ruiz-García *et al.*, 2012).

Methyl jasmonate treatment of harvested loquat fruit followed by cold storage for 35 days led to a reduced incidence of chilling injury symptoms, evident as decreased extractable juice and internal browning. The alleviation of such symptoms might be attributed to the substantial delay of increases in superoxide radical ($O_2^{-\cdot}$) production rate and H_2O_2 content and the higher activities of SOD, catalase and ascorbate peroxidase (Cao *et al.*, 2009).

7.5.6 Postharvest 1-MCP treatment

The application of 1-MCP is a success story in the apple fruit industry with beneficial effects in terms of quality retention by inhibition of climacteric ripening responses, whilst alleviating the chilling injury symptoms in cold-stored apple fruit. The effect of 1-MCP treatment on fruit phytochemical profile is another emerging area for future research. Within this area, Qiu *et al.* (2009) observed that the antioxidant capacity of 1-MCP-treated 'Sunrise' summer apples after short-term storage at a range of temperatures (5–22°C) scored higher values compared with controls. Furthermore, the application of 1-MCP at the optimum apple maturity stage has been proposed as a means of retaining key flavonoid compounds during storage and postharvest ripening (MacLean *et al.*, 2006).

Superficial scald is a cold-storage disorder of pome fruits attributed to oxidative stress. Application of 1-MCP or immersion in an antioxidant agent (diphenylamine) prevents or alleviates these symptoms. A recent study identified a key class of phytosterol metabolites and indicated that peel phytosteryl conjugate metabolism was affected by storage duration, oxidative stress, ethylene action/ripening and storage temperature (Rudell *et al.*, 2011). The efficacy of 1-MCP on superficial scald symptoms in Asian pears was attributed to the induction of catalase and peroxidase activities but not of SOD activity (Yazdani *et al.*, 2011). 1-MCP treatment proved to be particularly effective in 'Rocha' pear by alleviating cold-storage disorders (browning and superficial scald), but without affecting the free-radical scavenging activity or the levels of fruit AsA and glutathione contents. Therefore, it appears that the efficacy of 1-MCP is not directly related to its effects on antioxidant levels (Silva *et al.*, 2010).

Lately, the effect of 1-MCP on the antioxidant systems of 'Empire' apple fruit stored under controlled atmospheres for prolonged periods has been studied with special reference to the development of chilling injury symptoms, evident as internal browning; however, the data do not justify a direct role for antioxidant metabolism during the development of internal browning (Lee *et al.*, 2012).

1-MCP-treated apricot fruit cold stored for 3 weeks exhibited a greater resistance to oxidative stress, postulated as a function of the lower ion leakage values, higher antioxidant capacity and higher SOD and unspecific peroxidase activities (Egea *et al.*, 2010). The profile of individual anthocyanins and hydroxycinnamic acids in 'Lambert Compact' sweet cherry was not affected by cold storage or 1-MCP treatment, although colour changes were monitored (Mozetic *et al.*, 2006).

Application of 1-MCP inhibited the production of H_2O_2 and specific activities of antioxidant enzymes (catalase, SOD and ascorbate peroxidase) of cold-stored 'Tainong' mango fruit, suggesting its putative role in positively regulating the activated oxygen metabolic balance (Wang *et al.*, 2009a). However, a similar study in mango fruit showed that 1-MCP treatment led to decreased levels of H_2O_2 and lipid peroxidation, concomitant with increased activities and isozymes of catalase and SOD (Singh and Dwivedi, 2008).

7.5.7 Postharvest nitric oxide treatment

The free-radical gas nitric oxide has been applied recently as a postharvest treatment with the aim of retarding the ripening process, based on its antisenescence activity. Special attention was paid to the antioxidant response of 'Rojo Rito' peach fruit in response to nitric oxide treatment. The treated fruit were characterized by lower ethylene evolution and respiration rate and higher firmness retention, and a lower percentage of electrolyte leakage. These data provide supporting evidence that nitric oxide might have a beneficial effect on the oxidation equilibrium and the antioxidant capacity of peach fruit (Flores *et al.*, 2008). A more recent study, investigating the efficacy of nitric oxide treatment on qualitative attributes of 'Kensington Pride' mango fruit, showed a suppression of ethylene production and respiration rate and a delay of fruit softening, as well as alleviation of chilling injury symptoms. However, in terms of phytochemical properties, nitric oxide treatments did not seem to affect AsA or carotenoid content and the overall antioxidant capacity (Zaharah and Singh, 2011).

7.5.8 Other postharvest treatments

A recent review presented an overview of the effect of postharvest elicitors on phytochemical content and composition in horticultural produce, as an attempt to be an integral part of postharvest management (Schreiner and Huyskens-Keil, 2006). Such techniques may be applied to both ripening fruits destined for fresh consumption or to raw material intended to be used as functional foods and/or supplements.

Postharvest elicitors are segregated into physical (low temperature, heat treatment, ultraviolet and γ-irradiation, altered gas composition) and those of chemical origin; the latter includes signalling molecules implicated in the fruit-ripening process, such as ethylene, salicylic acid and methyl jasmonate.

Thermal treatments, alone or in combination with controlled atmosphere storage, are commonly applied against invasive pests, for preservation of fruit quality and for extension of market life. Such treatments did not adversely affect the polyphenol content in mango fruit, although a general decrease was observed during the progress of fruit ripening (Kim *et al.*, 2007). The effect of postharvest hot-air treatment on the antioxidant system of stored mature-green tomatoes has also been monitored, although without a clear picture emerging (Yahia *et al.*, 2007). It should be noted that the temperature applied in heat treatments is of critical importance in the efficacy of the treatment on both the qualitative profile and the antioxidant properties of the fruit commodity.

Processing (blanching, freezing, canning and cooking) generally has a negative impact on phytochemical concentrations but is dependent on temperature and processing time (Serrano *et al.*, 2011).

7.6 Molecular Breeding and Genetic Manipulation

In recent years, the biosynthetic pathways leading to the production of most bioactive compounds including antioxidants have been elucidated in higher plants. In addition to plant transformation, modern plant biotechnology has provided a variety of available genomic, transcriptomic, metabolomic and bioinformatic techniques. It is now possible to establish molecular markers linked to health-promoting substances for use in molecular breeding programmes to create new cultivars, hybrids and/or transgenic lines with improved nutritional content (Farre *et al.*, 2011; Goldman, 2011; Zhao and Shewry, 2011). It should be noted that this is a recent trend in breeding programmes, whereas the emphasis in the last decades of the 20th century was placed on the production of cultivars and hybrids possessing desirable appearance and disease and pest resistance coupled with increased yield and improved postharvest characteristics (Dorais *et al.*, 2008). More recently, breeding and biotechnological programmes have had an increased emphasis on increasing the accumulation of: (i) carotenoids including lycopene and xanthophylls in tomatoes (Tanaka and Ohmiya, 2008; Farre *et al.*, 2011; Zhao and Shewry, 2011); (ii) flavonoids, proanthocyanidins and anthocyanins in tomatoes (Willits *et al.*, 2005; Schijlen *et al.*, 2006; Butelli *et al.*, 2008; Gonzali *et al.*, 2009; Vogt, 2010), grape berries (Xie and Dixon, 2005; Ageorges *et al.*, 2006; Tanaka and Ohmiya, 2008; Terrier *et al.*, 2009; Zhao and Shewry, 2011) and other crops (Zhao and Shewry, 2011); (iii) vitamin C in different fruit crops (Davey *et al.*, 2006; Lippman *et al.*, 2007; Stevens *et al.*, 2007; Smirnoff, 2011; Bulley *et al.*, 2012); (iv) folates (Almeida *et al.*, 2011; Hanson and Gregory, 2011); (v) vitamin E (DellaPenna and Mene-Safrane, 2011); and (vi) other fruit bioactive molecules, like polyamines (Mehta *et al.*, 2002; Nambeesan *et al.*, 2010) and glucosinolates (Goldman, 2011). However, in addition to increasing the phytonutrient content in fruit, molecular breeding projects need also to consider the functionality and the bioavailability of the individual components in humans in combination with reducing the concentration of antinutritional compounds such as phytic acid and calcium oxalate (Goldman, 2011).

7.7 Conclusions and Perspectives

In order to fully exploit modern biotechnological and/or molecular breeding approaches, necessary prerequisites for improving the nutritional and health-promoting attributes of fruit include knowledge of the biosynthetic pathways at biochemical and molecular levels, as well as the regulatory steps controlling the rate of biosynthesis and the pool size of the individual compounds. In our opinion, the future for obtaining new cultivars and hybrids enriched with 'health-promoting substances' is connected with the development of molecular markers linked to useful traits. It is also essential to exploit all recourses and approaches as they become available, including potential systems/synthetic biology workflows applied to the production of target bioactive compounds in consumer-friendly hosts such as fruit crops. The discovery and use of tissue-specific promoters known to control biosynthesis and accumulation of bioactive compounds will lead to the efficient production of fruit and vegetable varieties that possess enhanced functional properties. The interplay between genotype and environment cannot be overlooked, so future investigations to close this research gap may include the following:

- Consideration of phytochemical characteristics of new fruit cultivars in addition to traditional breeding objectives of yield, market life and disease tolerance. An emerging research area is the investigation of phytochemical profiles of either traditional or 'forgotten' indigenous cultivars.
- Comparison of bioactive phytochemical content of fruit crops with distinct ripening profiles among their cultivars, e.g. suppressed-climacteric and climacteric plums; melting, non-melting and stony-hard peaches; European and Japanese pears; citrus hybrids.
- Studies of the possible synergistic effects among different classes of phytochemicals during fruit ripening.
- Avoid general statements regarding health-promoting properties of fruits. Results need justification, especially with regard to actual bioavailability of the phytochemicals within fruits.
- Detection, evaluation and interpretation of fruit antioxidant capacity are advised to be undertaken by more than one method due to the complexity of the different chemical compounds of the fruit.

Acknowledgements

The authors would like to thank Dr. Vlasios Goulas for lending his expertise in writing Section 7.3: Determination of Phytochemical Content and Antioxidant Potency. Critical reading of the manuscript by John Fellman is greatly appreciated. Work reported in this chapter was funded by grants to A.K.K: EU-SOL-FOOD-CT-2006-016214, GR-NUTRITOM/11Syn_3_480, GR-PYTHAGORAS – EPEAEK II and COST Action FA1106 'QualityFruit'.

References

Abushita, A.A., Daood, H.G. and Biacs, P.A. (2000) Change in carotenoids and antioxidant vitamins in tomato as a function of varietal and technological factors. *Journal of Agricultural and Food Chemistry* 48, 2075–2081.

Ageorges, A., Fernandez, L., Vialet, S., Merdinoglu, D., Terrier, N. and Romieu, C. (2006) Four specific isogenes of the anthocyanin metabolic pathway are systematically co-expressed with the red colour of grape berries. *Plant Science* 170, 372–383.

Águila Ruiz-Sola, M. and Rodríguez-Concepción, M. (2012) Carotenoid biosynthesis in Arabidopsis: a colorful pathway. *The Arabidopsis Book* 10, e0158.

Ali, K., Maltese, F., Fortes, A.M., Pais, M.S., Choi, Y.H. and Verpoorte, R. (2011). Monitoring biochemical changes during grape berry development in Portuguese cultivars by NMR spectroscopy. *Food Chemistry* 124, 1760–1769.

Alimohammadi, M., de Silva, K., Ballu, C., Ali, N. and Khodakovskaya, M.V. (2012) Reduction of inositol (1,4,5)-trisphosphate affects the overall phosphoinositol pathway and leads to modifications in light signaling and secondary metabolism in tomato plants. *Journal of Experimental Botany* 63, 825–835.

Almeida, J., Quadrana, L., Asís, R., Setta, N., de Godoy, F., Bermudez, L., Otaiza, S.N., Correa da Silva, J.V., Fernie, A.R., Carrari, F. and Rossi, M. (2011) Genetic dissection of vitamin E biosynthesis in tomato. *Journal of Experimental Botany* 62, 3781–3798.

Andersson, S.C., Rumpunen, K., Johansson, E. and Olsson, M.E. (2008) Tocopherols and tocotrienols in sea buckthorn (*Hippophae rhamnoides* L.) berries during ripening. *Journal of Agricultural and Food Chemistry* 56, 6701–6706.

Apak, R., Guclu, K., Demirata, B., Ozyurek, M., Celik, S.E., Bektasoglu, B., Berker, K.I. and Ozyurt, D. (2007) Comparative evaluation of various total antioxidant capacity assays applied to phenolic compounds with the CUPRAC assay. *Molecules* 12, 1496–1547.

Arnold, M. and Barbul, A. (2006) Nutrition and wound healing. *Plastic and Reconstructive Surgery* 117, 42S–58S.

Atkinson, N.J., Dew, T.P., Orfila, C. and Urwin, P.E. (2011) Influence of combined biotic and abiotic stress on nutritional quality parameters in tomato (*Solanum lycopersicum*). *Journal of Agricultural and Food Chemistry* 59, 9673–9682.

Azari, R., Tadmor, Y., Meir, A., Reuveni, M., Evenor, D., Nahon, S., Shlomo, H., Chen, L. and Levin, I. (2010) Light signaling genes and their manipulation towards modulation of phytonutrient content in tomato fruits. *Biotechnology Advances* 28, 108–118.

Bakkali, F., Averbeck, S., Averbeck, D. and Idaomar, M. (2008) Biological effects of essential oils – a review. *Food and Chemical Toxicology* 46, 446–475.

Bandoniene, D. and Murkovic, M. (2002) On-line HPLC-DPPH screening method for evaluation of radical scavenging phenols extracted from apples (*Malus domestica* L.). *Journal of Agricultural and Food Chemistry* 50, 2482–2487.

Battino, M., Beekwilder, J., Denoyes-Rothan, B., Laimer, M., McDougall, G.J. and Mezzetti, B. (2009) Bioactive compounds in berries relevant to human health. *Nutrition Reviews* 67, S145–S150.

Bramley, P.M. (2002) Regulation of carotenoid formation during tomato fruit ripening and development. *Journal of Experimental Botany* 53, 2107–2113.

Brand, A., Borovsky, Y., Meir, S., Rogachev, I., Aharoni, A. and Paran, I. (2012) pc8.1, a major QTL for pigment content in pepper fruit, is associated with variation in plastid compartment size. *Planta* 235, 579–588.

Bulley, S., Wright, M., Rommens, C., Yan, H., Rassam, M., Lin-Wang, K., Andre, C., Brewster, D., Karunairetnam, S., Allan, A.C. and Laing, W.A. (2012) Enhancing ascorbate in fruits and tubers through over-expression of the galactose pathway gene GDP-L-galactose phosphorylase. *Plant Biotechnology Journal* 10, 390–397.

Butelli, E., Titta, L., Giorgio, M., Mock, H.-P., Matros, A., Peterek, S., Schijlen, E.G.W.M., Hall, R.D., Bovy, A.G., Luo, J. and Martin, C. (2008) Enrichment of tomato fruit with health-promoting anthocyanins by expression of select transcription factors. *Nature Biotechnology* 26, 1301–1308.

Cao, S.F., Zheng, Y.H., Wang, K.T., Jin, P. and Rui, H.J. (2009) Methyl jasmonate reduces chilling injury and enhances antioxidant enzyme activity in postharvest loquat fruit. *Food Chemistry* 115, 1458–1463.

Castañeda-Ovando, A., Pacheco-Hernández, M.D.L., Páez-Hernández, M.E., Rodríguez, J.A. and Galán-Vidal, C.A. (2009) Chemical studies of anthocyanins: a review. *Food Chemistry* 113, 859–871.

Charisiadis, P., Primikyri, A., Exarchou, V., Tzakos, A. and Gerothanassis, I.P. (2011) Unprecedented ultra-high-resolution hydroxy group 1H NMR spectroscopic analysis of plant extracts. *Journal of Natural Products* 74, 2462–2466.

Chen, H., McCaig, C., Melotto, M., He, S.Y. and Howe, G.A. (2004) Regulation of plant arginase by wounding, jasmonate and the phytotoxin coronatine. *Journal of Biological Chemistry*, 44, 45998–46007.

Chung, M.Y., Vrebalov, J., Alba, R., Lee, J.M., McQuinn, R., Chung, J.D., Patricia Klein, P. and Giovannoni, J. (2010) A tomato (*Solanum lycopersicum*) *APETALA2/ERF* gene, *SlAP2a*, is a negative regulator of fruit ripening. *Plant Journal* 64, 936–947.

Crawford, N.M. (2006) Mechanisms for nitric oxide synthesis in plants. *Journal of Experimental Botany* 57, 471–478.

Cruz-Rus, E., Botella, M.A., Valpuesta, V. and Gomez-Jimenez, M.C. (2010). Analysis of genes involved in L-ascorbic acid biosynthesis during growth and ripening of grape berries. *Journal of Plant Physiology* 167, 739–748.

Cruz-Rus, E., Amaya, I., Sanchez-Sevilla, J.F., Botella, M.A. and Valpuesta, V. (2011) Regulation of L-ascorbic acid content in strawberry fruits. *Journal of Experimental Botany* 62, 4191–4201.

Davey, M.W., Van Montagu, M., Sanmartin, M., Kanellis, A.K., Smirnoff, N., Benzie, F.F., Strain, J.J., Fletcher, J. and Favell, D. (2000) Plant L-ascorbic acid: chemistry, function, metabolism, bioavailability and processing. *Journal of the Science of Food and Agriculture* 80, 825–860.

Davey, M.W., Franck, C. and Keulemans, J. (2004) Distribution, developmental and stress responses of antioxidant metabolism in *Malus*. *Plant, Cell and Environment* 27, 1309–1320.

Davey, M.W., Kenis, K. and Keulemans, J. (2006) Genetic control of fruit vitamin C contents. *Plant Physiology* 142, 335–343.

Davuluri, G.R., van Tuinen, A., Mustilli, A.C., Manfredonia, A., Newman, R., Burgess, D., Brummell, D.A., King, S.R., Palys, J., Uhlig, J., Pennings, H.M.J. and Bowler, C. (2004) Manipulation of *DET1* expression in tomato results in photomorphogenic phenotypes caused by post-transcriptional gene silencing. *Plant Journal* 40, 344–354.

Davuluri G.R., van Tuinen, A., Fraser, P.D., Manfredonia, A., Newman, R., Burgess, D., Brummell, D.A., King, S.R., Palys, J., Uhlig, J., Bramley, P.M., Pennings, H.M.J. and Bowler, C. (2005) Fruit specific RNAi-mediated suppression of *DET1* enhances carotenoid and flavonoid content in tomatoes. *Nature Biotechnology* 23, 890–895.

de Faria, A.F., Hasegawa, P.N., Chagas, E.A., Pio, R., Purgatto, E. and Mercadante, A.Z. (2009) Cultivar influence on carotenoid composition of loquats from Brazil. *Journal of Food Composition and Analysis* 22, 196–203.

de Nigris, F., Lerman, L.O., Williams-Ignarro, S., Sica, G., Lerman, A., Palinski, W., Ignarro, L.J. and Napoli, C. (2003) Beneficial effects of antioxidants and L-arginine on oxidation-sensitive gene expression and endothelial nitric oxide synthase activity at the sites of disturbed shear stress. *Proceedings of the National Academy of Sciences USA* 100, 1420–1425.

Delauney, A.J. and Verma, D.P.S. (1993) Proline biosynthesis and osmoregulation in plants. *Plant Journal* 4, 215–223.

DellaPenna, D. and Mene-Safrane, L. (2011) Vitamin E. *Advances in Botanical Research* 59, 179–227.

Diaz-Mula, H.M., Zapata, P.J., Guillen, F., Martinez-Romero, D., Castillo, S., Serrano, M. and Valero, D. (2009) Changes in hydrophilic and lipophilic antioxidant activity and related bioactive compounds during postharvest storage of yellow and purple plum cultivars. *Postharvest Biology and Technology* 51, 354–363.

Dorais, M., Ehret, D.L. and Papadopoulos, A.P. (2008) Tomato *Solanum lycopersicum* health components: from the seed to the consumer. *Phytochemistry Reviews* 7, 231–250.

Dumas, Y., Dadomo, M., Lucca, G.D., Grolier, P. and Di Lucca, G. (2003) Effects of environmental factors and agricultural techniques on antioxidant content of tomatoes. *Journal of the Science of Food and Agriculture* 83, 369–382.

Egea, I., Flores, F.B., Martinez-Madrid, M.C., Romojaro, F. and Sanchez-Bel, P. (2010) 1-Methylcyclopropene affects the antioxidant system of apricots (*Prunus armeniaca* L. cv. Bulida) during storage at low temperature. *Journal of the Science of Food and Agriculture* 90, 549–555.

Exarchou, V. and Gerothanasis, I.P (2006) NMR studies of antioxidant plant phenols: basic parameters and methodologies. In: Boskou D., Gerothanasis, I. and Kefalas, P. (eds) *Antioxidant Plant Phenols*. Research Signpost, Kerala, India, pp. 1–17.

Faria, A. and Calhau, C. (2011) The bioactivity of pomegranate: impact on health and disease. *Critical Reviews in Food Science and Nutrition* 51, 626–634.

Farre, G., Bai, C., Twyman, R.M., Capell, T., Christou, P. and Zhu, C. (2011) Nutritious crops producing multiple carotenoids – a metabolic balancing act. *Trends in Plant Science* 16, 532–540.

Fernie, A.R., Yaakov, T. and Zamir, D. (2006) Natural genetic variation for improving crop quality. *Current Opinion in Plant Biology* 9, 196–202.

Ferretti, G., Bacchetti, T., Belleggia, A. and Neri, D. (2010) Cherry antioxidants: from farm to table. *Molecules* 15, 6993–7005.

Fiorentino, A., D'Abrosca, B., Pacifico, S., Mastellone, C., Scognamiglio, M. and Monaco, P. (2009) Identification and assessment of antioxidant capacity of phytochemicals from kiwifruits. *Journal of Agricultural and Food Chemistry* 57, 4148–4155.

Fitzpatrick, T.B., Basset, G.J.C., Borel, P., Carrari, F., DellaPenna, D., Fraser, P.D., Hellmann, H., Osorio, S., Rothan, C., Valpuesta, V., Caris-Veyrat, C. and Fernie, A.R. (2012) Vitamin deficiencies in humans: can plant science help? *Plant Cell* 24, 395–414.

Flores, F.B., Sanchez-Bel, P., Valdenegro, M., Romojaro, F., Martinez-Madrid, M.C. and Egea, M.I. (2008) Effects of a pretreatment with nitric oxide on peach (*Prunus persica* L.) storage at room temperature. *European Food Research and Technology* 227, 1599–1611.

Flynn, N.E., Meininger, C.J., Haynes, T.E. and Wu, G. (2002) The metabolic basis of arginine nutrition and pharmacotherapy. *Biomedicine and Pharmacotherapy* 56, 427–438.

Fotopoulos, V. and Kanellis, A.K. (2013) Altered apoplastic ascorbate redox state in tobacco plants via ascorbate oxidase overexpression results in delayed dark-induced senescence in detached leaves. *Plant Physiology and Biochemistry* 73, 154–160.

Fotopoulos, V., Sanmartin, M. and Kanellis, A.K. (2006) Effect of ascorbate oxidase over-expression on ascorbate recycling gene expression in response to agents imposing oxidative stress. *Journal of Experimental Botany* 57, 3933–3943.

Fotopoulos, V., De Tullio, M.C., Barnes, J. and Kanellis, A.K. (2008) Altered stomatal dynamics in ascorbate oxidase over-expressing tobacco plants suggest a role for dehydroascorbate signalling. *Journal of Experimental Botany* 59, 729–737.

Frankel, E.N. and Meyer, A.S. (2000) The problem of using one-dimensional methods to evaluate multifunctional food and biological antioxidants. *Journal of the Science of Food and Agriculture* 80, 1925–1941.

Fraser, P.D., Enfissi E.M. and Bramley, P.M. (2009) Genetic engineering of carotenoid formation in tomato fruit and the potential application of systems and synthetic biology approaches. *Archives of Biochemistry and Biophysics* 483, 196–204.

Galanakis, C.M., Goulas, V., Tsakona, M., Manganaris, G.A. and Gekas, V. (2013) A knowledge base for the recovery of natural phenols with different solvents. *International Journal of Food Properties* 16, 382–396.

Galpaz, N., Ronen, G., Khalfa, Z., Zamir, D. and Hirschberg, J. (2006) A chromoplast-specific carotenoid biosynthesis pathway is revealed by cloning of the tomato white-flower locus. *Plant Cell* 18, 1947–1960.

Gao, L. and Mazza, G. (1995) Characterization, quantitation, and distribution of anthocyanins and colorless phenolics in sweet cherries. *Journal of Agricultural and Food Chemistry* 43, 343–346.

Gautier, H., Massot, C., Stevens, R., Serino, S. and Genard, M. (2009) Regulation of tomato fruit ascorbate content is more highly dependent on fruit irradiance than leaf irradiance. *Annals of Botany* 103, 495–504.

Genard, M., Bertin, N., Borel, C., Bussieres, P., Gautier, H., Habib, R., Lechaudel, M., Lecomte, A., Lescourret, F., Lobit, P. and Quilot, B. (2007) Towards a virtual fruit focusing on quality: modelling features and potential uses. *Journal of Experimental Botany* 58, 917–928.

George, B., Kaur, C., Khurdiya, D.S. and Kapoor, H.C. (2004) Antioxidants in tomato (*Lycopersicon esculentum*) as a function of genotype. *Food Chemistry* 84, 45–51.

Gil, M.I., Tomas-Barberan, F.A., Hess-Pierce, B. and Kader, A.A. (2002) Antioxidant capacities, phenolic compounds, carotenoids, and vitamin C contents of nectarine, peach, and plum cultivars from California. *Journal of Agricultural and Food Chemistry* 50, 4976–4982.

Giorio, G., Stigliani, A.L. and D'Ambrosio, C. (2008) Phytoene synthase genes in tomato (*Solanum lycopersicum* L.) – new data on the structures, the deduced amino acid sequences and the expression patterns. *FEBS Journal* 275, 527–535.

Giovannoni, J.J. (2004) Genetic regulation of fruit development and ripening. *Plant Cell* 16, S170–S180.

Goldman, I.L. (2011) Molecular breeding of healthy vegetables. *EMBO Reports* 12, 96–102.

Gonzali, S., Mazzucato, A. and Perata, P. (2009) Purple as a tomato: towards high anthocyanin tomatoes. *Trends in Plant Science* 14, 237–241.

Goulas, V. and Manganaris, G.A. (2011a) Exploring the phytochemical content and the antioxidant potential of *Citrus* fruits grown in Cyprus. *Food Chemistry* 131, 39–47.

Goulas, V. and Manganaris, G.A. (2011b) The effect of postharvest ripening on strawberry bioactive composition and antioxidant potential. *Journal of the Science of Food and Agriculture* 91, 1907–1914.

Goulas, V. and Manganaris, G.A. (2012) Towards an efficient protocol for the determination of triterpenic acids in olive fruit: a comparative study of drying and extraction methods. *Phytochemical Analysis* 23, 444–449.

Graßmann, J. (2005) Terpenoids as Plant Antioxidants. *Vitamins and Hormones* 72, 505–535.

Grillo, M.A. and Colombatto, S. (2004) Arginine revisited: mini review article. *Amino Acids* 26, 345–351.

Hancock, R.D., Walker, P.G., Simon, D.A., Pont, S.D.A., Marqui, N., Vivera, S., Gordon, S.L., Brennan, R.M. and Viola, R. (2007) L-ascorbic acid accumulation in fruit of *Ribes nigrum* occurs by in situ biosynthesis via the L-galactose pathway. *Functional Plant Biology* 34, 1080–1091.

Handa, A.K. and Mattoo, A.K. (2010) Differential and functional interactions emphasize the multiple roles of polyamines in plants. *Plant Physiology and Biochemistry* 48, 540–546.

Hanson, A.D. and Gregory, J.F. III (2011) Folate biosynthesis, turnover, and transport in plants. *Annual Review of Plant Biology* 62, 105–125.

Hasanuzzaman, M., Hossain, M.A. and Fujita, M. (2010) Selenium in higher plants: physiological role, antioxidant metabolism and abiotic stress tolerance. *Journal of Plant Science* 5, 354–375.

Havas, S., Heimendinger, J., Reynolds, K., Baranowski, T., Nicklas, T.A., Bishop, D., Buller, D., Sorensen, G., Beresford, S.A.A., Cowan, A. and Damron, D. (1994) 5 a day for better health: a new research initiative. *Journal of the American Dietetic Association* 94, 32–36.

Hayashi, T., Juliet, P.A., Matsui-Hirai, H., Miyazaki, A., Fukatsu A., Funami, J. *et al.* (2005) L-Citrulline and L-arginine supplementation retards the progression of high-cholesterol-diet-induced atherosclerosis in rabbits. *Proceedings of the National Academy of Sciences USA* 102, 13681–13686.

Ignat, I., Volf, I. and Popa, V.I. (2011) A critical review of methods for characterization of polyphenolic compounds in fruits and vegetables. *Food Chemistry* 126, 1821–1835.

Ilahy, R., Hdider, C., Lenucci, M.S., Tlili, I. and Dalessandro, G. (2011) Antioxidant activity and bioactive compound changes during fruit ripening of high-lycopene tomato cultivars. *Journal of Food Composition and Analysis* 24, 588–595.

Imai, T., Ban, Y., Terakami, S., Yamamoto, T. and Moriguchi, T. (2009) L-Ascorbate biosynthesis in peach: cloning of six L-galactose pathway-related genes and their expression during peach fruit development. *Physiologia Plantarum* 136, 139–149.

Ioannidi, E., Kalamaki, M.S., Pateraki, I., Engineer, C., Alexandrou, D., Giovannonni, J. and Kanellis, A.K. (2009) Expression profiling of ascorbic acid-related genes during tomato fruit development and ripening and in response to various stresses. *Journal of Experimental Botany* 60, 663–678.

Jimenez, A., Creissen, G., Kular, B., Firmin, J., Robinson, S., Verhoeyen, M. and Mullineaux, P. (2002) Changes in oxidative processes and components of the antioxidant system during tomato fruit ripening. *Planta* 214, 751–758.

Kalac, C. and Krausova, P. (2005) A review of dietary polyamines: formation, implications for growth and health and occurrence in foods. *Food Chemistry* 90, 219–230.

Kalamaki, M.S., Alexandrou, D., Lazari, D., Merkouropoulos, G., Fotopoulos, V., Pateraki, I., Aggelis, A., López Carrillo, A., Rubio-Cavetas, M. and Kanellis, A.K. (2009) Over-expression of a tomato *N-acetyl-glutamate synthase* gene in *Arabidopsis thaliana* results in increased tolerance in salt and drought stresses. *Journal of Experimental Botany* 60, 1859–1871.

Kim, H.-S., Park, S.J., Hyun, S.-H., Yang, S.-O., Lee, J., Auh, J.-H., Kim, J.-H., Cho, S.-M., Marriott, P.J. and Choi, H.-K. (2011) Biochemical monitoring of black raspberry (*Rubus coreanus* Miquel) fruits according to maturation stage by 1H NMR using multiple solvent systems. *Food Research International* 44, 1977–1987.

Kim, Y., Brecht, J.K. and Talcott, S.T. (2007) Antioxidant phytochemical and fruit quality changes in mango (*Mangifera indica* L.) following hot water immersion and controlled atmosphere storage. *Food Chemistry* 105, 1327–1334.

Kontogianni, V.G., Exarchou, V., Troganis, A. and Gerothanassis, I.P. (2009) Rapid and novel discrimination and quantification of oleanolic and ursolic acids in complex plant extracts using two-dimensional nuclear magnetic resonance spectroscopy-Comparison with HPLC methods. *Analytica Chimica Acta* 635, 188–195.

Kopsell, D.A. and Kopsell, D.E. (2006) Accumulation and bioavailability of dietary carotenoids in vegetable crops. *Trends in Plant Science* 11, 499–507.

Koyama, K., Ikeda, H., Poudel, P.R. and Goto-Yamamoto, N. (2012) Light quality affects flavonoid biosynthesis in young berries of Cabernet Sauvignon grape. *Phytochemistry* 78, 54–64.

Lansky, E.P. and Newman, R.A. (2007) *Punica granatum* (pomegranate) and its potential for prevention and treatment of inflammation and cancer. *Journal of Ethnopharmacology* 109, 177–206.

Larqué, E., Sabater-Molina, M. and Zamora, S. (2007) Biological significance of dietary polyamines. *Nutrition* 23, 87–95.

Lee, J., Cheng, L.L., Rudell, D.R. and Watkins, C.B. (2012) Antioxidant metabolism of 1-methylcyclopropene (1-MCP) treated 'Empire' apples during controlled atmosphere storage. *Postharvest Biology and Technology* 65, 79–91.

Lee, S.K. and Kader, A.A. (2000) Preharvest and postharvest factors influencing vitamin C content of horticultural crops. *Postharvest Biology and Technology* 20, 207–220.

Leskovar, D.I., Bang, H.J., Crosby, K., Maness, N., Franco, J.A. and Perkins-Veazie, P. (2004) Lycopene, carbohydrates, ascorbic acid, and yield components of diploid and triploid watermelon cultivars are affected by deficit irrigation. *Journal of Horticultural Science and Biotechnology* 79, 75–81.

Lewis, B. and Langkamp-Henken, B. (2000) Arginine enhances in vivo immune responses in young, adult and aged mice. *Journal of Nutrition* 130, 1827–1830.

Li, G., Zhang, X., You, J., Song, C., Sun, Z., Xia, L. and Suo, Y. (2011) Highly sensitive and selective pre-column derivatization high-performance liquid chromatography approach for rapid determination of triterpenes oleanolic and ursolic acids and application to *Swertia* species: optimization of triterpenic acids extraction and pre-column derivatization using response surface methodology. *Analytica Chimica Acta* 688, 208–218.

Li, M., Ma, F., Shang, P., Zhang, M., Hou, C. and Liang, D. (2009) Influence of light on ascorbate formation and metabolism in apple fruits. *Planta* 230, 39–51.

Li, M., Ma, F., Liang, D., Li, J. and Wang, Y. (2010) Ascorbate biosynthesis during early fruit development is the main reason for its accumulation in kiwi. *PLoS ONE* 5, e14281.

Lippman, Z.B., Semel, Y. and Zamir, D. (2007) An integrated view of quantitative trait variation using tomato interspecific introgression lines. *Current Opinion in Genetics and Development* 17, 545–542.

Lombardo, V.A., Osorio, S., Borsani, J., Lauxmann, M.A., Bustamante, C.A., Budde, C.O., Andreo, C.S., Lara, M.V., Fernie, A.R. and Drincovich, M.F. (2011) Metabolic profiling during peach fruit development and ripening reveals the metabolic networks that underpin each developmental stage. *Plant Physiology* 157, 1696–1710.

MacLean, D.D., Murr, D.P., DeEll, J.R. and Horvath, C.R. (2006) Postharvest variation in apple (*Malus × domestica* Borkh.) flavonoids following harvest, storage, and 1-MCP treatment. *Journal of Agricultural and Food Chemistry* 54, 870–878.

Malorgio, F., Diaz, K, Ferrante, A., Mensuali, A. and Pezzarossa, B. (2009) Effects of selenium addition on minimally processed leafy vegetables grown in floating system. *Journal of Science and Food and Agriculture* 89, 2243–2251.

Manach, C., Scalbert, A., Morand, C., Rémésy, C. and Jiménez, L. (2004) Polyphenols: food sources and bioavailability. *American Journal of Clinical Nutrition* 79, 727–747.

Manners, G.D. (2007) Citrus limonoids: analysis, bioactivity, and biomedical prospects. *Journal of Agricultural and Food Chemistry* 55, 8285–8294.

Martí, M.C., Camejo, D., Vallejo, F., Romojaro, F., Bacarizo, S., Palma, J.M., Sevilla, F. and Jiménez, A. (2011) Influence of fruit ripening stage and harvest period on the antioxidant content of sweet pepper cultivars. *Plant Foods and Human Nutrition* 66, 416–423.

Massot, C., Stevens, R., Genard, M., Longuenesse, J.J. and Gautier, H. (2012) Light affects ascorbate content and ascorbate-related gene expression in tomato leaves more than in fruits. *Planta* 235, 153–163.

Mattila, P., Hellström, J. and Törrönen, R. (2006) Phenolic acids in berries, fruits, and beverages. *Journal of Agricultural and Food Chemistry* 54, 7193–7199.

Mattoo, A.K., Minocha, S.C., Minocha, R. and Handa, A.K. (2010) Polyamines and cellular metabolism in plants: transgenic approaches reveal different responses to diamine putrescine versus higher polyamines spermidine and spermine. *Amino Acids* 38, 405–413.

Mehta, R.A., Cassol, T., Li, N., Ali, N., Handa, A.K. and Mattoo, A.K. (2002) Engineered polyamine accumulation in tomato enhances phytonutrient content, juice quality, and vine life. *Nature Biotechnology* 20, 613–618.

Mélendez-Martinez, A.J., Fraser P.D. and Bramley, P.M. (2010) Accumulation of health promoting phytochemicals in wild relatives of tomato and their contribution to *in vitro* antioxidant activity. *Phytochemistry* 71, 1104–1114.

Mellidou, I., Keulemans, J., Kanellis, A.K. and Davey, M.W. (2012) Regulation of fruit ascorbic acid concentrations during ripening in high and low vitamin C tomato cultivars. *BMC Plant Biology* 12, 239.

Meyer, M.D., Landahl, S., Donetti, M. and Terry, L.A. (2011) Avocado. In: Terry, L.A. (ed.) *Health-promoting Properties of Fruits and Vegetables*. CAB International, Wallingford, UK, pp. 27–50.

Milella, L., Caruso, M., Galgano, F., Favati, F., Padula, M.C. and Martelli, G. (2011) Role of the cultivar in choosing Clementine fruits with a high level of health-promoting compounds. *Journal of Agricultural and Food Chemistry* 59, 5293–5298.

Mozetic, B., Simcic, M. and Trebse, P. (2006) Anthocyanins and hydroxycinnamic acids of Lambert Compact cherries (*Prunus avium* L.) after cold storage and 1-methylcyclopropene treatment. *Food Chemistry* 97, 302–309.

Nagata, M. and Yamashita, I. (1992) Simple method for simultaneous determination of chlorophyll and carotenoids in tomato fruit. *Nippon Shokuhin Kogyo Gakkai-Shi* 39, 925–928.

Nambeesan, S., Datsenka, T., Ferruzzi, M.G., Malladi, A., Mattoo, A.K. and Handa, A.K. (2010) Overexpression of yeast spermidine synthase impacts ripening, senescence and decay symptoms in tomato. *Plant Journal* 63, 836–847.

Niederlander, H.A.G., van Beek, T.A., Bartasiute, A. and Koieva, I.I. (2008) Antioxidant activity assays on-line with liquid chromatography. *Journal of Chromatography A* 1210, 121–134.

Nishiyama, I., Yamashita, Y., Yamanaka, M., Shimohashi, A., Fukuda, T. and Oota, T. (2004) Varietal difference in vitamin C content in the fruit of kiwifruit and other *Actinidia* species. *Journal of Agricultural and Food Chemistry* 52, 5472–5475.

Obied, H.K., Allen, M.S., Bedgood, D.R., Prenzler, P.D. and Robards, K. (2005) Investigation of Australian olive mill waste for recovery of biophenols. *Journal of Agricultural and Food Chemistry* 53, 9911–9920.

Olsson, M.E., Ekvall, J., Gustavsson, K., Nilsson, J., Pillai, D., Sjoholm, I., Svensson, U., Akesson, B. and Nyman, M.G.L. (2004) Antioxidants, low molecular weight carbohydrates, and total antioxidant capacity in strawberry (*Fragaria* × *ananassa*): effects of cultivar, ripening, and storage. *Journal of Agricultural and Food Chemistry* 52, 2490–2498.

Palmer, R.M.J., Ashton, D. and Moncada, S. (1988) Vascular endothelial cells synthesize nitric oxide from L-arginine. *Nature* 333, 664–666.

Pantelidis, G.E., Vasilakakis, M., Manganaris, G.A. and Diamantidis, G. (2007) Antioxidant capacity, phenol, anthocyanin and ascorbic acid contents in raspberries, blackberries, red currants, gooseberries and cornelian cherries. *Food Chemistry* 102, 777–783.

Pateraki, I., Sanmartin, A.M., Kalamaki, M.S., Gerasopoulos, D. and Kanellis, A.K. (2004) Molecular characterization and expression studies during melon fruit development and ripening of L-galactono-1,4-lactone dehydrogenase. *Journal of Experimental Botany* 55, 1623–1633.

Pedrero, Z. and Madrid, Y. (2009) Novel approaches for selenium speciation in foodstuffs and biological specimens: a review. *Analytica Chimica Acta* 634, 135–152.

Perkins-Veazie, P., Collins, J.K., Pair, S.D. and Roberts, W. (2001) Lycopene content differs among red-fleshed watermelon cultivars. *Journal of the Science of Food and Agriculture* 81, 983–987.

Perkins-Veazie, P., Collins, J.K., Davis, A.R. and Roberts, W. (2006) Carotenoid content of 50 watermelon cultivars. *Journal of Agricultural and Food Chemistry* 54, 2593–2597.

Pezzarossa, B., Malorgio, F. and Tonutti, P. (1999) Effects of selenium uptake by tomato plants on senescence, fruit ripening and ethylene evolution. In: Kanellis, A.K., Chang, C., Klee, H., Bleecker, A.B., Pech, J.C. and Grierson, D. (eds) *Biology and Biotechnology of the Plant Hormone Ethylene II.* Kluwer, Dordrecht, pp. 275–276.

Pezzarossa, B., Remorini, D., Gentileb, M.L. and Massaib, R. (2012) Effects of foliar and fruit addition of sodium selenate on selenium accumulation and fruit quality. *Journal of Science and Food and Agriculture* 92, 781–786.

Proestos, C., Sereli, D. and Komaitis, M. (2006) Determination of phenolic compounds in aromatic plants by RP-HPLC and GC-MS. *Food Chemistry* 95, 44–52.

Pucciariello, C., Parlanti, S., Banti, V., Novi, G. and Perata, P. (2012) Reactive oxygen species-driven transcription in Arabidopsis under oxygen deprivation. *Plant Physiology* 159, 184–196.

Puerta-Gomez, A.F. and Cisneros-Zevallos, L. (2011) Postharvest studies beyond fresh market eating quality: phytochemical antioxidant changes in peach and plum fruit during ripening and advanced senescence. *Postharvest Biology and Technology* 60, 220–224.

Qiu, S.S., Lu, C.W., Li, X.H. and Toivonen, P.M.A. (2009) Effect of 1-MCP on quality and antioxidant capacity of *in vitro* digests from 'Sunrise' apples stored at different temperatures. *Food Research International* 42, 337–342.

Rabe, E. and Lovatt, C.J. (1986) Increased arginine biosynthesis during phosphorus deficiency, a response to the increased ammonia content of leaves. *Plant Physiology* 81, 774–779.

Radomski, M.W., Palmer, R.M.J. and Moncada, S. (1990) An L-arginine/nitric oxide pathway present in human platelets regulates aggregation, *Proceedings of the National Academy of Sciences USA* 87, 5193–5197.

Rice-Evans, C.A., Miller, N.J. and Paganga, G. (1997) Antioxidant properties of phenolic compounds. *Trends in Plant Science* 2, 152–159.

Rivera-Pastrana, D.M., Yahia, E.M. and Gonzalez-Aguilar, G.A. (2010) Phenolic and carotenoid profiles of papaya fruit (*Carica papaya* L.) and their contents under low temperature storage. *Journal of the Science of Food and Agriculture* 90, 2358–2365.

Rohrmann, J., Tohge, T., Alba, R., Osorio, S., Caldana, C., McQuinn, R., Arvidsson, S., Van Der Merwe, M.J., Riaño-Pachón, D.M., Mueller-Roeber, B., Fei, Z., Nesi, A.N., Giovannoni, J.J. and Fernie, A.R. (2011) Combined transcription factor profiling, microarray analysis and metabolite profiling reveals the transcriptional control of metabolic shifts occurring during tomato fruit development. *Plant Journal* 68, 999–1013.

Rudell, D.R., Buchanan, D.A., Leisso, R.S., Whitaker, B.D., Mattheis, J.P., Zhu, Y.M. and Varanasi, V. (2011) Ripening, storage temperature, ethylene action, and oxidative stress alter apple peel phytosterol metabolism. *Phytochemistry* 72, 1328–1340.

Ruiz-García, Y., Romero-Cascales, I., Gil-Muñoz, R., Fernández-Fernández, J.I., López-Roca, J.M. and Gómez-Plaza, E. (2012) Improving grape phenolic content and wine chromatic characteristics through the use of two different elicitors: methyl jasmonate versus benzothiadiazole. *Journal of Agricultural and Food Chemistry* 60, 1283–1290.

Sacchetti, G., Maietti, S., Muzzoli, M., Scaglianti, M., Manfredini, S., Radice, M. and Bruni, R. (2005) Comparative evaluation of 11 essential oils of different origin as functional antioxidants, antiradicals and antimicrobials in foods. *Food Chemistry* 91, 621–632.

Salunkhe, D.K., Bolin, H.R. and Reddy, N.R. (1991) *Storage, Processing, and Nutritional Quality of Fruits and Vegetables. Volume I. Fresh Fruits and Vegetables.* CRC Press, Boston, MA.

Sancho, L.E.G.G., Yahia, E.M. and Gonzalez-Aguilar, G.A. (2011) Identification and quantification of phenols, carotenoids, and vitamin C from papaya (*Carica papaya* L., cv. Maradol) fruit determined by HPLC-DAD-MS/MS-ESI. *Food Research International* 44, 1284–1291.

Sanmartin, I., Wanntorp, L. and Winkworth, R.C. (2007) West Wind Drift revisited: testing for directional dispersal in the Southern Hemisphere using event-based tree fitting. *Journal of Biogeography* 34, 398–416.

Sanmartin, M., Drogoudi, P.D., Lyons, T., Pateraki, I., Barnes, J. and Kanellis, A.K. (2003) Over-expression of ascorbate oxidase in the apoplast of transgenic tobacco results in altered ascorbate and glutathione redox states and increased sensitivity to ozone. *Planta* 216, 918–928.

Saura-Calixto, F. and Goni, I. (2009) Definition of the Mediterranean diet based on bioactive compounds. *Critical Reviews in Food Science and Nutrition* 49, 145–152.

Schijlen, E., de Vos C.H.R., Jonker, H., van den Broeck, H., Molthoff, J., van Tunen, A., Martens, S. and Bovy, A. (2006) Pathway engineering for healthy phytochemicals leading to the production of novel flavonoids in tomato fruit. *Plant Biotechnology Journal* 4, 433–444.

Schreiner, M. and Huyskens-Keil, S. (2006) Phytochemicals in fruit and vegetables: health promotion and postharvest elicitors. *Critical Reviews in Plant Sciences* 25, 267–278.

Seeram, N.P., Adams, L.S., Zhang, Y.J., Lee, R., Sand, D., Scheuller, H.S. and Heber, D. (2006) Blackberry, black raspberry, blueberry, cranberry, red raspberry, and strawberry extracts inhibit growth and stimulate apoptosis of human cancer cells in vitro. *Journal of Agricultural and Food Chemistry* 54, 9329–9339.

Serrano, M., Guillean, F.N, Martinez-Romero, D., Castillo, S. and Valero, D. (2005) Chemical constituents and antioxidant activity of sweet cherry at different ripening stages. *Journal of Agricultural and Food Chemistry* 53, 2741–2745.

Serrano, M., Díaz-Mula, H.D. and Valero, D. (2011) Antioxidant compounds in fruits and vegetables and changes during postharvest storage and processing. *Stewart Postharvest Review* 7, 1–10.

Silva, F.J.P., Gomes, M.H., Fidalgo, F., Rodrigues, J.A. and Almeida, D.P.F. (2010) Antioxidant properties and fruit quality during long-term storage of 'Rocha' pear: effects of maturity and storage conditions. *Journal of Food Quality* 33, 1–20.

Simonne, A.H. and Ritenour, M.A. (2011) Citrus. In: Terry, L.A. (ed.) *Health-promoting Properties of Fruits and Vegetables.* CAB International, Wallingford, UK, pp. 90–117.

Singh, R. and Dwivedi, U.N. (2008) Effect of ethrel and 1-methylcyclopropene (1-MCP) on antioxidants in mango (*Mangifera indica* var. Dashehari) during fruit ripening. *Food Chemistry* 111, 951–956.

Singh, R., Rastogi, S., and Dwivedi, U.N. (2010) Phenylpropanoid metabolism in ripening fruits. *Comprehensive Reviews in Food Science and Food Safety* 9, 398–416.

Smirnoff, N. (2011) Vitamin C: the metabolism and functions of ascorbic acid in plants. *Advances in Botanical Research*, 59, 107–177.

Sorensen, G., Stoddard, A., Peterson, K., Cohen, N., Hunt, M.K., Stein, E., Palombo, R. and Lederman, R. (1999) Increasing fruit and vegetable consumption through worksites and families in the Treatwell 5-a-Day study. *American Journal of Public Health* 89, 54–60.

Stevens, R., Buret, M., Duffe, P., Garchery, C., Baldet, P., Rothan, C. and Causse, M. (2007) Candidate genes and quantitative trait loci affecting fruit ascorbic acid content in three tomato populations. *Plant Physiology* 143, 1943–1953.

Stevens, R., Page, D., Gouble, B., Garchery, C., Zamir D. and Causse, M. (2008) Tomato fruit ascorbic acid content is linked with monodehydroascorbate reductase activity and tolerance to chilling stress. *Plant, Cell and Environment* 31, 1086–1096.

Struik, P. C., Yin, X. Y. and de Visser, P. (2005) Complex quality traits: now time to model. *Trends in Plant Science* 10, 513–516.

Szajdek, A. and Borowska, E.J. (2008) Bioactive compounds and health-promoting properties of berry fruits: a review. *Plant Foods for Human Nutrition* 63, 147–156.

Tabart, J., Kevers, C., Pincemail, J., Defraigne, J.-O. and Dommes, J. (2010) Evaluation of spectrophotometric methods for antioxidant compound measurement in relation to total antioxidant capacity in beverages. *Food Chemistry* 120, 607–614.

Tadmor, Y., Burger, J., Yaakov, I., Feder, A., Libhaber, S.E., Portnoy, V., Meir, A., Tzuri, G., Saar, U., Rogachev, I., Aharoni, A., Abeliovich, H., Schaffer, A.A., Lewinsohn, E. and Kutzir, N. (2010) Genetics of flavonoid, carotenoid, and chlorophyll pigments in melon fruit rinds. *Journal of Agricultural and Food Chemistry* 58, 10722–10728.

Takahara, K., Akashi, K. and Yokota, A. (2005) Purification and characterization of glutamate *N*-acetyltransferase involved in citrulline accumulation in wild watermelon. *FEBS Journal* 272, 5353–5364.

Tanaka, Y. and Ohmiya, A. (2008) Seeing is believing: engineering anthocyanin and carotenoid biosynthetic pathways. *Current Opinions in Biotechnology* 19, 190–197.

Terrier, N., Torregrosa, L., Ageorges, A., Vialet, S., Verries, C., Cheynier, V. and Romieu, C. (2009)

Ectopic expression of *VvMybPA2* promotes proanthocyanidin biosynthesis in grapevine and suggests additional targets in the pathway. *Plant Physiology* 149, 1028–1041.

Terry, L.A. (2011) *Health-promoting Properties of Fruits and Vegetables*. CAB International, Wallingford, UK.

Terry, L.A. and Thompson, A.K. (2011) Introduction. In: Terry, L.A. (ed) *Health-promoting Properties of Fruits and Vegetables*. CAB International, Wallingford, UK, pp. 27–50.

Tiburcio, A.F., Besford, R.T., Capell, T., Borrell, A., Testillano, P.S. and Risueño, M.C. (1994) Mechanisms of polyamine action during senescence responses induced by osmotic stress. *Journal of Experimental Botany* 45, 1789–1800.

Tlili, I., Hdider, C., Lenucci, M.S., Ilahy, R., Jebari, H. and Dalessandro, G. (2011) Bioactive compounds and antioxidant activities during fruit ripening of watermelon cultivars. *Journal of Food Composition and Analysis* 24, 923–928.

Tomas-Barberan, F.A., Gil, M.I., Cremin, P., Waterhouse, A.L., Hess-Pierce, B. and Kader, A.A. (2001) HPLC-DAD-ESIMS analysis of phenolic compounds in nectarines, peaches, and plums. *Journal of Agricultural and Food Chemistry* 49, 4748–4760.

Tripoli, E., La Guardia, M., Giammanco, S., Di Majo, D. and Giammanco, M. (2007) Citrus flavonoids: molecular structure, biological activity and nutritional properties: a review. *Food Chemistry* 104, 466–479.

Trojanowicz, M. (2011) Recent developments in electrochemical flow detections – a review: Part II. Liquid chromatography. *Analytica Chimica Acta* 688, 8–35.

Tulipani, S., Mezzetti, B., Capocas, F., Bompadre, S., Beekwilder, J., De Vos, C.H.R., Capanoglu, E., Bovy, A., and Battino, M. (2008) Antioxidants, phenolic compounds, and nutritional quality of different strawberry genotypes. *Journal of Agricultural and Food Chemistry* 56, 696–704.

Vallverdú-Queralt, A., Oms-Oliu, G., Odriozola-Serrano, I., Lamuela-Raventos, R.M., Martín-Belloso, O. and Elez-Martínez, P. (2012) Effects of pulsed electric fields on the bioactive compound content and antioxidant capacity of tomato fruit. *Journal of Agricultural and Food Chemistry* 60, 3126–3134.

Vicente, A.R., Manganaris, G.A., Sozzi, G.O. and Crisosto, C.H. (2009) Nutritional quality of fruits and vegetables. In: Florokowski, W.J., Shewfelt, R.L., Brueckner, B. and Prusia, S.E. (eds) *Postharvest Handling: A Systems Approach*. Academic Press, San Diego, CA, pp. 57–106.

Vigneault, C., Leblanc, D.I., Goyette, B. and Jenni, S. (2012) Engineering aspects of physical treatments to increase fruit and vegetable phytochemical content. *Canadian Journal of Plant Science* 92, 373–397.

Villa-Rodriguez, J.A., Molina-Corral, F.J., Ayala-Zavala, J.F., Olivas, G.I. and Gonzalez-Aguilar, G.A. (2011) Effect of maturity stage on the content of fatty acids and antioxidant activity of 'Hass' avocado. *Food Research International* 44, 1231–1237.

Vogt, T. (2010) Phenylpropanoid biosynthesis. *Molecular Plant* 3, 2–20.

Waller, J.C., Akhtar, T.A., Lara-Nunez, A., Gregory, J.F. III, McQuinn, R.P., Giovannoni, J.J. and Hanson, A.D. (2010) Developmental and feedforward control of the expression of folate biosynthesis genes in tomato fruit. *Molecular Plant* 3, 66–77.

Wang, B.G., Wang, J.H., Feng, X.Y., Lin, L., Zhao, Y.M. and Jiang, W.B. (2009a) Effects of 1-MCP and exogenous ethylene on fruit ripening and antioxidants in stored mango. *Plant Growth Regulation* 57, 185–192.

Wang, S.Y., Chen, C.T. and Wang, C.Y. (2009b) The influence of light and maturity on fruit quality and flavonoid content of red raspberries. *Food Chemistry* 112, 676–684.

White, P.J. and Broadley, M.R. (2009) Biofortification of crops with seven mineral elements often lacking in human diets – iron, zinc, copper, calcium, magnesium, selenium and iodine. *New Phytologist* 182, 49–84.

Willits, M.G., Kramer, C.M., Prata, R.T.N., De Luca, V., Potter, B.G., Stephens, J.C. and Graser, G. (2005) Utilization of the genetic resources of wild species to create a nontransgenic high flavonoid tomato. *Journal of Agricultural and Food Chemistry* 53, 1231–1236.

Wu, G., Meininger, C.J., Knabe, D.A., Bazer, F.W. and Rhoads, J.M. (2000) Arginine nutrition in development, health and disease. *Current Opinion in Clinical Nutrition and Metabolic Care* 3, 59–66.

Xie, D.Y. and Dixon, R.A. (2005) Proanthocyanidin biosynthesis: still more questions than answers? *Phytochemistry* 66, 2127–2144.

Xie, X.-B., Li, S., Zhang, R.-F., Zhao, J., Chen Y.-C., Zhao, Q., Yao Y.-X., You, C.-X., Zhang X.-S. and Hao, Y.-J. (2012) The bHLH transcription factor *MdbHLH3* promotes anthocyanin accumulation and fruit colouration in response to low temperature in apples. *Plant Cell Physiology* 35, 1884–1897.

Yahia, E.M., Soto-Zamora, G., Brecht, J.K.and Gardea, A. (2007) Postharvest hot air treatment effects on the antioxidant system in stored mature-green tomatoes. *Postharvest Biology and Technology* 44, 107–115.

Yazdani, N., Arzani, K., Mostofi, Y. and Shekarchi, M. (2011) α-Farnesene and antioxidative enzyme systems in Asian pear (*Pyrus serotina* Rehd.) fruit. *Postharvest Biology and Technology* 59, 227–231.

Yu, J., Wang, L., Walzem, R.L., Miller, E.G., Pike, L.M. and Patil, B.S. (2005) Antioxidant activity of citrus limonoids, flavonoids, and coumarins. *Journal of Agricultural and Food Chemistry* 53, 2009–2014.

Zaharah, S.S. and Singh, Z. (2011) Postharvest nitric oxide fumigation alleviates chilling injury, delays fruit ripening and maintains quality in cold-stored 'Kensington Pride' mango. *Postharvest Biology and Technology* 60, 202–210.

Zhao F.-J. and Shewry, P.R. (2011) Recent developments in modifying crops and agronomic practice to improve human health. *Food Policy* 96, S94–S101.

Zifkin, M., Jin, A., Ozga, J.A., Zaharia, L.I., Schernthaner, J.P., Gesell, A., Abrams, S.R., Kennedy, J.A. and Constabel, C.P. (2012) Gene expression and metabolite profiling of developing highbush blueberry fruit indicates transcriptional regulation of flavonoid metabolism and activation of abscisic acid metabolism. *Plant Physiology* 158, 200–224.

8 Vitamins in Fleshy Fruits

Pierre Baldet, Carine Ferrand and Christophe Rothan*
INRA, Villenave d'Ornon, France; and University of Bordeaux, Villenave d'Ornon, France

8.1 Introduction

Grains (rice, wheat, maize) and tubers (potato, cassava) constitute the staple foods in most human diets. They provide major nutrients required for sustaining a healthy and productive life in humans, such as polysaccharides, lipids and proteins. Cereal and tuber-based diets, however, need to be diversified by the addition of a variety of foodstuffs, such as legumes, vegetables and fruits, which will add micronutrients to the staple food. Micronutrients are essential dietary elements required in very small quantities and not synthesized by humans. They comprise minerals (e.g. selenium, zinc) and vitamins, deficiencies of which are responsible for numerous human diseases.

The crucial role in the human diet of fresh fruits and vegetables for preventing various diseases has been known for centuries. The well-known example of the discovery of the role of L-ascorbic acid, or vitamin C, illustrates how the link between a diet including fleshy fruits, the prevention of scurvy and the intake of vitamin C was made. Scurvy is a fatal disease that frequently affected sailors on long voyages, or soldiers, whose diet consisted mostly of bread, meat or beans. As early as 1737, a doctor from the Austrian armies named Kramer wrote a report indicating that intake of fruits and vegetables would prevent scurvy affecting the soldiers. Later, in 1747, a Scottish doctor named James Lind undertook one of the first controlled and scientific experiments carried out at that time, aimed at identifying which diet components could cure scurvy. His recommendations for including citrus fruits in the sailor's diet were initially ignored, but the British navy eventually added limes to the food stores on board, which effectively prevented scurvy. It took almost three centuries (1932) until the antiscorbutic factor, named vitamin C, was purified by Dr Szent-Gyorgyi from pepper fruit, which resulted in him being awarded a Nobel Prize in 1937. Meanwhile, whilst working on beriberi, a vitamin B1-deficiency disease, Casimir Funk (1912) developed the concept of vitamins. Vitamins refer to non-mineral micronutrients essential for human life: *vita* is the Latin word for life and, at that time, Funk thought that all vitamins were amines, which is not the case. The final 'e' was later dropped in English to acknowledge this fact.

Given their utmost importance in preventing a wide range of human diseases, decades of research have been devoted to vitamins since their first

* christophe.rothan@bordeaux.inra.fr

 Fruit Ripening: Physiology, Signalling and Genomics (eds P. Nath *et al.*)

discovery. This led to the identification of 13 organic micronutrients that are required in the human diet and which are therefore classified as vitamins (Fitzpatrick *et al.*, 2012). They comprise fat-soluble vitamins (vitamins A, D, E and K) and water-soluble vitamins (vitamin B complex: B1, B2, B3, B5, B6, B8, B9 and B12 and vitamin C). A wealth of studies has also underscored the prominent role played by micronutrients, including vitamins, in promoting human health. This led to the establishment of recommendations for minimum intake per day, the recommended daily allowance (RDA), for each vitamin in order to avoid vitamin deficiency (Table 8.1). The RDA may vary according to the needs of the individuals, which may depend on their age and physiological stage: for most vitamins, the requirements are not the same for children and adults or for lactating females and adult males. The RDA for several vitamins is regularly updated according to advances in food and nutrition research. As an example, the RDA for vitamin C was recently revised (in 2000) in the USA from the previous recommendation of 60 to 120 mg day^{-1} for breast-feeding women. The prevention of vitamin deficiency diseases, among others, led to the recent establishment of dietary guidelines by various national governments and by the FAO/WHO (2004). These guidelines recommend eating five or more servings of fruits and vegetables a day, including two to four servings of fruit (a fruit serving is equal to one piece of apple or a half cup of sliced fruit, for example). Whilst these objectives can easily be reached in western countries, in which a variety of fruits and vegetables are available for diversifying the staple diet, the intake of food with high micronutrient density is more problematic for populations subsisting mostly on grain- or tuber-based diets. As an alternative, food can be biofortified with vitamins. However, it has been shown, for example for vitamin C, that supplements are less efficient than vitamins taken from plant products (Inoue *et al.*, 2008), probably because of the synergistic effect between vitamins and other phytonutrients.

Due to the wide diversity of fleshy fruit species, the broad climatic conditions in which they can grow and the large variability in their micronutrient composition, a large panel of fleshy fruits can provide appreciable levels of vitamins in many parts of the world (Table 8.1). As an example, the recommended five servings of fruits and vegetables may provide >200 mg of vitamin C, which is the RDA proposed by some authors for preventing cancer (Levine *et al.*, 1996). Fruits such as mango, avocado and guava can be major sources of vitamin A, vitamins E and K, and vitamin C, respectively. In addition, fruits less rich in vitamins but available all year round in many countries, like tomato, may cover a significant part of the recommended intake for several vitamins. For a given species, wide variations in vitamin content are usually found by screening cultivated germplasm or related species, thus providing potential targets for improving fruit vitamin content by genetic means. As well as the widespread cultivated fleshy fruit species shown in Table 8.1, biochemical screening has revealed the high micronutrient density of other fleshy fruit species found in the wild and/or cultivated locally (e.g. Amazonian fruits; Rodriguez-Amaya, 1999; Justi *et al.*, 2000). These fruits can become important sources of vitamins locally or can be exploited for their nutritional value, as is the case for the Amazonian fruits acerola (*Malpighia glabra*) and camu-camu (*Myrceria dubia*), which are rich in vitamin C (Justi *et al.*, 2000; Badejo *et al.*, 2009). Additional factors affecting fruit nutritional value that need to be considered are which part of the fruit is edible (for example, vitamin C accumulates mostly in the peel in many fruit species), preharvest environmental conditions and cultural practices (for example, increasing light intensity will increase both provitamin A and vitamin C content, whilst high nitrogen may reduce

Table 8.1. Main vitamins in fruits: usual names, active forms, some fruit sources, content and recommended daily allowance (RDA). Data from the US Department of Agriculture (http://www.nal.usda.gov/fnic/foodcomp/search).

Vitamin/ common name	Metabolic forms (present in plants)	Main fruit sources	Fruit content (per 100 g)	RDA
Vitamin A/retinol	β-carotene, β-cryptoxanthin	Passion fruit Mango Guava Apple Tomato	907 RE 476 RE 309 RE 30 RE 5 RE	Children 400–700 μg RE, adult males 1000 μg RE, adult females 800 μg RE, lactating females 1200 μg RE
Vitamin E/ tocopherol, tocotrienol	α-Tocopherol	Avocado Mango Guava Kiwifruit	4.16 mg 2.32 mg 1.2 mg 1.01 mg	Children 6–11 mg, most adults 22.5 mg
Vitamin K/ phylloquinone	2-Methyl-3-phytyl-1,4-naphtoquinone	Avocado Blueberry Kiwifruit Tomato Mango	0.42 mg 0.03 mg 0.03 mg 0.01 mg 0.01 mg	Children 30–40 μg, adult males 70–80 μg, adult females 60–65 μg
Vitamin B1/ thiamin	Thiamine pyrophosphate, thiamine triphosphate	Breadfruit Avocado Mango Orange Guava	0.24 mg 0.13 mg 0.12 mg 0.11 mg 0.11 mg	Children 0.6–0.9 mg, adult males 1.2 mg, adult females 1.1 mg, lactating females 1.5 mg
Vitamin B2/ riboflavin	Flavine mononucleotide, flavine adenine dinucleotide	Passion fruit Avocado Mango Banana	0.31 mg 0.26 mg 0.12 mg 0.09 mg	Children 0.6–0.9 mg, adult males 1.2 mg, adult females 1.1 mg, lactating females 1.5 mg
Vitamin B3/ niacin (nicotinic acid, nicotinamide)	Nicotinamide adenine dinucleotide, nicotinamide adenine dinucleotide phosphate	Passion fruit Avocado Breadfruit Guava Mango	3.54 mg 3.49 mg 1.98 mg 1.79 mg 1.21 mg	Children 9–16 mg, adult males 16 mg, adult females 14 mg, lactating females 18 mg
Vitamin B5/ panthothenate	Panthothenate	Avocado Breadfruit Guava	2.79 mg 1.05 mg 0.74 mg	Children 2–4 mg, adults 5 mg, lactating females 6-7 mg
Vitamin B6/ pyridoxine	Pyridoxine phosphate, pyridoxamine phosphate, pyridoxal phosphate	Avocado Banana Passion fruit Mango	0.52 mg 0.43 mg 0.24 mg 0.23 mg	Children 0.6–1.3 mg, adults 1.3–1.7 mg, lactating females 2 mg
Vitamin B9/ folates	Tetrahydrofolates	Avocado Guava Strawberry	0.16 mg 0.08 mg 0.03 mg	Children 45–50 mg, adults 60 mg, lactating females 95 mg
Vitamin C/ ascorbate	Ascorbic acid, dehydroascorbic acid	Guava Strawberry Orange Kiwifruit Mango Tomato	376.7 mg 84.7 mg 69.7 mg 63.8 mg 57.3 mg 30 mg	Children 40–75 mg, adult males 90 mg, adult females 75 mg, lactating females 120 mg

RE, Retinol equivalent. 1 RE = 1 μg retinol, 2 μg β-carotene dissolved in oil.

vitamin C content), postharvest storage conditions (reviewed by Lee and Kader, 2000) and whether the product is consumed raw or transformed by cooking (vitamin C is heat labile, but cooking can increase the bioavailability of carotenoids).

In this chapter, we will focus on three vitamins for which fleshy fruits are important sources in the human diet: two fat-soluble vitamins, the provitamin A carotenoids, which are precursors of vitamin A (retinoids), and vitamin E (tocopherols), and the water-soluble vitamin C (ascorbate). State-of-the-art research on vitamins in plants and their potential benefits in promoting human health have been outlined in a recent review (Fitzpatrick *et al.*, 2012). The reader can also refer to two recent volumes of *Advances in Botanical Research* dedicated to vitamins in plants (Rébeillé and Douce, 2011a,b) for more detailed knowledge on specific vitamins. In the first part of the chapter, a general overview will be given for each vitamin: its physiological role in plants, the biosynthetic pathways in fleshy fruits and regulation by genetic factors, and environmental cues in terms of fruit development and during postharvest storage. In the second part, we will review the various strategies currently being undertaken to improve the vitamin content of fleshy fruits.

8.2 Provitamin A Biosynthesis and Regulation in Fleshy Fruits

Vitamin A comprises several compounds, among which the most abundant and active is retinol. In plants and fruits, several precursors of retinol known as provitamin A can be found. They mainly include α- and β-carotene and β-cryptoxanthin, which are orange-red carotenoids found in chloroplasts and chromoplasts. Plant carotenoids are C40 isoprenoids with polyene chains that may contain up to 15 conjugated double bonds. Carotenoids provide most of the provitamin A in the diet, with β-carotene being converted the most efficiently to vitamin A.

8.2.1 Role of vitamin A in humans and plants

Provitamin A carotenoid pigments found in fruits are embedded in chloroplasts and chromoplasts, which are complex cellular structures. During digestion, highly lipophilic carotenoids are freed from embedded food matrices and incorporated into lipid-containing water-miscible micellar solutions, a process that permits their passage into the lipid-rich membrane of intestinal mucosal cells. Fruit processing (e.g. thermal treatments, high-pressure or microwave preservation), which disrupts fruit chromoplasts and helps the constitution of emulsions, contributes to increase the bioaccessibility of provitamin A (Svelander *et al.*, 2011).

Vitamin A is essential to humans and is needed in small amounts for the visual system, in the retina of the eye, for normal development and growth, for differentiation of epithelial cells and for immune function. Vitamin A deficiency is frequent in economically deprived populations, in which it is the major cause of blindness and one of the important reasons for child mortality. Deficiency in developed countries, although less acute and frequent, can be responsible for night blindness. Other human diseases triggered by vitamin A deficiency are mostly linked to its effect on epithelial cells (e.g. increased sensitivity to pathogens, reduced immunity).

In plants, carotenoids are essential components of the photosynthetic system, in which they contribute to the stabilization of lipid membranes and to light harvesting for photosynthesis. Because of their chemical properties, they have a crucial function in protecting the photosynthetic apparatus against chlorophyll bleaching in intense light. In addition, carotenoids are precursors of the plant hormone abscisic acid (ABA). Carotenoids are also natural pigments acting as colouring agents in flowers and fruits, where their function is to attract animals to disseminate the seeds.

8.2.2 Provitamin A in fleshy fruits

Provitamin A carotenoids (α- and β-carotene, β-cryptoxanthin; Fig. 8.1) are found mostly in yellow and orange fruits, in which they can accumulate in the chromoplasts, plastids specialized in pigment synthesis and storage. Many widely consumed fleshy fruits such as passion fruit, mango and guava (Table 8.1) are rich in carotenoids and can be seasonally important sources of provitamin A in many countries. The fleshy mesocarp of palm fruit (*Elaeis guineensis*), the main oil crop in the world, also represents an interesting source of provitamin A (Tranbarger *et al.*, 2011) as it provides, at the same time, the provitamin A carotenoids and the lipids that increase their bioavailability. Many other fruit

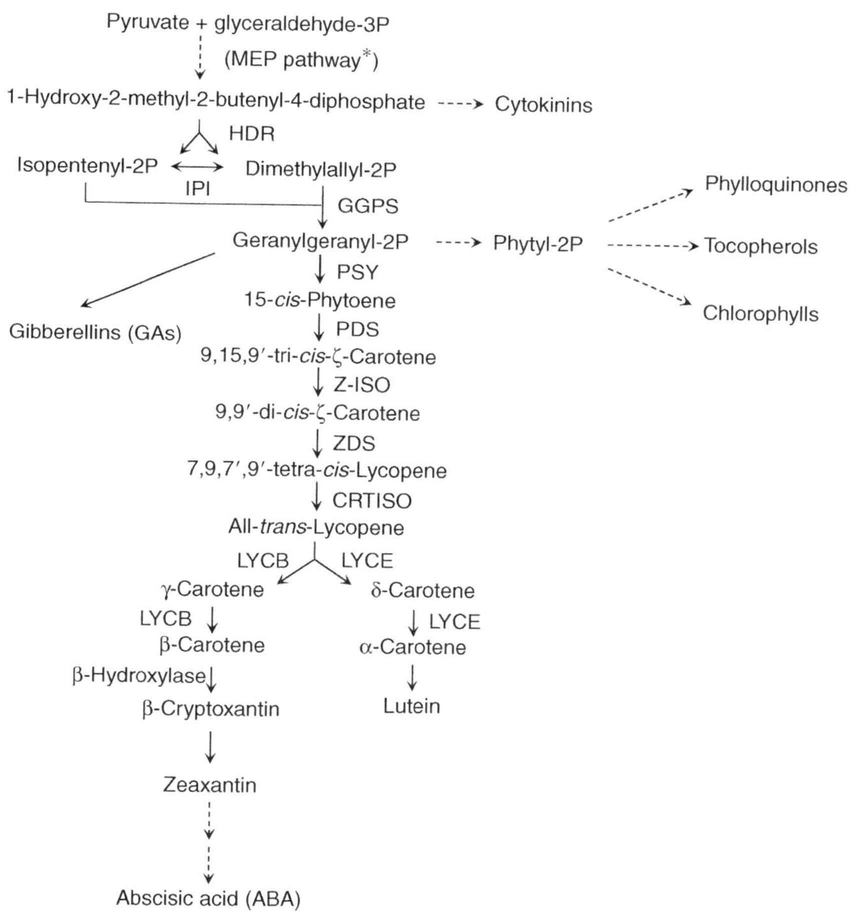

Fig. 8.1. The carotenoid biosynthetic pathway (provitamin A) in plants and related metabolisms: cytokinins, tocopherols, chlorophyll, phylloquinones (vitamin K), abscissic acid (ABA) and gibberellins (GA). The corresponding enzymes and metabolites are: HDR, 4-hydroxy-3-methylbut-2-enyl-diphosphate reductase; IPI, isopentenyl diphosphate isomerase; GGPS, geranylgeranyl pyrophosphate synthase; PSY, phytoene synthase; PDS, phytoene desaturase; Z-ISO, 15-*cis*-ζ-carotene isomerase; ZDS, ζ-carotene desaturase; CRTISO, carotenoid isomerase; LYCB, lycopene β-cyclase; LYCE, lycopene ε-cyclase. *, For the methylerythritol 4-phosphate (MEP) pathway, see detailed reactions in Fig. 8.2.

species are sources of provitamin A in the human diet. They include widespread species such as papaya, apricot, pepper and tomato and more local species produced in South-east Asia such as gac fruit (*Momordica cochinchinensis*) (Hyun *et al.*, 2012) and other species found in Latin America (Rodriguez-Amaya, 1999). Provitamin A rich fruits also contain usually other phytonutrients with potential health benefits, such as the red-coloured lycopene found in papaya (Schweiggert *et al.*, 2012) and in tomato (Fraser *et al.*, 2009). Tomato has been studied extensively because of the striking increase in carotenoids that takes place during fruit ripening and the many colour mutants available (Hirschberg, 2001). Thanks to the genomics tools currently available, carotenoid biosynthesis and regulation are also beginning to be well studied in less well-known species such as papaya (Devitt *et al.*, 2010), palm fruit (Tranbarger *et al.*, 2011) and gac (Hyun *et al.*, 2012).

8.2.3 Carotenoid biosynthetic pathway

Carotenoids are synthesized by nuclear-encoded enzymes within the plastids of the fruit, chloroplasts in photosynthetic tissues and chromoplasts in the ripening fruit (Hirschberg 2001; Fraser *et al.*, 2009). The sequence of reactions involved in carotenoid biosynthesis that lead to the formation of the major provitamin A carotenoids, α- and β-carotene and β-cryptoxanthin, is presented in Fig. 8.1.

In the well-studied tomato fruit (Fraser *et al.*, 2009), isopentenyl diphosphate (IPP), a C5 precursor from which the carotenoids and tocochromanols are derived, is synthesized mostly by the plastidial methylerythritol 4-phosphate (MEP) pathway (Figs 8.1 and 8.2). IPP is isomerized to dimethylallyl diphosphate by IPP isomerase (IPI). Three IPP molecules are then sequentially added to dimethylallyl diphosphate by geranylgeranyl diphosphate (GGDP) synthase (GGPS), producing GGDP, which constitutes the immediate precursor of carotenoids. The first committed step in the carotenoid pathway is the condensation head to tail of two GGDP (C20) molecules, which produces the non-coloured 15-*cis*-phytoene and is catalysed by phytoene synthase (PSY). The coloured carotenoid chromophore is then generated by a series of desaturation reactions, which extend the number of conjugated double bonds. The first desaturase enzyme is phytoene desaturase (PDS), which introduces double bonds in the molecule to form 9,15,9′-tri-*cis*-ζ-carotene, a yellow-green molecule with seven conjugated double bonds. This step is followed by an isomerization producing 9,9′-di-*cis*-ζ-carotene. A further desaturation catalysed by ζ-carotene desaturase (ZDS) forms 7,9,7′,9′-tetra-*cis*-lycopene (prolycopene), which is converted in the fruit by carotene isomerase (CRTISO) to *all-trans*-lycopene, a molecule with 11 conjugated double bonds. Lycopene is responsible for the red coloration of fruits such as tomato (Fraser *et al.*, 2009), red-fleshed papaya (Devitt *et al.*, 2010; Schweiggert *et al.*, 2012) and orange (Alquezar *et al.*, 2008). Cyclization of lycopene by β-cyclase (LCYB) and ε-cyclase (LCYE) forms the orange-coloured provitamin A cyclic carotenoids β-carotene (found for example in mango) and α-carotene. Further introduction of hydroxyl groups into β-carotene by β-hydroxylase results in the formation of the orange-coloured β-cryptoxanthin found, for example, in orange-fleshed papaya (Schweiggert *et al.*, 2012) and, ultimately, to the phytohormone ABA. Other precursors are shared between the carotenoid biosynthetic pathway and the phytohormones gibberellins (GAs) (Fig. 8.1) and cytokinins.

8.2.4 Regulation of carotenoids in fruits

Tomato has also been the model fruit for studying the regulation of carotenoid biosynthesis in fleshy fruits. Like many carotenoid-coloured fleshy fruits, during fruit ripening tomato undergoes a massive accumulation of carotenoids coordinated with other ripening-associated changes

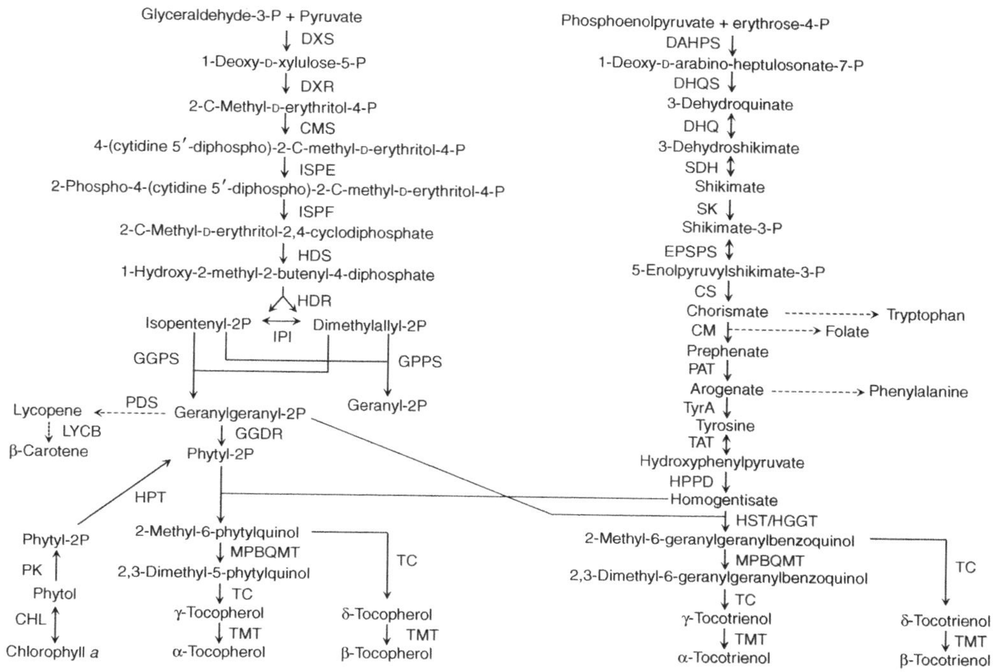

Fig. 8.2. Biosynthetic routes of vitamin E reactions in plants displaying the methylerythritol 4-phosphate (MEP), shikimate kinase (SK) and tocopherol core pathways as well as the VTE-related pathways (carotenoid, chlorophyll, tryptophan, phenylalanine and folate).The corresponding enzymes are: DXS, deoxy-D-xylulose-5- phosphate synthase; DXR, 2-C-methyl-D-erythritol 4-phosphate synthase; CMS, 2-C-methyl-D-erythritol 4-phosphate cytidylyltransferase; ISPE, 4-(cytidine 5′-diphospho)-2-C-methyl-D-erythritol kinase; ISPF, 2-C-methyl-D-erythritol 2,4-cyclodiphosphate synthase; HDS, 4-hydroxy-3-methylbut-2-enyl diphosphate synthase; HDR, 4-hydroxy-3-methylbut-2-enyl diphosphate reductase; IPI, isopentenyl diphosphate isomerase; GPPS, geranyl pyrophosphate synthase; GGPS, geranylgeranyl pyrophosphate synthase; GGDR, geranylgeranyl reductase; DAHPS, 3-deoxy-D-arabino-heptulosonate-7-phosphate synthase; DHQS, 3-dehydroquinate synthase; SDH, shikimate dehydrogenase; DHQ, 3-dehydroquinate dehydratase; EPSPS, 5-enolpyruvylshikimate-3-phosphate synthase; CS, chorismate synthase; CM, chorismate mutase; PAT, prephenate aminotransferase; TyrA, arogenate dehydrogenase; TAT, tyrosine aminotransferase; HPPD, 4-hydroxyphenylpyruvate dioxygenase; HTS/HGGT, homogentisate solanesyl transferase/homogentisate geranylgeranyl transferase; HPT (VTE2), homogentisate phytyl transferase; MPBQMT(VTE3), 2-methyl-6-phytyl-1,4-benzoquinone methyltransferase; TC (VTE1), tocopherol cyclase; γ-TMT (VTE4), γ-tocopherol methyl transferase; PK, phytol kinase; CHL, chlorophyllase; LYCB, lycopene β-cyclase. P, phosphate; 2P, diphosphate.

(see Part I). These include the transition from chloroplasts, which are present in the green tissues from developing tomato fruit until the onset of ripening, to carotenoid-accumulating chromoplasts in the coloured fruit. Carotenoids present in chloroplasts are mostly the carotene-derived xanthophylls, associated with photosynthesis, whilst in chromoplasts the acyclic carotenoids (phytoene, phytofluene, β-carotene, neurosporene and lycopene) accumulate. In tomato, the transcriptional upregulation of carotenoid biosynthesis genes is the major regulatory mechanism that takes place in fruit (Fraser *et al.*, 2009). There is a massive increase in transcripts for PSY and PDS, concomitant with a disappearance of the transcripts for the lycopene cyclases encoded by *LYCB* and *LYCE*. In addition, an alternative set of carotenoid biosynthetic genes encoding GGPS, PSY and CRTISO is specifically upregulated in the fruit to sustain the carotenoid accumulation. Remarkably, similar upregulation of carotenoid gene expression during fruit ripening has been found in other fruits, not only in pepper (Hugueney et al., 1996), another *Solanaceae* species, but also in satsuma mandarin (Ikoma et al., 2001), papaya (Devitt et al., 2010), palm fruit (Tranbarger et al., 2011) and gac (Hyun et al., 2012). In papaya, a chromoplast-specific lycopene cyclase expressed during fruit ripening leads to the formation of the provitamin A β-carotene and β-cryptoxanthin in the yellow-fleshed fruit, whilst a mutation reducing its functionality is probably responsible for lycopene accumulation in the red-fleshed cultivars (Devitt et al., 2010). Conversely in tomato, which normally accumulates lycopene, upregulation of a lycopene β-cyclase gene in the *Beta* (*B*) mutant results in β-carotene accumulation in fruits (Ronen et al., 2000).

Ripening is a complex process. In climacteric fruits, ethylene plays a key role in initiating and coordinating ripening-associated changes, including carotenoid accumulation. The strong upregulation of the PSY and PDS genes during tomato fruit ripening are under the control of ethylene, whilst regulation of the cyclase involved in β-carotene formation is ethylene independent (Alba et al., 2005). Light and temperature are among the main factors regulating carotenoid accumulation in the fruit (Gautier et al., 2008). The regulation of carotenoid biosynthesis by light has been well studied thanks to the large number of tomato mutants affected in the level and type of carotenoids in the fruit. Both phytochromes and cryptochromes are involved (Alba et al., 2000; Fraser et al., 2009). In addition, the involvement of the light signal transduction components in the regulation of carotenoid accumulation in tomato fruit has raised considerable interest in recent years (Lieberman et al., 2004; Liu et al., 2004; Wang et al., 2008; Enfissi et al., 2010). Both carotenoid gene transcription and plastid biogenesis are affected when this pathway is deregulated, leading to enhanced carotenoid accumulation in the fruit, together with increased levels of ascorbate and other phytonutrients (Enfissi et al., 2010). Increased plastid number due to enhanced plastid division has also been linked to the increased accumulation of carotenoids in a tomato ABA-deficient mutant (Galpaz et al., 2008), illustrating the close relationships between carotenoid and phytohormone biosynthesis and regulation.

8.3 Vitamin E Biosynthesis and Regulation in Fleshy Fruits

Vitamin E comprises several lipid-soluble compounds, termed tocochromanols, composed of a polar chromanol head group and a lipophilic polyprenyl side chain (Mène-Saffrané and DellaPenna, 2009). Tocopherols have a saturated phytil-derived side chain, whilst tocotrienols have an unsaturated geranylgeranyl-derived side chain. Tocopherol and tocotrienol each consist of four isoforms (α, β, γ and δ), which are differentiated by the position and number of the methyl groups on the chromanol ring. Tocochromanols are only found in photosynthetic organisms, and in plants are found exclusively in plastids. The most abundant tocochromanol found in plants is tocopherol, whilst tocotrienol is less widespread and is found mostly in monocot seeds and in fruits.

8.3.1 Role of vitamin E in humans and plants

Vitamin E deficiency is responsible for several human pathologies, but clinical evidence for vitamin E deficiencies is rare, except in conditions in which the metabolism of vitamin E is disturbed (FAO/WHO, 2004). Vitamin E can be obtained from plant and animal sources. In humans, as a lipid-soluble antioxidant, its major role is to protect polyunsaturated fatty acids and other cellular components (DNA, proteins) from oxidation by free radicals. Like provitamin A, vitamin E must be freed from the chloroplasts and enter a step of micelle formation before being adsorbed. The main forms retained in blood and tissues are the α- and, to a lower extent, the γ-form of tocopherol. In addition to its antioxidant activity, tocopherol is involved in the modulation of specific signalling pathways (reviewed by Galli and Azzi, 2010). In addition, minor forms of vitamin E such as tocotrienol may share with tocopherol several beneficial effects for human health such as the prevention of cancer. However, the prevention of cardiovascular disease and the anti-tumoural effect of tocopherol have been questioned recently and require further investigation (Galli and Azzi, 2010).

In plants, tocochromanols have an antioxidant function in photosynthetic membranes, by controlling lipid peroxidation. The main tocochromanol found in plants, tocopherol, has been shown to play a crucial role in the protection of plants against photo-oxidative stress (Falk and Munné-Bosch, 2010). In addition, tocopherol deficiency affects other plant functions, such as germination and photo-assimilate transport, suggesting additional but still ill-defined roles in plants. The function of tocotrienols is less clear, although they have been shown potentially to serve as antioxidants in photosynthetic membranes (Matringe *et al.*, 2008).

8.3.2 Vitamin E in fleshy fruits

The main plant sources of vitamin E in the human diet are by far the plant-derived oils, derived mostly from seeds. Oil extracted from the mesocarp of palm fruit, the major oil crop in the world and also rich in provitamin A, is very rich in both α-tocopherol (25.6 mg per 100 g) and γ-tocopherol (31.6 mg per 100 g), and in α-tocotrienol (14.3 mg per 100 g) (FAO/WHO, 2004). By comparison, sunflower oil content is 48.7, 5.1 and 0 mg per 100 g for α-tocopherol, γ-tocopherol and α-tocotrienol, respectively. The vitamin E content of other fleshy fruits is much lower, with avocado, mango, blackberry and raspberry having the highest vitamin E content (Table 8.1) (Chun *et al.*, 2006).

8.3.3 Vitamin E biosynthetic pathway

Whilst the biosynthetic pathway of vitamin E was proposed in the 1970s, the first enzyme of that pathway was cloned in 1998 (reviewed by Dellapenna, 2005; Mène-Saffrané and DellaPenna, 2009). Since then, vitamin E biosynthetic genes have been identified using a combination of genetic and genomic tools in the model plant species *Arabidopsis* and subsequently in many other plant species. More recently, vitamin E biosynthetic genes have also been identified in several fruit species including tomato (Almeida *et al.*, 2011), apple (Seo *et al.*, 2011), mango (Singh *et al.*, 2011) and palm fruit (Tranbarger *et al.*, 2011).

Tocochromanols are composed of a polar chromanol head group derived from a cytosolic aromatic amino acid pathway, the shikimate pathway, and by a lipophilic polyprenyl side chain, derived from the plastidial MEP pathway, which synthesizes the isoprenoid precursor IPP (Fig. 8.2). Tocopherols have a saturated side chain derived from phytyl diphosphate (PDP), produced from geranylgeranyl diphosphate (GGDP) or from chlorophyll *a*, whilst tocotrienols have an unsaturated side chain derived from GGDP. The committed step in head-group synthesis is the formation of homogentisate from hydroxyphenylpyruvate catalysed by 4-hydroxyphenylpyruvate dioxygenase (HPPD). Prenylation of homogentisate with either PDP or GGDP yields 2-methyl-6-phytylquinol (MPBQ) and 2-methyl-6-geranylgeranylbenzoquinol (MGGBQ), the first intermediates in tocopherol and tocotrienol synthesis, respectively. Subsequent steps of methylation of MPBQ and MGGBQ by 2-methyl-6-phytyl-1,4-benzoquinone methyltransferase (MPBQMT/VTE3), tocopherol cyclase (TC/VTE1) and γ-tocopherol methyltransferase (γ-TMT/VTE4) yield the differently methylated α-, β-, γ- and δ-forms of tocopherol and tocotrienols.

8.3.4 Regulation of tocochromanols in fruits

Regulation of tocochromanol biosynthesis in fruits remains poorly investigated. As mentioned in Section 8.2.3, tocochromanols and carotenoids share GGDP as a common precursor (Figs 8.1 and 8.2). As a consequence, manipulation of carotenoid biosynthetic pathways in the fruit will probably also alter tocochromanol levels in this organ. This was indeed the case in tomato fruit overexpressing a fruit PSY, which accumulated more α-tocopherol (Fraser *et al.*, 2007), although the mechanisms responsible for the increased vitamin E content in the fruit were unclear. Light regulation of tocopherol accumulation in the fruit through the transcriptional activation of tocopherol biosynthetic genes (*GGDR* and *γ-TMT*; see Fig. 8.2) is likely, as shown by *DET-1* defective tomato transgenics that accumulated two- to tenfold more tocopherol (Enfissi *et al.*, 2010). However, the *DET-1* mutation, which disturbs the light signal transduction pathway, also affects plastid biogenesis, carotenoid biosynthesis and other secondary metabolites, therefore opening up the possibility that tocopherol alterations are due to a more general effect on plastid compartment size or fruit metabolism.

Available data also indicate a developmental regulation of tocopherol biosynthesis during fruit ripening. In tomato, in which the main tocochromanols are α- and γ-tocopherol (Almeida *et al.*, 2011), as in most fruits (Chun *et al.*, 2006), the tocopherol content increases during fruit ripening (Enfissi *et al.*, 2010) and silencing of γ-TMT leads to substantial alterations of the tocopherol profiles of the fruits (Quadrana *et al.*, 2011). Likewise, in mango, tocopherol content increases during fruit ripening, together with that of carotenoids (Singh *et al.*, 2011). Tocopherol accumulation in mango (*Mangifera indica*) is correlated with increased transcripts for HPPD, which catalyses the committed step in chromanol head-group synthesis (Fig. 8.2). Furthermore, *MiHPPD* expression is ethylene inducible, as shown by promoter studies. Thus, although the regulation of vitamin E in fleshy fruits warrants further investigation, current data suggest that its accumulation during fruit ripening is developmentally controlled and submitted to hormonal regulation by ethylene in climacteric fruits.

8.4 Vitamin C Biosynthesis and Regulation in Fleshy Fruits

In general terms, vitamin C is used with reference to its nutritional virtues, whereas ascorbic acid refers to the purified compound. At physiological pH, ascorbic acid exists as monoanion form and is hence called ascorbate. This di-acid ($C_6H_8O_6$) contains an ene-diol group, which confers its reducing agent or antioxidant properties (Smirnoff, 2000). *In vivo*, the oxidation products of ascorbate are monodehydroascorbate (MDHA) and dehydroascorbate (DHA). As a whole, these comprise the ascorbate pool, which can be oxidized more or less according to the redox state of the cell. Eventually, if not recycled to ascorbate, DHA is degraded to produce intermediates such as oxalate, tartrate, and threonate (Green and Fry, 2005).

8.4.1 Role of ascorbate (vitamin C) in humans and plants

Humans, like a small number of mammals, are unable to synthesize ascorbate due to a mutation in the L-guluno-1,4-lactone oxidase gene corresponding to the last step of the biosynthetic pathway (Linster and Van Schaftingen, 2007). In humans, ascorbate is associated mainly with metabolism related to ageing, free radicals (mainly reactive oxygen and nitrogen species), redox homeostasis and carcinogenesis (Valko *et al.*, 2007). Numerous epidemiological studies have established a positive link between vitamin C content in food and/or plasma and health benefits, for example in the prevention of cardiovascular disease, cancer, influenza and other diseases (Blot *et al.*, 1993; Steinmetz and Potter, 1996), or an increase in iron bioavailability (López and Martos, 2004).

Ascorbate in plants has many cellular functions, linked mostly to the molecule's capacity to donate electrons. Ascorbate acts as a scavenger of the free radicals generated by photosynthesis, cellular respiration and abiotic stresses such as ozone and UV radiation (Conklin *et al.*, 1996; Noctor and Foyer, 1998; Smirnoff and Wheeler, 2000). This molecule is a cofactor for numerous enzymatic reactions such as those catalysed by oxygenases involved in the synthesis of flavonoids and alkaloids and hormones, such as ACC oxidase catalysing the last step of ethylene biosynthesis (see Part III) (Prescott and John, 1996). Ascorbate also has a role in plant growth and development (Arrigoni and De Tullio, 2002; Pastori *et al.*, 2003), including aspects such as flowering (Pavet *et al.*, 2005; Dowdle *et al.*, 2007). Finally, numerous studies have linked ascorbate to both biotic and abiotic stress tolerance in various species (Muckenschnabel *et al.*, 2002; Huang *et al.*, 2005; Kuzniak and Skłodowska, 2005; Yamamoto *et al.*, 2005; Stevens *et al.*, 2008). Besides the ascorbate biosynthetic pathway, the ascorbate recycling pathway also plays a crucial role, as both MDHA reductase (MDHAR) and DHA reductase (DHAR) have roles in stress tolerance (Eltayeb *et al.*, 2007; Stevens *et al.*, 2008; Gest *et al.*, 2010).

In fruit, in addition to its effect on stress tolerance, ascorbate may contribute to postharvest fruit quality (Davey and Keulemans, 2004; Malacrida *et al.*, 2006). Ascorbic acid levels have been linked to flesh browning in pear (Veltman *et al.*, 1999) and a quantitative trait locus (QTL) for flesh browning colocalizes with a QTL for oxidized ascorbate content in apple (Davey *et al.*, 2006). In tomato fruit, ascorbic acid content is linked to MDHAR activity and tolerance to chilling stress (Stevens *et al.*, 2008; Gest *et al.*, 2010).

8.4.2 Vitamin C in fleshy fruits

Ascorbate is present in all plant organs and in all cell compartments, including the nucleus. The concentrations of this molecule are highly variable in plants (Smirnoff, 2000; Zechmann *et al.*, 2011) and may vary within organs of the same plant. Photosynthetic tissues, in particular meristems, are often higher in ascorbate than roots, flowers and fruits (Loewus and Loewus, 1987). Fruit ascorbate levels are highly variable in diverse species (Table 8.1). For instance, citrus fruits are known for being vitamin C rich, their content being about 50 mg per 100 g of fresh weight (FW). Nevertheless, the highest vitamin C levels in fruits are found in two berries from South America, the camu-camu (2–3 g per 100 g FW) and acerola (1–2 g per 100 g FW). In addition, within the same genus, large disparities may be observed, as for example in tomato. Wild tomato species such as *Solanum pennellii* generally accumulate more ascorbate (50 mg per 100 g FW) than the large-fruited domesticated cultivars (10–20 mg per 100 g FW) of *Solanum lycopersicum* (Stevens *et al.*, 2007; Roselló *et al.*, 2011). Other major sources of natural variability are the cultural and environmental conditions, especially light (Massot *et al.*, 2012), and the postharvest losses during storage and processing, which can considerably reduce the vitamin C content (Lee and Kader, 2000; Franck *et al.*, 2003).

8.4.3 Vitamin C biosynthetic pathways

Whilst the biosynthesis of vitamin C was elucidated in the early 1960s in the animal kingdom (Chatterjee *et al.*, 1960), it was not until 1998 following work by the Smirnoff group that the first biosynthetic pathway in higher plants was proposed, known as the 'Smirnoff–Wheeler' pathway. It is now accepted that this pathway is the major *de novo* synthesis route for ascorbate (Wheeler *et al.*, 1998). Smirnoff and coworkers established that the direct precursor of ascorbate was L-galacto-1-4 lactone itself produced from GDP-D-mannose via oxidation of the comparatively rare sugar L-galactose (Wheeler *et al.*, 1998). This pathway was supported by feeding studies and by partial purification of a new enzyme, L-galactose dehydrogenase, which catalyses the NAD-dependent oxidation of L-galactose to L-galactono-1,4-lactone. Furthermore, clear evidence for this pathway was given by characterization of the ozone-sensitive and ascorbate-deficient *Arabidopsis vtc1* mutant that was later revealed to be defective in GDP-D-mannose pyrophosphorylase activity (Conklin *et al.*, 1996, 1999). The 'Smirnoff–Wheeler' pathway involves the conversion of D-mannose into ascorbate via a succession of L-galactose-containing intermediates, namely GDP-D-mannose, GDP-L-galactose, L-galactose-1-phosphate, L-galactose and L-galactono-1,4-lactone (Fig. 8.3) (Wheeler *et al.*, 1998; Conklin *et al.*, 1999, 2006; Bartoli *et al.*, 2000; Wolucka and Van Montagu, 2003; Dowdle *et al.*, 2007; Laing *et al.*, 2007). Based on radiotracer, biochemical and expression studies, three alternative routes to ascorbate have been proposed. The 'Smirnoff–Wheeler' pathway can be augmented through a 'pectin-scavenging' pathway. This pathway, initially described in strawberry fruit, initiates from D-galacturonic acid to produce an L-galactonic acid derivative via D-galacturonate reductase (Agius *et al.*, 2003). Alternatively, L-gulose (Wolucka and van Montagu, 2003) and *myo*-inositol (Lorence *et al.*, 2004) have also been proposed as intermediates of ascorbate biosynthesis pathways, which partially overlap with the animal pathway (Fig. 8.3). Although the main plant pathway is the 'Smirnoff–Wheeler' pathway, the prevalence of this pathway over the others is dependent on the fleshy fruit species and developmental stage (Lorence *et al.*, 2004; Laing *et al.*, 2007; Cruz-Rus *et al.*, 2010, 2011).

8.4.4 Regulation of vitamin C content in fruit

Unlike foliar tissues, few studies on ascorbate synthesis, recycling and regulation have been carried out in fruit. Nevertheless, fruits represent pertinent models to unravel this aspect of the ascorbate metabolism as they experience physiological and biochemical changes during development, ripening and post-harvest storage. Ascorbate can accumulate to very high levels in fruits (e.g. mango, kiwifruit, citrus fruits, strawberry, tomato), which are sink organs. In addition, fruits from the same genus display a large scale of ascorbate content, for example in tomato from 10 to more than 500 mg per 100 g FW (Galiana-Balaguer *et al.*, 2006), as well as in kiwifruit and related species where vitamin C content may range from 80 to 800 mg per 100 g FW (Bulley *et al.*, 2009). Many recent studies on fruits like tomato (Gautier *et al.*, 2009), kiwifruit (Bulley *et al.*, 2009), strawberry (Agius *et al.*, 2003), apple (Razavi *et al.*, 2005) and blackcurrant (Hancock *et al.*, 2007) have revealed that fruits are able to synthesize their own ascorbate. However, its translocation from source tissues to fruits has been described in kiwifruit (Bulley *et al.*, 2009) and tomato (Gautier *et al.*, 2009). Moreover, the recycling of ascorbate from MDHA and DHA significantly contributes to the regulation of ascorbate content (Stevens *et al*, 2008; Yin *et al.*, 2010). As in the leaf, light intensity influences ascorbate levels in fruits (Massot *et al.*, 2012).

However, little is known about the mechanisms underlying light regulation of ascorbate synthesis in fruits, even though this has been described since 1945 (Hamner *et al.*, 1945; McCollum, 1946; Madamba *et al.*, 1974; El-Gizawi *et al.*, 1993; Ioannidi *et al.*, 2009). Light regulation of the expression of ascorbate biosynthesis-related genes has been demonstrated in apple (Li *et al.*, 2009) and tomato (Massot *et al.*, 2012). In whole plants and leaves, recent studies have demonstrated that the GDP-L-galactose phosphorylase gene (Fig. 8.3) exerts most of the control of the flux through the synthesis pathway (Bulley *et al.*, 2009), and expression of this gene as well as the protein activity are often correlated with ascorbate content (reviewed by Linster and Clarke, 2008). The presence in strawberry fruit of the 'pectin-scavenging' pathway, which uses D-galacturonic acid to produce an L-galactonic acid derivative via D-galacturonate reductase, and ultimately

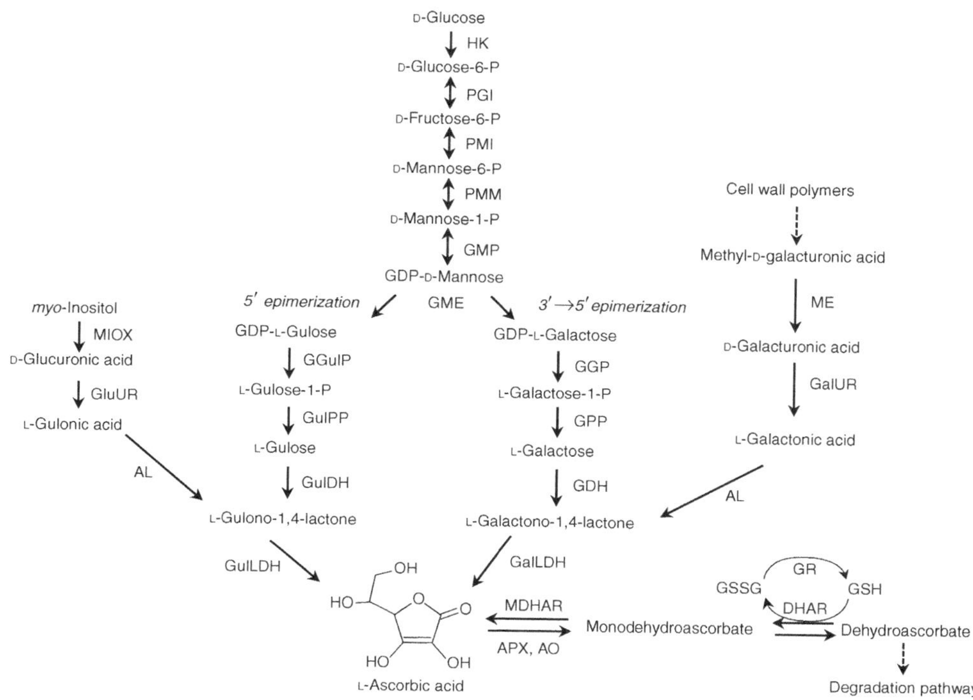

Fig. 8.3. Biosynthetic routes to L-ascorbic acid and the recycling reactions in plants. The corresponding enzymes and metabolites are: HK, hexokinase; PGI, D-glucose-6-phosphate isomerase; PMI, D-mannose-6-phosphate isomerase; PMM, phosphomannomutase; GMP, GDP-D-mannose pyrophosphorylase; GME, GDP-D-mannose-3′,5′-epimerase; GGP, GDP-L-galactose phosphorylase; GPP, L-galactose-1-phosphate phosphatase; GDH, L-galactose dehydrogenase; GalLDH, L-galactono-1,4-lactone dehydrogenase; ME, methylesterase; GalUR, D-galacturonate reductase; AL, aldono-lactonase; GGulP, GDP-L-gulose phosphorylase; GulPP, L-gulose-1-P phosphatase; GulDH, L-gulose dehydrogenase; GulLDH, L-gulono-1,4-lactone dehydrogenase; MIOX, *myo*-inositol oxygenase; GluUR, D-glucuronate reductase; AL, aldono-lactonase; APX, ascorbate peroxidase; AO, ascorbate oxidase; MDHAR, monodehydroascorbate reductase; DHAR, dehydroascorbate reductase; GR, glutathione reductase; GSSG, oxidized glutathione; GSH, reduced glutathione. P, phosphate.

ascorbate (Agius *et al.*, 2003), is remarkable as D-galacturonic acid is produced by the degradation of pectins from the cell wall during fruit ripening (see Tucker, Chapter 4, this volume). In addition, two *Arabidopsis* transcription factors, the F-box E3 protein AMR1 (ascorbic acid mannose pathway regulator 1) and the ERF98 (ethylene response factor 98) protein, were suggested to regulate the expression of the D-mannose/L-galactose pathway genes at the transcriptional level and to modulate ascorbate levels in the leaf in response to abiotic stress (Zhang *et al.*, 2009, 2012). Such ascorbate biosynthetic and regulatory genes are putative targets for understanding the regulation and manipulation of the levels of vitamin C in fleshy fruit species.

8.5 Enhancing Vitamin Content in Fleshy Fruits

Carotenoids, tocochromanols (tocopherols and tocotrienols) and ascorbate are among the main non-enzymatic antioxidants found in plants, in which they play crucial roles in plant development and metabolism and in protection against various stresses. They are therefore under tight regulation by environmental conditions (Roselló *et al.*, 2011; Massot *et al.*, 2012), especially light, which may exert a strong stress effect on plants. Temperatures during fruit growth may also affect fruit content in vitamin C and provitamin A (Gautier *et al.*, 2008). As a consequence, climatic conditions and cultural practices can be adapted to modify the fruit contents in provitamin A, vitamin E and vitamin C. This can be achieved by modulation of the amount and quality of light during fruit growth and ripening, for example by cultivating plants in different growing seasons or at different altitudes, or through the control of irradiance by light-emitting diodes in the greenhouse (Ma *et al.*, 2012) or photoselective nets in open fields (Shahak *et al.*, 2006). These strategies can be applied commercially to increase fruit provitamin A (e.g. β-cryptoxanthin in citrus fruits; Ma *et al.*, 2012), as they will improve the nutritional value of the fruit and its visual appearance at the same time. Modulating fruit vitamin E and C content by technological means can be more challenging, unless the increase in nutritional value is sufficient to bring significant added value to the fruit product. Fruit vitamin content can also be controlled at subsequent steps, during postharvest storage and/or during fruit processing. Storage can lead to substantial losses in vitamins, especially in vitamin C (Stevens *et al.*, 2008; Cruz-Rus *et al.*, 2011). The various methods used to extend fruit shelf-life will have an impact on the fruit micronutrient content, for example in strawberry and mango (Rivera-Pastrana *et al.*, 2010; Yang *et al.*, 2010; reviewed by Lee and Kader, 2000), especially when their synthesis is regulated by hormones and/or fruit ripening. Likewise, cooking or thermal treatments and subsequent storage of the processed product will result in significant reductions in some vitamins, such as the provitamin A β-carotene, vitamin E and vitamin C (Abushita *et al.*, 2000; Koh *et al.*, 2012), although the processed fruit product may still retain significant amounts of vitamins (e.g. 45% of the initial content of vitamin C in tomato paste; Abushita *et al.*, 2000). As an alternative to conventional thermal treatments, new methods such as high-pressure or microwave preservation will help retain the initial content of vitamins in processed products (reviewed by Barrett and Lloyd, 2012). In addition, disruption of fruit chromoplasts by processing and constitution of emulsions may increase the bioaccessibility of carotene (Svelander *et al.*, 2011).

In recent years, most efforts have been focused on the genetic improvement of vitamin content in fruits, and especially in tomato, a model fleshy fruit; whilst tomato is only moderately rich in carotenoids and vitamins E and C, this fruit species has a widespread consumption throughout the world, therefore contributing in a large way to human dietary nutrition. Other well-studied species are mango (rich in carotenoids and vitamin E), papaya (carotenoids), palm fruit (carotenoids and vitamin E), apricot (carotenoids), strawberry (vitamin C), kiwifruit (vitamin C) and apple (vitamin C). Species such as mango represent seasonally important sources of vitamins in some tropical countries whose inhabitants suffer chronically from vitamin deficiency (FAO/WHO, 2004).

Improving vitamin content in fruit through genetic means is a realistic goal because: (i) our understanding of plant biosynthesis pathways and of the regulation of carotenoids and vitamins E and C has progressed considerably in recent years (Hirschberg, 2001; Linster and Clarke, 2008; Fraser *et al.*, 2009; Mène-Saffrané and DellaPenna, 2009; Almeida *et al.*, 2011; Fitzpatrick *et al.*, 2012); and (ii) genomic tools, such as marker-saturated genetic maps, genome sequences and other genetic resources, are now available for most crop species, allowing easier identification of alleles that improve vitamin content and their introduction into elite varieties.

8.5.1 Enhancing fruit vitamin content by genetic engineering

Provitamin A

Tomato fruit has been the model of choice for the discovery of important steps in the carotenoid biosynthetic pathway and for the genetic engineering of carotenoids in plants. More than 20 studies aimed at enhancing fruit carotenoid levels have been conducted in this species in recent years. Most of these studies have been thoroughly presented and discussed in a recent excellent review on genetic engineering of carotenoid formation in tomato fruit (Fraser *et al.*, 2009). A two- to fourfold increase in total carotenoids or in β-carotene was generally obtained by ectopic expression, overexpression or silencing of genes encoding various enzymes from the carotenoid pathway or proteins involved in light signalling, originating from tomato, other plants and bacteria. Notable achievements were a greater than 30-fold increase in β-carotene triggered by overexpression of β-lycopene cyclase (Fig. 8.1) from tomato, although there was a corresponding reduction in lycopene (D'Ambrosio *et al.*, 2004), the production of β-cryptoxanthin and zeaxanthin by coexpressing β-lycopene cyclase from *Arabidopsis* and β-carotene hydroxylase from pepper (Dharmapuri *et al.*, 2002), and concomitant and strong increases in β-carotene (tenfold), lycopene (fourfold) and flavonoids obtained by specifically silencing the *DET-1* gene involved in light signalling using fruit-specific promoters (Davuluri *et al.*, 2005).

Vitamin E

In recent years, many biotechnological attempts to increase vitamin E have been carried out successfully in plants (Mène-Saffrané and DellaPenna, 2009). In oilseed crops, which are the main source of vitamin E in the human diet, metabolic engineering of vitamin E in *Arabidopsis*, canola and soybean has been done by expressing both *Synechocystis* bifunctional prephenate dehydrogenase (*tyrA*), a feedback insensitive enzyme, and the plant HPPD genes. This led to up to 1.8–2.6-fold increases in tocochomanol levels in the seeds (Karunanandaa *et al.*, 2005). Likewise, expression of the yeast (*Saccharomyces cerevisiae*) prephenate dehydrogenase gene (*PDH*) in tobacco overexpressing the *Arabidopsis HPPD* gene resulted in a massive accumulation of tocotrienols in the leaves (Rippert *et al.*, 2004). Using a similar strategy, based on the stable coexpression in tomato of the yeast *PDH* and *Arabidopsis HPPD* genes under the control of the *SlPPC2* fruit-specific promoter (Fernandez *et al.*, 2009; Guillet *et al.*, 2012), our group recently obtained a threefold increase in tocotrienol content in tomato fruit (unpublished results). In parallel, also in tomato, overexpression of homogentisate phytyl transferase (HPT; Fig. 8.2) isolated from apple fruit led to approximately 1.8–3.6-fold and approximately 1.6–2.9-fold increases in tomato leaf of α-tocopherol and γ-tocopherol, respectively, whilst the levels of α-tocopherol and γ-tocopherol in the fruit increased up to 1.7- and 3.1-fold, respectively (Seo *et al.*, 2011).

Vitamin C

Success has been achieved by the expression of plant vitamin C biosynthesis or regulatory genes in *Arabidopsis* and crop species (Ishikawa *et al.*, 2006; Bulley *et al.*, 2009; Zhang *et al.*, 2009). Increases in vitamin content were usually two- to threefold, reaching 12-fold when the newly discovered GDP-L-galactose phosphorylase (VTC2/VTC5) enzyme, which catalyses the first committed step in the L-galactose ascorbate biosynthetic pathway (Fig. 8.3), was transiently coexpressed with GDP-D-mannose epimerase (GME) in tobacco (Bulley *et al.*, 2009). In fruit, the stable overexpression of GME led to a slight increase in ascorbate content in tomato (1.6-fold increase in red fruit; Zhang *et al.*, 2011). The most promising results were obtained recently through the stable overexpression of *VTC2* in tomato and strawberry, which increased the fruit vitamin C content by two- to sixfold (Bulley *et al.*, 2012).

8.5.2 Enhancing fruit vitamin content by breeding

Most fleshy fruit species present considerable variation in vitamin content. Fruit ascorbate content of wild tomato species accessions can be as low as 11 mg per 100 g FW or reach >500 mg per 100 g FW (Galiana-Balaguer *et al.*, 2006), whilst in kiwifruit and related species, vitamin C ranges from 80 to 800 mg per 100 g FW (Bulley *et al.*, 2009). Similarly wide variations can be found for carotenoids or vitamin E in existing cultivars or in wild related species or accessions of cultivated fleshy fruits species such as tomato (Schauer *et al.*, 2005; Almeida *et al.*, 2011), pepper (Brand *et al.*, 2012), melon (Cuevas *et al.*, 2009), papaya (Schweigert *et al.*, 2012) and *Citrus* species (Fanciullino *et al.*, 2006). For several fruit species, the genomic regions controlling the quantitative variations in vitamin content (QTLs) have been located on genetic maps using introgression lines, advanced backcrossing and recombinant inbred line populations derived from crosses between genotypes with contrasting vitamin content (cultivated varieties or wild related species). To date, the most extensive studies have been carried out in tomato in which, for example, metabolite profiling of 76 introgression lines issued from crosses between *S. esculentum* and *S. pennellii* allowed the identification of up to 889 QTLs controlling the variations in several metabolites linked to fruit sensorial and nutritional quality, including vitamin E and vitamin C (Schauer *et al.*, 2006). QTLs controlling fruit vitamin content have been detected in a large number of species such as provitamin A in melon (Cuevas *et al.*, 2009) and pepper (Brand *et al.*, 2012), vitamin E in tomato (Almeida *et al.*, 2011), and vitamin C in tomato (Stevens *et al.*, 2007), strawberry (Zorrilla-Fontanesi *et al.*, 2011) and apple (Davey *et al.*, 2006). Once QTLs have been mapped, candidate gene approaches, positional cloning and/or association genetics can be used to identify the locus responsible for the vitamin variation. Using a combination of QTL and candidate gene mapping, components of the light signal transduction pathway responsible for variations in provitamin A levels and in overall fruit nutritional quality have been identified in tomato (Lieberman *et al.*, 2004; Liu *et al.*, 2004). Using a similar approach, Almeida *et al.*, (2011) identified 16 candidate genes putatively affecting the various tocopherol isoforms found in tomato fruit. Likewise, Stevens *et al.* (2007, 2008) identified an allelic form of MDHAR, which co-segregated with a major vitamin C QTL and explained more than 80% of the variation in reduced ascorbic acid levels following storage. A candidate gene approach has also been successful in less-studied species such as papaya, in which an allelic form of lycopene β-cyclase was shown to control fruit colour and provitamin A content (Devitt *et al.*, 2010). Positional cloning of loci responsible for variations in vitamin levels have been reported for provitamin A (Ronen *et al.*, 2000; Isaacson *et al.*, 2002) but not yet for vitamins E and C in fleshy fruit species. The rapid acquisition of genome sequences and high-density-marker genetic maps in fleshy fruit species will considerably improve the speed and efficiency of the identification of genes underlying fruit vitamin QTLs and, subsequently, the transfer of the alleles of interest to elite varieties.

8.6 Conclusion

In the near future, analysis of cultivated fruit varieties, introgression lines from wild related species, mutants and transgenic lines with contrasting vitamin content, using a combination of the currently available genomic tools (metabolomics, transcriptomics and proteomics), should allow the discovery of new target genes for enhancing the fruit levels of vitamin A (Enfissi *et al.*, 2010; Tranbarger *et al.*, 2011; Hyun *et al.*, 2012; Schweiggert *et al.*, 2012), vitamin E (Almeida *et al.*, 2011) and vitamin C (Stevens *et al.*, 2007; Bulley *et al.*, 2009; Garcia *et al.*, 2009; Cruz-Rus *et al.*, 2011) through biotechnological means or marker-assisted breeding. Although current studies are focused mostly on model fleshy fruit species like tomato, these strategies should soon become relevant to species for which wide genetic diversity exists in terms of vitamin content (e.g. papaya; Schweiggert *et al.*, 2012) and/or that are easily genetically transformable. The overwhelming development of new and cheap sequencing strategies allowing a complete inventory of expressed genes and deciphering of whole-genome sequences will help considerably in advancing these studies.

However, one must keep in mind that alterations of provitamin A, vitamin E and vitamin C biosynthesis may have profound repercussions on whole-plant physiology and on fruit quality. As an example, the isoprenoid and carotenoid pathways are involved in the formation not only of provitamin A and vitamin E but also of several major plant hormones such as gibberellin and ABA (Fig. 8.1) and of carotenoid-derived volatiles (Lewinsohn *et al.*, 2005; Mathieu *et al.*, 2009; Vogel *et al.*, 2010). In addition to increasing the carotenoid level, the constitutive overexpression of the *Psy-1* gene in tomato affected plant vigour and isoprenoid-derived hormones in vegetative tissues, as well as fruit chlorophyll and tocopherol levels (Fraser *et al.*, 2009). Conversely, fruit-specific silencing of the *SlNCED1* gene involved in ABA biosynthesis, which effectively reduced ABA accumulation in tomato fruit, also led to the enhancement of tomato fruit lycopene and β-carotene levels (Sun *et al.*, 2012). Silencing of GME, a key enzyme of ascorbate biosynthesis, affected both tomato fruit vitamin content and firmness (Gilbert *et al.*, 2009), thereby demonstrating the tight connections between vitamin C and cell-wall biosynthesis and between fruit nutritional and sensorial quality. For genetic engineering of fruit vitamin levels, it is therefore essential to use fruit-specific promoters, as was done for the *DET1* and *CUL4* genes involved in light signalling (Wang *et al.*, 2008; Enfissi *et al.*, 2010) and for tocotrienol engineering in tomato fruit (our unpublished results), or, when using natural or artificially induced genetic variability, to find alleles that have no detrimental effects on plant and fruit.

References

Abushita, A.A., Daood, H.G. and Biacs, P.A. (2000) Change in carotenoids and antioxidant vitamins in tomato as a function of varietal and technological factors. *Journal of Agricultural and Food Chemistry* 48, 2075–2081.

Agius, F., Gonzalez-Lamothe, R., Caballero, J.L., Munoz-Blanco, J., Botella, M.A. and Valpuesta, V. (2003) Engineering increased vitamin C levels in plants by overexpression of a D-galacturonic acid reductase. *Nature Biotechnology* 21, 177–181.

Alba, R., Cordonnier-Pratt, M.-M. and Pratt, L.H. (2000) Fruit-localized phytochromes regulate lycopene accumulation independently of ethylene production in tomato. *Plant Physiology* 123, 363–370.

Alba, R., Payton, P., Fei, Z.J., McQuinn, R., Debbie, P., Martin, G.B., Tanksley, S.D. and Giovannoni, J.J. (2005) Transcriptome and selected metabolite analyses reveal multiple points of ethylene control during tomato fruit development. *Plant Cell* 17, 2954–2965.

Almeida, J., Quadrana, L., Asís, R., Setta, N., de Godoy, F., Bermúdez, L., Otaiza, S.N., Corrêa da Silva, J.V., Fernie, A.R., Carrari, F. and Rossi, M. (2011) Genetic dissection of vitamin E biosynthesis in tomato. *Journal of Experimental Botany* 62, 3781–3798.

Alquezar, B., Rodrigo, M.J. and Zacarias, L. (2008) Regulation of carotenoid biosynthesis during fruit maturation in the red-fleshed orange mutant Cara Cara. *Phytochemistry* 69, 1997–2007.

Arrigoni, O. and De Tullio, M.C. (2002) Ascorbic acid: much more than just an antioxidant. *Biochimica et Biophysica Acta* 1569, 1–9.

Badejo, A.A., Fujikawa, Y. and Esaka, M. (2009) Gene expression of ascorbic acid biosynthesis related enzymes of the Smirnoff–Wheeler pathway in acerola (*Malpighia glabra*). *Journal of Plant Physiology* 166, 652–660.

Barrett, D.M. and Lloyd, B. (2012) Advanced preservation methods and nutrient retention in fruits and vegetables. *Journal of the Science of Food and Agriculture* 92, 7–22.

Bartoli, C.G., Pastori, G.M. and Foyer, C.H. (2000) Ascorbate biosynthesis in mitochondria is linked to the electron transport chain between complexes III and IV. *Plant Physiology* 123, 335–344.

Blot, W.J., Li, J.Y., Taylor, P.R., Guo, W., Dawsey, S., Wang, G.Q., Yang, C.S., Zheng, S.F., Gail, M., Li, G.Y., Yu, Y., Liu, B.-Q., Tangrea, J., Sun, Y.-H., Liu, F., Fraumeni, J.F., Zhang Y.-H. Jr. and Li, B. (1993) Nutrition intervention trials in Linxian, China: supplementation with specific vitamin/mineral combinations, cancer incidence, and disease-specific mortality in the general population. *Journal of the National Cancer Institute* 85, 1483–1492.

Brand, A., Borovsky, Y., Meir, S., Rogachev, I., Aharoni, A. and Paran, I. (2012) *pc8.1*, a major QTL for pigment content in pepper fruit, is associated with variation in plastid compartment size. *Planta* 235, 579–588.

Bulley, S., Wright, M., Rommens, C., Yan, H., Rassam, M., Lin-Wang, K., Andre, C., Brewster, D., Karunairetnam, S., Allan, A.C. and Laing, W.A. (2012) Enhancing ascorbate in fruits and tubers through over-expression of the L-galactose pathway gene GDP-L-galactosephosphorylase. *Plant Biotechnology Journal* 10, 390–397.

Bulley, S.M., Rassam, M., Hoser, D., Otto, W., Schünemann, N., Wright, M., MacRae, E., Gleave, A. and Laing, W. (2009) Gene expression studies in kiwifruit and gene over-expression in *Arabidopsis* indicates that GDP-L-galactose guanyltransferase is a major control point of vitamin C biosynthesis. *Journal of Experimental Botany* 60, 765–778.

Chatterjee, I.B., Chatterjee, G.C., Ghosh, N.C., Ghosh, J.J. and Guha, B.C. (1960) Biological synthesis of L-ascorbic acid in animal tissues: conversion of D-glucuronolactone and L-gulonolactone into L-ascorbic acid. *Biochemical Journal* 76, 279–292.

Chun, J., Lee, J., Ye, L., Exler, J. and Eitenmiller, R.R. (2006) Tocopherol and tocotrienol contents of raw and processed fruits and vegetables in the United States diet. *Journal of Food Composition and Analysis* 19, 196–204.

Conklin, P.L., Williams, E.H. and Last, R.L. (1996) Environmental stress sensitivity of an ascorbic acid-deficient Arabidopsis mutant. *Proceeding of the National Academy of Sciences USA* 93, 9970–9974.

Conklin, P.L., Norris, S.R., Wheeler, G.L., Williams, E.H., Smirnoff, N. and Last, R.L. (1999) Genetic evidence for the role of GDP-mannose in plant ascorbic acid (vitamin C) biosynthesis. *Proceedings of the National Academy of Sciences USA* 96, 4198–4203.

Conklin, P.L., Gatzek, S., Wheeler, G.L., Dowdle, J., Raymond, M.J., Rolinski, S., Isupov, M., Littlechild, J.A. and Smirnoff, N. (2006) *Arabidopsis thaliana* VTC4 encodes L-galactose-1-P phosphatase, a plant ascorbic acid biosynthetic enzyme. *Journal of Biological Chemistry* 281, 15662–15670.

Cruz-Rus, E., Botella, M.A., Valpuesta, V. and Gomez-Jimenez, M.C. (2010) Analysis of genes involved in L-ascorbic acid biosynthesis during growth and ripening of grape berries. *Journal of Plant Physiology* 167, 739–748.

Cruz-Rus, E., Amaya, I., Sánchez-Sevilla, J.F., Botella, M.A. and Valpuesta, V. (2011) Regulation of L-ascorbic acid content in strawberry fruits. *Journal of Experimental Botany* 62, 4191–4201.

Cuevas, H.E., Staub, J.E., Simon, P.W. and Zalapa, J.E (2009) A consensus linkage map identifies genomic regions controlling fruit maturity and β-carotene-associated flesh colour in melon (*Cucumis melo* L.). *Theoretical and Applied Genetics* 119, 741–756.

D'Ambrosio, C., Giorio, C., Marino, G., Merendino, I., Petrozza, A., Salfi, L., Stigliani, A.L. and Cellini, F. (2004) Virtually complete conversion of lycopene into β-carotene in fruits of tomato plants transformed with the tomato lycopene β-cyclase (*tlyc-b*) cDNA. *Plant Science* 166, 207–214.

Davey, M.W. and Keulemans, J. (2004) Determining the potential to breed for enhanced antioxidant status in *Malus*: mean inter- and intravarietal fruit vitamin C and glutathione contents at harvest and their evolution during storage. *Journal of Agricultural and Food Chemistry* 52, 8031–8038.

Davey, M.W., Kenis, K. and Keulemans, J. (2006) Genetic control of fruit vitamin C contents. *Plant Physiology* 142, 343–351.

Davuluri, G.R., van Tuinen, A., Fraser, P.D., Manfedonia, A., Newman, R., Burgess, D., Brummell, D.A., King, S.R., Palys, J., Uhlig, J., Bramley, P.M., Pennings, H.M.J. and Bowler, C. (2005) Fruit-specific RNAi-mediated suppression of DET1 enhances carotenoid and flavonoid content in tomatoes. *Nature Biotechnology* 23, 890–895.

DellaPenna, D. (2005) Progress in the dissection and manipulation of vitamin E synthesis. *Trends in Plant Science* 10, 574–579.

Devitt, L.C., Fanning, K., Dietzgen, R.G. and Holton, T.A. (2010) Isolation and functional characterisation of a lycopene β-cyclase gene that controls fruit colour of papaya (*Carica papaya* L.). *Journal of Experimental Botany* 61, 33–39.

Dharmapuri, S., Rosati, C., Pallara, P., Aquilani, R., Bouvier, F., Camara, B. and Giuliano, G. (2002) Metabolic engineering of xanthophyll content in tomato fruits. *FEBS Letters* 519, 30–34.

Dowdle, J., Ishikawa, T., Gatzek, S., Rolinski, S. and Smirnoff, N. (2007) Two genes in *Arabidopsis thaliana* encoding GDP-L-galactose phosphorylase are required for ascorbate biosynthesis and seedling viability. *Plant Journal* 52, 673–689.

El-Gizawi, A.M., Abdallah, M.M.F., Gomaa, H.M. and Mohamed, S.S. (1993) Effect of different shading levels on tomato plants. 2. Yield and fruit quality. *Acta Horticulturae* 323, 349–354.

Eltayeb, A.E., Kawano, N., Badawi, G.H., Kaminaka, H., Sanekata, T., Shibahara, T., Inanaga, S. and Tanaka, K. (2007) Overexpression of monodehydroascorbate reductase in transgenic tobacco confers enhanced tolerance to ozone, salt and polyethylene glycol stresses. *Planta* 225, 1255–1264.

Enfissi, E.M., Barneche, F., Ahmed, I., Lichtlé, C., Gerrish, C., McQuinn, R.P., Giovannoni, J.J., Lopez-Juez, E., Bowler, C., Bramley, P.M. and Fraser, P.D. (2010) Integrative transcript and metabolite analysis of nutritionally enhanced DE-ETIOLATED1 downregulated tomato fruit. *Plant Cell* 22, 1190–1215.

Falk, J. and Munné-Bosch, S. (2010) Tocochromanol functions in plants: antioxidation and beyond. *Journal of Experimental Botany* 61, 1549–1566.

Fanciullino, A.L., Dhuique-Mayer, C., Luro, F., Casanova, J., Morillon, R. and Ollitrault, P. (2006) Carotenoid diversity in cultivated citrus is highly influenced by genetic factors. *Journal of Agricultural and Food Chemistry* 54, 4397–4406.

FAO/WHO (2004) Vitamin and mineral requirements in human nutrition, 2nd edn <http://www.who.int/nutrition/publications/micronutrients/9241546123/en/index.html>.

Fernandez, A.I., Viron, N., Alhagdow, M., Karimi, M., Jones, M., Amsellem, Z., Sicard, A., Czerednik, A., Angenent, G., Grierson, D., May, S., Seymour, G., Eshed, Y., Lemaire-Chamley, M., Rothan, C. and Hilson, P (2009) Flexible tools for gene expression and silencing in tomato. *Plant Physiology* 151, 1729–1740.

Fitzpatrick, T.B, Basset, G.J., Borel, P., Carrari, F., Dellapenna, D., Fraser, P.D., Hellmann, H., Osorio, S., Rothan, C., Valpuesta, V., Caris-Veyrat, C. and Fernie, A.R. (2012) Vitamin deficiency in humans: can plant sciences help? *Plant Cell* 24, 395–414.

Franck, C., Baetens, M., Lammertyn, J., Verboven, P., Davey, M.W. and Nicolaï, B.M. (2003) Ascorbic acid concentration in Cv. conference pears during fruit development and postharvest storage. *Journal of Agricultural and Food Chemistry* 51, 4757–4763.

Fraser, P.D., Enfissi, E.M., Halket, J.M., Truesdale, M.R., Yu, D.M., Gerrish, C. and Bramley, P.M. (2007) Manipulation of phytoene levels in tomato fruit: effects on isoprenoids, plastids, and intermediary metabolism. *Plant Cell* 19, 3194–3211.

Fraser, P.D., Enfissi, E.M. and Bramley, P.M. (2009) Genetic engineering of carotenoid formation in tomato fruit and the potential application of systems and synthetic biology approaches. *Archives of Biochemistry and Biophysics* 483, 196–204.

Funk, C. (1912) The etiology of the deficiency diseases. *Journal of State Medicine* 20, 341–368.

Galiana-Balaguer, L., Roselló, S. and Nuez, F. (2006) Characterisation and selection of balanced sources of variability for breeding tomato (*Lycopersicon*) internal quality. *Genetic Resources in Crop Evolution* 53, 907–923.

Galli, F. and Azzi, A. (2010) Present trends in vitamin E research. *Biofactors* 36, 33–42.

Galpaz, N., Wang, Q., Menda, N., Zamir, D. and Hirschberg, J. (2008) Abscisic acid deficiency in the tomato mutant high-pigment 3 leading to increased plastid number and higher fruit lycopene content. *Plant Journal* 53, 717–730.

Garcia, V., Stevens, R., Gil, L., Gilbert, L., Gest, N., Petit, J., Faurobert, M., Maucourt, M., Deborde, C., Moing, A., Poessel, J.L., Jacob, D., Bouchet, J.-P., Giraudel, J.-L., Gouble, B., Page, D., Alhagdow, M., Massot, C., Gautier, H., Lemaire-Chamley, M., de Daruvar, A., Rolin, D., Usadel, B., Lahaye, M., Causse, M., Baldet, P. and Rothan, C. (2009) An integrative genomic approach for deciphering the complex interactions between ascorbate metabolism and fruit growth and composition in tomato. *Compte-Rendus Biologies* 332, 1007–1021.

Gautier, H., Diakou-Verdin, V., Bénard, C., Reich, M., Buret, M., Bourgaud, F., Poëssel, J.-L., Caris-Veyrat, C. and Génard, M. (2008) How does tomato quality (sugar, acid, and nutritional quality) vary with ripening stage, temperature, and irradiance? *Journal of Agricultural and Food Chemistry* 56, 1241–1250.

Gautier, H., Massot, C., Stevens, R., Sérino, S. and Génard, M. (2009) Regulation of tomato fruit ascorbate content is more highly dependent on fruit irradiance than leaf irradiance. *Annals of Botany* 103, 495–504.

Gest, N., Page, D., Birtic, S., Gouble, B., Gilbert, L., Garchery, C., Causse, M. and Stevens, R. (2010) Response of the fruit antioxidant system to the post-chilling period in two different tomato lines. *Functional Plant Science and Biotechnology* 4, 76–83.

Gilbert, L., Alhagdow, M., Nunes-Nesi, A., Quemener, B., Guillon, F., Bouchet, B., Faurobert, M., Gouble, B., Page, D., Garcia, V., Petit, J., Stevens, R., Causse, M., Fernie, A.R., Lahaye, M., Rothan, C. and Baldet, P. (2009) GDP-D-mannose 3,5-epimerase (GME) plays a key role at the intersection of ascorbate and non-cellulosic cell-wall biosynthesis in tomato. *Plant Journal* 60, 499–508.

Green, M.A. and Fry, S.C. (2005) Vitamin C degradation in plant cells via enzymatic hydrolysis of 4-*O*-oxalyl-L-threonate. *Nature* 433, 83–87.

Guillet, C., Aboul-Soud, M.A., Le Menn, A., Viron, N., Pribat, A., Germain, V., Just, D., Baldet, P., Rousselle, P., Lemaire-Chamley, M. and Rothan, C. (2012) Regulation of the fruit-specific PEP carboxylase *SlPPC2* promoter at early stages of tomato fruit development. *PLoS One* 7, e36795.

Hamner, K.C., Bernstein, L. and Maynard, L.A. (1945) Effects of light intensity, day length, temperature, and other environmental factors on the ascorbic acid content of tomatoes. *Journal of Nutrition* 29, 85–97.

Hancock, R.D., Walker, P.G., Pont, S.D.A., Marquis, N., Vivera, S., Gordon, S.L., Brennan, R.M. and Viola, R. (2007) L-Ascorbic acid accumulation in fruit of *Ribes nigrum* occurs by *in situ* biosynthesis via the L-galactose pathway. *Functional Plant Biology* 34, 1080–1091.

Hirschberg, J. (2001) Carotenoid biosynthesis in flowering plants. *Current Opinion in Plant Biology* 4, 210–218.

Huang, C., He, W., Guo, J., Chang, X., Su, P. and Zhang, L. (2005) Increased sensitivity to salt stress in an ascorbate-deficient *Arabidopsis* mutant. *Journal of Experimental Botany* 56, 3041–3049.

Hugueney, P., Bouvier, F., Badillo, A., Quennemet, J., d'Harlingue, A. and Camara, B. (1996) Developmental and stress regulation of gene expression for plastid and cytosolic isoprenoid pathways in pepper fruits. *Plant Physiology* 111, 619–626.

Hyun, T.K., Rim, Y., Jang, H.-J., Kim, C.H., Park, J., Kumar, R., Lee, S., Kim, B.C., Bhak, J., Nguyen-Quoc, B., Kim, S.-W., Lee, S.Y. and Kim, J.-Y. (2012) De novo transcriptome sequencing of *Momordica cochinchinensis* to identify genes involved in the carotenoid biosynthesis. *Plant Molecular Biology* 79, 413–427.

Ikoma, Y., Komatsu, A., Kita, M., Ogawa, K., Omura, M., Yano, M. and Moriguchi, T. (2001) Expression of a phytoene synthase gene and characteristic carotenoid accumulation during citrus fruit development. *Physiologia Plantarum* 111, 232–238.

Inoue, T., Komoda, H., Uchida, T. and Node, K. (2008) Tropical fruit camu-camu (*Myrciariadubia*) has anti-oxidative and anti-inflammatory properties. *Journal of Cardiology* 52, 127–132.

Ioannidi, E., Kalamaki, M.S., Engineer, C., Pateraki, I., Alexandrou, D., Mellidou, I., Giovannonni, J. and Kanellis, A.K. (2009) Expression profiling of ascorbic acid-related genes during tomato fruit development and ripening and in response to stress conditions. *Journal of Experimental Botany* 60, 663–678.

Isaacson, T., Ronen, G., Zamir, D. and Hirschberg, J. (2002) Cloning of *tangerine* from tomato reveals a carotenoid isomerase essential for the production of β-carotene and xanthophylls in plants. *Plant Cell* 14, 333–342.

Ishikawa, T., Dowdle, J. and Smirnoff, N. (2006) Progress in manipulating ascorbic acid biosynthesis and accumulation in plants. *Physiologia Plantarum* 126, 343–355.

Justi, K.C., Visentainer, J.V., Evelázio de Souza, N. and Matsushita, M. (2000) Nutritional composition and vitamin C stability in stored camu-camu (*Myrciaria dubia*) pulp. *Archivos Latinoamericanos de Nutrición* 50, 405–408.

Karunanandaa, B., Qi, Q., Hao, M., Baszis, S.R., Jensen, P.K., Wong, Y.H., Jiang, J., Venkatramesh, M., Gruys, K.J., Moshiri, F., Post-Beittenmiller, D., Weiss, J.D. and Valentin, H.E. (2005) Metabolically engineered oilseed crops with enhanced seed tocopherol. *Metabolic Engineering* 7, 384–400.

Koh, E., Charoenprasert, S. and Mitchell, A.E. (2012) Effects of industrial tomato paste processing on ascorbic acid, flavonoids and carotenoids and their stability over one-year storage. *Journal of the Science of Food and Agriculture* 92, 23–28.

Kuzniak, E. and Skłodowska, M. (2005) Fungal pathogen-induced changes in the antioxidant systems of leaf peroxisomes from infected tomato plants. *Planta* 222, 192–200.

Laing, W.A., Wright, M.A., Cooney, J. and Bulley, S.M. (2007) The missing step of the L-galactose pathway of ascorbate biosynthesis in plants, an L-galactoseguanyltransferase, increases leaf ascorbate content. *Proceedings of the National Academy of Sciences USA* 104, 9534–9539.

Lee, S.K. and Kader, A.A. (2000) Preharvest and postharvest factors influencing vitamin C content of horticultural crops. *Postharvest Biology and Technology* 20, 207–220.

Levine, M., Conry-Cantilena, C., Wang, Y., Welch, R.W., Washko, P.W., Dhariwal, K.R., Park, J.B., Lazarev, A., Graumlich, J.F., King, J. and Cantilena, L.R. (1996) Vitamin C pharmacokinetics in healthy volunteers: evidence for a recommended dietary allowance. *Proceedings of the National Academy of Sciences USA* 93, 3704–3709.

Lewinsohn, E., Sitrit, Y., Bar, E., Azulay, Y., Meir, A., Zamir, D. and Tadmor, Y. (2005) Carotenoid pigmentation affects the volatile composition of tomato and watermelon fruits, as revealed by comparative genetic analyses. *Journal of Agricultural and Food Chemistry* 53, 3142–3148.

Li, M., Ma, F., Shang, P., Zhang, M., Hou, C. and Liang, D. (2009) Influence of light on ascorbate formation and metabolism in apple fruits. *Planta* 230, 39–51.

Lieberman, M., Segev, O., Gilboa, N., Lalazar, A. and Levin, I. (2004) The tomato homolog of the gene encoding UV-damaged DNA binding protein 1 (DDB1) underlined as the gene that causes the *high pigment-1* mutant phenotype. *Theoretical and Applied Genetics* 108, 1574–1581.

Linster, C.L. and Clarke, S.G. (2008) L-Ascorbate biosynthesis in higher plants: the role of VTC2. *Trends in Plant Science* 13, 567–573.

Linster, C.L. and Van Schaftingen, E. (2007) Vitamin C. Biosynthesis, recycling and degradation in mammals. *Federation of European Biochemical Societies Letters* 274, 1–22.

Liu, Y., Roof, S., Ye, Z., Barry, C., van Tuinen, A., Vrebalov, J., Bowler, C. and Giovannoni, J. (2004) Manipulation of light signal transduction as a means of modifying fruit nutritional quality in tomato. *Proceedings of the National Academy of Sciences USA* 101, 9897–9902.

Loewus, F.A. and Loewus, M.W. (1987) Biosynthesis and metabolism of ascorbic-acid in plants. *Critical Reviews in Plant Sciences* 5, 101–119.

López, M.A. and Martos, F.C. (2004) Iron availability: an updated review. *International Journal of Food Sciences and Nutrition* 55, 597–606.

Lorence, A., Chevone, B.I., Mendes, P. and Nessler, C.L. (2004) Myo-inositol oxygenase offers a possible entry point into plant ascorbate biosynthesis. *Plant Physiology* 134, 1200–1205.

Ma, G., Zhang, L., Kato, M., Yamawaki, K., Kiriiwa, Y., Yahata, M., Ikoma, Y. and Matsumoto, H. (2012) Effect of blue and red LED light irradiation on β-cryptoxanthin accumulation in the flavedo of citrus fruits. *Journal of Agricultural and Food Chemistry* 60, 197–201.

Madamba, L.S.P., Morales, E.R. and Villanueva, T.G. (1974) Some factors affecting the ascorbic acid contents of tomato fruits. *Philippine Journal of Nutrition* 27, 7–12.

Malacrida, C., Valle, E.M. and Boggio, S.B. (2006) Postharvest chilling induces oxidative stress response in the dwarf tomato cultivar Micro-Tom. *Physiologia Plantarum* 127, 10–18.

Massot, C., Stevens, R., Genard, M., Longuenesse, J.J. and Gautier, H. (2012) Light affects ascorbate content and ascorbate-related gene expression in tomato leaves more than in fruits. *Planta* 235, 153–163.

Mathieu, S., Cin, V.D., Fei, Z., Li, H., Bliss, P., Taylor, M.G., Klee, H.J. and Tieman, D.M. (2009) Flavour compounds in tomato fruits: identification of loci and potential pathways affecting volatile composition. *Journal of Experimental Botany* 60, 325–337.

Matringe, M., Ksas, B., Rey, P. and Havaux, M. (2008) Tocotrienols, the unsaturated forms of vitamin E, can function as antioxidants and lipid protectors in tobacco leaves. *Plant Physiology* 147, 764–778.

McCollum, J. (1946) Effect of sunlight exposure on the quality of constituents of tomato fruits. *Proceedings of the American Society of Horticultural Science* 48, 413–416.

Mène-Saffrané, L. and DellaPenna, D. (2009) Biosynthesis, regulation and functions of tocochromanols in plants. *Plant Physiology and Biochemistry* 48, 301–309.

Muckenschnabel, I., Goodman, B.A., Williamson, B., Lyon, G.D. and Deighton, N. (2002) Infection of leaves of Arabidopsis thaliana by Botrytis cinerea: changes in ascorbic acid, free radicals and lipid peroxidation products. *Journal of Experimental Botany* 53, 207–214.

Noctor, G. and Foyer, C.H. (1998) Ascorbate and glutathione: keeping active oxygen under control. *Annual Review of Plant Physiology and Plant Molecular Biology* 49, 249–279.

Pastori, G.M., Kiddle, G., Antoniw, J., Bernard, S., Veljovic-Jovanovic, S., Verrier, P.J., Noctor, G. and Foyer, C.H. (2003) Leaf vitamin C contents modulate plant defense transcripts and regulate genes that control development through hormone signalling. *Plant Cell* 5, 939–951.

Pavet, V., Olmos, E., Kiddle, G., Mowla, S., Kumar, S., Antoniw, J., Alvarez, M.E. and Foyer, C.H. (2005) Ascorbic acid deficiency activates cell death and disease resistance responses in Arabidopsis. *Plant Physiology* 139, 1291–1303.

Prescott, A.G. and John, P. (1996) Dioxygenases: molecular structure and role in plant metabolism. *Annual Review of Plant Physiology and Plant Molecular Biology* 47, 245–271.

Quadrana, L., Rodriguez, M.C., Lopez, M., Bermudez, L., Nunes-Nesi, A., Fernie, A.R., Descalzo, A., Asis, R., Rossi, M., Asurmendi, S. and Carrari, F. (2011) Coupling virus-induced gene silencing to exogenous green fluorescence protein expression provides a highly efficient system for functional genomics in Arabidopsis and across all stages of tomato fruit development. *Plant Physiology* 156, 1278–1291.

Razavi, F., Keulemans, J. and Davey, M.W. (2005) A study of the L-ascorbate biosynthetic capacity of apple fruit. *Communications in Agricultural Applied Biological Science* 70, 213–216.

Rébeillé, F. and Douce, R. (eds) (2011a) Biosynthesis of vitamins in plants, Part A: vitamins A, B1, B2, B3, B5. *Advances in Botanical Research*, Vol. 58. Elsevier Shannon, Ireland.

Rébeillé, F. and Douce, R., (eds) (2011b) Biosynthesis of vitamins in plants, Part B: vitamins B6, B7, B9, C, D, E. *Advances in Botanical Research*, Vol. 59. Elsevier Shannon, Ireland.

Rippert, P., Scimemi, C., Dubald, M. and Matringe, M. (2004) Engineering plant shikimate pathway for production of tocotrienol and improving herbicide resistance. *Plant Physiology* 134, 92–100.

Rivera-Pastrana, D.M., Yahia, E.M. and González-Aguilar, G.A. (2010) Phenolic and carotenoid profiles of papaya fruit (*Carica papaya* L.) and their contents under low temperature storage. *Journal of the Science of Food and Agriculture* 90, 2358–2365.

Rodriguez-Amaya, D.B. (1999) Latin American food sources of carotenoids. *Archivos Latinoamericanos de Nutrición* 49, 74–84.

Ronen, G., Carmel-Goren, L., Zamir, D. and Hirschberg, J. (2000) An alternative pathway to β-carotene formation in plant chromoplasts discovered by map-based cloning of *Beta* and *old-gold* colour mutations in tomato. *Proceedings of the National Academy of Sciences USA* 97, 11102–11107.

Roselló, S., Adalid, A.M., Cebolla-Cornejo, J. and Nuez, F. (2011) Evaluation of the genotype, environment and their interaction on carotenoid and ascorbic acid accumulation in tomato germplasm. *Journal of Agricultural and Food Chemistry* 9, 1014–1021.

Schauer, N., Zamir, D. and Fernie, A.R. (2005) Metabolic profiling of leaves and fruit of wild species tomato: a survey of the *Solanum lycopersicum* complex. *Journal of Experimental Botany* 56, 297–307.

Schauer, N., Semel, Y., Roessner, U., Gur, A., Balbo, I., Carrari, F., Pleban, T., Perez-Melis, A., Bruedigam, C., Kopka, J., Willmitzer, L., Zamir, D. and Fernie, A.R. (2006) Comprehensive metabolic profiling and phenotyping of interspecific introgression lines for tomato improvement. *Nature Biotechnology* 24, 447–454.

Schweiggert, R.M., Steingass, C.B., Esquivel, P. and Carle, R. (2012) Chemical and morphological characterisation of Costa Rican papaya (*Carica papaya* L.) hybrids and lines with particular focus on their genuine carotenoid profiles. *Journal of Agricultural and Food Chemistry* 60, 2577–2585.

Seo, Y.S., Kim, S.J., Harn, C.H. and Kim, W.T. (2011) Ectopic expression of apple fruit homogentisatephytyltransferase gene (MdHPT1) increases tocopherol in transgenic tomato (*Solanum lycopersicum* cv. Micro-Tom) leaves and fruits. *Phytochemistry* 72, 321–329.

Shahak, Y., Ratner, K., Giller, Y.E., Zur, N., Or, E., Gussakovsky, E.E., Stern, R., Sarig, P., Raban, E., Harcavi, E., Doron, I. and Greenblat-Avron, Y. (2006) Improving solar energy utilisation, productivity and fruit quality in orchards and vineyards by photoselective netting. ISHS *Acta Horticulturae* 772: XXVII International Horticultural Congress – IHC2006: International Symposium on Enhancing Economic and Environmental Sustainability of Fruit Production in a Global Economy.

Singh, R.K., Ali, S.A., Nath, P. and Sane, V.A. (2011) Activation of *p*-hydroxyphenylpyruvate dioxygenase leads to increased tocopherol levels during ripening in mango. *Journal of Experimental Botany* 62, 3375–3385.

Smirnoff, N. (2000) Ascorbic acid: metabolism and functions of a multi-facetted molecule. *Current Opinion in Plant Biology* 3, 229–235.

Smirnoff, N. and Wheeler, G.L. (2000) Ascorbic acid in plants: biosynthesis and function. *Critical Reviews in Biochemistry and Molecular Biology* 35, 291–314.

Steinmetz, K.A. and Potter, J.D. (1996) Vegetables, fruit, and cancer prevention: a review. *Journal of the American Dietetic Association* 96, 1027–1039.

Stevens, R., Buret, M., Duffe, P., Garchery, C., Baldet, P., Rothan, C. and Causse, M. (2007) Candidate genes and quantitative trait loci affecting fruit ascorbic acid content in three tomato populations. *Plant Physiology* 143, 1943–1953.

Stevens, R., Page, D., Gouble, B., Garchery, C., Zamir, D. and Causse, M. (2008) Tomato fruit ascorbic acid content is linked with monodehydroascorbate reductase activity and tolerance to chilling stress. *Plant, Cell and Environment* 31, 1086–1096.

Sun, L., Yuan, B., Zhang, M., Wang, L., Cui, M., Wang, Q. and Leng, P. (2012) Fruit-specific RNAi-mediated suppression of *SlNCED1* increases both lycopene and β-carotene contents in tomato fruit. *Journal of Experimental Botany* 63, 3097–3108.

Svelander, C.A., Lopez-Sanchez, P., Pudney, P.D., Schumm, S. and Alminger, M.A. (2011) High pressure homogenisation increases the in vitro bioaccessibility of α- and β-carotene in carrot emulsions but not of lycopene in tomato emulsions. *Journal of Food Science* 76, 215–225.

Tranbarger, T.J., Dussert, S., Joët, T., Argout, X., Summo, M., Champion, A., Cros, D., Omore, A., Nouy, B. and Morcillo, F. (2011) Regulatory mechanisms underlying oil palm fruit mesocarp maturation, ripening, and functional specialisation in lipid and carotenoid metabolism. *Plant Physiology* 156, 564–584.

Valko, M., Leibfritz, D., Moncola, J., Cronin, M.T.D., Mazura, M. and Telser, J. (2007) Free radicals and antioxidants in normal physiological functions and human disease. *International Journal of Biochemistry and Cell Biology* 39, 44–84.

Veltman, R.H., Sanders, M.G., Persijn, S.T., Peppelenbos, H.W. and Oosterhaven, J. (1999) Decreased ascorbic acid levels and brown core development in pears (*Pyrus communis* L. cv. Conference). *Physiologia Plantarum* 107, 39–45.

Vogel, J.T., Tieman, D.M., Sims, C.A., Odabasi, A.Z., Clark, D.G. and Klee, H.J. (2010) Carotenoid content impacts flavor acceptability in tomato (*Solanum lycopersicum*). *Journal of the Science of Food and Agriculture* 90, 2233–2240.

Wang, S., Liu, J., Feng, Y., Niu, X., Giovannoni, J. and Liu, Y. (2008) Altered plastid levels and potential for improved fruit nutrient content by downregulation of the tomato DDB1-interacting protein CUL4. *Plant Journal* 55, 89–103.

Wheeler, G.L., Jones, M.A. and Smirnoff, N. (1998) The biosynthetic pathway of vitamin C in higher plants. *Nature* 393, 365–369.

Wolucka, B.A. and Van Montagu, M. (2003) GDP-mannose 3′,5′-epimerase forms GDP-L-gulose, a putative intermediate for the *de novo* biosynthesis of vitamin C in plants. *Journal of Biological Chemistry* 278, 47483–47490.

Yamamoto, A., Bhuiyan, M.N., Waditee, R., Tanaka, Y., Esaka, M., Oba, K., Jagendorf, A.T. and Takabe, T. (2005) Suppressed expression of the apoplastic ascorbate oxidase gene increases salt tolerance in tobacco and *Arabidopsis* plants. *Journal of Experimental Botany* 56, 1785–1796.

Yang, F.M., Li, H.M., Li, F., Xin, Z.H., Zhao, L.Y., Zheng, Y.H. and Hu, Q.H. (2010) Effect of nano-packing on preservation quality of fresh strawberry (*Fragaria ananassa* Duch. cv Fengxiang) during storage at 4 degrees C. *Journal of Food Science* 75, 236–240.

Yin, L., Wang, S., Eltayeb, A.E., Uddin, M.I., Yamamoto, Y., Tsuji, W., Takeuchi, Y. and Tanaka, K. (2010) Overexpression of dehydroascorbate reductase, but not monodehydroascorbate reductase, confers tolerance to aluminum stress in transgenic tobacco. *Planta* 231, 609–621.

Zechmann, B., Stumpe, M. and Mauch, F. (2011) Immunocytochemical determination of the subcellular distribution of ascorbate in plants. *Planta* 233, 1–12.

Zhang, C., Liu, J., Zhang, Y., Cai, X., Gong, P., Zhang, J., Wang, T., Li, H. and Ye, Z. (2011) Overexpression of *SlGMEs* leads to ascorbate accumulation with enhanced oxidative stress, cold, and salt tolerance in tomato. *Plant Cell Reporter* 30, 389–398.

Zhang, W., Lorence, A., Gruszewski, H.A., Chevone, B.I. and Nessler, C.L. (2009) *AMR1*, an Arabidopsis gene that coordinately and negatively regulates the mannose/L-galactose ascorbic acid biosynthetic pathway. *Plant Physiology* 150, 942–950.

Zhang, Z., Wang, J., Zhang, R. and Huang, R. (2012) The ethylene response factor AtERF98 enhances tolerance to salt through the transcriptional activation of ascorbic acid synthesis in Arabidopsis. *Plant Journal* 71, 273–287.

Zorrilla-Fontanesi, Y., Cabeza, A., Domínguez, P., Medina, J.J., Valpuesta, V., Denoyes-Rothan, B., Sánchez-Sevilla, J.F. and Amaya, I. (2011) Quantitative trait loci and underlying candidate genes controlling agronomical and fruit quality traits in octoploid strawberry (*Fragaria* × *ananassa*). *Theoretical and Applied Genetics* 123, 755–778.

9 Polyphenols

Agnès Ageorges, Véronique Cheynier* and Nancy Terrier
INRA, Montpellier cedex, France

9.1 Introduction

Polyphenols are a large class of plant secondary metabolites, ubiquitous in plants and structurally diverse. The earlier definition of polyphenols, proposed by Bate-Smith and Swain (1962), implied the ability to precipitate alkaloids and proteins from solution, while many recent papers refer to all phenolic compounds as polyphenols. In fact, the term polyphenols should be restricted to plant phenolic compounds 'derived exclusively from the shikimate derived phenylpropanoid and/or the polyketide pathway(s), featuring more than one phenolic ring and being devoid of any nitrogen-based functional group in their most basic structural expression', as stated recently by Quideau *et al.* (2011). This definition covers several groups, including flavonoids, hydroxystilbenes, lignans and benzoic acid derivatives such as gallotannins and ellagitannins. Wide structural diversity is encountered within each group, and especially the flavonoid family, comprising over 8000 molecules (Andersen and Markham, 2006).

Phenolic compounds are found throughout the plant kingdom, but polyphenols are represented mostly in vascular plants, as reviewed recently (Lattanzio *et al.*, 2012). Some polyphenols are rather common, while others are restricted to certain botanical families. Phenolic profiles are genetically determined and commonly used for taxonomic studies, providing evidence of phylogenic proximities and markers in varietal authentication. For instance, the relative abundance of anthocyanin diglucosides distinguishes berries (and wines) of non-*vinifera* or hybrid grape varieties from those of *Vitis vinifera* cultivars. Polyphenols are also differently distributed within the plant, in both time and space, in relation to their particular roles.

Fruit polyphenol composition has been studied extensively, in particular with respect to food quality and nutrition issues. Indeed, polyphenols contribute to major organoleptic properties of fruits, including colour, sensitivity to enzymatic browning and taste properties such as bitterness and astringency. They also attract considerable interest for their potential role in the health beneficial effects associated with dietary consumption of fruits and vegetables.

After reviewing the structures and properties of major fruit polyphenols, this chapter will discuss fruit polyphenol composition and sources of variations, and highlight some of the biosynthetic and regulatory processes.

* cheynier@supagro.inra.fr

 Fruit Ripening: Physiology, Signalling and Genomics (eds P. Nath *et al.*)

9.2 Structures and Occurrence of Major Fruit Polyphenols

Polyphenols are found in most fruit species. Fruit polyphenols are represented primarily by flavonoids, including some widespread families such as anthocyanins, flavan-3-ols and flavonols, and rather uncommon groups such as flavanones and dihydrochalcones. Major non-flavonoid polyphenols in fruits are ellagitannins and ellagic acid conjugates. Other families, such as stilbenoids and lignans, often mentioned for their potential health effects, are also encountered in fruits, in minor quantities. Information available in the literature on the occurrence and quantities of phenolic compounds in common foods has recently been compiled in the Phenol-Explorer database (Neveu *et al.*, 2010). Specific information on fruit flavonoids can also be found in the USDA flavonoid database (http://www.ars.usda.gov/ba/bhnrc/ndl).

9.2.1 Flavonoids

Flavonoids share a common structure (C6–C3–C6) consisting of two aromatic rings (A and B) and an oxygen heterocycle (C). They are subdivided into several subclasses that differ by the oxidation state of the heterocycle and position of the B-ring, as illustrated in Fig. 9.1(a–h).

Anthocyanins

Anthocyanins are the pigments of most red, blue and black fruits and are rather common. In a recent screening, they were detected in 14 fruits out of 25 (Wu and Prior, 2005). Fruit anthocyanins are based on six aglycones (anthocyanidins) that differ by the number of hydroxyl groups on their B-ring and their methylation pattern (Fig. 9.1a) and are glycosylated in the 3-position. Pelargonidin is often missing, while cyanidin derivatives are ubiquitous and delphinidin derivatives are found in some species (e.g. grape, blueberry and cranberry). Methylation is commonly encountered in both the cyanidin and delphinidin series. Further diversity arises from the nature of the sugar (monosaccharide, often glucose but also galactose, xylose, rhamnose and arabinose, or disaccharide, such as rutinose, sambubiose and sophorose) and from additional substituents (glycosylation in the 5-position, acylation of the sugar with aliphatic (e.g. acetic, malic and oxalic) or hydroxycinnamic (*p*-coumaric, caffeic) acids). Anthocyanin profiles are extremely variable and are characteristic of each plant species (Mazza and Miniati, 1993; Wu and Prior, 2005), with only two anthocyanins in apples and peaches, and up to 31 in 'Concord' grape. Quantitatively, large varietal differences are also observed: for example, anthocyanin content in grape berries ranges from zero in white cultivars to 5 g kg^{-1} in the teinturier cultivar 'Alicante Bouschet' (Mattivi *et al.*, 2006).

Flavonols

Flavonols (Fig. 9.1b) are widespread in fruits and are structurally diverse, differing by the B-ring substitution pattern, and the nature and position of sugar substituents, which can also be acylated. According to the USDA database, flavonols are present only in small quantities in fruits, but concentrations in the range 10–50 mg per 100 g are found in some berries.

Flavan-3-ols

Flavan-3-ols (Fig. 9.1c) comprise monomers, oligomers and polymers, called proanthocyanidins (PAs, syn. condensed tannins). The presence of asymmetric carbons in the saturated C-ring gives rise to several isomers (2,3-*trans* (2R,3R; 2S,3R) and 2,3-*cis* (2R,3S; 2S,3S), designated by the prefix epi-). In PAs, flavan-3-ol units are linked by C4–C6 or C4–C8 linkages in the B-type series, plus C7–O–C2 or C5–O–C2 bonds in the A-type series. PAs are present in most fruits, except citrus, cantaloupe, watermelon and tomato (Gu *et al.*, 2003; Hellstrom *et al.*, 2009). Procyanidins, consisting of catechin and epicatechin units, are the most widespread,

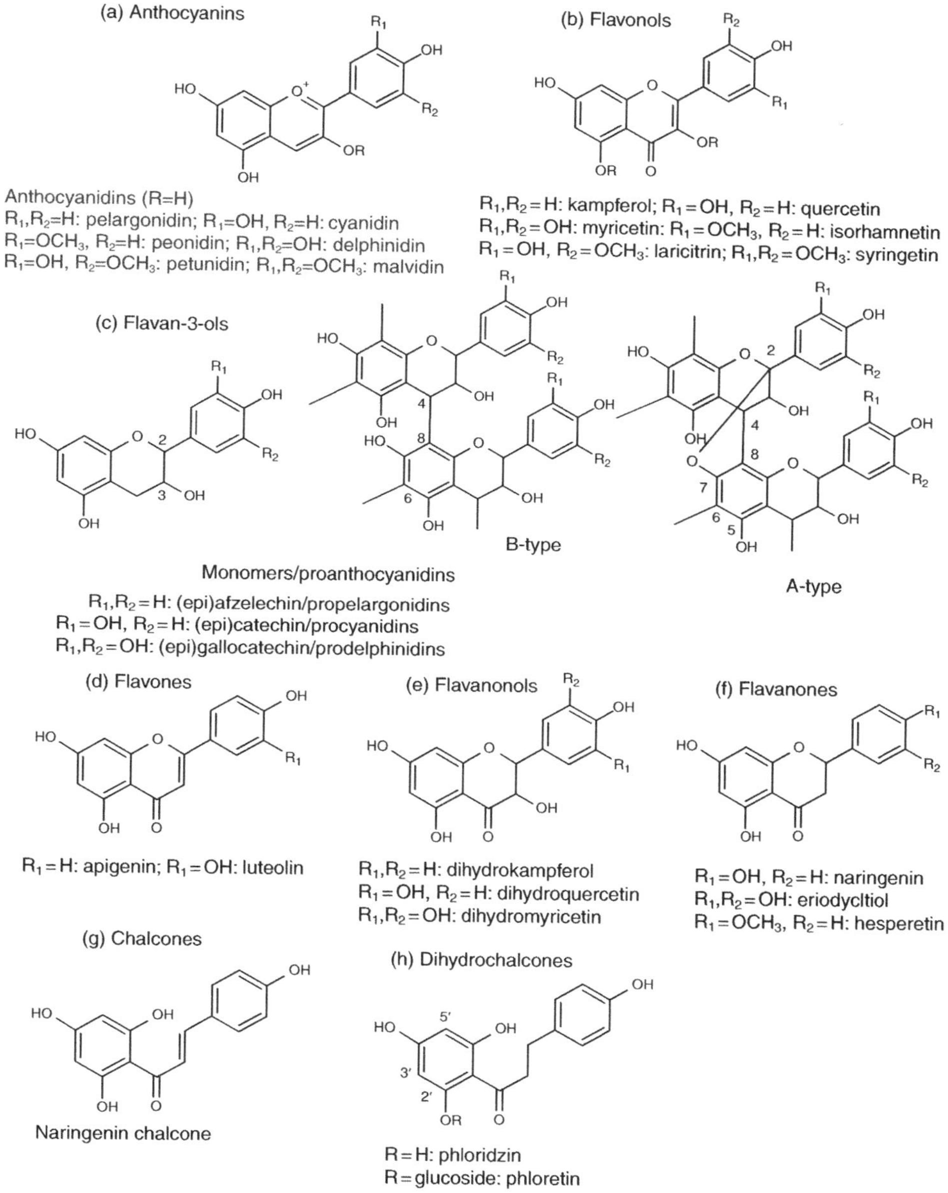

Fig. 9.1. Structures of the major fruit flavonoids.

but strawberry and raspberry fruit also contain (epi)afzelechin units, and blackcurrant, persimmon, grape and banana contain (epi)gallocatechin units. Flavan-3-ol glycosides are found in cherry, kiwi and various nuts and galloylated derivatives in grape and persimmon fruit. Most fruits contain only B-type PAs, but A-type ones are encountered in cranberry, plum, avocado and groundnut (Gu *et al.*, 2003).

PA-rich fruits include persimmons and most berries, whereas citrus fruits are devoid of them. Quantities ranging from 4 mg (in banana and kiwi) to 540 mg (in chockeberries) per 100 g of fresh weight (FW) have been determined by normal-phase high-performance liquid chromatography (HPLC) analysis (Gu *et al.*, 2004). However, PA content is often underestimated. Methods relying upon acid-catalysed depolymerization in the presence of a nucleophile followed by HPLC analysis showed that they are the most abundant polyphenols in common fruits such as apple (Guyot *et al.*, 2002; Wojdylo *et al.*, 2008), grape (Mané *et al.*, 2007) and strawberry (Buendia *et al.*, 2010), accounting for 71–90% of total phenolic compounds in apple (Vrhovsek *et al.*, 2004) and 60–93% in grape (Mané *et al.*, 2007). Large varietal differences are also reported and contents ranging from 1.3 to 19.8 g kg^{-1} of dry weight (DW) have been determined in apple varieties (Wojdylo *et al.*, 2008). Moreover, PAs differ in terms of the average degree of polymerization (aDP). For instance, aDP values in apples varied from 3.2 to 28.7 (Wojdylo *et al.*, 2008), and higher values (up to 100) have been reported in cider apples (Guyot *et al.*, 2001).

In addition, PAs are only partly extracted by the usual solvents. The unextractable fraction can be determined by performing acid-catalysed depolymerization directly on the plant material (Guyot *et al.*, 2001) or on the residue after extraction (Downey *et al.*, 2003; Verries *et al.*, 2008; Hellstrom *et al.*, 2009). The proportion of extractable PAs reported by Hellstrom *et al.* (2009) varied from 0% in banana to 90–95% in apple, although much lower values (10%) were obtained for apples in another study (Arranz *et al.*, 2009). The poor extraction rate of banana PAs may be related to their high aDP, as higher-molecular-weight PAs adsorb to plant cell-wall material (Le Bourvellec and Renard, 2005). Differences in sample preparation (freeze drying versus freezing), in extraction and analysis protocols, and in the development stage (green versus ripe) may explain the large discrepancy between the PA levels reported in banana (cv. 'Cavendish') by different authors: 420–500 mg per 100 g FW, extractable (Uclés Santos *et al.*, 2010); 64 mg per 100 g fresh weight, unextractable (Hellstrom *et al.*, 2009); and 4 mg per 100 g FW, extractable (Gu *et al.*, 2004).

Other flavonoids

Flavonoids also include minor groups such as flavones, flavanones (syn. dihydroflavones), flavanonols (syn. dihydroflavonols), chalcones and dihydrochalcones.

Flavones (Fig. 9.1d) and flavanonols (Fig. 9.1e) are seldom present in fruits, but small amounts of flavones (7-glycosides and 6- and/or 8-C-glucosides of apigenin and luteolin) are found in citrus fruits (Gil-Izquierdo *et al.*, 2001) and dihydroflavonol 3-rhamnosides in grapes (Trousdale and Singleton, 1983; Vitrac *et al.*, 2002).

Fruit flavanones (Fig. 9.1f) are usually encountered as 7-*O*-diglycosides (e.g. naringin (naringenin 7-*O*-neohesperidoside), narirutin (naringenin 7-*O*-rutinoside) and hesperidin (hesperetin 7-*O*-rutinoside)), along with smaller amounts of C-glycosides. Naringenin derivatives have been detected in tomato (Slimestad and Verheul, 2005; Vallverdu-Queralt *et al.*, 2010). However, flavanones are present in large amounts only in citrus fruits. Flavanone compositions are specific for the different *Citrus* species (Rouseff *et al.*, 1987). Major flavanone glycosides are eriocitrin (eriodictyol 7-*O*-rutinoside) and hesperidin in lemon and lime, and naringin and narirutin in grapefruit (Mouly *et al.*, 1994). Sweet oranges are a rich source of hesperidin and narirutin and contain polymethoxyflavones (Gil-Izquierdo *et al.*, 2001). Flavanone glycosides are present in citrus juices in substantial amounts (up to about 700–800 mg l^{-1}) in hand-squeezed juice (Gil-Izquierdo *et al.*, 2001).

Chalcones (Fig. 9.1g) are a specific group of flavonoids in which the C-ring is open (hence the different C numbering). Naringenin chalcone is an upstream intermediate in the flavonoid biosynthesis pathway and does not usually accumulate in plants but has been detected in tomato fruit (Hunt and Baker, 1980). It is the major flavonoid in cherry tomato, ranging from 7 to 28 mg per 100 g FW (Slimestad and Verheul, 2005). Dihydrochalcones (Fig. 9.1h) have been detected in tomato (namely the 3′,5′ C-diglucosides of phloretin (Slimestad *et al.*, 2008) and C-diglycosides of phloridzin (phloretin 2′-O-glucoside) (Vallverdu-Queralt *et al.*, 2010)) but this family is mostly represented in apples, with phloridzin, phloretin-xyloglucoside and 3-hydroxyphloridzin accounting together for 2–6% of the total phenolic composition (Vrhovsek *et al.*, 2004).

Flavonoid reaction products

All flavonoids are highly reactive. Their reactions are well documented in food transformation processes, and there is some evidence of their occurrence *in vivo*. Some changes have been attributed to oxidation, especially for PA in seeds (Kennedy *et al.*, 2000), and to formation of covalent linkages between PAs and plant cell walls (Matthews *et al.*, 1997; Kennedy *et al.*, 2001). The seeds of the transparent testa *Arabidopsis* mutant, *tt10*, missing a laccase gene, contains much higher PA levels and a higher ratio of flavonol monomers to flavonol dimers than the brown wild-type seeds, confirming that seed coat browning results from flavonoid oxidation (Pourcel *et al.*, 2005), as postulated in grape seeds (Kennedy *et al.*, 2000). Various types of anthocyanin-derived pigments, including flavanol–anthocyanin adducts, pyranoanthocyanins and anthocyanin polymers, have also been detected in fruits such as strawberry (Andersen *et al.*, 2004; Fossen *et al.*, 2004), blackcurrants (Lu *et al.*, 2000), grapes (Vidal *et al.*, 2004a,b; Gonzalez-Paramas *et al.*, 2006) and cranberries (Tarascou *et al.*, 2011), suggesting that they can form *in planta*. Condensation products of flavan-3-ols with acetaldehyde have been detected in persimmon (Tanaka *et al.*, 1994). Similar products in which flavan-3-ols are linked through methylmethine bridges, often called 'ethyl bridges', to anthocyanins and flavonols have also been detected in cranberry extracts (Tarascou *et al.*, 2011).

9.2.2 Hydrolysable tannins

Ellagitannins and gallotannins, that is, multiple esters of hexahydroxydiphenic acid group(s) or gallic acid groups, respectively, with a polyol (usually glucose), belong to the hydrolysable tannin class. Hydrolysis of ellagitannins releases hexahydroxydiphenic acid (Fig. 9.2a), which spontaneously lactonizes to ellagic acid (Fig. 9.2b), while hydrolysis of gallotannins yields gallic acid (Fig. 9.2c). These reactions are commonly used to detect and quantify them. Their occurrence in fruits is rather limited. Gallotannins have been reported in mangoes and pomegranate. Ellagic acid conjugates and ellagitanins are found in strawberry, pomegranates, *Rubus* species (e.g. raspberries, blackberries and cloudberries), muscadine grapes and in some nuts (e.g. hazelnuts, walnuts and pecans) (Clifford and Scalbert, 2000; Törrönen, 2009). When screened in 33 fruits consumed in Finland, ellagitannins were detected only in five species of berries, with contents ranging from 1 mg per 100 g FW (in sea buckthorn) to 330 mg per 100 g FW (in cloudberry) (Koponen *et al.*, 2007). Major ellagitannins in *Rubus* species (Mullen *et al.*, 2003) and in strawberry (Buendia *et al.*, 2010) are the dimeric sanguiin H-6 (Fig. 9.2d) and the trimeric lambertianin. Ellagitannins are abundant (up to 1.9 g l^{-1} found in juice due

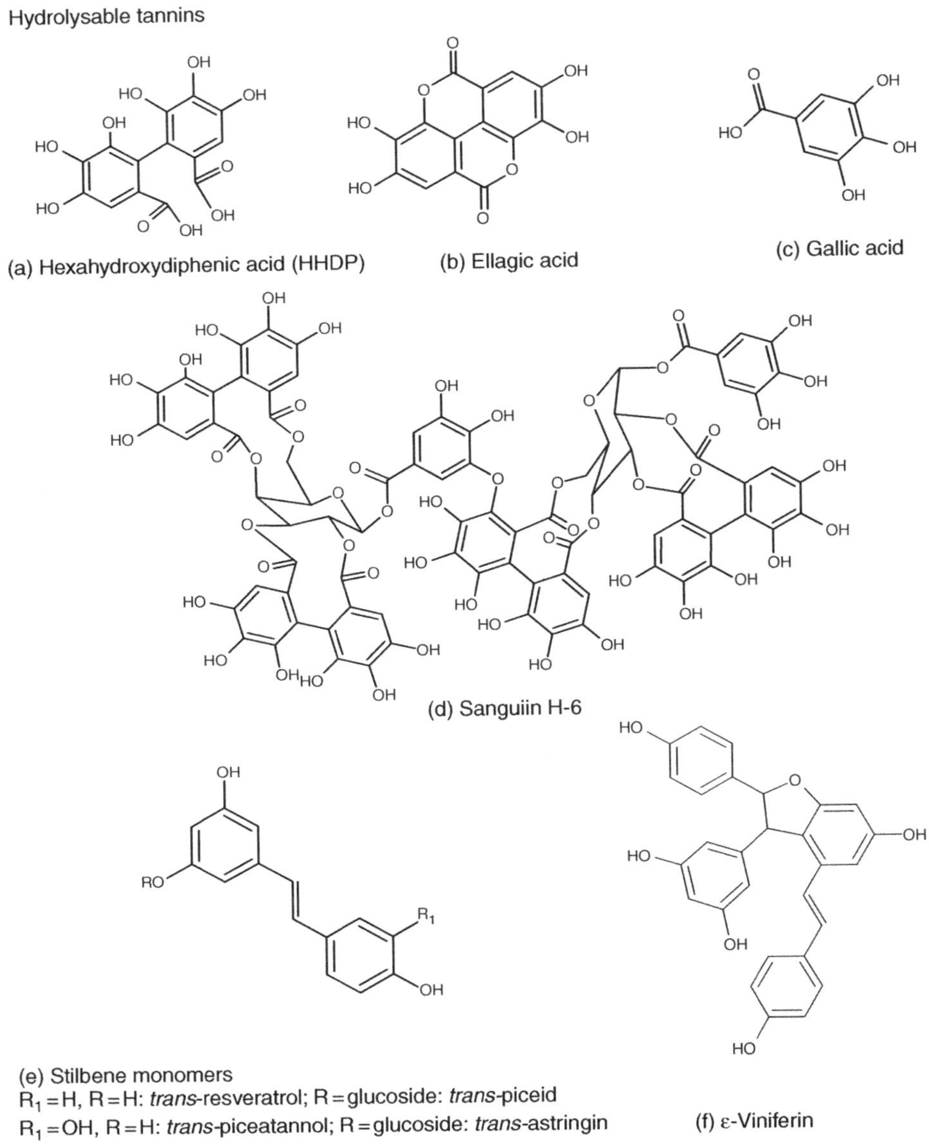

Fig. 9.2. Structure of some non-flavonoid polyphenols found in fruits.

to extraction from the rind) in pomegranate where they are represented mostly by punicalagins and ellagic acid (Gil *et al.*, 2000). Rather high levels (360–910 mg kg^{-1}) of ellagitannins have also been reported in muscadine grape (Lee *et al.*, 2005).

9.2.3 Other polyphenols

Stilbenoids, often called stilbenes, are polyhydroxystilbenes (C6–C2–C6), that is, hydroxy and methoxy derivatives of stilbene (1,2-diphenylethylene), as well as their glycosides and polymers. Stilbenes have been reported in grapes and berries including blackcurrant, redcurrant, lingonberry, strawberry (Ehala *et al.*, 2005), bilberry and blueberry (Moze *et al.*, 2011). Major representatives are resveratrol, piceatannol and their 3-*O*-β-D-glucosides (R=glucoside), piceid and astringin (Fig. 9.2e), found in the *trans* and *cis* forms. A variety of oligostilbenes are also found in grape, including (+)-ε-viniferin, a resveratrol cylic dehydrodimer (Fig. 9.2f), other dimers derived from it, and several trimers (e.g. α-viniferin), and tetramers such as β-viniferin and hopeaphenols (Takaya *et al.*, 2002; Amira-Guebailia *et al.*, 2006). Other minor polyphenol classes, such as isoflavones, coumestans and lignans, are attracting much interest because of their phyto-oestrogenic properties (Kuhnle *et al.*, 2007). They occur mostly in foods as glycosides but can be hydrolysed by intestinal enzymes. Their major dietary sources are legumes (soybean and chickpea for isoflavones, legume sprouts for coumestans) and cereals for lignans, but low amounts have also been found in fruits that contribute about 30% of the total phytoestrogen intake in western diets (Carmichael *et al.*, 2011). Major fruit sources of isoflavones and lignans are citrus fruits, although higher concentrations of lignans are found in apricot, watermelon, blackcurrant and gooseberry (a few 100 μg per 100 g DW; Kuhnle *et al.*, 2007). Citrus fruits also contribute small amounts of coumestrol (15 μg per 100 g DW; Kuhnle *et al.*, 2007), accounting for about half the dietary intake (Carmichael *et al.*, 2011).

9.3 Polyphenols and Fruit Quality

9.3.1 Colour properties

Polyphenols comprise a variety of pigments: red and blue anthocyanins, yellow flavonols and products formed by oxidative reactions that are responsible for the brown colour of seed coats, as explained previously under ‘Flavonoid reaction products’.

Anthocyanins are usually represented (and analysed) under their flavylium cation form, whose colour shifts from red to purple as the number of B-ring substituents increases. However, this form is prevalent only in very acidic conditions because hydration and deprotonation reactions convert flavylium ions to colourless hemiketal forms and blue quinoidal bases as the pH is increased. Pigment stabilization is provided by regulation of the pH of intracellular compartments such as the vacuole and by self-association of anthocyanins, association with other molecules (co-pigmentation), or complexation with metal ions, as reviewed recently (Yoshida *et al.*, 2012). Co-pigmentation, involving hydrophobic π–π interactions of the anthocyanin flavylium with other planar structures (Goto and Kondo, 1991), results in colour enhancement and a slight bathochromic shift from red to purple. In particular, flavonols, which by themselves are light-yellow pigments, are known to act as co-pigments and enhance anthocyanin colour.

Some of the anthocyanin-derived pigments listed under ‘Flavonoid reaction products’ also show particular colour properties. In particular, among pigments detected in fruits, pyranoanthocyanins are orange (Fulcrand *et al.*, 1996), while products resulting from condensation of anthocyanins with aldehydes are purple, presumably because of intra-molecular co-pigmentation (Escribano-Bailon *et al.*, 1996; Dueñas *et al.*, 2006).

9.3.2 Taste properties

Some polyphenols contribute to taste properties, especially bitterness and astringency. It should be emphasized that bitterness is a taste, mediated by interaction with taste receptors, while astringency is a tactile perception, usually ascribed to interaction with (and eventually precipitation of) salivary proteins.

Numerous phenolic compounds such as oleoropein, the bitter principle of olive, and most phenolic acids exhibit bitterness.

Among polyphenols, flavanones, flavonols and flavan-3-ols have been reported to taste bitter.

Flavanones, and especially naringin, are responsible for the bitter flavour of grapefruit and sour orange (Horowitz and Gentili, 1969; Rouseff, 1990). Replacement of the neohesperidose (2-*O*-α-L-rhamnosyl-β-D-glucose) by its isomer rutinose (6-*O*-α-L-rhamnosyl-β-D-glucose) results in tasteless compounds, while naringenin 7-glucoside is only mildly bitter. Members of the flavanone family have other important taste properties. Flavanone aglycones such as eriodictyol and homoeriodictyol have been reported to mask the bitterness of caffeine without exhibiting intrinsic flavour (Ley *et al.*, 2005). Moreover, opening of the C-ring of flavanone neohesperidosides to form the dihydrochalcones generates extremely sweet compounds.

Flavonols have also been described as bitter and astringent in water, 5% ethanol, and beer (Dadic and Belleau, 1973; Delcour *et al.*, 1984). Flavonols extracted from blackcurrant (Schwarz and Hofmann, 2007) and bilberry (Laaksonen *et al.*, 2010) were also perceived as bitter and astringent. However, supplementing the bilberry juice with flavonol extracts did not modify its sensory profile (Laaksonen *et al.*, 2010), and there is no report of their contribution to fruit taste.

Flavan-3-ols taste less bitter and more astringent as their chain length increases (Lea and Arnold, 1978). Interactions of tannins with proteins increase with the number of phenolic groups, and thus with DP and galloylation, and so does astringency (Vidal *et al.*, 2003a). Larger PAs (beyond DP8 or so) have classically been considered insoluble and thus unable to exhibit astringency (Lea, 1990), but other studies have established that larger polymers are highly astringent (Vidal *et al.*, 2003a). Most persimmon cultivars show intense astringency and require postharvest treatments (with CO_2, ethanol or ethylene) to reduce it (Kato, 1990). Two mechanisms have been proposed. The first involves condensation of PAs with acetaldehyde (Tanaka *et al.*, 1994) released from seeds during development (Sugiura and Tomana, 1983). However, the products of catechin condensation with acetaldehyde were perceived to be as astringent as PAs of equivalent chain length and were more bitter (Vidal *et al.*, 2004c). The other mechanism involves increased solubilization of pectins that can interact with PAs (Taira *et al.*, 1998) and reduce astringency when added to procyanidin solutions (Vidal *et al.*, 2003b). Indeed, postharvest deastringency treatments induce genes encoding xyloglucan endotransglycosylase/hydrolase that catalyse degradation of xyloglucans, the major hemicellulosic polysaccharides in plant cell walls (Nakatsuka *et al.*, 2011).

9.3.3 Health benefits

Dietary polyphenols are attracting great interest for their potential beneficial health effects. This assumption arises from epidemiological studies relating higher consumption of plant-based foods and beverages to a lower incidence of certain degenerative diseases. Health effects of polyphenols are often attributed to their antioxidant activity, involving a variety of mechanisms (i.e. reduction or scavenging of reactive oxygen species, chelation of transition metal ions and inhibition of enzymes involved in oxidative stress) (Dangles, 2012). Numerous *in vitro* studies have shown antioxidant activity of phenolic compounds, but scientific evidence is not yet conclusive and more human studies are required (Törrönen, 2009). Indeed, the relevance of such *in vitro* studies is questionable because of the poor bioavailability and extensive metabolism of phenolic compounds in the digestive process (Williamson and Stalmach, 2012). This applies in particular to the potential health benefits of stilbenes, and especially of resveratrol, suggested by over 4000 *in vitro* studies, much highlighted by media coverage. Although resveratrol shows potential to reduce the incidence of chronic diseases, clinical trials are still needed to establish its effects in humans (Smoliga *et al.*, 2011). Moreover, normal dietary intake is unlikely to exert any protective effect, because of the small quantities consumed (Manach *et al.*, 2006). However, the antioxidant action may be exerted in the digestive tract (Halliwell *et al.*, 2005), and catabolites formed by microbial transformation in the colon may be absorbed and also show biological activities.

The biological effects of polyphenols may implicate other mechanisms such as antimicrobial properties, an impact on intestinal flora and modulation of cell signalling pathways (Manach *et al.*, 2009). In particular, a health claim has been approved for the use of A-type procyanidins from cranberries for prevention of urinary tract infection. Finally, other minor polyphenol classes, including isoflavones (which are abundant in legumes and especially soybean), coumestans and lignans, are reported to show phyto-oestrogenic properties (Kuhnle *et al.*, 2007).

9.4 Fruit Polyphenol Composition

9.4.1 Factors affecting fruit phenolic composition

Several factors have been reported to affect the fruit polyphenolic composition.

Temperature

In persimmon, cold stress has been demonstrated to stimulate PA accumulation in fruit when applied 1 week after bloom (WAB) but has no significant effect when applied 5 weeks after bloom. The most significant impact was measured on trihydroxylated subunits (gallocatechin and epigallogatechin), rather than on dihydroxylated units (catechin and epicatechin or galloylated units). On the other hand, no significant correlation between temperature and PA accumulation was detected on grape berries (Cohen *et al.*, 2012), despite a similar early application of thermal stress (10–12 days after anthesis).

Ellagitannins concentration was significantly increased with temperature in raspberry (Remberg *et al.*, 2010), but this result was not confirmed by McDougall *et al.* (2011).

Low temperature stimulates anthocyanin accumulation by upregulating the expression of biosynthetic genes in apple and pear (Ubi *et al.*, 2006; Steyn *et al.*, 2009), in grape (Yamane *et al.*, 2006) and also in orange (Crifò *et al.*, 2011). In mature oranges, the response to cold treatment is strictly observed in blood oranges but not in common oranges. Furthermore, in apple, the environmental temperatures also modulate the expression of regulatory genes (Ban *et al.*, 2007; Lin-Wang *et al.*, 2010).

Light

Several studies have reported that light exposure stimulates accumulation of piceid (Adrian *et al.*, 2000) and of PAs, flavonols and anthocyanin in grape berries (Cortell and Kennedy, 2006; Koyama *et al.*, 2012). Howewer, the impact of light exclusion on PA biosynthesis is limited (Koyama *et al.*, 2012). This increase more specially affects B-ring trihydroxylated compounds (Cortell and Kennedy, 2006; Fujita *et al.*, 2007). In grape berries, flavonol biosynthesis is also induced by light (Downey *et al.*, 2004; Matus *et al.*, 2009), even during developmental stages when flavonols are normally not synthesized. Light also enhances anthocyanin synthesis in many other fruit species such as apple (Kim *et al.*, 2003), pear (Steyn *et al.*, 2004) and peach (Kataoka and Beppu, 2004; Tsuda *et al.*, 2004). Nevertheless, this response appears to depend on cultivars. In Syrah berries, anthocyanin synthesis is not influenced by shading (Downey *et al.*, 2004), whereas Cabernet Sauvignon grape berries contain fewer anthocyanins when shaded from sunlight (Jeong *et al.*, 2004). Similarly, tomato fruit from different cultivars are also differentially affected by light exposure. In most cases, the MYB transcription factor appears to be the primary determinant of fruit pigmentation in response to light, such as in apple,

Chinese bayberry and red-skinned pear (Takos *et al.*, 2006a; Feng *et al.*, 2010; Niu *et al.*, 2010).

Hormones

In strawberry, the central role of abscisic acid (ABA) was recently demonstrated directly, as RNA interference-mediated silencing of an ABA receptor gene led to the inhibition of anthocyanin production (Jia *et al.*, 2011). Grape berries treated with ABA at the onset of ripening showed an increased accumulation of anthocyanins associated with increased expression of the anthocyanin structural genes tested and of the regulatory gene *VvMYBA1* (Jeong *et al.*, 2004). By contrast, in grape, there is little effect of ABA on PA synthesis and gene expression (Koyama *et al.*, 2010; Lacampagne *et al.*, 2010).

Auxins appear to have the opposite effect to ABA, delaying anthocyanin accumulation in grape berries (Davies *et al.*, 1997). The treatment of berries at veraison with the auxin 2,4-dichlorophenoxyacetic acid reduced the anthocyanin level in berries. Similar results were seen when naphthaleneacetic acid was sprayed on to Cabernet Sauvignon berries at veraison (Jeong *et al.*, 2004).

Ethephon, an analogue of ethylene, increases stilbene accumulation when sprayed on grapevine (Belhadj *et al.*, 2008a), and, in groundnuts, stilbene biosynthetic genes are induced by ethylene (Chung *et al.*, 2001). Methyl jasmonate was shown to stimulate stilbene and anthocyanin production by grapevine cell culture (Belhadj *et al.*, 2008b). Spraying with the ethylene-releasing compound 2-chloroethylphosphonic acid enhanced anthocyanin accumulation and expression of several structural genes in grape (El-Kereamy *et al.*, 2003), but ethylene did not affect expression of VvMYBA transcription factors (Tipa-Umphon *et al.*, 2007).

Hydric status

Water stress can greatly increase the content and alter the anthocyanin composition in fruits during ripening (Ojeda *et al.*, 2002; Castellarin *et al.*, 2007; Ollé *et al.*, 2011). In particular, under water deficit stimulating hydroxylation and methoxylation of the flavonoid B-ring, the anthocyanin profile of water-stressed berries shifts towards purple/blue pigments (Castellarin *et al.*, 2007). Conversely, the flavan-3-ol and flavonol composition of grape berry is only slightly affected by water stress (Kennedy *et al.*, 2002; Ojeda *et al.*, 2002; Castellarin *et al.*, 2007).

Biotic stress

Phenolics are said to be involved in defence against biotic stresses. Induction of the stilbene pathway after pathogen infection is one of the main responses described and has been particularly studied on grapevine, where accumulation of stilbenes is significantly induced after infection with powdery mildew (Fung *et al.*, 2008), downy mildew (Adrian *et al.*, 1997) or grey mould (Langcake and McCarthy, 1979). Bilberry infection by a fungal endophyte (*Paraphaeosphaeria* sp.) and a pathogen (*Botrytis cinerea*) resulted in increased concentrations of flavan-3-ols and quercetin derivatives (Koskimäki *et al.*, 2009). Applications of elicitors like benzothiadiazole and methyl jasmonate have been described as promoting the accumulation of stilbenes, anthocyanins, flavonols and PAs of grape skins (Iriti *et al.*, 2004, 2005; Ruiz-Garcia *et al.*, 2012). In strawberry fruits, benzothiadiazole induced the accumulation of flavan-3-ols, anthocyanins and kaempferol derivatives (Hukkanen *et al.*, 2007).

9.4.2 Changes during development

Fruit polyphenol composition varies throughout growth and ripening. In most cases, unripe fruits have higher levels of phenolic compounds than ripe fruits. PAs are in general synthesized during the early fruit development stages, for example in grape berries (Kennedy *et al.*, 2000), strawberries (Carbone *et al.*, 2009) and blueberries (Zifkin *et al.*, 2012). PA levels usually decrease thereafter when expressed per FW, but remain almost constant when expressed per fruit. Ellagitannins exhibit the same accumulation profile as PAs, and constantly decreased in strawberry when expressed per fresh weight (Williner *et al.*, 2003). Accumulation of stilbenes in healthy grape berries is restricted to ripening stages (Gatto *et al.*, 2008). In grape, flavonol synthesis occurs at two distinct periods during berry development (Downey *et al.*, 2003). The first synthesis phase occurs early in the inflorescence and the second after veraison. Accumulation of anthocyanin begins during ripening in most fruits, and then stabilizes or decreases slightly towards the harvesting stage (Boss *et al.*, 1996; Lister *et al.*, 1996). While phenolics are markers of development stages, the molecular circuits connecting their accumulation to the ripening process are poorly understood.

9.4.3 Distribution and role in fruit tissues

The phenolic compounds, like most secondary metabolites, present a very uneven distribution within the fruit, and for the same species, depending on variety. In young unripe fruit skin, while seeds are still immature, it is generally assumed that PAs function as feeding deterrents thanks to their astringency and bitterness (Wrangham and Waterman, 1983). Skin PAs may also protect the fruits against pathogens, fungi and viruses due to their protein-binding properties (Treutter, 2006). In seeds, PAs are often present in the seed coat, in agreement with their protective role (Debeaujon *et al.*, 2000). To date, there are few reports of the localization of PAs in fruit: in grape (Cadot *et al.*, 2011), apple (Lees *et al.*, 1995) and blueberry (Zifkin *et al.*, 2012), higher concentrations of PAs were detected in the skin and the seed coat, but parenchyma is the major contributor of PAs in apple (Guyot *et al.*, 1998), and PAs were also found in grape berry pulp (Mané *et al.*, 2007; Verries *et al.*, 2008). At maturity, ellagitannins are found mostly (about 80%) in achenes (Williner *et al.*, 2003). Stilbenes were also found in much higher concentration in achenes than in the receptacle of strawberries (Wang *et al.*, 2007). Localization of stilbenes in grape berry skin is in accordance with their role as a barrier against pathogens (Fornara *et al.*, 2008). Ellagitannins were detected in equivalent concentrations in the peel and mesocarp of pomegranate (Fischer *et al.*, 2011) and only in trace amounts in arils (Gil *et al.*, 2000). Anthocyanins accumulate in the skins and, in some species or varieties, in the pulp of red fruits, in accordance with their roles as pigments to attract animals for seed dispersion and protectants against UV irradiation (Winkel-Shirley, 2001). Similarly, flavonols are mainly detected in fruit skins, which is consistent with their roles as UV filters (Winkel-Shirley, 2002).

9.5 Biosynthesis Pathways

9.5.1 General and particular pathways

Most of the phenolics described in this chapter, except hydrolysable tannins, derive from the general phenylpropanoid with phenylalanine as substrate (Fig. 9.3), already described in numerous reviews. The biosynthetic pathway of flavonoids was characterized first in parsley (Kreuzaler *et al.*, 1983) and then in maize, petunia (Holton and Cornish, 1995) and *Arabidopsis* (Shirley *et al.*, 1995). More recently, this pathway has also been described in fruit crops such as apple (Takos *et al.*, 2006b; Espley *et al.*, 2007), bilberry (Jaakola *et al.*, 2002) and grapevine (Boss *et al.*, 1996). These compounds derive from a common precursor, the flavanone naringenin (Fig. 9.1f). This intermediate is subsequently hydroxylated by flavanone-3β-hydroxylase to dihydroflavonol (Fig. 9.3). This part of the pathway is common to the biosynthesis of PAs, anthocyanins and flavonols. Hydroxylation catalysed by flavonoid 3′ hydroxylase and flavonoid 3′5′ hydroxylase gives rise to the different B-ring

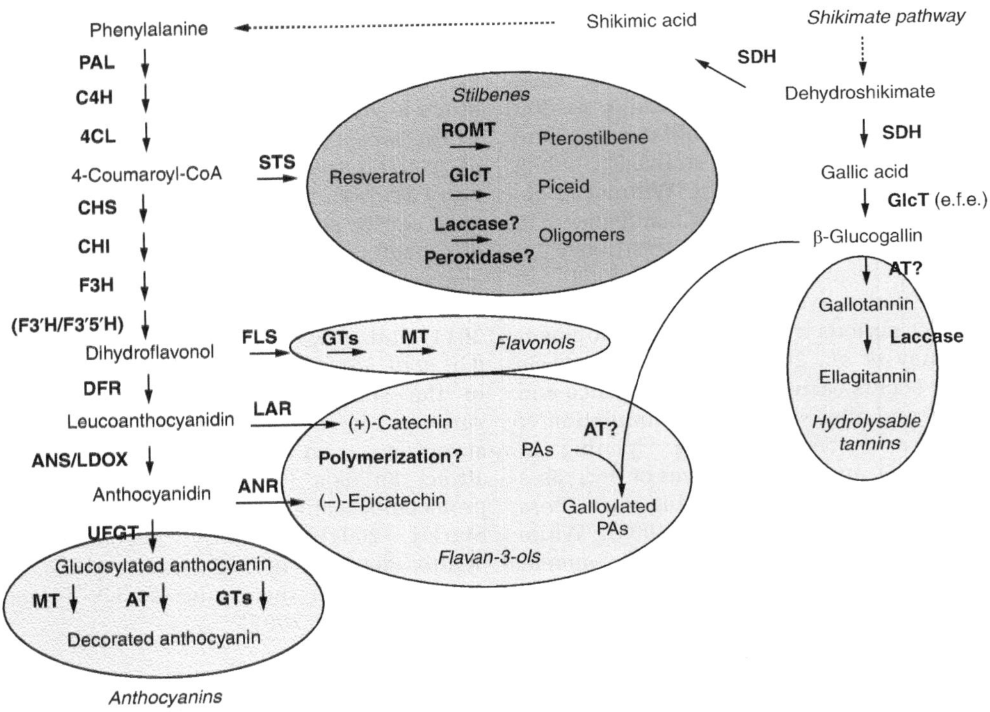

Fig. 9.3. Schematic representation of the biosynthesis pathways of phenolic compounds. Enzyme names are shown in bold. PAL, phenylalanine ammonia lyase; C4H, cinnamate 4-hydroxylase; 4CL, 4-coumarate CoA ligase; CHS, chalcone synthase; CHI, chalcone isomerase; F3H, flavonone 3-hydroxylase; F3′H, flavonoid 3′-hydroxylase; F3′5′H, flavonoid 3′,5′-hydroxylase; FLS, flavonol synthase; DFR, dihydroflavonol 4-reductase; ANS, anthocyanidin synthase; LDOX, leucoanthocyanidin dioxygenase; LAR, leucoanthocyanidin reductase; ANR, anthocyanidin reductase; UFGT, UDP-glucose:flavonoid 3-*O*-glucosyl transferase; AT, acyltransferase; MT, methyltransferase; GTs, glycosyltransferase; ROMT, resveratrol-*O*-methyltransferase; GlcT, glucosyltransferase; e.f.e., ester-forming enzyme; SDH, shikimate dehydrogenase.

substitutions. The dihydroflavonol is then oxidized to flavonol via flavonol synthase (FLS) or reduced by dihydroflavonol reductase to leucoanthocyanidin and then to anthocyanidin, by leucoanthocyanidin dioxygenase (also called anthocyanidin synthase). Further reactions such as glycosylation, methylation and acylation lead to flavonol derivatives and anthocyanins. In particular 3-*O*-glycosylation of the anthocyanidin (e.g. glucosylation, catalysed by UDP-glucose:flavonoid 3-*O*-glucosyltransferase (UFGT)), stabilizing anthocyanidins, is the first step in anthocyanin biosynthesis (Harborne and Williams, 2000). Glycosylation is specific both for the sugar and for the substitution position. The sequence of methylation, glycosylations and acylations is variable. In contrast to the well-conserved main pathway, these steps are family, species and even variety dependent, and are believed to provide flavonoids with unique properties.

Concerning the PA pathway, anthocyanidin reductase (ANR), first described in *Arabidopsis*, catalyses the formation of epicatechin from anthocyanidin (Xie *et al.*, 2003), whereas leucoanthocyanidin reductase, first identified in *Lotus*, catalyses the formation of catechin from leucoanthocyanidin (Tanner *et al.*, 2003). In fruits, these compounds were isolated in grapevine (Bogs *et al.*, 2005), apple (Takos *et al.*, 2006b), pear (Fischer *et al.*, 2007), strawberry (Almeida *et al.*, 2007), kaki (Ikegami *et al.*, 2007; Akagi *et al.*, 2009a) and blueberry (Zifkin *et al.*, 2012). ANR is suspected to produce catechin in addition to epicatechin (Akagi *et al.*, 2009a; Gargouri *et al.*, 2009; Han *et al.*, 2012). Mechanisms of PA polymerization are still unknown in fruits and in model plants (Terrier *et al.*, 2009a).

Biosynthesis of hydrolysable tannins is much less understood. The origin of gallic acid (GA) was unclear until recently. Labelling experiments revealed that it derives from an intermediate product of the shikimate pathway, probably dehydroshikimate (Werner *et al.*, 1997, 2004). Dehydroshikimate was successfully reduced to GA by an enzymatic extract of *Betula* leaves (Ossipov *et al.*, 2003), but the enzyme was not identified. Recently, Muir *et al.* (2011) demonstrated that shikimate dehydrogenase, an enzyme from the shikimate pathway essential for aromatic amino acid synthesis, is also able to catalyse GA production.

The following step, esterification of GA and glucose to yield β-glucogallin, was observed with oak enzymatic extract (Gross, 1983). Gallotannins from di- to pentagalloylglucose are formed by successive position-specific steps, using β-glucogallin as both acyl donor and acyl acceptor. The enzymatic transfer of the galloyl moiety to one of the galloyl hydroxyls (transacylation forming a meta-depside group), leading to more complex gallotannin, was demonstrated with extract from oak and sumach leaves (Gross *et al.*, 1990; Gross and Denzel, 1991). Several enzymes with different substrate specificities have been isolated and may cooperate in synthesizing the various gallotannins found in these plants (Niemetz and Gross, 2005). Further oxidation steps catalysed by laccases are then involved in the formation of the hexahydroxydiphenoyl groups of ellagitannins (Niemetz and Gross, 2003a,b). However, the proteins have not been molecularly identified.

β-Glucogallin is probably common to PA galloylation and gallotannin biosynthesis. Recently, Terrier *et al.* (2009b) identified glucosyltransferases induced in parallel with PA synthesis in grapevine hairy roots. These glucosyltransferases are able to form β-glucogallin, and it was hypothesized that this compound is an intermediate for PA galloylation (Khater *et al.*, 2012). Serine carboxy peptidase-like proteins (SCPLs) have been identified during transcriptomic screening comparing grape or persimmon samples differing in their PA content (Ikegami *et al.*, 2007; Akagi *et al.*, 2009a; Terrier *et al.*, 2009b). As both fruits contain galloylated PA and SCPLs are able to catalyse transacylation with glucose esters as acyl donors (Strack and Mock, 1993; Steffens, 2000; Milkowski and Strack, 2004), these genes represent good candidates for the enzymes involved in the second step of galloylation.

Stilbene synthase is the first enzyme specific for stilbene biosynthesis. The genes encoding those enzymes were isolated first from groundnuts and then from grapes (Schröder *et al.*, 1988; Melchior and Kindl, 1991). They belong to the superfamily polyketide synthases, like chalcone synthases. Hall and DeLuca (2007) suggested that a bifunctionnal glucosyltransferase could catalyse the glucosylation of resveratrol to form piceid, despite the optimum pH for this reaction being quite alkaline (pH 9) and its rate much lower than that of its other activity (i.e. forming glucose esters with phenolic acids). A resveratrol *O*-methyltransferase gene isolated from *V. vinifera* was characterized, and the corresponding enzyme was found to be able to catalyse resveratrol methylation to yield pterostilbene both *in vitro* and *in planta* (Schmidlin *et al.*, 2008). Several hypotheses have been proposed to explain the oxidative polymerization of stilbenes: by laccase-like stilbene oxidases from the pathogens (Breuil *et al.*, 1999) or by host peroxidases localized in the vacuole, cell wall or apoplast (Ros Barcelo *et al.*, 2003).

9.5.2 Regulation

Of the phenolics mentioned in this chapter, only regulators of flavonoid have been described. Flavonoid synthesis is assumed to be controlled by a complex of two transcription factors belonging to the R2R3 MYB and basic helix–loop–helix (bHLH) families, associated with a WD-repeat protein as demonstrated in *Arabidopsis* (Lepiniec *et al.*, 2006). These MYB–bHLH–WD40 complexes are responsible for the regulation of anthocyanin, PA and flavonol biosynthesis in a variety of species and tissues, including flowers and fruits (Allan *et al.*, 2008; Dubos *et al.*, 2010). There appears to be considerable redundancy for the bHLH cofactors in particular, while the MYB protein usually provides specificity (Broun, 2005; Feller *et al.*, 2011).

Most information available on the regulation of flavonoid biosynthesis in fleshy fruits has been obtained from studies with grapevine and apple, as a result of sequence homologies with model plants and the availability of genome sequences (Jaillon *et al.*, 2007; Velasco *et al.*, 2010).

MYB factors

Some of the identified MYB factors, such as VvMYB5a and VvMYB5b in grape, appear to be 'general regulators', activating different parts of the flavonoid pathway (Deluc *et al.*, 2006, 2008). *VvMYB5a* and *VvMYB5b* are expressed in both skins and seeds, at green stage and during ripening, respectively. Both induce anthocyanin and PA accumulation when expressed ectopically in tobacco. Other MYBs are more specific of one branch of the flavonoid pathway.

PAS. In blueberry, a MYB factor called *VcMYBPA1* is expressed during the green stage when PAs accumulate (Zifkin *et al.*, 2012). It activates the promoter of poplar *ANR* but not that of apple anthocyanidin-3-*O*-glycosyltransferase, and would therefore appear to be specific for the PA gene. However, this is inconsistent with its continuing expression after PA synthesis has stopped. In kaki, two MYB factors, *DkMYB2* and *DkMYB4*, regulated the PA pathway when overexpressed in callus (Akagi *et al.*, 2009b, 2010). However, the phenological expression patterns of *DkMYB2* did not correlate with PA accumulation (Akagi *et al.*, 2009b). *DkMYB2* was induced by wounding and activated the PA pathway (Akagi *et al.*, 2010). *DkMYB4* exhibited much lower expression in fruit from non-astringent (NA) cultivars than in astringent (A) ones (Akagi *et al.*, 2009b). In addition, its expression and PA accumulation were affected by low temperatures in NA-type cultivars but not in the A-type cultivar (Akagi *et al.*, 2011). In grape, *VvMYBPA1* and *VvMYBPA2*, controlling the PA pathway, were identified (Bogs *et al.*, 2007; Terrier *et al.*, 2009b). They are expressed in parallel with PA accumulation during the early stages of grape berry development, more specifically in berry seeds and skin, respectively. Both activate the PA pathway when overexpressed ectopically in grapevine hairy roots.

ANTHOCYANINS. The anthocyanin synthesis pathway in red grape cultivars is controlled by a MYBA complex locus (including VvMYBA1 and VvMYBA2) (Kobayashi *et al.*, 2002, 2004). These genes are inactive in some white grape cultivars due to a retrotransposon insertion in the promoter region of *VvMYBA1* (Kobayashi *et al.*, 2004) and non-conservative substitutions in the coding region of *VvMYBA2* (Walker *et al.*, 2007). *VvMYBA1* and *VvMYBA2* regulate anthocyanin pigmentation of berry skins predominantly by controlling UFGT (Boss *et al.*, 1996) but also other genes encoding putative targets involved in vacuolar transport and methylation of anthocyanins (Cutanda-Perez *et al.*, 2009). In apple (*Malus domestica*), *MdMYB1* and *MdMYBA* control anthocyanin synthesis in red-skinned apple cultivars (Takos *et al.*, 2006a), whereas *MdMYB10* controls red pigmentation in both skin and flesh of fruit and in foliage (Espley *et al.*, 2007). Putative orthologues of the apple *MdMYB10* have been identified in many other fruit species from the *Rosaceae* family, such as pears (*PyMYB10*, Feng *et al.*, 2010), plums, cherries, peaches, raspberries and strawberries (Lin-Wang *et al.*, 2010). The expression of these *MYB* genes correlates with anthocyanin accumulation during fruit ripening and is higher in anthocyanin-rich varieties (Feng *et al.*, 2010). *FaMYB1* from strawberry (*Fragaria ananassa*), the only known repressor of anthocyanin biosynthesis in fruit species, is present at high transcript levels only at ripe fruit stages, and may balance the levels of anthocyanin pigments in the latter stages of strawberry fruit maturation (Lin-Wang *et al.*, 2010). Other MYB regulators controlling anthocyanin pigmentation in fruits have been identified in *Solanaceae*, such as tomato and pepper. In tomato (*Solanum lycopersicum*), three *R2R3-MYB* genes control anthocyanin and flavonoid levels in fruit skin: *LeMYB12* (Adato *et al.*, 2009) and the two tightly linked genes *LeANT1* (Sapir *et al.*, 2008) and *LeAN2* (Mes *et al.*, 2008).

FLAVONOLS. Few studies are available on the regulation of flavonol synthesis in fruit species. In grape, a putative flavonol regulator was identified (Matus *et al.*, 2008), given its close homology to AtMYB12, which controls *FLS* expression in *Arabidopsis thaliana* (Mehrtens *et al.*, 2005; Stracke *et al.*, 2007). It was functionally validated and called *VvMYBF1* (Czemmel *et al.*, 2009). Its expression in the berry was strongly reduced as a result of shading (Matus *et al.*, 2009) and was induced by light (Czemmel *et al.*, 2009). Unlike other MYB transcription factors, MYB12 does not require a bHLH partner for promoter activation (Mehrtens *et al.*, 2005).

Other members of the regulatory complex

In grape, two WD40-type (WDR1 and WDR2) and two bHLH-type (MYC1 and MYCA1) proteins involved in regulation of the flavonoid pathway have been identified (Hichri *et al.*, 2010; Matus *et al.*, 2010). WDR1 and WDR2 are expressed in skin and seeds throughout berry development with a higher expression level during ripening. However, only ectopic expression of WDR1 in wild-type *Arabidopsis* led to anthocyanin overproduction in leaves and shoot (Matus *et al.*, 2010). The expression pattern of *MYCA1* and *WDR1* compared with those of *ANR* and *UFGT* suggest that MYCA1 could be a putative regulator of both PA and anthocyanin synthesis, whereas WDR1 would be more specific for anthocyanins (Bogs *et al.*, 2005; Matus *et al.*, 2010). By yeast two-hybrid assay, it was verified that MYC1 interacts physically with MYB5a, MYB5b, MYBPA1 and MYBA (Hichri *et al.*, 2010), suggesting the non-specifity of bHLH proteins towards particular branches of the flavonoid pathway. In apple, a WD40 protein, MdTTG1, was recently identified (Brueggemann *et al.*, 2010). This protein is able to complement the corresponding *A. thaliana* mutant with respect to anthocyanin and PA production.

Other regulating events

Other regulators affecting the flavonoid pathway have been identified in fruits. Basic leucine zipper transcription factors directly affect the expression of some flavonoid pathway genes such as *FLS* in *Arabidopsis* and grape (Hartmann *et al.*, 2005; Czemmel *et al.*, 2009; Hichri *et al.*, 2011). *DkbZip5* identified in persimmon (*Diospyros kaki*), mediates the effect of ABA signalling on PA accumulation through regulation of seasonal *DkMyb4* expression (Akagi *et al.*, 2012). The existence of a link between the regulatory genes controlling fruit ripening and the downstream anthocyanin pathway has been described recently (Jaakola *et al.*, 2010). A MADS-box transcription factor, *VmTDR4*, from bilberry (*Vaccinium myrtillus* L.) appears to control anthocyanin accumulation during fruit ripening by direct or indirect control of the expression of *a MYB* factor controlling the anthocyanin pathway (Jaakola *et al.*, 2010). Regulation can also occur through competition between the different branches of the pathway. Thus, UFGT may be an important branching-point enzyme that drives the pathway towards anthocyanins. In the absence of UFGT activity, the flux may be redirected towards other flavonoid branches, such as the PA branch. Indeed, reduced levels of anthocyanins induced by downregulation of UFGT in strawberry were accompanied by an increase in epiafzelechin (Griesser *et al.*, 2008).

This section has highlighted the large array of transcription factors regulating flavonoid biosynthesis in fruits. This may appear to be redundancy but probably reflects the necessity of numerous genes for the fine-tuning of this pathway, each regulator having a tissue, stage, biotic and abiotic condition-specific expression.

9.6 Conclusion

Polyphenols are key components of fruit quality, responsible for important organoleptic properties. They are also believed to be beneficial to human health, but the precise mechanisms involved need further investigation. Fruits and berries are major dietary sources of polyphenols, and their composition varies qualitatively and quantitatively among cultivars, making this character a potential breeding target. By studying polyphenols biosynthetic pathways and the genetic factors involved, the amount of these compounds in the fruit can be optimized to improve nutritional quality for human consumption. Polyphenols are also involved in plant defence against biotic and abiotic stresses. A better understanding of the regulation processes triggering their biosynthesis in response to environmental threats will be of great help for selection (based on genetic markers) and/or bioengineering of fruit crops showing increased UV, heat or drought tolerance and resistance to pathogens. The increasing number of genome sequences available for agronomical plants will enable investigation of these complex and challenging pathways. However, this will also require the development of analytical methods for high-throughput phenotyping of plant polyphenols, and in particular of tannins, which are abundant in fruits and are qualitatively important but are often neglected because of the difficulties in extracting and analysing them.

References

Adato, A., Mandel, T., Mintz-Oron, S., Venger, I., Levy, D., Yativ, M., Domınguez, E., Wang, Z., De Vos, R.C., Jetter, R., Schreiber, L., Heredia, A., Rogachev, I. and Aharoni, A. (2009) Fruit-surface flavonoid accumulation in tomato is controlled by a *SlMYB12*-regulated transcriptional network. *PLoS Genetics* 5, e1000777.

Adrian, M., Jeandet, A.C., Douillet-Breuil, L., Tesson, R. and Bessis, E. (2000) Stilbene content of mature *Vitis vinifera* berries in response to UV-C elicitation. *Journal of Agricultural and Food Chemistry* 48, 6103–6105.

Adrian, M.X., Daire, P., Jeandet, A.C., Breuil, L.A., Weston, R., Bessis, E. and Boudon, M. (1997) Comparisons of stilbene synthase activity (resveratrol amounts and stilbene synthase mRNAs levels) in grapevines treated with biotic and abiotic phytoalexin inducers. *American Journal of Enology and Viticulture* 48, 394–395.

Akagi, T., Ikegami, A., Suzuki, Y., Yoshida, J., Yamada, M., Sato, A. and Yonemori, K. (2009a) Expression balances of structural genes in shikimate and flavonoid biosynthesis cause a difference in proanthocyanidin accumulation in persimmon (*Diospyros kaki* Thunb.) fruit. *Planta* 230, 899–915.

Akagi, T., Ikegami, A., Tsujimoto, T., Kobayashi, S., Sato, A., Kono, A. and Yonemori, K. (2009b) DkMyb4 is a Myb transcription factor involved in proanthocyanidin biosynthesis in persimmon fruit. *Plant Physiology* 151, 2028–2045.

Akagi, T., Ikegami, A. and Yonemori, K. (2010) DkMyb2 wound-induced transcription factor of persimmon (*Diospyros kaki* Thunb.) contributes to proanthocyanidin regulation. *Planta* 232, 1045–1059.

Akagi, T., Tsujimoto, T., Ikegami, A. and Yonemori, K. (2011) Effects of seasonal temperature changes on *DkMyb4* expression involved in proanthocyanidin regulation in two genotypes of persimmon (*Diospyros kaki* Thunb.) fruit. *Planta* 233, 883–894.

Akagi, T., Katayama-Ikegami, A., Kobayashi, S., Sato, A., Kono, A. and Yonemori, K. (2012) Seasonal abscisic acid signal and a basic leucine zipper transcription factor, DkbZIP5, regulate proanthocyanidin biosynthesis in persimmon fruit. *Plant Physiology* 158, 1089–1102.

Allan, A.C., Hellens, R.P. and Laing, W.A. (2008) MYB transcription factors that colour our fruit. *Trends in Plant Science* 13, 99–102.

Almeida, J.R., D'Amico, E., Preuss, A., Carbone, F., de Vos, C.H., Deiml, B., Mourgues, F., Perrotta, G., Fischer, T.C., Bovy, A.G., Martens, S. and Rosati, C. (2007) Characterization of major enzymes and genes involved in flavonoid and proanthocyanidin biosynthesis during fruit development in strawberry (*Fragaria* × *ananassa*). *Archives of Biochemistry and Biophysics* 465, 61–71.

Amira-Guebailia, H., Chira, K., Richard, T., Mabrouk, T., Furiga, A., Vitrac, X., Monti, J.P., Delaunay, J.C. and Mérillon, J.M. (2006) Hopeaphenol: the first resveratrol tetramer in wines from North Africa. *Journal of Agricultural and Food Chemistry* 54, 9559–9564.

Andersen, O. and Markham, K. (2006) *Flavonoids: Chemistry, Biochemistry and Applications*. CRC Press, Boca Raton, FL.

Andersen, O.M., Fossen, T., Torskangerpoll, K., Fossen, A. and Hauge, U. (2004) Anthocyanin from strawberry (*Fragaria ananassa*) with the novel aglycon, 5-carboxypyranopelargonidin. *Phytochemistry* 65, 405–410.

Arranz, A., Saura-Calixto, F., Shaha, S. and Kroon, P.A. (2009) High contents of nonextractable polyphenols in fruits suggest that polyphenol contents of plant foods have been underestimated. *Journal of Agricultural and Food Chemistry* 57, 7298–7303.

Ban, Y., Honda, C., Hatsuyama, Y., Igarashi, M., Hideo Bessho, H. and Moriguchi, T. (2007) Isolation and functional analysis of a MYB transcription factor gene that is a key regulator for the development of red colouration in apple skin. *Plant Cell Physiology* 48, 958–970.

Bate-Smith, E. and Swain, T. (1962) Flavonoid compounds. In: Mason, H. and Florkin, A. (eds) *Comparative Biochemistry*, Vol. 3. Academic Press, New York, pp. 755–809.

Belhadj, A., Telef, N., Cluzet, S., Bouscaut, J., Corio-Costet, M.F. and Merillon, J.M. (2008a) Ethephon elicits protection against *Erysiphe necator* in grapevine. *Journal of Agricultural and Food Chemistry* 56, 5781–5787.

Belhadj, A., Telef, N., Saigne, C., Cluzet, S., Barrieu, F., Hamdi, S. and Merillon, J.M. (2008b) Effect of methyl jasmonate in combination with carbohydrates on gene expression of PR proteins, stilbene and anthocyanin accumulation in grapevine cell cultures. *Plant Physiology and Biochemistry* 46, 493–499.

Bogs, J., Downey, M.O., Harvey, J.S., Ashton, A.R., Tanner, G.J. and Robinson, S.P. (2005) Proanthocyanidin synthesis and expression of genes encoding leucoanthocyanidin reductase and anthocyanidin reductase in developing grape berries and grapevine leaves. *Plant Physiology* 139, 652–663.

Bogs, J., Jaffé, F.W., Takos, A.M., Walker, A.R. and Robinson, S.P. (2007) The grapevine transcription factor VvMYBPA1 regulates proanthocyanidin synthesis during fruit development. *Plant Physiology* 143, 1347–1361.

Boss, P.K., Davies, C. and Robinson, S.P. (1996) Analysis and the expression of anthocyanin pathway genes in developing *Vitis vinifera* L. cv Shiraz grape berries and the implications for pathway regulation. *Plant Physiology* 11, 1059–1066.

Breuil, A.C., Jeandet, P., Chopin, F., Adrian, M., Pirio, N., Meunier, P. and Bessis, R. (1999) Characterization of a pterostilbene dehydromer produced by a laccase of *Botrytis cinerea*. *Phytopathology* 89, 298–302.

Broun, P. (2005) Transcriptional control of flavonoid biosynthesis: a complex network of conserved regulators involved in multiple aspects of differentiation in *Arabidopsis*. *Current Opinion in Plant Biology* 8, 272–279.

Brueggemann, J., Weisshaar, B. and Sagasser, M. (2010) A WD40-repeat gene from *Malus* × *domestica* is a functional homologue of *Arabidopsis thaliana TRANSPARENT TESTA GLABRA1*. *Plant Cell Reports* 29, 285–294.

Buendia, B., Gil, M.I., Tudela, J.A., Gady, A.L., Medina, J.J., Soria, C., Lopez, J.M. and Tomas-Barberan, F.A. (2010) HPLC-MS analysis of proanthocyanidin oligomers and other phenolics in 15 strawberry cultivars. *Journal of Agricultural and Food Chemistry* 58, 3916–3926.

Cadot, Y., Chevalier, M. and Barbeau, G. (2011) Evolution of the localisation and composition of phenolics in grape skin between veraison and maturity in relation to water availability and some climatic conditions. *Journal of the Science of Food and Agriculture* 91, 1963–1976.

Carbone, F., Preuss, A., De Vos, R.C., D'Amico, E., Perrotta, G., Bovy, A.G., Martens, S. and Rosati, C. (2009) Developmental, genetic and environmental factors affect the expression of flavonoid genes, enzymes and metabolites in strawberry fruits. *Plant, Cell and Environment* 32, 1117–1131.

Carmichael, S., Gonzalez-Feliciano, A., Ma, C., Shaw, G.M. and Cogswell, M.E. (2011) Estimated dietary phytoestrogen intake and major food sources among women during the year before pregnancy. *Nutrition Journal* 10, 105–113.

Castellarin, S.D., Pfeiffer, A., Sivilotti, P., Degan, M., Peterlunger, E. and Di Gaspero, G. (2007) Transcriptional regulation of anthocyanin biosynthesis in ripening fruits of grapevine under seasonal water deficit. *Plant, Cell and Environment* 30, 1381–1389.

Chung, I.M., Park, M.R., Rehman, S. and Yun, S.J. (2001) Tissue specific and inducible expression of resveratrol synthase gene in peanut plants. *Molecular Cell* 12, 353–359.

Clifford, M. and Scalbert, A. (2000) Ellagitannins – nature, occurrence and dietary burden. *Journal of the Science of Food and Agriculture* 80, 1118–1125.

Cohen, S.D., Tarara, J.M., Gambetta, G.A., Matthews, M.A. and Kennedy, J.A. (2012) Impact of diurnal temperature variation on grape berry development, proanthocyanidin accumulation, and the expression of flavonoid pathway genes. *Journal of Experimental Botany* 63, 2655–2665.

Cortell, J. and Kennedy, J.A. (2006) Effect of shading on accumulation of flavonoid compounds in *Vitis vinifera* L. Pinot Noir fruit and extraction in a model system. *Journal of Agricultural and Food Chemistry* 54, 8510–8520.

Crifò, T., Puglisi, I., Petrone, G., Recupero, G.R. and Lo Piero, A.R. (2011) Expression analysis in response to low temperature stress in blood oranges: implication of the flavonoid biosynthetic pathway. *Gene* 476, 1–9.

Cutanda-Perez, M.C., Ageorges, A., Gomez, C., Vialet, S., Terrier, N., Romieu, C. and Torregrosa, L. (2009) Ectopic expression of *VlmybA1* in grapevine activates a narrow set of genes involved in anthocyanin synthesis and transport. *Plant Molecular Biology* 69, 633–648.

Czemmel, S., Stracke, R., Weisshaar, B., Cordon, N., Harris, N.N., Walker, A.R., Robinson, S.P. and Bogs, J. (2009) The grapevine R2R3-MYB transcription factor VvMYBF1 regulates flavonol synthesis in developing grape berries. *Plant Physiology* 151, 1513–1530.

Dadic, M. and Belleau, G. (1973) Polyphenols and beer flavor. *Proceedings of the American Society of Brewing Chemists* 4, 107–114.

Dangles, O. (2012) Antioxidant activity of plant phenols: chemical mechanisms and biological significance. *Current Organic Chemistry* 16, 692–714.

Davies, C., Boss, P.K. and Robinson, S.P. (1997) Treatment of grape berries, a nonclimacteric fruit with a synthetic auxin, retards ripening and alters the expression of developmentally regulated genes. *Plant Physiology* 115, 1155–1161.

Debeaujon, I., Lèon-Kloosterziel, K.M. and Koornneef, M. (2000) Influence of the testa on seed dormancy, germination, and longevity in Arabidopsis. *Plant Physiology* 122, 403–414.

Delcour, J.A., Vandenberghe, M.M., Corten, P.F. and Dondeyne, P. (1984) Flavor thresholds of polyphenolics in water. *American Journal of Enology and Viticulture* 35, 134–136.

Deluc, L., Barrieu, F., Marchive, C., Lauvergeat, V., Decendit, A., Richard, T., Carde, J.P., Merillon, J.M., and Hamdi, S. (2006) Characterization of a grapevine R2R3-MYB transcription factor that regulates the phenylpropanoid pathway. *Plant Physiology* 140, 499–511.

Deluc, L., Bogs, J., Walker, A.R., Ferrier, T., Decendit, A., Merillon, J., Robinson, S.P. and Barrieu, F. (2008) The transcription factor VvMYB5b contributes to the regulation of anthocyanin and proanthocyanidin biosynthesis in developing grape berries. *Plant Physiology* 147, 2041–2053.

Downey, M., Harvey, J., and Robinson, S.P. (2003) Synthesis of flavonols and expression of flavonol synthase genes in the developing grape berries of Shiraz and Chardonnay (*Vitis vinifera* L.). *Australian Journal of Grape and Wine Research* 9, 110–121.

Downey, M.O., Harvey, J.S. and Robinson, S.R. (2004) The effect of bunch shading on berry development and flavonoid accumulation in Shiraz grapes. *Australian Journal of Grape and Wine Research* 10, 55–73.

Dubos, C., Stracke, R., Grotewold, E., Weisshaar, B., Martin, C. and Lepiniec, L. (2010) MYB transcription factors in *Arabidopsis*. *Trends in Plant Science* 15, 573–581.

Dueñas, M., Salas, E., Cheynier, V., Dangles, O. and Fulcrand, H. (2006) UV-Visible spectroscopic investigation of the 8–8-methylmethine catechin-malvidin 3-Glucoside pigments in aqueous solution: structural transformations and molecular complexation with chlorogenic acid. *Journal of Agricultural and Food Chemistry* 54, 189–196.

Ehala, S., Vaher, M. and Kaljurand, M. (2005) Characterization of phenolic profiles of Northern European berries by capillary electrophoresis and determination of their antioxidant activity. *Journal of Agricultural and Food Chemistry* 53, 6484–6490.

El-Kereamy, A., Chervin, C., Roustan, J.P., Cheynier, V., Souquet, J.M., Moutounet, M., Raynal, J., Ford, C.M., Latche, A., Pech, J.C. and Bouzayen, M. (2003) Exogenous ethylene stimulates the long term expression of genes related to the anthocyanin synthesis in grape berries. *Physiologia Plantarum* 119, 1–8.

Escribano-Bailon, M., Dangles, O. and Brouillard, R. (1996) Coupling reactions between flavylium ions and catechin. *Phytochemistry* 41, 1583–1592.

Espley, R.V., Hellens, R.P., Putterill, J., Stevenson, D.E., Kutty-Amma, S. and Allan, A.C. (2007) Red colouration in apple fruit is due to the activity of the MYB transcription factor, MdMYB10. *Plant Journal* 49, 414–427.

Feller, A., Machemer, K., Braun, E.L. and Grotewold, E. (2011) Evolutionary and comparative analysis of MYB and bHLH plant transcription factors. *Plant Journal* 66, 94–116.

Feng, S., Wang, Y., Yang, S., Xu, Y. and Chen, X. (2010) Anthocyanin biosynthesis in pears is regulated by a R2R3-MYB transcription factor PyMYB10. *Planta* 232, 245–255.

Fischer, T.C., Gosch, C., Pfeiffer, J., Halbwirth, H., Halle, C., Stich, K. and Forkmann, G. (2007) Flavonoid genes of pear (*Pyrus communis*). *Trees* 21, 521–529.

Fischer, U.A., Reinhold, C. and Kammerer, D.R. (2011) Identification and quantification of phenolic compounds from pomegranate (*Punica granatum* L.) peel, mesocarp, aril and differently produced juices by HPLC-DAD-ESI/MS^n. *Food Chemistry* 127, 807–821.

Fornara, V.E., Onelli, F., Sparvoli, M., Rossoni, R., Aina, G., Marino, S. and Citterio, F. (2008) Localization of stilbene synthase in *Vitis vinifera* L. during berry development. *Protoplasma* 233, 83–93.

Fossen, T., Rayyan, S. and Andersen, O.M. (2004) Dimeric anthocyanins from strawberry (*Fragaria ananassa*) consisting of pelargonidin 3-glucoside covalently linked to four flavan-3-ols. *Phytochemistry* 65, 1421–1428.

Fujita, A., Soma, N., Goto-Yamamoto, N., Mizuno, A., Kiso, K. and Hashizume, K. (2007) Effect of shading on proanthocyanidin biosynthesis in the grape berry. *Journal of Japanese Society of Horticultural Science* 76, 112–119.

Fulcrand, H., Cameira Dos Santos, P.J., Sarni-Manchado, P., Cheynier, V. and Favre-Bonvin, J. (1996) Structure of new anthocyanin-derived wine pigments. *Journal of the Chemical Society Perkin Transactions 1* 7, 735–739.

Fung, R.W., Gonzalo, M., Fekete, C., Kovacs, L.G., He, Y., Marsh, E., McIntyre, L.M., Schachtman, D.P. and Qiu, W. (2008) Powdery mildew induces defense-oriented reprogramming of the transcriptome in a susceptible but not in a resistant grapevine. *Plant Physiology* 146, 236–249.

Gargouri, M., Manigand, C., Mauge, C., Granier, T., Langlois d'Estaintot, B., Cala, O., Pianet, I., Bathany, K., Chaudiere, J. and Gallois, B. (2009) Structure and epimerase activity of anthocyanidin reductase from *Vitis vinifera*. *Acta Crystallography* 65, 989–1000.

Gatto, P., Vrhovsek, U., Muth, J., Segala, C., Romualdi, C., Fontana, P., Pruefer, D., Stefanini, M., Moser, C., Mattivi, F. and Velasco, R. (2008) Ripening and genotype control stilbene accumulation in healthy grape. *Journal of Agricultural and Food Chemistry* 56, 11773–11785.

Gil, M.I., Tomas-Barberan, F.A., Hess-Pierce, B., Holcroft, D.M. and Kader, A.A. (2000) Antioxidant activity of pomegranate juice and its relationship with phenolic composition and processing. *Journal of Agricultural and Food Chemistry* 48, 4581–4589.

Gil-Izquierdo, A., Gil, M., Ferreres, F. and Tomas-Barberan, F.A. (2001) *In vitro* availability of flavonoids and other phenolics in orange juice. *Journal of Agricultural and Food Chemistry* 49, 1035–1041.

Gonzalez-Paramas, A.M., Lopes da Silva, F., Martin-Lopez, P., Macz-Pop, G., Gonzalez-Manzano, S., Alcalde-Eon, C., Perez-Alonso, J.J., Escribano-Bailon, M.T., Rivas-Gonzalo, J.C. and Santos-Buelga, C. (2006) Flavanol–anthocyanin condensed pigments in plant extracts. *Food Chemistry* 94, 428–436.

Goto, T. and Kondo, T. (1991) Structure and molecular stacking of anthocyanins. Flower color variation. *Angewandte Chemie International Edition* 30, 17–33.

Griesser, M., Hoffmann, T., Bellido, M.L., Rosati, C., Fink, B., Kurtzer, R., Aharoni, A., Muñoz-Blanco, J. and Schwab, W. (2008) Redirection of flavonoid biosynthesis through the down-regulation of an anthocyanidin glucosyltransferase in ripening strawberry fruit. *Plant Physiology* 146, 1528–1539.

Gross, G.G. (1983) Synthesis of mono-, di- and trigalloyl- β-D-glucose by β-glucogallin-dependent galloyltransferases from oak leaves. *Zeitschrift für Naturforschung* 38, 519–523.

Gross, G.G. and Denzel, K. (1991) Biosynthesis of gallotannins. β-Glucogallin-dependent galloylation of 1,6-digalloylglucose to 1,2,6-trigalloylglucose. *Zeitschrift für Naturforschung* 46, 389–394.

Gross, G.G., Denzel, K. and Schilling, G. (1990) Enzymatic-synthesis of di-*O*-phenylcarboxyl-β-D-glucose esters by an acyltransferase from oak leaves. *Zeitschrift für Naturforschung* 45, 37–41.

Gu, L., Kelm, M., Hammerstone, J.F., Beecher, G., Holden, J., Haytowitz, D. and Prior, R.L. (2003) Screening of foods containing proanthocyanidins and their structural characterization using LC-MS/MS and thiolytic degradation. *Journal of Agricultural and Food Chemistry* 51, 7513–7521.

Gu, L., Kelm, M., Hammerstone, J.F., Beecher, G., Holden, J., Haytowitz, D., Gebhardt, S. and Prior, R.L. (2004) Concentrations of proanthocyanidins in common foods and estimations of normal consumption. *Journal of Nutrition* 134, 613–617.

Guyot, S., Marnet, N., Laraba, D., Sanoner, P. and Drilleau, F.F. (1998) Reversed-phase HPLC following thiolysis for quantitative estimation and characterization of the four main classes of phenolic compounds in different tissue zones of a french cider apple variety (*Malus domestica* var. Kermerrien). *Journal of Agricultural and Food Chemistry* 46, 1698–1705.

Guyot, S., Marnet, N. and Drilleau, J.F. (2001) Thiolysis-HPLC characterization of apple procyanidins covering a large range of polymerisation states. *Journal of Agricultural and Food Chemistry* 49, 14–20.

Guyot, S., Le Bourvellec, C., Marnet, N. and Drilleau, J.F. (2002) Procyanidins are the most abundant polyphenols in dessert apples at maturity. *Lebensmittel-Wissenschaft und Technologie* 35, 289–291.

Hall, D. and DeLuca, V. (2007) Mesocarp localization of a bi-functional resveratrol/hydroxycinnamic acid glucosyltransferase of Concord grape (*Vitis labrusca*). *Plant Journal* 49, 579–591.

Halliwell, B., Rafter, J. and Jenner, A. (2005) Health promotion by flavonoids, tocopherols, tocotrienols, and other phenols: direct or indirect effects? Antioxidant or not? *American Jouranl of Clinical Nutrition* 81, 268S–276S.

Han, Y., Vimolmangkang, S., Soria-Guerra, R.E. and Korban, S.S. (2012) Introduction of apple *ANR* genes into tobacco inhibits expression of both *CHI* and *DFR* genes in flowers, leading to loss of anthocyanin. *Journal of Experimental Botany* 63, 2437–2447.

Harborne, J.B. and Williams, C.A. (2000) Advances in flavonoid research since 1992. *Phytochemistry* 55, 481–504.

Hartmann, U., Sagasser, M., Mehrtens, F., Stracke, R. and Weisshaar, B. (2005) Differential combinatorial interactions of *cis*-acting elements recognized by R2R3-MYB, BZIP, and bHLH factors control light-responsive and tissue-specific activation of phenylpropanoid biosynthesis genes. *Plant Molecular Biology* 57, 155–171.

Hellstrom, J., Torronen, A.R. and Mattila, P.H. (2009) Proanthocyanidins in common food products of plant origin. *Journal of Agricultural and Food Chemistry* 57, 7899–7906.

Hichri, I., Heppel, S.C., Pillet, J., Léon, C., Czemmel, S., Delrot, S., Lauvergeat, V. and Bogs, J. (2010) The basic helix–loop–helix transcription factor MYC1 is involved in the regulation of the flavonoid biosynthesis pathway in grapevine. *Molecular Plant* 3, 509–523.

Hichri, I., Barrieu, F., Bogs, J., Kappel, C., Delrot, S. and Lauvergeat, V. (2011) Recent advances in the transcriptional regulation of the flavonoid biosynthetic pathway. *Journal of Experimental Botany* 62, 2465–2483.

Holton, T.A. and Cornish, E.C. (1995) Genetics and biochemistry of anthocyanin biosynthesis. *Plant Cell* 7, 1071–1083.

Horowitz, R.M. and Gentili, B. (1969) Taste and structure in phenolic glycosides. *Journal of Agricultural and Food Chemistry* 17, 696–700.

Hukkanen, A.T., Kokko, H.I., Buchala, A.J., McDougall, G.J., Stewart, D., Karenlampi, S.O. and Karjalainen, R.O. (2007) Benzothiadiazole induces the accumulation of phenolics and improves resistance to powdery mildew in strawberries. *Journal of Agricultural and Food Chemistry* 55, 1862–1870.

Hunt, G. and Baker, E. (1980) Phenolic constituents of tomato fruit cuticles. *Phytochemistry* 19, 1415–1419.

Ikegami, A., Eguchi, S., Kitajima, A., Inoue, K. and Yonemori, K. (2007) Identification of genes involved in proanthocyanidin biosynthesis of persimmon (*Diospyros kaki*) fruit. *Plant Science* 172, 1037–1047.

Iriti, M., Rossoni, M., Borgo, M. and Faoro, F. (2004) Benzothiadiazole enhances resveratrol and anthocyanin biosynthesis in grapevine, meanwhile improving resistance to *Botrytis cinerea*. *Journal of Agricultural and Food Chemistry* 52, 4406–4413.

Iriti, M., Rossoni, M., Borgo, M., Ferrara, L. and Faoro, F. (2005) Induction of resistance to gray mold with benzothiadiazole modifies amino acid profile and increases proanthocyanidins in grape: primary versus secondary metabolism. *Journal of Agricultural and Food Chemistry* 53, 9133–9139.

Jaakola, L., Maatta, K., Pirttila, A., Torronen, R., Karenlampi, S. and Hohtola, A. (2002) Expression of genes involved in anthocyanin biosynthesis in relation to anthocyanin, proanthocyanidin, and flavonol levels during bilberry fruit development. *Plant Physiology* 130, 729–739.

Jaakola, L., Poole, M., Jones, M.O., Kämäräinen-Karppinen, T., Koskimäki, J., Hohtola, A., Häggman, H., Fraser, P., Manning, K., King, G., Thomson, H. and Seymour, G. (2010) A SQUAMOSA MADS box gene involved in the regulation of anthocyanin accumulation in bilberry fruits. *Plant Physiology* 153, 1619–1629.

Jaillon, O., Aury, J.M., Noel, B., Policriti, A., Clepet, C., Casagrande, A., Choisne, N., Aubourg, S., Vitulo, N., Jubin, C., Vezzi, A., Legeai, F., Hugueney, P., Dasilva, C., Horner, D., Mica, E., Jublot, D., Poulain, J., Bruyère, C., Billault, A., Segurens, B., Gouyvenoux, M., Ugarte, E., Cattonaro, F., Anthouard, V., Vico, V., Del Fabbro, C., Alaux, M., Di Gaspero, G., Dumas, V., Felice, N., Paillard, S., Juman, I., Moroldo, M., Scalabrin, S., Canaguier, A., Le Clainche, I., Malacrida, G., Durand, E., Pesole, G., Laucou, V., Chatelet, P., Merdinoglu, D., Delledonne, M., Pezzotti, M., Lecharny, A., Scarpelli, C., Artiguenave, F., Pè, M.E., Valle, G., Morgante, M., Caboche, M., Adam-Blondon, A.F., Weissenbach, J., Quétier, F., Wincker, P. and the French-Italian Public Consortium for Grapevine Genome Characterization (2007) The grapevine genome sequence suggests ancestral hexaploidization in major angiosperm phyla. *Nature* 449, 463–467.

Jeong, S.T., Goto-Yamamoto, N., Kobayashi, S. and Esaka, M. (2004) Effects of plant hormones and shading on the accumulation of anthocyanins and the expression of anthocyanin biosynthetic genes in grape berry skins. *Plant Science* 167, 247–252.

Jia, H.F., Chai, Y.M., Li, C.L., Lu, D., Luo, J.J., Qin, L. and Shen, Y.Y. (2011) Abscisic acid plays an important role in the regulation of strawberry fruit ripening. *Plant Physiology* 157, 188–199.

Kataoka, I. and Beppu, K. (2004) UV irradiance increases development of red skin color and anthocyanins in 'Hakuho' peach. *Hortscience* 39, 1234–1237.

Kato, K. (1990) Astringency removal and ripening in persimmons treated with ethanol and ethylene. *Horticultural Science* 25, 205–207.

Kennedy, J.A., Matthews, M.A. and Waterhouse, A.L. (2000) Changes in grape seed polyphenols during fruit ripening. *Phytochemistry* 55, 77–85.

Kennedy, J.A., Hayasaka, Y., Vidal, S., Waters, E.J. and Jones, G.P. (2001) Composition of grape skin proanthocyanidins at different stages of berry development. *Journal of Agricultural and Food Chemistry* 49, 5348–5355.

Kennedy, J.A., Matthews, M.A. and Waterhouse, A.L. (2002) Effect of maturity and vine water status on grape skin and wine flavonoids. *American Journal of Enology and Viticulture* 53, 268–274.

Khater, F., Fournand, D., Vialet, S., Meudec, E., Cheynier, V. and Terrier, N. (2012) Identification and functional characterization of cDNAs coding for hydroxybenzoate/hydroxycinnamate glucosyltransferases co-expressed with genes related to proanthocyanidin biosynthesis. *Journal of Experimental Botany* 63, 1201–1214.

Kim, S.H., Lee, J.R., Hong, S.T., Yoo, Y.K., An, G. and Kim, S.R. (2003) Molecular cloning and analysis of anthocyanin biosynthesis genes preferentially expressed in apple skin. *Plant Science* 165, 403–413.

Kobayashi, S., Ishimaru, M., Hiraoka, K. and Honda, C. (2002) *Myb*-related genes of the Kyoho grape (*Vitis labruscana*) regulate anthocyanin biosynthesis. *Planta* 215, 924–933.

Kobayashi, S., Goto-Yamamoto, N. and Hirochika, H. (2004) Retrotransposon-induced mutations in grape skin color. *Science* 304, 982.

Koponen, J., Happonen, A., Mattila, P.H. and Törrönen, A.R. (2007) Contents of anthocyanins and ellagitannins in selected foods consumed in Finland. *Journal of Agricultural and Food Chemistry* 55, 1612–1619.

Koskimäki, J.J., Hokkanen, J., Jaakola, L., Suorsa, M., Tolonen, A., Mattila, S., Pirttilä, A.M. and Hohtola, A. (2009) Flavonoid biosynthesis and degradation play a role in early defense responses of bilberry (*Vaccinium myrtillus*) against biotic stress. *European Journal of Plant Pathology* 125, 629–640.

Koyama, K., Sadamatsu, K. and Goto-Yamamoto, N. (2010) Abscisic acid stimulated ripening and gene expression in berry skins of the Cabernet Sauvignon grape. *Functional and Integrative Genomics* 10, 367–381.

Koyama, K., Ikeda, H., Poudel, P.R. and Goto-Yamamoto, N. (2012) Light quality affects flavonoid biosynthesis in young berries of Cabernet Sauvignon grape. *Phytochemistry* 78, 54–64.

Kreuzaler, F., Ragg, H., Fautz, E., Kuhn, D.N. and Hahlbrock, K. (1983) UV-induction of chalcone synthase mRNA in cell suspension cultures of *Petroselinum hortense. Proceedings of the National Academy of Sciences USA* 80, 2591–2593.

Kuhnle, G., Dell'Aquila, C., Low, Y.L., Kussmaul, M. and Bingham, S.A. (2007) Extraction and quantification of phytoestrogens in foods using automated solid-phase extraction and LC/MS/MS. *Analytical Chemistry* 79, 9234–9239.

Laaksonen, O., Sandell, M. and Kallio, K. (2010) Chemical factors contributing to orosensory profiles of bilberry (*Vaccinium myrtillus*) fractions. *European Food Research and Technology* 231, 271–285.

Lacampagne, S., Gagné, S. and Gény, L. (2010) Involvement of abscisic acid in controlling the proanthocyanidin biosynthesis pathway in grape skin: new elements regarding the regulation of tannin composition and leucoanthocyanidin reductase (LAR) and anthocyanidin reductase (ANR) activities and expression. *Journal of Plant Growth Regulation* 29, 81–90.

Langcake, P. and McCarthy, W.V. (1979) The relationship of resveratrol production to infection of grapevine leaves by *Botrytis cinerea. Vitis* 18, 244–253.

Lattanzio, V., Cardinali, A. and Linsalata, V. (2012) Plant phenolics: a biochemical and physiological perspective. In: Cheynier, V., Sarni-Machado, P. and Quideau, S. (eds) *Recent Advances in Polyphenol Research*, Vol. 3. Wiley-Blackwell, Oxford, pp. 1–39.

Le Bourvellec, C. and Renard, C.M.G.C. (2005) Non-covalent interaction between procyanidins and apple cell wall material. Part II: Quantification and impact of cell wall drying. *Biochimica Biophysica Acta* 1725, 1–9.

Lea, A.G.H. (1990) Bitterness and astringency: the procyanidins of fermented apple ciders. In: Rouseff, R.L. (ed.) *Bitterness in Foods and Beverages.* Elsevier, Amsterdam, pp. 123–143.

Lea, A.G.H. and Arnold, G.M. (1978) The phenolics of ciders: bitterness and astringency. *Journal of the Science of Food and Agriculture* 29, 478–483.

Lee, J., Johnson, J. and Talcott, S.T. (2005) Identification of ellagic acid conjugates and other polyphenolics in muscadine grapes by HPLC-ESI-MS. *Journal of Agricultural and Food Chemistry* 53, 6003–6010.

Lees, G.L., Wall, K.M., Beveridge, T.H. and Suttill, N.H. (1995) Localisation of condensed tannins in apple fruit, pulp and seeds. *Canadian Journal of Botany* 73, 1897–1904.

Lepiniec, L., Debeaujon, I., Routaboul, J.M., Baudry, A., Pourcel, L., Nesi, N. and Caboche, M. (2006) Genetics and biochemistry of seed flavonoids. *Annual Review of Plant Biology* 57, 405–430.

Ley, J., Krammer, G., Reinders, G., Gatfield, I.L. and Bertram, H.J. (2005) Evaluation of bitter masking flavanones from herba santa (*Eriodictyon californicum* (H. & A.) Torr., Hydrophyllaceae). *Journal of Agricultural and Food Chemistry* 53, 6061–6066.

Lin-Wang, K., Bolitho, K., Grafton, K., Kortstee, A., Karunairetnam, S., McGhie, T.K., Espley, R.V., Hellens, R.P. and Allan, A.C. (2010) An R2R3 MYB transcription factor associated with regulation of the anthocyanin biosynthetic pathway in Rosaceae. *BMC Plant Biology* 10, 50.

Lister, C.E., Lancaster, J.E. and Walker, J.R. (1996) Developmental changes in enzymes of flavonoid biosynthesis in the skins of red and green apple cultivars. *Journal of the Science of Food and Agriculture* 71, 313–320.

Lu, Y., Sun, Y. and Foo, L.Y. (2000) Novel pyranoanthocyanins from black current seeds. *Tetrahedron Letters* 41, 5975–5979.

Manach, C., Scalbert, A., Remesy, C. and Morand, C. (2006) Consommation et disponibilité des polyphénols. In: Sarni-Manchado, P. and Cheynier, V. (eds) *Les Polyphénols en Agroalimentaire.* Lavoisier, Paris, pp. 361–380.

Manach, C., Hubert, J., Llorach, R. and Scalbert, A. (2009) The complex links between dietary phytochemicals and human health deciphered by metabolomics. *Molecular Nutrition and Food Research* 3, 1303–1315.

Mané, C., Souquet, J.M., Olle, D., Verries, C., Veran, F., Mazerolles, G., Cheynier, V. and Fulcrand, H. (2007) Optimization of simultaneous flavanol, phenolic acid, and anthocyanin extraction from grapes using an experimental design: application to the characterization of Champagne grape varieties. *Journal of Agricultural and Food Chemistry* 55, 7224–7233.

Matthews, S., Mila, I., Scalbert, A. and Donnelly, D.M.X. (1997) Extractable and non-extractable proanthocyanidins in bark. *Phytochemistry* 45, 405–410.

Mattivi, F., Guzzon, R., Vrhovsek, U., Stefanini, M. and Velasco, R. (2006). Metabolite profiling of grape: flavonols and anthocyanins. *Journal of Agricultural and Food Chemistry* 54, 7692–7702.

Matus, J.T., Aquea, F. and Arce-Johnson, P. (2008) Analysis of the grape MYB R2R3 subfamily reveals expanded wine quality-related clades and conserved gene structure organization across *Vitis* and *Arabidopsis* genomes. *BMC Plant Biology* 8, 83.

Matus, J.T., Loyola, R., Vega, A., Peña-Neira, A., Bordeu, E., Arce-Johnson, P. and Alcalde, J.A. (2009) Post-veraison sunlight exposure induces MYB-mediated transcriptional regulation of anthocyanin and flavonol synthesis in berry skins of *Vitis vinifera*. *Journal of Experimental Botany* 60, 853–867.

Matus, J.T., Poupin, M., Cañon, P., Bordeu, E., Alcalde, J. and Arce-Johnson, P. (2010) Isolation of WDR and bHLH genes related to flavonoid synthesis in grapevine (*Vitis vinifera* L.). *Plant Molecular Biology* 72, 607–620.

Mazza, G. and Miniati, E. (1993) *Anthocyanins in Fruits, Vegetables and Grains*. CRC Press, Boca Raton, FL.

McDougall, G.J., Martinussen, I., Junttila, O., Verrall, S. and Stewart, D. (2011) Assessing the influence of genotype and temperature on polyphenol composition in cloudberry (*Rubus chamaemorus* L.) using a novel mass spectrometric method. *Journal of Agricultural and Food Chemistry* 59, 10860–10868.

Mehrtens, F., Kranz, H., Bednarek, P. and Weisshaar, B. (2005) The Arabidopsis transcription factor MYB12 is a flavonol-specific regulator of phenylpropanoid biosynthesis. *Plant Physiology* 138, 1083–1096.

Melchior, F. and Kindl, H. (1991) Coordinate- and elicitor-dependent expression of stilbene synthase and phenylalanine ammonia-lyase genes in *Vitis* cv. Optima. *Archives of Biochemistry and Biophysics* 288, 552–557.

Mes, P.J., Boches, P., Myers, J.R. and Durst, R. (2008) Characterization of tomatoes expressing anthocyanin in the fruit. *Journal of the American Society for Horticultural Science* 133, 262–269.

Milkowski, C., and Strack, D. (2004) Serine carboxypeptidase-like acyltransferases. *Phytochemistry* 66, 517–524.

Mouly, P., Arzouyan, C., Gaydou, E.M. and Estienne, J.M. (1994) Differentiation of citrus juices by factorial discriminant analysis using liquid chromatography of flavanone glycosides. *Journal of Agricultural and Food Chemistry* 42, 70–79.

Moze, S., Polak, T., Gasperlin, L., Koron, D., Vanzo, A., Ulrih, N.P. and Abram, V. (2011) Phenolics in Slovenian bilberries (*Vaccinium myrtillus* L.) and blueberries (*Vaccinium corymbosum* L.). *Journal of Agricultural and Food Chemistry* 59, 6998–7004.

Muir, R.M., Ibáñez, A.M., Uratsu, S.L., Ingham, E.S., Leslie, C.A., McGranahan, G.H., Batra, N., Goyal, S., Joseph, J., Jemmis, E.D. and Dandekar, A.M. (2011) Mechanism of gallic acid biosynthesis in bacteria (*Escherichia coli*) and walnut (*Juglans regia*). *Plant Molecular Biology* 75, 555–565.

Mullen, W., Yokota T., Lean, M.E.J. and Crozier, A. (2003) Analysis of ellagitannins and conjugates of ellagic acid and quercetin in raspberry fruits by MV-MSn. *Phytochemistry* 64, 617–624.

Nakatsuka, A., Maruo, T., Shibashi, C., Ueda, Y., Kobayashi, N., Yamagishi, M. and Itamura, H. (2011) Expression of genes encoding xyloglucan endotransglycosylase/hydrolase in 'Saijo' persimmon fruit during softening after deastringency treatment. *Postharvest Biology and Technology* 62, 89–92.

Neveu, V., Perez-Jiménez, J., Vos, F., Crespy, V., du Chaffaut, L., Mennen, L., Knox, C., Eisner, R., Cruz, J., Wishart, D. and Scalbert, A. (2010) Phenol-Explorer: an online comprehensive database on polyphenol contents in foods. Database, doi: 10.1093/database/bap024.

Niemetz, R. and Gross, G.G. (2003a) Oxidation of pentagalloylglucose to the ellagitannin, tellimagrandin II, by a phenol oxidase from *Tellima grandiflora* leaves. *Phytochemistry* 62, 301–306.

Niemetz, R. and Gross, G.G. (2003b) Ellagitannin biosynthesis: laccase-catalyzed dimerization of tellimagrandin II to cornusiin E in *Tellima grandiflora*. *Phytochemistry* 64, 1197–1201.

Niemetz, R. and Gross, G.G. (2005) Enzymology of gallotannin and ellagitannin biosynthesis. *Phytochemistry* 66, 2001–2011.

Niu, S.S., Xu, C.J., Zhang, W.S., Zhang, B., Li, X., Lin-Wang, K., Ferguson, I.B., Allan, A.C. and Chen, K.S. (2010) Coordinated regulation of anthocyanin biosynthesis in Chinese bayberry (*Myrica rubra*) fruit by a R2R3 MYB transcription factor. *Planta* 231, 887–899.

Ojeda, H., Andary, C., Kraeva, E., Carbonneau, A. and Deloire, A. (2002) Influence of pre- and postveraison water deficit on synthesis and concentration of skin phenolic compounds during berry growth of *Vitis vinifera* cv. Shiraz. *American Journal of Enology and Viticulture* 53, 261–267.

Ollé, D., Guiraud, J.L., Souquet, J.M., Terrier, N., Ageorges, A., Cheynier, V. and Verries, C. (2011) Effect of pre- and post-veraison water deficit on proanthocyanidin and anthocyanin accumulation during Shiraz berry development. *Australian Journal of Grape and Wine Research* 17, 90–100.

Ossipov, V., Salminen, J., Ossipova, S., Haukioja, E. and Pihlaja, K. (2003) Gallic acid and hydrolyzable tannins are formed in birch leaves from an intermediate compound of the shikimate pathway. *Biochemical Systematics and Ecology* 31, 3–16.

Pourcel, L., Routaboul, J., Kerhoas, L., Caboche, M., Lepiniec, L. and Debeaujon, I. (2005) *TRANSPARENT TESTA10* encodes a laccase-like enzyme involved in oxidative polymerization of flavonoids in *Arabidopsis* seed coat. *Plant Cell* 17, 2966–2980.

Quideau, S., Deffieux, D., Douat-Casassus, C. and Pouységu, L. (2011) Plant polyphenols: chemical properties, biological activities, and synthesis. *Angewandte Chemie International Edition* 50, 586–621.

Remberg, S.F., Sonsteby, A., Aaby, K. and Heide, O.M. (2010) Influence of postflowering temperature on fruit size and chemical composition of Glen Ample raspberry (*Rubus idaeus* L.). *Journal of Agricultural and Food Chemistry* 58, 9120–9128.

Ros Barcelo, A., Pomar, F., Lopez-Serrano, M. and Pedreno, M.A. (2003) Peroxidase: a multi-functional enzyme in grapevines. *Functional Plant Biology* 30, 577–591.

Rouseff, R.L. (1990) Bitterness in food products: an overview. In: Rousseff, R.L. (ed.) *Bitterness in Foods and Beverages*. Elsevier, Amsterdam, pp. 1–12.

Rouseff, R.L., Martin, S.F. and Youtsey, C.O. (1987) Quantitative survey of narirutin, hesperidin and neohesperidin in citrus. *Journal of Agricultural and Food Chemistry* 35, 1027–1030.

Ruiz-Garcia, Y., Romero-Cascales, I., Gil-Munoz, R., Fernandez-Fernandez, J.I., Lopez-Roca, J.M. and Gomez-Plaza, E. (2012) Improving grape phenolic content and wine chromatic characteristics through the use of two different elicitors: methyl jasmonate versus benzothiadiazole. *Journal of Agricultural and Food Chemistry* 60, 1283–1290.

Sapir, M., Oren-Shamir, M., Ovadia, R., Reuveni, M., Evenor, D., Tadmor, Y., Nahon, S., Shlomo, H., Chen, L., Meir, A. and Levin, I. (2008) Molecular aspects of *Anthocyanin* fruit tomato in relation to *high pigment-1*. *Journal of Heredity* 99, 292–303.

Schmidlin, L., Poutaraud, A., Claudel, P., Mestre, P., Prado, E., Santos-Rosa, M., Wiedemann-Merdinoglu, S., Karst, F., Merdinoglu, D. and Hugueney, P. (2008) A stress-inducible resveratrol *O*-methyltransferase involved in the biosynthesis of pterostilbene in grapevine. *Plant Physiology* 148, 1630–1639.

Schröder, G., Brown, J.W.S. and Schröder, J. (1988) Molecular analysis of resveratrol synthase: cDNA, genomic clones and relationship with chalcone synthase. *European Journal of Biochemistry* 172, 161–169.

Schwarz, B. and Hofmann, T. (2007) Isolation, structure determination, and sensory activity of mouth-drying and astringent nitrogen-containing phytochemicals isolated from red currants (*Ribes rubrum*). *Journal of Agricultural and Food Chemistry* 55, 1405–1410.

Shirley, B.W., Kubasek, W., Storz, G., Bruggemann, E., Koornneef, M., Ausubel, F.M. and Goodman, H.M. (1995) Analysis of *Arabidopsis* mutants deficient in flavonoid biosynthesis. *Plant Journal* 8, 659–671.

Slimestad, R. and Verheul, M. (2005) Seasonal variations in the level of plant constituents in greenhouse production of cherry tomatoes. *Journal of Agricultural and Food Chemistry* 53, 3114–3119.

Slimestad, R., Fossen, T. and Verheul, M. (2008) The flavonoids of tomatoes. *Journal of Agricultural and Food Chemistry* 56, 2436–2441.

Smoliga, J.M., Baur, J.A. and Hausenblas, H.A. (2011) Resveratrol and health – a comprehensive review of human clinical trials. *Molecular Nutrition and Food Research* 55, 1129–1141.

Steffens, J.C. (2000) Acyltransferases in protease's clothing. *Plant Cell* 12, 1253–1255.

Steyn, W.J., Holcroft, D.M, Wand, S.J.E. and Jacobs, G. (2004) Regulation of pear color development in relation to activity of flavonoid enzymes. *Journal of the American Society for Horticultural Science* 129, 1–6.

Steyn, W.J., Wand, S.J., Jacobs, G., Rosecrance, R.C. and Roberts, S.C. (2009) Evidence for a photoprotective function of low-temperature-induced anthocyanin accumulation in apple and pear peel. *Physiologia Plantarum* 136, 461–472.

Strack, D. and Mock, H.P. (1993) Hydroxycinnamic acids and lignins. *Plant Biochemistry* 9, 45–97.

Stracke, R., Ishihara, H., Huep, G., Barsch, A., Mehrtens, F., Niehaus, K. and Weisshaar, B. (2007) Differential regulation of closely related R2R3-MYB transcription factors controls flavonol accumulation in different parts of the *Arabidopsis thaliana* seedling. *Plant Journal* 20, 660–677.

Sugiura, A. and Tomana, T. (1983) Relationships of ethanol production by seeds of different types of Japanese persimmons and their tannin content. *Horticultural Science* 18, 319–321.

Taira, S., Ono, M. and Matsumoto, M. (1998) Reduction of persimmon astringency by complex formation between pectin and tannins. *Postharvest Biology and Technology* 12, 265–271.

Takaya, Y., Yan, K., Terashima, K., He, Y.H. and Niwa, M. (2002). Biogenetic reactions on stilbene tetramers from Vitaceaeous plants. *Tetrahedron* 58, 9265–9271.

Takos, A.M., Jaffe, F.W., Jacob, S.R., Bogs, J., Robinson, S.P. and Walker, A.R. (2006a) Light-induced expression of a MYB gene regulates anthocyanin biosynthesis in red apples. *Plant Physiology* 142, 1216–1232.

Takos, A.M., Ubi, B.E., Robinson, S.P. and Walker, A.R. (2006b) Condensed tannin biosynthesis genes are regulated separately from other flavonoid biosynthesis genes in apple fruit skin. *Plant Science* 170, 487–499.

Tanaka, T., Takahashi, R. Kouno, I. and Nonaka, G.I. (1994) Chemical evidence for the de-astringency (insolubilisation of tannins) of persimmon fruit. *Journal of Chemical Society Perkin Transactions* 1, 3013–3022.

Tanner, G.J., Francki, K.T., Abrahams, S., Watson, J.M., Larkin, P.J. and Ashton, A.R. (2003) Proanthocyanidin biosynthesis in plants. Purification of legume leucoanthocyanidin reductase and molecular cloning of its cDNA. *Journal of Biological Chemistry* 278, 31647–31656.

Tarascou, I., Mazauric, J.P., Meudec, E., Souquet, J.M., Cunningham, D., Nojeim, S., Cheynier, V. and Fulcrand, H. (2011) Characterisation of genuine and derived cranberry proanthocyanidins by LC-ESI-MS. *Food Chemistry* 128, 802–810.

Terrier, N., Olle, D., Verries, C. and Cheynier, V. (2009a) Biochemical and molecular aspects of flavan-3-ol synthesis during berry development. In: Roubelakis-Angelakis, K.A. (ed.) *Grapevine Molecular Physiology & Biotechnology*. Springer Book, New York, USA, pp. 365–388.

Terrier, N., Torregrosa, L., Ageorges, A., Vialet, S., Verriès, C., Cheynier, V. and Romieu, C. (2009b) Ectopic expression of VvMybPA2 promotes proanthocyanidin biosynthesis in *Vitis vinifera* L. and suggests additional targets in the pathway. *Plant Physiology* 149, 1028–1041.

Tipa-Umphon, A., Roustan, J.P. and Chervin, C. (2007) The stimulation by ethylene of the UDP glucose-flavonoid 3-*O*-glucosyltransferase (UFGT) in grape tissues is independent from the MybA transcription factors. *Vitis* 46, 210–211.

Törrönen, R. (2009) Sources and health effects of dietary ellagitannins. In: Quideau, S. (ed.) *Chemistry and Biology of Ellagitannins – an Underestimated Class of Bioactive Plant Polyphenols*. World Scientific, Singapore, pp. 298–319.

Treutter, D. (2006) Significance of flavonoids in plant resistance: a review. *Environmental Chemistry Letters* 4, 147–157.

Trousdale, E. and Singleton, V.L. (1983) Astilbin and engeletin in grapes and wines. *Phytochemistry* 22, 619–620.

Tsuda, T., Yamaguchi, M., Honda, C. and Moriguchi, T. (2004) Expression of anthocyanin biosynthesis genes in the skin of peach and nectarine fruit. *Journal of the American Society for Horticultural Science* 129, 857–862.

Ubi, B.W., Honda, C., Bessho, H., Kondo, S., Wada, M., Kobayashi, S. and Moriguchi, T. (2006) Expression analysis of anthocyanin biosynthetic genes in apple skin: effect of UV-B and temperature. *Plant Science* 170, 571–578.

Uclés Santos, J., Bakry, F. and Brillouet, J.M. (2010) A preliminary chemotaxonomic study on the condensed tannins of green banana flesh in the *Musa* genus. *Biochemical Systematics and Ecology* 38, 1010–1017.

Vallverdu-Queralt, A., Jauregui, O., Medina-Remon, A., Andres-Lacueva, C. and Lamuela-Ravento, R. (2010) Improved characterization of tomato polyphenols using liquid chromatography/electrospray ionization linear ion trap quadrupole Orbitrap mass spectrometry and liquid chromatography/electrospray ionization tandem mass spectrometry. *Rapid Communications in Mass Spectrometry* 24, 2986–2992.

Velasco, R., Zharkikh, A., Affourtit, J., Dhingra, A., Cestaro, A., Kalyanaraman, A., Fontana, P., Bhatnagar, S.K., Troggio, M. *et al.* (2010) The genome of the domesticated apple (*Malus domestica* Borkh.). *Nature Genetics* 42, 833–839.

Verries, C., Guiraud, J.L., Souquet, J.M, Vialet, S., Terrier, N. and Ollé, D. (2008) Validation of an extraction method on whole pericarp of grape berry (*Vitis vinifera* L. cv. Shiraz) to study biochemical and molecular aspects of flavan-3-ol synthesis during berry development. *Journal of Agricultural and Food Chemistry* 56, 5896–8904.

Vidal, S., Francis L., Guyot, S., Marnet, N., Kwiatkowski, M., Gawel, R., Cheynier, V. and Waters, E.J. (2003a) The mouth-feel properties of grape and apple proanthocyanidins in a wine-like medium. *Journal of the Science of Food and Agriculture* 83, 564–573.

Vidal, S., Courcoux, P., Francis, L., Gawel, R., Waters, E. and Cheynier, V. (2003b) Use of an experimental design approach for evaluation of key wine components on mouth-feel perception. *Food Quality and Preference* 15, 209–217.

Vidal, S., Hayasaka, Y., Meudec, E., Cheynier, V. and Skouroumounis, G. (2004a) Fractionation of grape anthocyanin classes using multilayer coil countercurrent chromatography with step gradient elution. *Journal of Agricultural and Food Chemistry* 52, 713–719.

Vidal, S., Meudec, E., Cheynier, V., Skouroumounis, G. and Hayasaka, Y. (2004b) Mass spectrometric evidence for the existence of oligomeric anthocyanins in grape skins. *Journal of Agricultural and Food Chemistry* 52, 7144–7151.

Vidal, S., Francis, L., Kwiatkowski, M., Cheynier, V. and Waters, E. (2004c) Taste and mouth-feel properties of different types of tannin-like polyphenolic compounds and anthocyanins in wine. *Analytica Chimica Acta* 513, 57–65.

Vitrac, X., Monti, J., Vercauteren, J., Deffieux, G. and Merillon, J.M. (2002) Direct liquid chromatographic analysis of resveratrol derivatives and flavanonols in wines with absorbance and fluorescence detection. *Analytica Chimica Acta* 458, 103–110.

Vrhovsek, U., Rigo, A., Tonon, D. and Mattivi, F. (2004) Quantitation of polyphenols in different apple varieties. *Journal of Agricultural and Food Chemistry* 52, 6532–6538.

Walker, A.R., Lee, E., Bogs, J., McDavid, D.A.J., Thomas, M.R. and Robinson, S.P. (2007) White grapes arose through the mutation of two similar and adjacent regulatory genes. *Plant Journal* 49, 772–785.

Wang, C.T., Chen, C.Y., Wang, P. and Chen, Y. (2007) Resveratrol content in strawberry fruit is affected by preharvest conditions. *Journal of Agricultural and Food Chemistry* 55, 8269–8274.

Werner, I., Bacher, A. and Eisenreich, W. (1997) Retrobiosynthetic NMR studies with 13C-labelled glucose. Formation of gallic acid in plants and fungi. *Journal of Biological Chemistry* 272, 25474–25482.

Werner, R.A., Rossmann, A., Schwarz, C., Bacher, A., Schmidt, H.L. and Eisenreich, W. (2004) Biosynthesis of gallic acid in *Rhus typhina*: discrimination between alternative pathways from natural oxygen isotope abundance. *Phytochemistry* 65, 2809–2813.

Williamson, G. and Stalmach, A. (2012) Absorption and metabolism of dietary chlorogenic acids and procyanidins. In: Cheynier, V., Sarni-Manchado, P. and Quideau, S. *Recent Advances in Polyphenol Research*, Vol. 3. Wiley and Blackwell, London, pp. 209–222.

Williner, M.R., Pirovani, M.E. and Güemes, D.R. (2003) Ellagic acid content in strawberries of different cultivars and ripening stages. *Journal of the Science of Food and Agriculture* 83, 842–845.

Winkel-Shirley, B. (2001) Flavonoid biosynthesis. A colorful model for genetics, biochemistry, cell biology, and biotechnology. *Plant Physiology* 126, 485–493.

Winkel-Shirley, B. (2002) Biosynthesis of flavonoids and effects of stress. *Current Opinion in Plant Biology* 5, 218–223.

Wojdylo, A., Oszmianski, J. and Laskowski, P. (2008) Polyphenolic compounds and antioxidant activity of new and old apple varieties. *Journal of Agricultural and Food Chemistry* 56, 6520–6530.

Wrangham, R.W. and Waterman, P.G. (1983) Condensed tannins in fruits eaten by chimpanzees. *Biotropica* 15, 217–222.

Wu, L.C. and Prior, R. (2005) Systematic identification and characterization of anthocyanins by HPLC-ESI-MS/MS in common foods in the United States: fruits and berries. *Journal of Agricultural and Food Chemistry* 53, 2589–2599.

Xie, D.Y., Sharma, S.B., Paiva, N.L., Ferreira, D. and Dixon, R.A. (2003) Role of anthocyanidin reductase, encoded by *BANYULS* in plant flavonoid biosynthesis. *Science* 299, 396–399.

Yamane, T., Jeong, S.T., Goto-Yamamoto, N., Koshita, Y. and Kobayashi, K. (2006) Effects of temperature on anthocyanin biosynthesis in grape berry skins. *American Journal of Enology and Viticulture* 57, 54–59.

Yoshida, K., Oyama, K. and Kondo, T. (2012) Chemistry of flavonoids in color development. In: Cheynier, V., Sarni-Manchado, P. and Quideau, S. *Recent Advances on Polyphenol Research*, Vol. 3. Wiley and Blackwell, London, pp. 99–129.

Zifkin, M., Jin, A., Ozga, J.A., Zaharia, L.I., Schernthaner, J.P., Gesell, A., Abrams, S.R., Kennedy, J.A. and Constabel, C.P. (2012) Gene expression and metabolite profiling of developing highbush blueberry fruit indicates transcriptional regulation of flavonoid metabolism and activation of abscisic acid metabolism. *Plant Physiology* 158, 200–224.

10 Ethylene Biosynthesis

Donald Grierson*

Laboratory of Fruit Quality Biology/The State Agriculture Ministry Laboratory of Horticultural Plant Growth, Development and Quality Improvement, Zhejiang University, Hangzhou, China; Plant and Crop Sciences Division, School of Biosciences, University of Nottingham, Loughborough, UK

10.1 Introduction

Research during the first few decades of the 20th century showed that hydrocarbon gases in the environment influence plant growth, development and fruit ripening. Once it was realized that ethylene was the key molecule in this process, and that plants produce it themselves, it was recognized as a *bona fide* hormone. This stimulated interest in determining the pathway of ethylene biosynthesis and led, ultimately, to the discovery of the enzymes, genes and regulatory factors that control ethylene production and action at different stages in the life cycle. All plants produce ethylene, but increased ethylene production occurs at many stages of development, particularly in response to developmental signals (e.g. flower development and sex determination, abscission, fruit ripening, leaf senescence), hormones (e.g. auxin, cytokinin, ethylene) and environmental influences (e.g. infection, wounding, chilling drought, UV light, oxidative stresses such as SO_2 and O_3) (Abeles *et al.*, 1992). This raises the question: how are the observed increases in ethylene production regulated so precisely in different situations? As we shall see, there are two key enzymes required for ethylene synthesis (ACC synthase and ACC oxidase) and there is a range of mechanisms controlling their transcriptional and post-transcriptional regulation.

Classic research by Mapson, Lieberman and colleagues (Lieberman *et al.*, 1965, 1966; reviewed by Lieberman, 1979) showed that the amino acid methionine was a precursor of ethylene in plants, and that the CH_2 groups of methionine form carbons 1 and 2 of ethylene (Fig. 10.1). The three essential reactions are: activation of L-methionine (Met) to form *S*-adenosylmethionine (AdoMet); conversion of AdoMet to 1-aminocyclopropane-1-carboxylic acid (ACC), a non-protein amino acid; and conversion of ACC to ethylene, as follows.

The first reaction is catalysed by *S*-adenosylmethionine synthetase (EC 2.45.1.6), which requires Mg^{2+} and K^+ (or another monovalent cation):

$$\text{Met} + \text{ATP} \rightarrow \text{AdoMet} + \text{PP}_i + \text{P}_i \qquad (1)$$

* donald.grierson@nottingham.ac.uk

 Fruit Ripening: Physiology, Signalling and Genomics (eds P. Nath *et al.*)

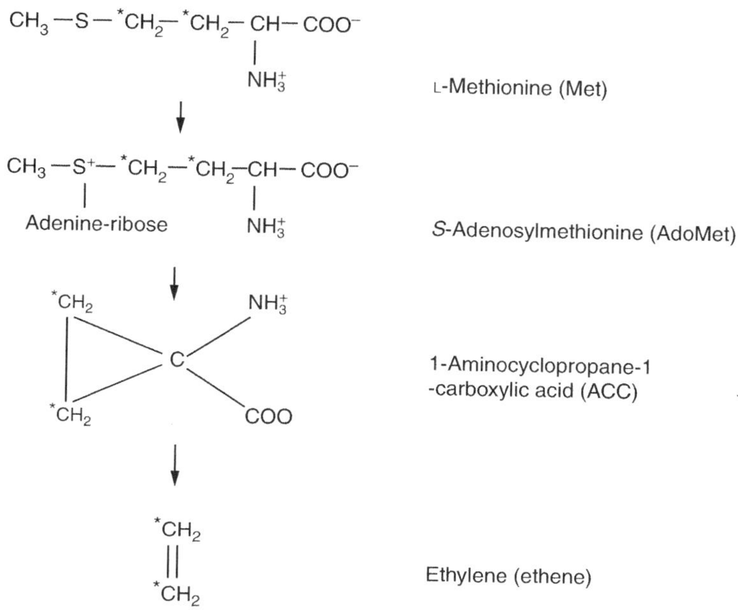

Fig. 10.1. Biosynthesis of ethylene from methionine. The two CH_2 groups of methionine, which form ethylene, are marked with asterisks. Note that plants convert methionine into 1-amino-cyclopropane-1-carboxylic acid to produce ethylene, whereas bacteria utilize 2-oxo-4-methylthiobutyric acid formed from methionine as an intermediate in ethylene synthesis. From Primrose (1979) and Yang and Hoffman (1984).

The second reaction is catalysed by ACC synthase (ACS; EC 4.4.1.14), which requires pyridoxal phosphate:

The last reaction is catalysed by ACC oxidase (ACO) (EC 1.14.17.4), which requires oxygen, Fe^{2+}, CO_2 (supplied as bicarbonate) and ascorbate:

Methionine and *S*-adenosylmethionine are important constituents of all living cells. Reactions (2) and (3), however, are specific for ethylene biosynthesis in plants, and are catalysed by ACS and ACO. These two enzymes require a number of factors for activity, and ACO, for example, can only function in the presence of O_2. The requirement for O_2 for ethylene synthesis was exploited by Adams and Yang (1979) to study the biosynthetic pathway. They showed that when apple tissue was placed under N_2, ethylene synthesis was inhibited and radioactive carbon tracer from methionine accumulated in an unknown compound. When O_2 was restored to the tissue, ethylene production rapidly resumed and the radioactive carbon appeared in ethylene. This suggested that the unknown compound, which Adams and Yang later identified as ACC, was a precursor, and that its conversion to ethylene required O_2. The enzyme that catalyses this reaction (ACO) could, at the time, only be studied in tissue slices, and it was originally referred to as 'ethylene-forming enzyme' (EFE). Although Yang studied the stereospecificity of the reaction, it was more than a decade before the enzyme was fully characterized, as discussed below, and named ACC oxidase (ACO).

In addition to their roles in ethylene synthesis, methionine is required for protein synthesis, and *S*-adenosylmethionine functions as a methyl donor in general metabolism. In order to maintain high rates of ethylene production, without depleting the cellular pools of these compounds, the methylthioadenosine formed during ethylene biosynthesis is recycled in a series of reactions to form methionine (Fig. 10.2). In addition, the cyanide produced during the conversion of ACC to ethylene is detoxified by β-cyanoalanine synthase, forming β-cyanoalanine. Without this, the consumption of ripening fruits and cut vegetables would be potentially hazardous (Fig. 10.2)! ACC can also be converted to malonyl-ACC, which is stored in the vacuole, where it is thought to be unavailable for ethylene synthesis. Recent developments in ethylene synthesis, signalling and responses have been reviewed by Lin *et al.* (2009). For a full appreciation of how the ethylene pathway fits into the general metabolism, including the synthesis of another group of important growth compounds, the polyamines, readers are recommended to consult Urstenbinder and Sauter (2012) and Harpaz-Saad *et al.* (2012).

The ACC pathway is believed to be the source of most, if not all, ethylene produced by plants and its control holds

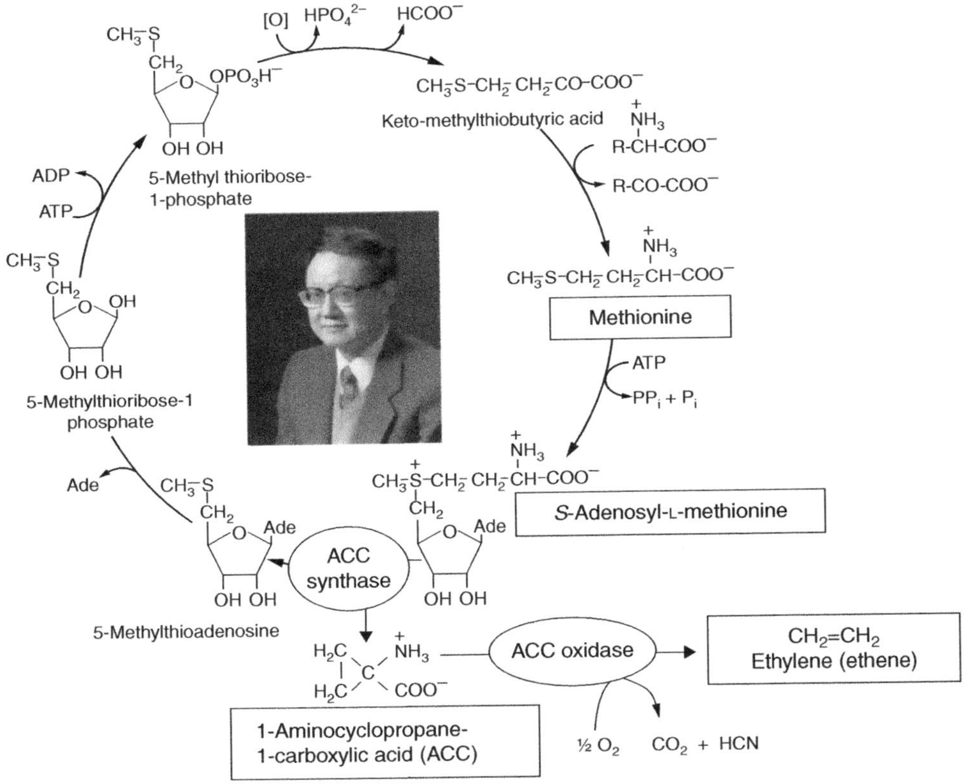

Fig. 10.2. The ethylene biosynthesis cycle. The reactions and enzymes involved in ethylene synthesis, methionine resynthesis, cyanide detoxification and malonyl-ACC production surrounding a photograph of the late Shang Fa Yang (Yang and Hoffman, 1984; Bradford, 2008). Enzymes and substrates/products referred to in the text are in boxes. Modified from Bradford (2008) and reproduced with permission.

S-Adenosyl-L-methionine → 1-aminocyclopropane-1-carboxylate + methylthioadenosine (2)

1-Aminocyclopropane-1-carboxylate + ascorbate + O_2 → ethylene + HCN + dehydroascorbate + CO_2 + $2H_2O$ (3)

the key to understanding the regulation of ethylene production during plant growth and development. In bacteria, however, there is clear evidence that ethylene biosynthesis occurs via the intermediate 2-oxo-4-methylthiobutyric acid and not ACC (Primrose, 1979).

Frequently, the ethylene literature states that ACS is the regulatory step in ethylene biosynthesis. This is partly true, but transcriptional regulation of ACO is also an important control point for ethylene synthesis, as explained below. The demonstration that *ACO1* mRNA appears rapidly after wounding indicates that early conclusions that ACO is always constitutive, and that control is only exerted at the level of ACS, are not entirely correct (discussed below). In addition, there are multiple *ACO* and *ACS* genes controlled by multiple transcription factors (TFs). These gene families can generate multiple isoforms of ACO and ACS and some have different half-lives and regulation. In principle, the pathway could be controlled at the level of transcription, translation or enzyme modification/protein degradation of either ACS or ACO. In fact, both enzymatic steps are transcriptionally regulated, and ACS activity is also controlled post-translationally. It was necessary to purify and study the enzymes, clone and sequence their genes, and investigate mutants with altered ethylene evolution in order to reveal and understand these complex control mechanisms.

10.2 Identification of Genes for ACO and ACS

ACS is a pyridoxal phosphate-requiring enzyme. It was first studied in homogenates of tissue that synthesized large amounts of ethylene and was purified by several groups using conventional protein purification procedures (Bleecker *et al.*, 1986; Nakajima and Imaseki, 1986; Nakajima *et al.*, 1988). An antibody raised against ACS was used to identify the mRNA translation product, which enabled the amino acid sequence to be predicted and provided probes for gene identification (Sato *et al.*, 1991; Van der Straeten *et al.*, 1992). Cloning of the genes showed that ACS is a member of a multigene family, related to the aminotransferase class of proteins, which form dimers *in vivo*. In *Arabidopsis*, *ACS* genes encode a range of active homo- and heterodimer isoforms of ACS with different kinetic properties (Yamagami *et al.*, 2003; Tsuchisaka and Theologis, 2004). These isoforms can have different affinities for pyridoxal phosphate and show structural and regulatory differences. It is believed that this range of ACS forms can function in different cellular environments and substrate concentrations. The crystal structure of ACS from apple (Capitani *et al.*, 1999) showed that the amino acids in the active site are identical to those of chicken mitochondrial aminotransferase. ACS binds the pyridoxal phosphate cofactor at a critical lysine residue (Lys278 in tomato ACS), which is conserved in other ACS isoforms (Yip *et al.*, 1990).

There was a widespread belief that the final step in the biosynthesis of ethylene by EFE required membrane integrity (Yang and Hoffman 1984; see also John, 1991) because, when attempts were made to solubilize the activity, it disappeared. Yang had studied the activity in tissue slices and used analogues to study the stereospecificity of the reaction (Hoffman *et al.*, 1982), but there was a lack of appreciation of the requirements of EFE (now called ACO). The problem was resolved when the author's group, after a systematic search, identified a cDNA clone from a tomato fruit ripening cDNA library, TOM13, encoding a mRNA that was expressed both in ripening fruit and rapidly in wounded leaves, situations where large amounts of ethylene are produced (Smith *et al.*, 1986; Holdsworth *et al.*, 1987) (Figs 10.3 and 10.4). Later, it was shown that this mRNA is also expressed in senescing leaves (Davies and Grierson, 1989) where again there is a requirement for ethylene synthesis.

Sequencing of the mRNA showed that it appeared to encode a soluble protein, and

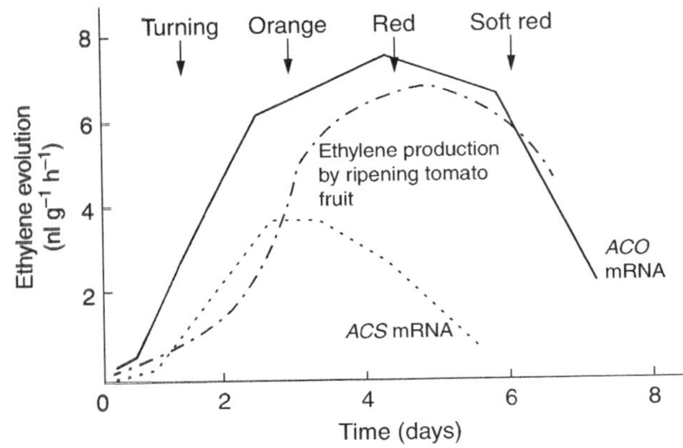

Fig. 10.3. Ethylene evolution and *ACS* and *ACO* mRNA levels during tomato fruit (*Lycopersicon esculentum*) ripening. During ripening, ethylene synthesis is autocatalytic: its production is initiated in a few cells and this recruits other cells to switch on ethylene synthesis, so that a sustained burst of ethylene production moves throughout the fruit. The turning stage of ripening is when the flesh is beginning to change from green to orange. The increase in ethylene biosynthesis (dotted-dashed line) is closely paralleled by increases in mRNA for *LeACS2* and *LeACS4* (dotted line), and *LeACO1*, *LeACO3* and *LeACO4* (solid line), as discussed in the text. High ethylene production levels continue for days, without feedback inhibition, and production then declines slowly as the fruit become fully ripe. Compare this with the kinetics of ethylene production in wounded leaves in Fig 10.4. The figure shows a diagrammatic representation of the results of Holdsworth *et al.* (1987), Barry *et al.*, (1996, 2000) and Nakatsuka *et al.* (1998).

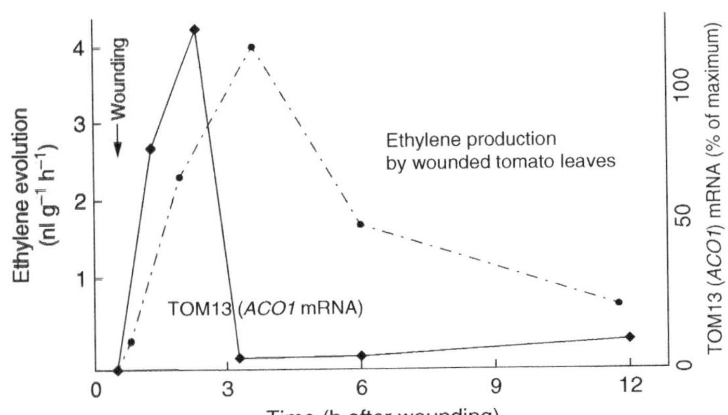

Fig. 10.4. Ethylene evolution and levels of TOM13 (*ACO1*) mRNA in wounded tomato leaves. Wounding initiates a rapid burst of ethylene production within 20 min of wounding (cutting) (dotted-dashed line). This is preceded by a rise in TOM13 (*ACO1*) mRNA (solid line). Ethylene and ACO1 mRNA start declining within a few minutes and continue to decline for a few hours to reach the basal levels. Redrawn from Holdsworth *et al.* (1987).

inhibiting its expression in transgenic tomato plants, using an antisense TOM13 gene-silencing construct, led to inhibition of ethylene biosynthesis (Hamilton *et al.*, 1990) (Fig. 10.5). By expressing it in yeast, TOM13 was shown to encode a protein that converted ACC to ethylene with the stereospecificity predicted by Yang (Hamilton *et al.*, 1991), and the identity of TOM13 was confirmed by expressing a

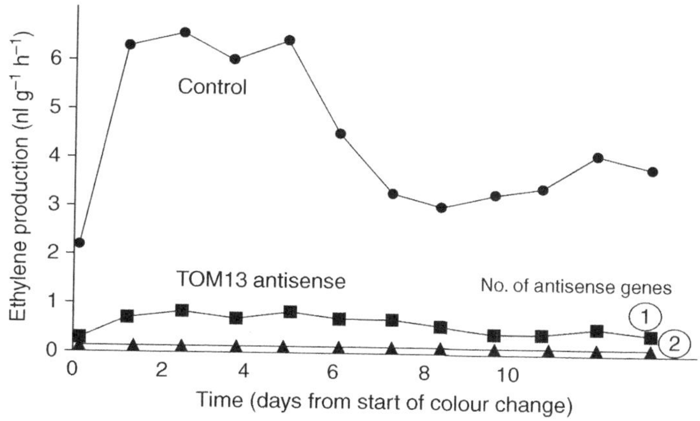

Fig. 10.5. Reduction in ethylene synthesis by inhibiting TOM13 (*ACO1*) expression in ripening tomato, by expressing antisense genes in transgenic plants. The control value for ethylene production (compare with Fig. 10.3) is reduced by over 85% with one antisense TOM13 (*ACO1*) gene (heterozygote) and over 95% with two antisense genes (homozygote). Redrawn from Hamilton *et al.* (1990).

related cDNA in *Xenopus* oocytes (Spanu *et al.*, 1991). The similarity between the TOM13 (*ACO*) predicted amino acid sequence (Holdsworth *et al.*, 1987) and flavanone-3-hydroxylase suggested that the enzyme would require, among other things, anaerobic extraction (Ververidis and John, 1991) and Fe(II) and ascorbate (see Hamilton *et al.*, 1991). Armed with this information, it was relatively easy to purify soluble EFE (ACO) and study its properties in detail.

ACO is a member of the non-haem iron oxygenase/oxidase superfamily of enzymes that utilizes Fe(II) and ascorbate rather than oxoglutarate as cofactors (Schofield and Zhang, 1999). It requires O, and also CO_2 (from bicarbonate), which activates the enzyme (Zhang *et al.*, 2004). Without CO_2, ACO is rapidly inactivated. Early work by Holdsworth *et al.* (1988) identified at least three genes in tomato, and it is now known that there are six related tomato sequences, although the catalytic activity of only three has been confirmed *in vitro* (Bidonde *et al.*, 1998). From the crystal structure of *Petunia* ACO, it has been deduced that it forms a complex with Fe(II), coordinated by His177, Asp179 and His234 (Zhang *et al.*, 2004). The oxidation of ACC occurs concomitantly with the reduction of O_2, presumed to be by ascorbate, to generate CO_2, cyanide and two molecules of H_2O (Bassan *et al.*, 2006).

The importance of ethylene in controlling fruit ripening and leaf senescence was demonstrated by directly inhibiting the expression of tomato *ACO1* and *ACS2* in transgenic tomato plants (Hamilton *et al.*, 1990; Oeller *et al.*, 1991; Picton *et al.*, 1993; John *et al.*, 1995). Lack of ethylene synthesis in these plants and fruit led to partial or complete inhibition of ripening and could be reversed by supplying ethylene externally (Oeller *et al.*, 1991). Other attempts were also made to inhibit ethylene in transgenic plants. The DNAP company developed a transgenic variety (called Endless Summer) in which a truncated *ACS* gene caused silencing of the endogenous gene, thus limiting ACC, and therefore ethylene, production during fruit ripening. Monsanto produced a transgenic tomato expressing a bacterial ACC deaminase gene that reduced the amount of ACC available for ethylene synthesis, whereas Agritope used an *S*-adenosylmethionine hydrolase from bacteriophage T3 to reduce the level of a precursor to ACC. None of these ventures has been commercially successful so far, partly for patent reasons, partly due to consumer resistance and also because, although reducing ethylene biosynthesis does reduce the rate of ripening, overripening and deterioration, it also inhibits flavour development, which makes this particular technological approach unattractive to consumers (Ayub *et al.*, 1996; Flores *et al.*, 2002).

10.3 Differential Expression of *ACO* and *ACS* Genes: System 1 and System 2 Ethylene

All plant parts synthesize small amounts of ethylene all the time, including unripe fruits. This basal level is approximately 0.05 nl g^{-1} h^{-1} in tomato, but can vary with the variety, temperature and other environmental conditions. In climacteric fruits, there is a ripening-related burst of ethylene biosynthesis, accompanied by the respiratory climacteric. This may reach 10 nl g^{-1} h^{-1} in tomato but can be much higher in other fruits, such as banana. Before ripening, ethylene production increases rapidly if the fruit or leaves are cut, or subjected to some other injury, but this is generally reduced again after a few hours (Fig. 10.4), because at this stage of development ethylene biosynthesis is subject to feedback inhibition. At the onset of ripening, however, a burst of autocatalytic ethylene production begins (i.e. ethylene stimulates its own synthesis; there is no feedback inhibition) and this stimulates ripening. McMurchie *et al.* (1972) proposed the existence of two systems (system 1 and system 2) involved in ethylene biosynthesis. System 1 functions during normal vegetative growth, is autoinhibited by ethylene and is responsible for producing the basal levels of ethylene that are synthesized by all plant tissues. System 2 operates during ripening of climacteric fruit, during senescence and in some other situations. The molecular and biochemical data accumulated subsequently support this proposal, and we now know that both systems utilize different isoforms of ACS and ACO, which are regulated differently.

ACO and *ACS* genes are differentially expressed in different plant organs at specific times, particularly during fruit development and ripening (Figs 10.6 and 10.7). The expression of members of the *ACS* multigene family in ripening tomato fruit and senescing flowers was demonstrated by Rottmann *et al.* (1991). *ACO1* mRNA (when it was still referred to as TOM13) was shown very early on to accumulate within minutes of cutting or wounding plant material and during ripening (Smith *et al.*, 1986; Holdsworth *et al.*, 1987; Blume and Grierson, 1997), leaf senescence (Davies and Grierson, 1989; Picton *et al.*, 1993; John *et al.*, 1995; Blume and Grierson, 1997) and following fungal infection (Blume and Grierson, 1997). Barry *et al.* (1996) demonstrated that mRNA from *ACO1*, *ACO2* and *ACO3* accumulated in a variety of developmental situations, including fruit ripening. *ACO* genes also show differential expression in flower petals in response to fertilization (Llop-Tous *et al.*, 2000).

A gene called *E8*, which encodes an Fe(II) oxygenase with some similarity to *ACO* genes, was reported to modulate *LeACO1* and *LeACS2* expression and ethylene production in tomato (Kneissl and Deikman, 1996), but the mechanism remains to be elucidated. The normal accumulation of *ACO1* and other mRNAs during tomato ripening is prevented by heat shock at 35°C (Picton and Grierson, 1988), which can occur under field conditions. Ozone, on the other hand, induces rapid accumulation of *ACS2* and *ACO* mRNA, and the induced ethylene biosynthesis is linked to oxidative stress and induction of cell death (Tuomainen *et al.*, 1997; Moeder *et al.*, 2002). Barry *et al.* (2000) proposed that, in tomato, system 1 ethylene involves the expression of *LeACS1A* and *LeACS6*, and that during the transition from system 1 to system 2, the *rin* (or *LeMADS-RIN*) gene, which is mutated in the *ripening inhibitor* (*rin*) mutant of tomato, enhances expression of *LeACS1A* and induction of *LeACS4* and that the maintenance of system 2 ethylene production is due to the ethylene-dependent induction of *LeACS2* (Barry

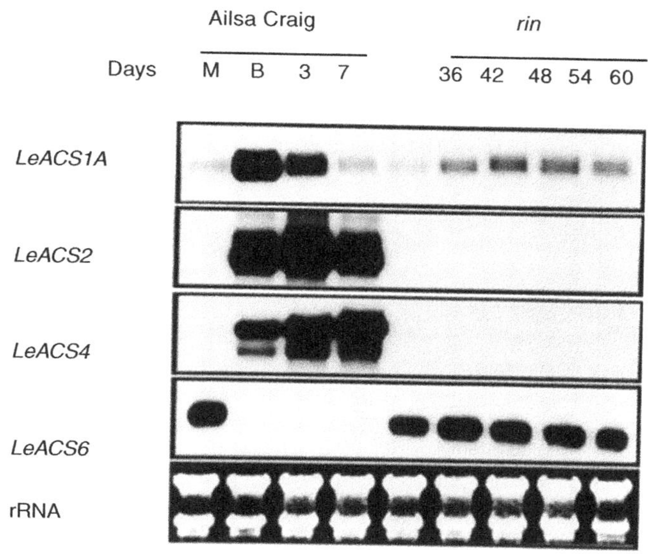

Fig. 10.6. Expression of different *ACS* genes during ripening of tomato. Note that *LeACS6* is expressed before ripening and then rapidly decreases in amount. *LeACS1A*, on the other hand, is expressed early in ripening, together with *LeACS2* and *LeACS4*, but the latter two mRNAs persist for longer. In fruit of the *ripening inhibitor* (*rin*) mutant, which fails to ripen even over a period of several weeks, *LeACS2* and *LeACS4* are not upregulated and *LeACS6* is not repressed, because there is no normal ethylene response in the mutant. M, mature green fruit, prior to ripening onset; B, breaker fruit (first sign of colour change). Days indicate days after breaking. Hybridization signals to fruit RNA samples were detected by X-ray film. *LeACS1A* and *LeACS6* expression was much lower than that of the other *ACS* genes detected. Exposure times to the film (a measure of signal strength) were as follows: *LeACS1A*, 7 days; *LeACS2*, 20 h; *LeACS4*, 20 h; *LeACS6*, 16 days. From Barry *et al.* (2000).

et al., 2000). A modified version of this scheme, proposed by Yokotani *et al.* (2009), is shown in Fig. 10.7.

Auxins, gibberellins and cytokinins all affect ethylene production in particular tissues and only a few examples will be discussed here. *ACO1* expression increased when beech (*Fagus sylvatica* L.) seeds were treated with gibberellic acid 3 (GA_3) or ethephon (which generates ethylene in plants) (Calvo *et al.*, 2004). The stimulation by ethephon was reversed by the GA biosynthesis inhibitor paclobutrazol, suggesting that ethephon affects GA production. Auxin increases the rate of ethylene production and RNA synthesis in seedlings (Grierson *et al.*, 1982) and induces a subset of *ACS* genes (Tsuchisaka and Theologis, 2004; Salman-Minkov *et al.*, 2008). The expression of auxin-induced *OsACO2* in rice (*Oriza sativa*) seedlings is partially inhibited by ethylene, while ethylene induction of *OsACO3* transcription is completely blocked by auxin (Chae *et al.*, 2000).

10.4 Transcriptional Regulation of *ACO* and *ACS* Genes

Given the existence of many *ACO* and *ACS* gene family members, and considering the range of developmental, environmental and hormonal factors that influence their expression, it is important to identify the TFs that regulate the transcription of individual genes in order to understand the complexities of controlling ethylene biosynthesis. Progress has been slow in this area, but four TFs have now been identified

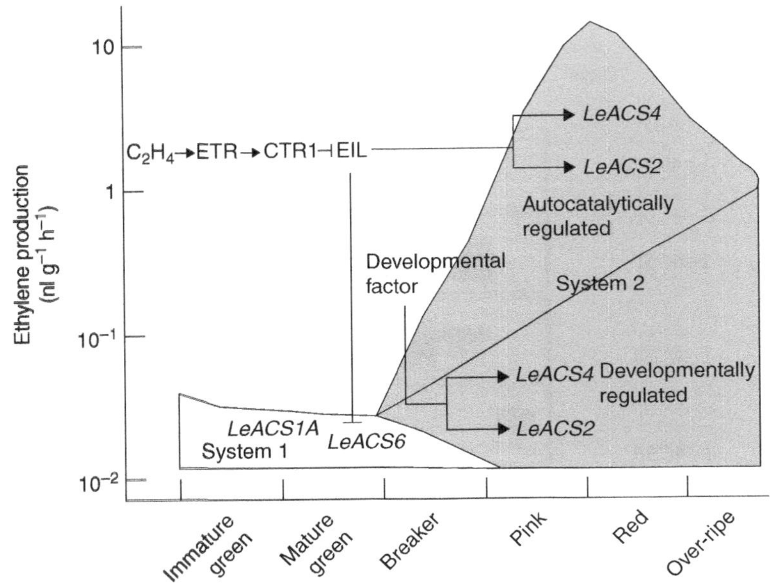

Fig. 10.7. Expression of different *LeACS* and *LeACO* genes related to system 1 and system 2 ethylene production in developing and ripening tomato. The contribution of each isoform to total ethylene evolution is indicated for each stage of development and ripening. Note that developmental factors are shown promoting upregulation of *LeACS4* and *LeACS2*. The ethylene response pathway (ethylene (C2H4) → ethylene receptor (ETR) proteins → constitutive triple response (CTR) proteins → EIN3-like (EIL) proteins) is shown inhibiting *LeACS6* and promoting transcription of *LeACS4* and *LeACS2*. From Yokotani *et al.* (2009); see also Barry *et al.* (2000).

(Fig. 10.8), using a combination of genetic, molecular and biochemical procedures. The *rin* tomato mutant fails to express some ethylene biosynthetic genes (Fig. 10.6) and does not ripen. The TF gene responsible, *LeMADS-RIN*, was identified (Vrebalov *et al.*, 2002) and loss of activity was shown to be the result of the *rin* mutation. Fujisawa *et al.* (2011) subsequently showed that the normal RIN protein binds to the *LeACS2* gene promoter and causes transcription of the gene. A second MADS-box TF, TOMATO AGAMOUS-LIKE1 (TAGL1), has also been shown to bind the *LeACS2* promoter (Itkin *et al.*, 2009) (Fig. 10.8). It is known that MADS-box proteins can form homo- and heterodimers and higher-order complexes, but it is not clear whether RIN and TAGL1 operate separately or in combination. A different type of TF, a homeodomain-leucine zipper (HD-Zip) protein named LeHB-1, was discovered by Lin *et al.* (2008) during a search for TFs that bind the *LeACO1* promoter. LeHB-1 binds to a putative binding site in the tomato *LeACO1* gene and appears to activate transcription at the onset of ripening, as inhibiting LeHB-1 expression in tomato fruit leads to inhibition of ripening (Lin *et al.*, 2008). Interestingly, there are also putative LeHB-1-binding sites in the gene promoters of *LeACO2* and *LeMADS_RIN*, and the promoter for the tomato cell-wall-modifying enzyme polygalacturonase (Lin *et al.*, 2008). RIN itself activates *LeACS2*, so RIN and LeHB-1 appear to be crucial for switching on system 2 ethylene synthesis and ripening. Interestingly, TAGL1 expression was enhanced in the presence of the ethylene perception/action inhibitor 1-methylcyclopropene and appears not to be regulated by ethylene or RIN (Itkin *et al.*, 2009).

Members of a large superfamily of plant-specific TFs, called APETALA2/ ethylene responsive factor (AP2/ERF),

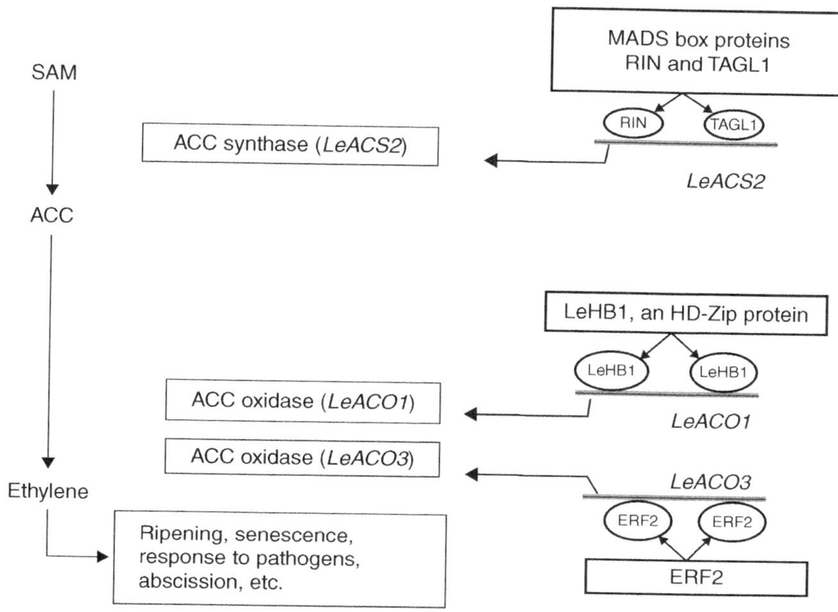

Fig. 10.8. Transcription factors regulating *LeACS2, LeACO1* and *LeACO3* and enhanced ethylene synthesis in ripening tomato fruit. The RIN protein was identified as the product of the *rin* gene in the non-ripening *rin* mutant of tomato (Vrebalov *et al.*, 2002) and the normal protein was shown to bind the *LeACO2* promoter (Ito *et al.*, 2008). The role of TOMATO AGAMOUS-LIKE 1 (TAGL1) was determined by Itkin *et al.* (2009). Identification and functional analysis of LeHB1, a homeobox protein, was performed by Lin *et al.* (2008). The role of the ethylene response factor LeERF2 was proposed by Zhang *et al.* (2009).

operate downstream of the ethylene signalling pathway (ethylene (C_2H_4) → ethylene receptor (ETR) proteins → constitutive triple response (CTR) proteins → EIN3-like (EIL) proteins) and have diverse expression patterns and DNA-binding capacity to the GCC box and other sequences in the promoters of genes they regulate (Tournier *et al.*, 2003; Pirrello *et al.*, 2012). In addition to ripening, they can be involved in responses to wounding, biotic and salt stress, anaerobiosis and signalling pathways involving brassinosteroids, ethylene, jasmonic acid and salicylic acid. One of these (LeERF2) interacts with the GCC box in the promoter of dehydration-responsive element in the promoter of *LeACO3*, resulting in transcriptional activation of the gene. Inhibiting ERF2 expression by gene silencing was shown to reduce ethylene synthesis (Zhang *et al.*, 2009).

10.5 Post-translational Control of ACS

Some forms of ACS, such as those induced by wounding or auxin, have a short half-life (i.e. they are unstable and rapidly degraded), and the demonstration that LeACS2, but not LeACS4, is phosphorylated (Tatsuki and Mori, 2001) hinted at a mechanism for regulating the stability of some isoforms. Generally, three ACS types are recognized, types 1, 2 and 3, which differ in their structure and regulation (Table 10.1). Type 1 ACSs are phosphorylated by mitogen-activated protein kinase 6 (MAPK6) (Liu and Zhang, 2004), and probably also by calcium-dependent protein kinase (Hernandez Sebastia *et al.*, 2004); in the absence of phosphorylation, they are rapidly degraded by the 26S proteasome pathway for protein degradation (Joo *et al.*, 2008). Type 2 isoforms have only one phosphorylation

Table 10.1. Classification and regulation of ACS protein isoforms.

ACS type	Properties	Examples in *Arabidopsis*	Examples in tomato
Type 1	Extended C-terminal domain, multiple phosphorylation sites for mitogen-activating protein kinase (MAPK) and (probably) calcium-dependent protein kinases (CDPKs); unphosphorylated proteins are rapidly degraded by the ubiquitin 26S proteasome system	AtACS1, AtACS2, AtACS6	LeACS1A, LeACS1B, LeACS2, LeACS6
Type 2	Shorter C terminus and phosphorylated by CDPK only, at a single site; type 2 isoforms are recognized and interact with proteins such as ATO1, EOL1 and EOL2, which leads to their degradation by the ubiquitin 26S proteasome	AtACS9, AtACS8, AtACS5, AtACS4	LeACS8, LeACS7, LeACS3
Type 3	No known phosphorylation sites	AtACS11, AtACS7	LeACS5, LeACS4

site, and type 3 have no known phosphorylation sites (Table 10.1). Mutations affecting the C termini of type 2 isoforms ACS5 and ACS6 from the *eto2* and *eto3* mutants of *Arabidopsis* caused dominant overproduction of ethylene (Chae *et al.*, 2003), indicating regulation involved the C termini of the proteins. The recessive mutation *eto1*, which also causes ethylene overproduction, revealed a class of regulatory proteins (ETO1, EOL1 and EOL2) that interact specifically with type 2 isoforms in their C termini and regulate their degradation (Wang *et al.*, 2004; Yoshida *et al.*, 2005; Christians *et al.*, 2009). Cytokinins and brassinosteroids can increase ethylene synthesis (Woeste *et al.*, 1999; Yi *et al.*, 1999; Arteca and Arteca, 2008) by increasing ACS5 protein stability (Chae *et al.*, 2003), and each may affect type 2 ACS protein stability independently (Hansen *et al.*, 2009). Controlling ACS protein stability provides a means of tightly controlling ethylene synthesis. There is little evidence that ACO activity is regulated post-transcriptionally, but it has not been studied in the same detail as ACS.

10.6 Conclusion

Ethylene biosynthesis, perception and response are essential for full ripening of many, if not all, climacteric fruits, and it is important to understand the biological and technological control of ethylene synthesis, as this offers the opportunity to influence fruit storage, ripening and fruit quality. Intriguingly, ethylene controls its own synthesis in some situations, including ripening. Control of ethylene production in response to stress factors and the molecular regulation of these responses should be a fertile area for future research. ACS and ACO are both actively regulated at different stages of development at either the mRNA or protein level, or both. Regulation of mRNA concentrations is achieved partly by transcriptional control of families of *ACS* and *ACO* genes with differential expression patterns. This transcriptional regulation involves MADS-box, ERF and HD-zip TFs and it is predicted more TFs that participate in different situations will be discovered as our understanding of the regulation of ethylene biosynthesis improves. There is sparse information about the control of mRNA turnover, but this may also represent a potential control point. Interplay between different hormones makes an important contribution to developmental and environmental regulation of ethylene production, and an important mechanism is post-translational control of the stability of type 2 ACS isoforms.

References

Abeles, F.B., Morgan, P.W. and Saltveit, M.E. Jr (1992) *Ethylene in Plant Biology*, 2nd edn. Academic Press, New York.

Adams, D.O. and Yang, S.F. (1979) Ethylene biosynthesis: Identification of 1-aminocyclopropane-l-carboxylic acid as an intermediate in the conversion of methionine to ethylene. *Proceedings of the National Academy of Sciences USA* 76, 170–174.

Arteca, R. and Arteca, J.M. (2008) Effects of brassinosteroid, auxin, and cytokinin on ethylene production in *Arabidopsis thaliana* plants. *Journal of Experimental Botany* 59, 3019–3026.

Ayub, R., Guis, M., Ben Amor, M., Gillot, L., Roustan, J.-P., Latché, A., Bouzayen, M. and Pech, J.-C. (1996) Expression of ACC oxidase antisense gene inhibits ripening of cantaloupe melon fruits. *Nature Biotechnology* 14, 862–866.

Barry, C.S., Blume, B., Bouzayen, M., Cooper, W., Hamilton, A.J. and Grierson, D. (1996) Differential expression of the 1-aminocyclopropane-1-carboxylate oxidase gene family of tomato. *Plant Journal* 9, 525–535.

Barry, C.S., Llop-Tous, M.I. and Grierson, D. (2000) The regulation of 1-aminocyclopropane-1-carboxylic acid synthase gene expression during the transition from system-1 to system-2 ethylene synthesis in tomato. *Plant Physiology* 123, 979–986.

Bassan, A., Borowski, T., Schofield, C.J. and Siegbahn, P.E. (2006) Ethylene biosynthesis by 1-aminocyclopropane-1-carboxylic acid oxidase: a DFT study. *Chemistry* 12, 8835–8846.

Bidonde, S., Ferrer, M.A., Zegzouti, H., Ramassamy, S., Latché, A., Pech, J.C., Hamilton, A.J., Grierson, D. and Bouzayen, M. (1998) Expression and characterization of three tomato 1-aminocyclopropane-1-carboxylate oxidase cDNAs in yeast. *European Journal of Biochemistry* 253, 20–26.

Bleecker, A.B., Kenyon, W.H., Somerville, S.C. and Kende, H. (1986) Use of monoclonal antibodies in the purification and characterization of 1-aminocyclopropane-1-carboxylate synthase, an enzyme in ethylene biosynthesis. *Proceedings of the National Academy of Sciences USA* 83, 7755–7759.

Blume, B. and Grierson, D. (1997) Expression of ACC oxidase promoter-GUS fusions in tomato and *Nicotiana plumbaginifolia* regulated by developmental and environmental stimuli. *Plant Journal* 12, 731–746.

Bradford, K.J. (2008) Shang Fa Yang: pioneer in plant ethylene biochemistry. *Plant Science* 175, 2–7.

Calvo, A.P., Nicolás, C., Nicolás, G. and Rodrıguez, D. (2004) Evidence of a cross-talk regulation of a GA 20-oxidase (FsGA20ox1) by gibberellins and ethylene during the breaking of dormancy in *Fagus sylvatica* seeds. *Physiologia Plantarum* 120, 623–630.

Capitani, G., Hohenester, E., Feng, L., Storici, P., Kirsch, J.F. and Jansonius, J.N. (1999) Structure of 1-aminocyclopropane-1-carboxylate synthase, a key enzyme in the biosynthesis of the plant hormone ethylene. *Journal of Molecular Biology* 294, 745–756.

Chae, H.S., Cho, Y.G., Park, M.Y., Lee, M.C., Eun, M.Y., Kang, B.G. and Kim, W.T. (2000) Hormonal cross-talk between auxin and ethylene differentially regulates the expression of two members of the 1-aminocyclopropane-1-carboxylate oxidase gene family in rice (*Oryza sativa* L.). *Plant and Cell Physiology* 41, 354–362.

Chae, H.S., Faure, F. and Kieber, J.J. (2003). The *eto1*, *eto2*, and *eto3* mutations and cytokinin treatment increase ethylene biosynthesis in Arabidopsis by increasing the stability of ACS protein. *Plant Cell* 15, 545–559.

Christians, M.J., Gingerich, D.J., Hansen, M., Binder, B.M., Kieber, J.J. and Vierstra, R.D. (2009) The BTB ubiquitin ligases ETO1, EOL1 and EOL2 act collectively to regulate ethylene biosynthesis in Arabidopsis by controlling type-2 ACC synthase levels. *Plant Journal* 57, 332–345.

Davies, K.M. and Grierson, D. (1989) Identification of cDNA clones for tomato (*Lycopersicon esculentum*) mRNAs that accumulate during fruit ripening and leaf senescence in response to ethylene. *Planta* 179, 73–80.

Flores, F., El Yahyaoui, F., de Billerbeck, G., Romojaro, F., Latché, A., Bouzayen, M., Pech, J.-C. and Ambid, C. (2002) Role of ethylene in the biosynthetic pathway of aliphatic ester aroma volatiles in Charentais Cantaloupe melons. *Journal of Experimental Botany* 53, 201–206.

Fujisawa, M., Nakano, T. and Ito, Y. (2011) Identification of potential target genes for the tomato fruit-ripening regulator RIN by chromatin immunoprecipitation. *BMC Plant Biology* 11, 26.

Grierson, D., Kear, R.J., Thompson, J.R. and Garcia-Mora, R. (1982) Stimulation of in vitro RNA synthesis by pre-treating plants with auxins is due to auxin-induced ethylene production. *Zeitschrift für Pflanzenphysiologie* 107, 419–426.

Hamilton, A., Lycett, G.W. and Grierson, D. (1990) Antisense gene that inhibits synthesis of the hormone ethylene in transgenic plants. *Nature* 346, 284–287.

Hamilton, A.J., Bouzayen, N. and Grierson, D. (1991) Identification of a tomato gene for the ethylene forming enzyme by expression in yeast. *Proceedings of the National Academy of Sciences USA* 88, 7434–7437.

Hansen, M., Chae, H.S. and Kieber, J.J. (2009) The regulation of ACS protein stability by cytokinin and brassinosteroid. *Plant Journal* 57, 606–614.

Harpaz-Saad, S., Yoon, G.M., Mattoo, A. and Kieber, J.J. (2012) The formation of ACC and competition between polyamines and ethylene for SAM. In: McManus, M.T. (ed.) *Annual Plant Reviews*, Vol. 44. *The Plant Hormone Ethylene.* Wiley Blackwell.

Hernandez Sebastia, C., Hardin, S.C., Clouse, S.D., Kieber. J.J. and Huber, S.C. (2004) Identification of a new motif for CDPK phosphorylation *in vitro* that suggests ACC synthase may be a CDPK substrate. *Archives of Biochemistry and Biophysics* 428, 81–91.

Hoffman, N., Yang, S.F., Ichihara, A. and Sakamura, S. (1982) Stereospecific conversion of 1-aminocyclopropanecarboxylic acid to ethylene by plant tissues. Conversion of stereoisomers of 1-amino-2-ethylcyclopropanecarboxylic acid to 1-butene. *Plant Physiology* 70, 195–199.

Holdsworth, M.J., Bird, C.R., Ray, J., Schuch, W. and Grierson, D. (1987) Structure and expression of an ethylene-related mRNA from tomato. *Nucleic Acids Research* 15, 731–739.

Holdsworth, M.J., Schuch, W. and Grierson, D. (1988) Organisation and expression of a wound/ripening-related small multigene family from tomato. *Plant Molecular Biology* 11, 81–88.

Itkin, M., Seybold, H., Breite, D., Rogachev, I., Meir, S. and Aharoni, A. (2009) TOMATO AGAMOUS-LIKE 1 is a component of the fruit ripening regulatory network. *Plant Journal* 60, 1081–1095.

Ito, Y., Kitagawa, M., Ihashi, H., Yabe, K., Kimbara, J., Yasuda, J., Ito, H., Inakuma, T., Hiroi, S. and Kasumi, T. (2008) DNA-binding specificity, transcriptional activation potential, and the rin mutation effect for the tomato fruit-ripening regulator RIN. *Plant Journal* 55, 212–223.

John, I., Drake, R., Farrell, A., Cooper, W., Lee, P., Horton, P. and Grierson, D. (1995) Delayed leaf senescence in ethylene-deficient ACC oxidase antisense tomato plants molecular and physiological analysis. *Plant Journal* 7, 483–490.

John, P. (1991) How plant molecular biologists revealed a surprising relationship between two enzymes, which took an enzyme out of a membrane where it was not located, and put it into the soluble phase where it could be studied. *Plant Molecular Biology Reporter* 9, 192–194.

Joo, S., Liu, Y., Lueth, A. and Zhang, S. (2008) MAPK phosphorylation induced stabilization of ACS6 protein is mediated by the non-catalytic C-terminal domain, which also contains the *cis*-determinant for rapid degradation by the 26S proteasome pathway. *Plant Journal* 54, 129–140.

Kneissl, M.L. and Deikman, J. (1996) The tomato *E8* gene influences ethylene biosynthesis in fruit but not in flowers. *Plant Physiology* 112, 537–547.

Lieberman, M. (1979) Biosynthesis and action of ethylene. *Annual Review of Plant Physiology* 30, 533–591.

Lieberman, M., Kunishi, A.T., Mapson, L.W. and Wardale, D.A. (1965) Ethylene production from methionine. *Biochemical Journal* 97, 449–459.

Lieberman, M., Kunishi, A., Mapson, L.W. and Wardale, D.A. (1966) Stimulation of ethylene production in apple tissue slices by methionine. *Plant Physiology* 41, 376–382.

Lin, Z., Hong, Y., Yin, M., Li, C., Zhang, G.K. and Grierson, D. (2008) A tomato HB-zip homeobox protein, LeHB-1, plays an important role in floral organogenesis and ripening. *Plant Journal* 55, 301–310.

Lin, Z., Zhong, S. and Grierson, D. (2009) Recent advances in ethylene research. *Journal of Experimental Botany* 60, 3311–3336.

Liu, Y. and Zhang, S. (2004) Phosphorylation of 1-aminocyclopropane-1-carboxylic acid synthase by MPK6, a stress-responsive mitogen activated protein kinase, induces ethylene biosynthesis in Arabidopsis. *Plant Cell* 16, 3386–3399.

Llop-Tous, I., Barry, C. and Grierson, D. (2000) Regulation of ethylene biosynthesis in response to pollination in tomato flowers. *Plant Physiology* 123, 971–978.

McMurchie, E.J., McGlasson, W.B. and Eaks, I.L. (1972) Treatment of fruit with propylene gives information about the biogenesis of ethylene. *Nature* 237, 235–236.

Moeder, W., Barry, C.S., Tauriainen, A.A., Betz, C., Tuomainen, J., Utrianen, M., Grierson, D., Sandermann, H., Langerbartels, C. and Kangasjarvi, J. (2002) Ethylene synthesis regulated by biphasic induction of 1-aminocyclopropane-1-carboxylic acid synthase and 1-aminocyclopropane-1-carboxylic acid oxidase genes is required for hydrogen peroxide accumulation and cell death in ozone-exposed tomato. *Plant Physiology* 130, 1918–1926.

Nakajima, N. and Imaseki, H. (1986) Purification and properties of 1-aminocyclopropane-1-carboxylate synthase of mesocarp of *Cucurbita maxima* Duch. fruits. *Plant Cell Physiology* 27, 969–980.

Nakajima, N., Nakagawa, N. and Imaseki, H. (1988) Molecular size of wound-induced 1-aminocyclopropane-1-carboxylate synthase from *Cucurbita maxima* Duch. and change of translatable mRNA of the enzyme after wounding. *Plant Cell Physiology* 29, 989–998.

Nakatsuka, A., Murachi, S., Okunishi, H., Shiomi, S., Nakano, R., Kubo, Y. and Inaba, A. (1998) Differential expression and internal feedback regulation of 1-aminocyclopropane-1-carboxylate synthase, 1-aminocyclopropane-1-carboxylate oxidase, and ethylene receptor genes in tomato fruit during development and ripening. *Plant Physiology* 18, 1295–1305.

Oeller, P.W., Lu, M.W., Taylor, L.P., Pike, D.A. and Theologis, A. (1991) Reversible inhibition of tomato fruit senescence by antisense RNA. *Science* 254, 437–439.

Picton, S. and Grierson, D. (1988) Inhibition of expression of tomato-ripening genes at high temperature. *Plant, Cell and Environment* 11, 265–272.

Picton, S., Barton, S.L., Bouzayen, M., Hamilton, A.J. and Grierson, D. (1993) Altered fruit ripening and leaf senescence in tomatoes expressing an antisense ethylene-forming enzyme transgene. *Plant Journal* 3, 469–481.

Pirrello, J., Prasad, B.C.N, Zhang, W., Chen, K., Mila, I., Zouine, M., Latché, A., Pech, J.C., Ohme-Takagi, M., Regad, F., and Bouzayen, M. (2012) Functional analysis and binding affinity of tomato ethylene response factors provide insight on the molecular bases of plant differential responses to ethylene. *BMC Plant Biology* 12, 190.

Primrose, S.B. (1979) Ethylene and agriculture: the role of the microbe. *Journal of Applied Bacteriology* 46, 1–25.

Rottmann, W.H., Peter, G.F., Oeller, P.W., Keller, J.A., Shen, N.F., Nagy, B.P., Taylor, L.P., Campbell, A.D. and Theologis, A. (1991) 1-Aminocyclopropane-1-carboxylate synthase in tomato is encoded by a multigene family whose transcription is induced during fruit and floral senescence. *Journal of Molecular Biology* 12, 937–961.

Salman-Minkov, A., Levi, A., Wolf, S. and Trebitsh, T. (2008) ACC synthase genes are polymorphic in watermelon (*Citrullus* spp.) and differentially expressed in flowers and in response to auxin and gibberellin. *Plant and Cell Physiology* 49, 740–750.

Sato, T., Oeller, P.W. and Theologis, A. (1991) The l-aminocyclopropane-1-carboxylate synthase of *Cucurbita*. Purification, properties, expression in *Escherichia coli*, and primary structure determination by DNA sequence analysis. *Journal of Biological Chemistry* 266, 3752–3759.

Schofield, C.J. and Zhang, Z. (1999) Structural and mechanistic studies on 2-oxoglutarate-dependent oxygenases and related enzymes. *Current Opinion in Structural Biology* 9, 722–731.

Smith, C.J.S., Slater, A. and Grierson, D. (1986) Rapid appearance of an mRNA correlated with ethylene synthesis encoding a protein of molecular weight 35,000. *Planta* 168, 94–100.

Spanu, P., Reinhardt, D. and Boller, T. (1991) Analysis and cloning of the ethylene-forming enzyme from tomato by functional expression of its messenger-RNA in *Xenopus laevis* oocytes. *EMBO Journal* 10, 2007–2013.

Tatsuki, M. and Mori, H. (2001) Phosphorylation of tomato 1-aminocyclopropane-1-carboxylic acid synthase, LE-ACS2, at the C-terminal region. *Journal of Biological Chemistry* 276, 28051–28057.

Tournier, B., Sanchez-Ballesta, M.T., Jones, B., Pesquet, E., Regad, F., Latché, A., Pech, J.C. and Bouzayen, M. (2003) New members of the tomato ERF family show specific expression pattern and diverse DNA-binding capacity to the GCC box element. *FEBS Letters* 550, 149–154.

Tsuchisaka, A. and Theologis, A. (2004) Unique and overlapping expression patterns among the Arabidopsis 1-amino-cyclopropane-1-carboxylate synthase gene family members. *Plant Physiology* 136, 2982–3000.

Tuomainen, J., Betz, C., Kangasjarvi, J., Ernst, D., Yin, Z-H., Langebartels, C. and Sandermann, H. Jr (1997) Ozone induction of ethylene emission in tomato plants: regulation by differential accumulation of transcripts for the biosynthetic enzymes. *Plant Journal* 12, 1151–1162.

Urstenbinder, K.B. and Sauter, M. (2012) Early events in the ethylene biosynthetic pathway – regulation of the pools of methionine and *S*-adenosylmethionine. In: McManus, M.T. (ed.) *Annual Plant Reviews*, Vol. 44. *The Plant Hormone Ethylene*. Wiley Blackwell.

Van Der Straeten, D., Rodrigues-Pousada, R.A., Villarroel, R., Hanley, S., Goodman, H.M. and Montague, M.V. (1992) Cloning, genetic mapping, and expression analysis of an *Arabidopsis thaliana* gene that encodes 1-aminocyclopropane-1-carboxylate synthase. *Proceedings of the National Academy of Sciences USA* 89, 9969–9973.

Ververidis, P. and John, P. (1991) Complete recovery in vitro of ethylene-forming enzyme activity. *Phytochemistry* 30, 725–727.

Vrebalov, J., Ruezinsky, D., Padmanabhan, V., White, R., Medrano, D., Drake, R., Schuch, W. and Giovannoni, J. (2002) A MADS-box gene necessary for fruit ripening at the tomato ripening-inhibitor (*rin*) locus. *Science* 296, 343–346.

Wang, K.L., Yoshida, H., Lurin, C. and Ecker, J.R. (2004) Regulation of ethylene gas biosynthesis by the *Arabidopsis* ETO1 protein. *Nature* 428, 945–950.

Woeste, K.E., Ye, C. and Kieber, J.J. (1999) Two Arabidopsis mutants that overproduce ethylene are affected in the posttranscriptional regulation of 1-aminocyclopropane-1-carboxylic acid synthase. *Plant Physiology* 119, 521–530.

Yamagami, T., Tsuchisaka, A., Yamada, K., Haddon, W.F., Harden, L.A. and Theologis, A. (2003) Biochemical diversity among the 1-aminocyclopropane-1-carboxylate synthase isozymes encoded by the Arabidopsis gene family. *Journal of Biological Chemistry* 278, 49102–49112.

Yang, S.F. and Hoffman, N.E. (1984) Ethylene biosysnthesis and its regulation in higher plants. *Annual Review of Plant Physiology* 35, 155–189.

Yi, H.C., Joo, S., Nam, K.H., Lee, J.S., Kang, B.G. and Kim, W.T. (1999) Auxin and brassinosteroid differentially regulate the expression of three members of the 1-aminocyclopropane-1-carboxylate synthase gene family in mung bean (*Vigna radiata* L.). *Plant Molecular Biology* 41, 443–454.

Yip, W.K., Dong, J.G., Kenny, J.W., Thompson, G.A. and Yang, S.F. (1990) Characterization and sequencing of the active site of 1-aminocyclopropane-1-carboxylate synthase. *Proceedings of the National Academy of Sciences USA* 87, 7930–7934.

Yokotani, N., Nakano, R., Imanishi, S., Nagata, M., Inaba, A. and Kubo, Y. (2009) Ripening-associated ethylene biosynthesis in tomato fruit is autocatalytically and developmentally regulated. *Journal of Experimental Botany* 60, 3433–3442.

Yoshida, H., Nagata, M., Saito, K., Wang, K.L. and Ecker, J.R. (2005) *Arabidopsis* ETO1 specifically interacts with and negatively regulates type 2 1-aminocyclopropane-1-carboxylate synthases. *BMC Plant Biology* 5, 14.

Zhang, Z., Ren, J.S., Clifton, I.J. and Schofield, C.J. (2004) Crystal structure and mechanistic implications of 1-aminocyclopropane-1-carboxylic acid oxidase – the ethylene-forming enzyme. *Chemistry and Biology* 11, 1383–1394.

Zhang, Z., Zhang, H., Quan, R., Wang, X.-C. and Huang, R. (2009) Transcriptional regulation of the ethylene response factor LeERF2 in the expression of ethylene biosynthesis genes controls ethylene production in tomato and tobacco. *Plant Physiology* 150, 365–377.

11 Ethylene Perception and Signalling in Ripening Fruit

Rahul Kumar and Arun K. Sharma*
Department of Plant Molecular Biology, University of Delhi South Campus, New Delhi, India

11.1 Introduction

Fruits are an indispensable part of the human diet as they are a rich source of vitamins, minerals, antioxidants, sugars, fibres and flavour compounds. Accumulation of the majority of these chemicals/products starts once ripening commences in fruits. The triggering of the ripening initiates a phase change in fruit development, which is accompanied by a dramatic shift in primary and secondary metabolism. These coordinated changes result in the accumulation of carotenoids, flavour and volatile components, antioxidants and sugars in the ripening fruits and add to the nutritional quality of the fruits. In plants, the bright colours and complex aromas act as attractants and help in seed dispersal, while in humans these pigments with antioxidative properties are involved in prevention of certain diseases, such as cancer and heart attack (Fraser and Bramley, 2004). Therefore, fruit development and ripening have received considerable scientific scrutiny, and efforts have been made to investigate and improve our understanding of the molecular and genetic changes underlying ripening, especially in the case of fleshy fruits. Endogenously produced ethylene is one of the most important factors influencing various aspects of ripening in many fleshy fruits. Based on the ripening physiology, fruits have been classified into two categories: climacteric and non-climacteric fruits. The ripening of climacteric fruits, such as tomato, banana, peach and apple, involves a sharp increase in respiration and ethylene production at the onset of ripening. These two attributes have different patterns in non-climacteric fruits, such as strawberry, grape and citrus. None the less, occurrence of similar biochemical events during ripening in non-climacteric fruits indicates that coordination between ethylene-dependent and ethylene-independent regulatory events is important during fruit maturation (Alba *et al.*, 2005; Cara and Giovannoni, 2008; Klee and Giovannoni, 2011; Kumar *et al.*, 2012b).

Due to the availability of advanced cytological, genetic and physical maps and a wealth of other genomic resources, tomato has been used as the system of choice for studying fleshy fruit development and ripening for decades (Foolad, 2007; Yano *et al.*, 2007; Fatima *et al.*, 2008; Gupta *et al.*, 2009). Furthermore, the availability of several non-ripening mutants, altered in either ethylene perception or downstream signal transduction of tomato, has also

* arun@genomeindia.org

facilitated a better understanding of the molecular and biochemical mechanisms behind ethylene-dependent aspects of ripening (Lanahan *et al.*, 1994; Wilkinson *et al.*, 1995; Thompson *et al.*, 1999). As tomato is the most studied plant for fruit ripening, the major emphasis in this chapter is on the knowledge accumulated using this model system, although significant insights obtained from other plants are also discussed. Although the botanical name of tomato has been changed recently from *Lycopersicum esculentum* to *Solanum lycopersicum*, the genes characterized before the name change have been named using the old prefix 'Le' in the main text.

11.2 Ethylene Signal Transduction

Ethylene plays a pivotal role during various developmental events and environmental responses in plants. Its role in ripening has received the most attention (Yang and Hoffman, 1984; Mattoo and Suttle, 1991; Abeles *et al.*, 1992). Discussions on the ethylene biosynthesis pathway and its regulation are described in detail by Grierson (Chapter 10, this volume). Much knowledge about ethylene signalling components in plants has been obtained using various ethylene-response mutants of *Arabidopsis* (Guo and Ecker, 2003). A high level of similarity was found in various components of ethylene signalling in *Arabidopsis* and tomato. This knowledge has served as a starting point for the identification of its components in other plants as well. During ethylene signalling, any change at the phenotype level can be regulated at three steps of this pathway: first, at the level of ethylene perception; secondly, during transduction of this signal to downstream regulators; and thirdly, at the level of expression of ethylene-responsive genes (see Plate 6).

11.2.1 Ethylene receptors

The receptor proteins responsible for perception of ethylene exist as disulfide-linked dimers and exhibit similarity to bacterial two-component regulators. These proteins are endoplasmic reticulum-associated integral membrane proteins. The receptor proteins possess two to three domains, including the N-terminal sensor, a kinase domain in the middle and the C-terminal receiver domain. The sensor domain is involved in ethylene perception, dimerization and binding to the copper cofactor, while the kinase domain is involved in autophosphorylation. Transfer of this phosphate group to an aspartate residue on the receiver domain of either the same or another ethylene receptor protein activates the receiver domain and results in the initiation of ethylene signalling. Detailed characterization of five ethylene-receptor genes in *Arabidopsis* has revealed their negative regulatory nature during ethylene signalling and provided the basic framework needed for their characterization during fruit ripening in tomato. When ethylene is absent, these receptors actively suppress ethylene responses, while during its availability, receptors bind to ethylene causing removal of this suppression and leading to the ethylene responses.

Seven ethylene-receptor proteins (LeETR1–7) have been identified in tomato and the first five of these have been shown to bind ethylene. LeETR7 is the least studied among the tomato receptor proteins (Klee and Giovannoni, 2011). The ethylene receptor NR (Never-Ripe) was the first receptor to be characterized during fruit ripening (Lanahan *et al.*, 1994). Later, the *ETR1* gene of *Arabidopsis* was used as a heterologous probe to screen and identify additional receptor proteins in tomato (Wilikinson *et al.*, 1995; Zhou *et al.*, 1996). Based on their structural similarity, tomato ethylene-receptor proteins have been divided into two subgroups. LeETR1, LeETR2 and LeETR3 proteins possess three N-terminal membrane-spanning domains and a conserved kinase domain, while LeETR4, LeETR5 and LeETR6 proteins are characterized by the presence of an additional transmembrane domain. Furthermore, the kinase domain of members of the latter subgroup does not possess one or more of the catalytic subdomains, including the autophosphorylated domain. In addition, the LeETR3 receptor protein differs structurally from the remaining receptor proteins in that it lacks the receptor domain. Although these genes are expressed ubiquitously in tomato, they have been found to respond differently to different external cues or at different stages of fruit development. *LeETR1* and *LeETR2* genes are expressed ubiquitously in all tissues, while expression levels of the *LeETR3–6* genes increase in reproductive tissues such as fruits and flowers (Klee and Giovannoni, 2011). These receptor genes have been utilized to manipulate ripening in tomato; however, in many cases, no effect of their altered expression was observed on fruit ripening, indicating a degree of functional redundancy among them (Zhou *et al.*, 1996). For example, knockdown antisense transgenic lines of *LeETR1* and *LeETR3* did not show any change in the ripening behaviour of their fruits. Increased expression of *LeETR4* in the knockdown antisense *LeETR3* lines was found to compensate for reduced expression of the *LeETR3* gene (Tieman *et al.*, 2000). Furthermore, functions of LeETR4 and LeETR6 proteins have been implicated in altering the time of fruit ripening in tomato. These proteins exhibited ethylene-induced degradation by the 26S proteasome pathway. Transgenic plants with attenuated expression of these genes resulted in early fruit ripening, suggesting that regulation at the protein level is a major determinant of the ripening initiation programme. The authors proposed that, as the receptors are negative regulators of ethylene signalling, depletion of receptors levels would result in a progressive increase in hormone sensitivity causing an early ripening phenotype. Thus, it was suggested that these receptors themselves may act as regulators of ripening initiation (Kevany *et al.*, 2007).

Additionally, other tissue-specific regulators of ethylene signalling have been identified in tomato. *Green-Ripe* (*Gr*) is a dominant non-ripening mutant in tomato. Fruits from *Gr* mutants do not fully ripen and exhibit reduced ethylene responsiveness during ripening. Mutation in this gene results in overaccumulation of otherwise normal GR protein in fruit. This is suggested to be a membrane protein of unknown function that negatively regulates ethylene responses only in fruits. Mutation in its *Arabidopsis* homologue, *REVERSION-TO-ETHYLENE SENSITIVITY1* (*RTE1*), was found to restore the ethylene insensitivity phenotype caused by *etr1-2* gain-of-function mutations. Together, evidence from GR and its *Arabidopsis* homologue (RTE1) suggests that these proteins interact with ethylene receptors to alter/modify their action in an as-yet-unknown way. In the case of the *Gr* mutation, it has been suggested that the GR protein can interfere with the normal activity of related proteins in tissue-dependent ethylene responses unique to tomato fruit (Barry and Giovannoni, 2006).

11.2.2 Constitutive triple response (CTR) proteins

In tomato, four *CTR* genes (*LeCTR1–4*) are members of a small multigene gene family (Adams-Phillips *et al.*, 2004). These MAP kinase kinase kinase (MAP3K) genes are present downstream of the ethylene receptors and act as negative regulators of ethylene responses. They interact physically with the receptor proteins and form a signalling complex (Cara and Giovannoni, 2008). These genes were identified either through heterologous hybridizations using the *Arabidopsis CTR1* gene as a probe or in a differential display. The *LeCTR1*, *LeCTR3* and *LeCTR4* genes are close homologues of *Arabidopsis CTR1*. These genes have been used successfully in complementing the *Arabidopsis ctr1* mutation, although with different efficacies, suggesting that all three *CTR* genes may be functional in tomato and could have a degree of functional redundancy. Similar to the ethylene receptors, these genes also exhibit tissue-specific transcript accumulation. Expression of *LeCTR1* is induced in an ethylene-dependent manner but not that of *LeCTR3* and *LeCTR4*, suggesting that these three genes might be subjected to both transcriptional and post-translational regulation during ripening (Lin *et al.*, 1998; Leclercq *et al.*, 2002).

11.2.3 Ethylene Insensitive 2 (EIN2)

In the ethylene signalling cascade, EIN2, a member of the Nramp family protein of metal transporters, is present downstream of the *CTR* genes (Roman *et al.*, 1995). Expression of the *EIN2* gene is ethylene independent and does not exhibit substantial changes at different stages of fruit development and ripening. Transgenic plants with attenuated *EIN2* expression show a delayed ripening phenotype. In addition, inhibited transcript accumulation of ethylene- and ripening-related genes suggests that this gene is a positive regulator of ethylene-mediated responses during tomato fruit development and ripening. This protein has been suggested as a converging point for the signalling pathways of various phytohormones, including auxin, abscisic acid, jasmonic acid and ethylene, indicating that these hormonal pathways can cross-talk to each other during various developmental aspects in plants (Cara and Giovannoni, 2008).

11.2.4 EIN3-like (EIL) proteins

A family of *trans*-acting proteins, namely EIL proteins, is present downstream of EIN2. These proteins act as transcription regulators and control the expression of another class of transcription factors, the ethylene-response factor (ERFs). EILs recognize specific motifs known as ethylene-response elements (EREs) present in the promoters of various ripening-associated genes. In tomato, four *EIL* genes constitute a small multigene family (Tieman *et al.*, 2001; Yokotani *et al.*, 2009). The transcript accumulation of *LeEIL1*, *LeEIL2* and *LeEIL3* genes is ethylene independent, while *LeEIL4* exhibits a ripening-associated increase during ripening. Reduced expression levels of *LeEIL1–3* resulted in ethylene insensitivity, indicating that these proteins are positive regulators of ethylene responses (Cara and Giovannoni, 2008). Yokotani *et al.* (2009) reported that, in transgenic plants with attenuated transcript accumulation of *LeEIL* genes, expression of the *LeACS2* and *LeACS4* genes exhibited a limited increase during fruit ripening that was not abolished even after 1-methylcyclopropene treatment. This study concluded that ethylene production is developmentally regulated in an ethylene-independent manner during fruit ripening in tomato. EIL proteins have also been found to control the expression of genes associated with ethylene biosynthesis, such as 1-aminocyclopropane-1-carboxylic acid oxidase (*ACO*) genes, directly by interacting with their promoters in kiwifruit and melon fruits. It has been suggested that EIL proteins could also be involved in the regulation of ethylene biosynthesis (Yin *et al.*, 2010). Recently, analysis of phosphorylation-dependent regulation of ethylene responses at the level of EIN3/EIL has led to the identification of a phosphorylation site in the tomato EIL1 protein. This phosphorylation region, named EPR1 (EIN3/EIL phosphorylation region 1), was found to be essential for its transcriptional activity. Mutation in this region prevented phosphorylation of EIL1, which in turn led to the complete loss of its transactivation potential and finally abolished the ethylene constitutive responses. Furthermore, functional EPR1 was found to be necessary for mediating the dimerization step of EIL1 proteins, indicating that this region might represent a molecular mechanism for the regulation of EIN3/EIL activity. In conclusion, the authors proposed a model where phosphorylation within EPR1 caused dimerization of EIL1 protein, and the homodimer complex then bound to the target genes allowing transcriptional activation of ethylene-regulated genes (Li *et al.*, 2011). Moreover, two EBF (EIN3-binding F-box) proteins have been found to regulate fruit development in tomato by negative regulation of the ethylene signalling pathway. These proteins were found to be associated with the turnover of EIN3/EIL proteins as they mediated the degradation of LeEILs via the ubiquitin/26S proteasome pathway (Yang *et al.*, 2010).

11.2.5 ERF proteins

ERF proteins are present further downstream of the EIL proteins in the ethylene signalling cascade. ERFs contain a highly conserved DNA-binding domain (AP2) consisting of 58 or 59 aa. These proteins regulate ethylene responses by binding to the GCC box promoter elements of the ethylene-responsive genes. To date, several *ERF* genes (*ERF1–6*, *3b*, *Pti4-6* and *AP2a*) have been characterized in tomato (Tournier *et al.*, 2003; Klee and Giovannoni, 2011; Lee *et al.*, 2011). Among the genes encoding these ERFs, *LeERF2*, *LeERF3b*, *Pti4*, *SlAP2a* and *SlERF6* exhibit ripening-associated transcript accumulation in tomato fruits. Expression of *LeERF2* and *Pti4* was found to be reduced in tomato ripening mutants, such as *ripening-inhibitor* (*rin*), *Never-ripe* (*Nr*) and *non-ripening* (*nor*), while increased transcript accumulation of *LeERF3b* was observed in transgenic plants with attenuated *ACO* expression and in *Nr* mutant fruit. Antisense transgenic lines of the *LeERF2* gene did not exhibit any visible change in fruit phenotype (Pirrello *et al.*, 2006). Antisense tomato fruits with inhibited expression of *LeERF1* genes have been shown to exhibit a longer shelf-life compared with wild-type fruits (Li *et al.*, 2007). Genome-wide analysis to identify the full complement of *ERF* genes encoded by tomato resulted in the identification of 85 *SlERF* genes. A comprehensive analysis of their expression profiles demonstrated that a total of 57 genes were differentially expressed during the temporal stages of tomato fruit development and ripening (Sharma *et al.*, 2010). Recently, two ERFs (*SlAP2a* and *SlERF6*) were found to negatively affect fruit ripening in tomato. Higher transcript accumulation levels of the *LeACS2*, *LeACS4* and *LeACO1* genes in RNA interference (RNAi) tomato plants with reduced expression of *SlAP2a* suggested that this gene acts as a negative regulator of ethylene production during ripening (Chung *et al.*, 2010). RNAi plants with reduced expression of *SlERF6* exhibited enhanced production of ethylene and carotenoids, demonstrating an important role for SlERF6 in ripening (Lee *et al.*, 2011).

As in tomato, several genes encoding components of ethylene signalling have been found to be differentially expressed during maturation and ripening in non-climacteric fruits, such as loquat (*Eriobotrya japonica* Lindl.), watermelon (*Citrullus lanatus* Thunb.), grapevine (*Vitis* species) and strawberry (*Fragaria* × *ananassa)*, as well as other climacteric fruits such as kiwifruit (*Actinidia deliciosa*), apple (*Malus domestica*) and oil palm (*Elaeis guineensis*), suggesting that the ethylene signalling pathway is well conserved across plant species (Wang *et al.*, 2010; Klee and Giovannoni, 2011).

11.3 Interaction of Ethylene with Other Hormones

Recent studies on the transcriptome of tomato fruit have suggested the possibility of a highly complex interactive hormonal network operating during fruit ripening. In addition to ethylene-related genes, genes related to other hormones such as auxin, jasmonic acid, abscisic acid are differentially regulated during ripening in tomato (Osorio *et al.*, 2011). Comparison of fruit trancriptomes between wild-type fruit and *rin* mutants showed that auxin-related genes formed the second most represented subcategory, after ethylene-related genes, in the hormone responses category during ripening (Kumar *et al.*, 2012b). Study of *cis*-regulatory elements for the presence of auxin-responsive elements in the upstream regulatory region of differentially expressed genes resulted in the identification of 34 such genes, suggesting that these might be regulated by auxin-response factors (ARFs) directly during ripening in tomato (Kumar *et al.*, 2011). These observations have been further supported by the analyses of expression profiles of genes involved in early auxin responses, such as ARFs and Gretchen Hagen3 (GH3). These studies have resulted in the identification of four ARFs (*SlARF3*, *SlARF5*, *SlARF6* and *SlARF13*) and two GH3 genes (*SlGH3*-1 and *SlGH3*-2) with ripening-associated expression (Kumar *et al.*, 2011, 2012a). In addition, the role of abscisic acid in the regulation of fruit ripening alone and/or in relation to ethylene is also being studied (Sun *et al.*, 2012). Together, these studies give support to the observation that other hormones, besides ethylene, are also involved in the regulation of fruit ripening.

11.4 Ethylene-regulated Transcriptional Aspects of Fruit Ripening

In addition to our improved knowledge of the components of ethylene signalling, numerous ethylene-regulated genes that are responsible for the outcome of the final ripened fruit phenotype have been identified across the plant species, including tomato (Cara and Giovannoni, 2008). Along with the coding sequence, the upstream regulatory sequences of several such genes have also been characterized for their ethylene responsiveness during ripening. Initially, ethylene-regulated aspects of ripening were studied using both ripening mutants and antisense transgenic plants altered in ethylene production/responses or by upregulating expression of genes related to these aspects. It has been clearly shown that ethylene controls ripening by regulating the expression of genes at the mRNA or protein level (Cara and Giovannoni, 2008). Some of the best-characterized genes identified in this manner include genes involved in ethylene biosynthesis, such as 1-aminocyclopropane-1-carboxylic acid synthases (*ACSs*) and *ACOs*, genes encoding proteins related to cell-wall dynamics such as polygalacturonase (*PG*), Expansin1 (*LeEXP1*) and pectin methyl esterase (*PME*), genes controlling carotenoid biosynthesis and volatile production such as phytoene synthase (*PSY*), lipoxygenases (*TomloxA*, *TomloxB* and *TomloxC*) and some other ethylene-responsive genes with unidentified functions such as *E4* and *E8* (Cara and Giovannoni, 2008). Some of the other less-well-characterized ethylene-responsive genes include *ER24*, *ER49*, *ER50* and *ER68* and the gene encoding an enzyme of the Rab GTPase family, *LeRab11a* (Zegzouti *et al.*, 1999; Lu *et al.*, 2001). Transcriptome analyses of non-ripening mutants with altered ethylene responses have revealed multiple ethylene-associated events during tomato ripening (Alba *et al.*, 2005; Osorio *et al.*, 2011; Kumar *et al.*, 2012b). Recently, two microRNAs, miR828 and miR1917, whose targets are EIN2 and the serine/threonine protein kinase CTR1, were identified in tomato fruit, suggesting their role in the regulation of ethylene signalling during ripening (Zuo *et al.*, 2012).

Characterization of promoter regions of several ethylene-induced and ripening-related genes have provided some insights into their transcriptional regulation. These studies resulted in the identification of several motifs that contribute to their ethylene responsiveness. For example, two repeat regions and multiple EREs (A(A/T)TTCAAA) along with a stress-related motif (TCATCTTCTT) were found in the promoters of the *ACO1*, *2A11* and *E4* genes. Promoter analysis of two ethylene-responsive genes (*E4* and *E8*) with contrasting expression patterns in response to ethylene resulted in the identification of motifs responsible for the tissue-specific and developmentally regulated responses of *E8*. *Cis*-element analysis of the promoters of ripening-associated genes, including these two genes, has also led to the identification of a similar region necessary for their ethylene responsiveness in several such genes (Cara and Giovannoni, 2008; Kumar *et al.*, 2012b).

11.5 Conclusions and Prospects

The initiation and subsequent coordinated manifestation of ripening events is a highly complex phenomenon that depends on several internal factors, including the type and category of fruit, developmental regulation and complex interactive hormonal networks, as well as numerous external factors including light, temperature, and biotic and abiotic stress. Ethylene is one of the main determinants of ripening in climacteric fruits. As most of the ethylene signalling components, upstream of the ERF proteins and above the levels of ERFs, have already been characterized during ripening, the current emphasis is on characterizing additional ripening-related ERFs in tomato. Characterization of these transcription factors and their interactions with targets is likely to result in a better understanding of the molecular mechanisms of ethylene signalling, underlying ripening in fleshy fruits. Several genomics tools such as microarray resources have already been utilized to gain an insight into the ethylene-dependent aspects of ripening in tomato (Alba *et al.*, 2005; Osorio *et al.*, 2011; Kumar *et al.*, 2012b). Furthermore, the identification of fruit ripening-specific small RNAs, including microRNAs and small interfering RNAs, will help in further elucidation of the post-transcriptional regulation of expression of ethylene-related ripening-specific genes (Zuo *et al.*, 2012). The recently completed tomato genome sequencing project will also help in finding the chromosomal locations of ripening-specific genes, and will result in identification of various common regulatory sequences present in the promoters of ripening-related genes that will ultimately help in unravelling the regulatory mechanism of their spatial/temporal expression. The availability of such genomic resources for other fruit species will promote opportunities to understand various aspects of fruit development and ripening and to apply this knowledge for crop improvement.

Acknowledgements

The authors thank the Department of Biotechnology, Government of India, for financial support. R.K. acknowledges CSIR, India, for the fellowship granted during his tenure as a research fellow.

References

Abeles, F.B., Morgan, P.W. and Saltveit, M.E. Jr (1992) *Ethylene in Plant Biology*, 2nd edn. Academic Press, New York.

Adams-Phillips, L., Barry, C., Kannan, P., Leclercq, J., Bouzayen, M. and Giovannoni, J.J. (2004) Evidence that CTR1-mediated ethylene signal transduction in tomato is encoded by a multigene family whose members display distinct regulatory features. *Plant Molecular Biology* 54, 387–404.

Alba, R., Payton, P., Fei, Z., McQuinn, R., Debbie, P., Martin, G.B., Tanksley, S.D. and Giovannoni, J.J. (2005) Transcriptome and selected metabolite analyses reveal multiple points of ethylene control during tomato fruit development. *Plant Cell* 17, 2954–2965.

Barry, C.S. and Giovannoni, J.J. (2006) Ripening in the tomato *Green-ripe* mutant is inhibited by ectopic expression of a protein that disrupts ethylene signaling. *Proceedings of the National Academy of Sciences USA* 103, 7923–7928.

Cara, B. and Giovannoni, J.J. (2008) Molecular biology of ethylene during tomato fruit development and maturation. *Plant Science* 175, 106–113.

Chung, M.Y., Vrebalov, J., Alba, R., Lee, J., McQuinn, R., Chung, J.D., Klein, P. and Giovannoni, J. (2010) A tomato (*Solanum lycopersicum*) *APETALA2*/ERF gene, *SlAP2a*, is a negative regulator of fruit ripening. *Plant Journal* 64, 936–947.

Fatima, T., Rivera-Dominguez, M., Troncoso-Rojas, R., Tiznado-Hernandez, M.-E., Handa, A.K. and Mattoo, A.K. (2008) Tomato. In: Kole, C. and Hall, T.C. (eds) *Compendium of Transgenic Crop Plants: Transgenic Vegetable Crops*, Vol. 6. Blackwell Publishing, Oxford, pp. 1–46.

Foolad, M.R. (2007) Genome mapping and molecular breeding of tomato. *Integrative Journal of Plant Genomics* 2007, 64358.

Fraser, P.D. and Bramley, P.M. (2004) The biosynthesis and nutritional uses of carotenoids. *Progress in Lipid Research* 43, 228–265.

Guo, H. and Ecker, J.R. (2003) Plant responses to ethylene gas are mediated by SCF(EBF1/EBF2)-dependent proteolysis of EIN3 transcription factor. *Cell* 115, 667–677.

Gupta, V., Mathur, S., Solanke, A.U., Sharma, M.K., Kumar, R., Vyas, S., Khurana, P., Khurana, J.P., Tyagi, A.K. and Sharma, A.K. (2009) Genome analysis and genetic enhancement of tomato. *Critical Reviews in Biotechnology* 29, 152–181.

Kevany, B.M., Tieman, D.M., Taylor, M.G., Cin, V.D. and Klee, H.J. (2007) Ethylene receptor degradation controls the timing of ripening in tomato fruit. *Plant Journal* 51, 458–467.

Klee, H.J. and Giovannoni, J.J. (2011) Genetics and control of tomato fruit ripening and quality attributes. *Annual Review of Genetics* 45, 41–59.

Kumar, R., Tyagi, A.K. and Sharma, A.K. (2011) Genome-wide analysis of auxin response factor (ARF) gene family from tomato and analysis of their role in flower and fruit development. *Molecular Genetics and Genomics* 285, 245–260.

Kumar, R., Agarwal, P., Tyagi, A.K. and Sharma, A.K. (2012a) Genome-wide analysis of the auxin-responsive *GH3* gene family in tomato (*Solanum lycopersicum*). *Molecular Genetics and Genomics* 287, 221–235.

Kumar, R., Sharma, M.K., Kapoor, S., Tyagi, A.K. and Sharma, A.K. (2012b) Transcriptome analysis of *rin* mutant fruit and *in silico* analysis of promoters of differentially regulated genes provides insight into LeMADS-RIN-regulated ethylene-dependent as well as ethylene-independent aspects of ripening in tomato. *Molecular Genetics and Genomics* 287, 189–203.

Lanahan, M.B., Yen, H.C., Giovannoni, J.J. and Klee, H.J. (1994) The never ripe mutation blocks ethylene perception in tomato. *Plant Cell* 6, 521–530.

Leclercq, J., Adams-Phillips, L.C., Zegzouti, H., Jones, B., Latche, A., Giovannoni, J.J., Pech, J.C. and Bouzayen, M. (2002) *LeCTR1*, a tomato *CTR1*-like gene, demonstrates ethylene signaling ability in *Arabidopsis* and novel expression patterns in tomato. *Plant Physiology* 130, 1132–1142.

Lee, J.M., Joung, J.G., McQuinn, R., Chung, M.Y., Fei, Z., Tieman, D., Klee, H. and Giovannoni, J. (2011) Combined transcriptome, genetic diversity and metabolite profiling in tomato fruit reveals that the ethylene response factor *SlERF6* plays an important role in ripening and carotenoid accumulation. *Plant Journal* 70, 191–204.

Li, J., Li, Z., Tang, L., Yang, Y., Zouine, M. and Bouzayen, M. (2011) A conserved phosphorylation site regulates the transcriptional function of ETHYLENE-INSENSITIVE3-like1 in tomato. *Journal of Experimental Botany* 63, 427–439.

Li, Y., Zhu, B., Xu, W., Zhu, H., Chen, A., Xie, Y., Shao, Y. and Luo, Y. (2007) LeERF1 positively modulated ethylene triple response on etiolated seedling, plant development and fruit ripening and softening in tomato. *Plant Cell Reports* 26, 1999–2008.

Lin, Z., Hackett, R.M., Payton, S. and Grierson, D. (1998) A tomato sequence, TCTR2 (accession no. AJ005077) encoding an *Arabidopsis* CTR1 homologue. *Plant Physiology* 117, 1126.

Lu, C., Zainal, Z., Tucker, G.A. and Lycett, G.W. (2001) Developmental abnormalities and reduced fruit softening in tomato plants expressing an antisense Rab11 GTPase gene. *Plant Cell* 13, 1819–1833.

Mattoo, A.K. and Suttle, J.C. (eds) (1991) *The Plant Hormone Ethylene*, CRC Press, Boca Raton, Florida.

Osorio, S., Alba, R., Damasceno, C.M., Lopez-Casado, G., Lohse, M., Zanor, M.I., Tohge, T., Usadel, B., Rose, J.K., Fei, Z., Giovannoni, J.J. and Fernie, A.R. (2011) Systems biology of tomato fruit development: combined transcript, protein, and metabolite analysis of tomato transcription factor (*nor, rin*) and ethylene receptor (*Nr*) mutants reveals novel regulatory interactions. *Plant Physiology* 157, 405–425.

Pirrello, J., Jaimes-Miranda, F., Sanchez-Ballesta, M.T., Tournier, B., Khalil-Ahmad, Q., Regad, F., Latche, A., Pech, J.C. and Bouzayen, M. (2006) Sl-ERF2, a tomato ethylene response factor involved in ethylene response and seed germination. *Plant and Cell Physiology* 47, 1195–1205.

Roman, G., Lubarsky, B., Kieber, J.J., Rothenberg, M. and Ecker, J.R. (1995) Genetic analysis of ethylene signal transduction in *Arabidopsis thaliana*: five novel mutant loci integrated into a stress response pathway. *Genetics* 139, 1393–1409.

Sharma, M.K., Kumar, R., Solanke, A.U., Sharma, R., Tyagi, A.K. and Sharma, A.K. (2010) Identification, phylogeny, and transcript profiling of ERF family genes during development and abiotic stress treatments in tomato. *Molecular Genetics and Genomics* 284, 455–475.

Sun, L., Sun, Y., Zhang, M., Wang, L., Ren, J., Cui, M., Wang, Y., Ji, K., Li, P., Li, Q., Chen, P., Dai, S., Duan, C., Wu, Y. and Leng, P. (2012) Suppression of 9-*cis*-epoxycarotenoid dioxygenase, which encodes a key enzyme in abscisic acid biosynthesis, alters fruit texture in transgenic tomato. *Plant Physiology* 158, 283–298.

Thompson, A.J., Tor, M., Barry, C.S., Vrebalov, J., Orfila, C., Jarvis, M.C., Giovannoni, J.J., Grierson, D. and Seymour, G.B. (1999) Molecular and genetic characterization of a novel pleiotropic tomato-ripening mutant. *Plant Physiology* 120, 383–390.

Tieman, D.M., Taylor, M.G., Ciardi, J.A. and Klee, H.J. (2000) The tomato ethylene receptors NR and LeETR4 are negative regulators of ethylene response and exhibit functional compensation within a multigene family. *Proceedings of the National Academy of Sciences USA* 97, 5663–5668.

Tieman, D.M., Ciardi, J.A., Taylor, M.G. and Klee, H.J. (2001) Members of the tomato *LeEIL* (*EIN3-like*) gene family are functionally redundant and regulate ethylene responses throughout plant development. *Plant Journal* 26, 47–58.

Tournier, B., Sanchez-Ballesta, M.T., Jones, B., Pesquet, E., Regad, F., Latche, A., Pech, J.C. and Bouzayen, M. (2003) New members of the tomato ERF family show specific expression pattern and diverse DNA-binding capacity to the GCC box element. *FEBS Letters* 550, 149–154.

Wang, P., Zhang, B., Li, X., Xu, C., Yin, X., Shan, L., Ferguson, I. and Chen, K. (2010) Ethylene signal transduction elements involved in chilling injury in non-climacteric loquat fruit. *Journal of Experimental Botany* 61, 179–190.

Wilkinson, J.Q., Lanahan, M.B., Yen, H.C., Giovannoni, J.J. and Klee, H.J. (1995) An ethylene-inducible component of signal transduction encoded by *never-ripe*. *Science* 270, 1807–1809.

Yang, S.F. and Hoffman, N.E. (1984) Ethylene biosynthesis and its regulation in higher plants. *Annual Reviews in Plant Physiology* 35, 155–190.

Yang, Y., Wu, Y., Pirrello, J., Regad, F., Bouzayen, M., Deng, W. and Li, Z. (2010) Silencing *Sl-EBF1* and *Sl-EBF2* expression causes constitutive ethylene response phenotype, accelerated plant senescence, and fruit ripening in tomato. *Journal of Experimental Botany* 61, 697–708.

Yano, K., Aoki, K. and Shibata, D. (2007) Genomic databases for tomato. *Plant Biotechnology* 24, 17–25.

Yin, X.R, Allan, A.C., Chen, K.S. and Ferguson, I.B. (2010) Kiwifruit EIL and ERF genes involved in regulating fruit ripening. *Plant Physiology* 153, 1280-1292.

Yokotani, N., Nakano, R., Imanishi, S., Nagata, M., Inaba, A. and Kubo, Y. (2009) Ripening-associated ethylene biosynthesis in tomato fruit is autocatalytically and developmentally regulated. *Journal of Experimental Botany* 60, 3433–3442.

Zegzouti, H., Jones, B., Frasse, P., Marty, C., Maitre, B., Latch, A., Pech, J.C. and Bouzayen, M. (1999) Ethylene-regulated gene expression in tomato fruit: characterization of novel ethylene-responsive and ripening-related genes isolated by differential display. *Plant Journal* 18, 589–600.

Zhou, D., Kalaitzis, P., Mattoo, A.K. and Tucker, M.L. (1996) The mRNA for an ETR1 homologue in tomato is constitutively expressed in vegetative and reproductive tissues. *Plant Molecular Biology* 30, 1331–1338.

Zuo, J., Zhu, B., Fu, D., Zhu, Y., Ma, Y., Chi, L., Ju, Z., Wang, Y., Zhai, B. and Luo, Y. (2012) Sculpting the maturation, softening and ethylene pathway: the influences of microRNAs on tomato fruits. *BMC Genomics* 13, 7.

12 Other Hormonal Signals during Ripening

Christopher Davies* and Christine Böttcher
CSIRO Plant Industry, Glen Osmond, SA, Australia

12.1 Introduction

Ask any plant biologist which hormone is involved in fruit ripening and the answer will almost inevitably be 'ethylene'. The role of ethylene during fruit development has been much discussed, and the case for it being pivotal in climacteric ripening is well established (see Grierson, Chapter 10, and Kumar and Sharma, Chapter 11, this volume). This simple molecule has dominated the research effort into the control of fruit ripening. This is partly because of its rather obvious effects on the ripening of some fruit and partly because it coordinates the ripening of many commercially important fruits that can also serve as model species for study, such as tomato. However, ethylene is far from being the only hormonal influence on fruit ripening. There is increasing interest in other hormones that deserve our attention with regard to the control of ripening in both climacteric and non-climacteric fruits. This chapter will outline the role of other hormones that may affect fleshy fruit ripening. It is not intended to be an exhaustive review of the literature; rather, we seek to explore the role of 'non-ethylene' ripening control with the use of relevant examples.

In our minds, the critical criteria required to make a convincing case for a hormone's involvement in the control of ripening are as follows: (i) if the hormone in question is an inhibitor of ripening, then its level, or sensitivity to it, should decrease before ripening initiation; if it is a promoter of ripening, the levels, or sensitivity to it, should increase before or coincide with, the initiation of ripening; (ii) in most cases, it would be expected that exogenous hormones, or inhibitors of hormone synthesis or perception, affect fruit development in line with their proposed action. Manipulation of the hormonal status or response through transgenesis should also have effects consistent with the proposed role. During the following discussions, it will become clear that, for a number of hormones, not all of the above criteria will be met.

We have not included a detailed discussion of the voluminous data concerning treatment of fruits with hormones and plant growth regulators that may affect ripening. This is not merely a question of available space due to the large numbers of studies (for example, one review of jasmonate application in horticultural crops quotes 381 papers; Rohwer and Erwin, 2008)). Although useful, this data

* christopher.davies@csiro.au

must be viewed carefully due to the complexity of these experiments. The complexity is due to there being a wide range of fruit species with a range of growth and ripening modes, variable sensitivity to hormonal treatments depending on growth conditions and stage, considerable variation in experimental design and procedures (e.g. depending on the species, the response of the fruit to hormone application can differ depending on whether the fruit is on or off the plant) and the number of candidate hormones to be tested, which is further complicated by interactions between them.

12.2 Regulators of Ripening, Both Positive and Negative and Others of Uncertain Status

12.2.1 Abscisic acid (ABA)

ABA is most commonly thought of as being associated with embryogenesis, seed maturation and abiotic stress responses (Nambara and Marion-Poll, 2005), but a role in the control of fruit ripening can be added to this list. ABA levels in a range of fruits, both climacteric and non-climacteric, appear to follow the same basic pattern: they are low prior to ripening and then increase at about the time ripening commences (e.g. strawberry, Chen *et al.*, 2011; avacado, Chernys and Zeevaart, 2000; grape, Davies *et al.*, 1997; cherry, Kondo and Gemma, 1993; orange, Rodrigo *et al.*, 2006; tomato, Sun *et al.*, 2012). In a number of these examples, ABA levels were high at flowering and during early fruit development and then decreased prior to the increase coincident with ripening. This pattern of accumulation is consistent with a role for ABA as a positive regulator of ripening. Interestingly, in tomato, the increase in ABA levels peaks before the peak in ethylene levels (Sun *et al.*, 2012), suggesting that ABA may have some role in modulating ethylene accumulation.

The transcript levels of a number of genes either involved in ABA synthesis or signalling or putatively involved in ABA responses have been shown to change during fruit development in response to ABA application and in association with ripening. Increases in the expression of 9-*cis*-epoxycarotenoid dioxygenase (NCED), which catalyses the key step in ABA synthesis, concomitant with an increase in ABA levels have been reported in both climacteric and non-climacteric fruits (e.g. Deluc *et al.*, 2007; Ren *et al.*, 2010; Sun *et al.*, 2012). Microarray experiments, particularly in grape, have been used to study changes in ABA-related gene expression in fruits during ripening and in response to stress (Deluc *et al.*, 2007, 2009; Grimplet *et al.*, 2007; Pilati *et al.*, 2007; Koyama *et al.*, 2010; Fortes *et al.*, 2011). Changes in protein levels following ABA treatment have also been observed in grapes (Giribaldi *et al.*, 2010).

The list of genes involved in the above studies and their putative functions is too long to discuss in detail here, but a few examples, which tie in with the effects of ABA application to developing fruits, deserve particular mention. In many cases, the application of ABA to unripe fruit has promoted ripening, or at least some aspect of it. The timing of ABA application is crucial for effect, as it is with other hormones that can alter fruit ripening. ABA application has frequently been reported to increase the activity of the flavonoid pathway, leading to increased levels of anthocyanins and other flavonoids related to fruit quality. This increase has been associated with ABA-induced increases in the transcript levels of genes encoding biosynthetic enzymes and transcription factors (Ban *et al.*, 2003; Jeong *et al.*, 2004). ABA application has also resulted in increased sugar uptake and accumulation and increased expression of enzymes thought to be associated with these processes such as invertases (Pan *et al.*, 2005). Interactions between ABA and sugars have been reported (Rook *et al.*, 2006) and an example is given in another part of this chapter.

In addition to the positive effects of ABA on fruit ripening, inhibitors of ABA biosynthesis have been shown to delay climacteric and non-climacteric fruit ripening. For example, in tomato, while ABA application was shown to enhance ripening, including inducing ethylene biosynthesis gene expression, treatment with fluridone and nordihydroguaiaretic acid delayed ripening (Zhang *et al.*, 2009a). Fluridone treatment also delayed strawberry fruit ripening (Jia *et al.*, 2011).

The most convincing evidence for ABA being involved in the ripening of both climacteric and non-climacteric fruits comes from recent experiments using transient or stably transformed strawberry and tomato fruit. Bastías *et al.* (2011) overexpressed and downregulated an ABA response element-binding factor in tomato. They found higher levels of the sugars glucose and fructose in fruit from the overexpression lines and elevated expression of genes encoding a vacuolar invertase and sucrose synthase. ABA levels were increased in immature green and red-ripe fruit, while ethylene levels were elevated in red-ripe fruit alone. In strawberries (*Fragaria* × *ananassa*), Chai *et al.* (2011) used virus-induced gene silencing to downregulate the expression of a putative ABA receptor, FaPYR1. This delayed fruit ripening and increased ABA content by more than twofold. ABA application did not overcome the ripening inhibition, and the expression of some ABA-responsive genes was downregulated. These data suggest that the *FaPYR1* gene encodes an ABA receptor and that ABA positively regulates strawberry ripening. In further experiments on strawberry using virus-induced gene silencing, Jia *et al.* (2011) downregulated *NCED* expression, which resulted in uncoloured fruit with lower ABA levels. Coloration could be restored by exogenous ABA. They also downregulated the expression of a putative ABA receptor gene, *FaCHLH/ABAR*, and again produced uncoloured fruit that did not colour with applied ABA. These fruit also had elevated ABA levels and reduced sugar content, and there were alterations in the expression of ABA-related and sugar-responsive genes. These results indicated that FaCHLH/ABAR acts as an ABA receptor and that ABA can promote ripening. The downregulation of *NCED* in tomato fruit provides support for ABA being involved in another important aspect of fruit ripening – changes in cell walls that relate to fruit texture. An RNA interference construct with the fruit-specific *E8* gene promoter was used to specifically downregulate *NCED* expression (Sun *et al.*, 2012). This resulted in a significant reduction in ABA levels and a reduction in the transcript levels of a number of genes encoding proteins involved in cell-wall metabolism, including polygalaturonase, pectin methyl esterase, β-galactosidase, xyloglucan endotransglycosylase, endo-1,4-β-cellulase and an expansin. This resulted in firmer fruit with higher pectin levels (Sun *et al.*, 2012). In summary, ABA appears to act as a positive regulator of ripening, or some aspects of it, in a range of fruits.

12.2.2 Brassinosteroids (BRs)

The role of BRs in plant growth is well established, but more recently many other functions have been defined (Gudesblat and Russinova, 2011). Among these functions, and like ABA, BRs have been shown to be involved in the response of plants to various stresses. Interestingly, studies in a small range of fruit species indicate that BRs may also play a role in the control of fruit ripening, another property shared with ABA. Indications that BRs could be involved in the control of fruit ripening first came from tomato. Tomato pericarp discs treated with BRs exhibited an increase in ripening-related parameters, such as increased lycopene accumulation, a reduction in chlorophyll levels, a decrease in ascorbic acid levels and an increase in sugar levels (Vardhini and Rao, 2002). As seen with the application of some other hormones, such as auxin, to climacteric fruits, advancement of ripening by BRs may operate through induction of ethylene evolution. High levels of endogenous BRs have been reported during early tomato development (Montoya *et al.*, 2005), but their role in the control of ripening is still unclear and we could find no information regarding the effects of BR mutants on tomato ripening. Exogenous BRs also seem to affect passionfruit ripening, but the response varied depending on the frequency and timing of treatment, and the increase in fruit number due to treatments at flowering was the most dramatic effect (Gomes *et al.*, 2006).

Further evidence of a possible involvement of BRs in ripening comes from gene expression studies. The expression of a BR-6-oxidase gene was increased in red-skinned apples when compared with green-skinned apples, but there was no other data to support a role for BRs in apple ripening (Han *et al.*, 2011). Bombarely *et al.* (2010) sequenced expressed sequence tags from strawberry fruit and used quantitative PCR to confirm that expression of the BR receptor gene and a gene encoding a BR signalling component was upregulated in the receptacle in the red stage, but there was no statistical analysis to support this observation.

The most convincing evidence that BRs play a role in fruit ripening comes from grape berries. Symons *et al.* (2006) showed that the levels of castasterone were elevated in flowers and young berries, decreased prior to veraison and then increased sharply at veraison, consistent with a role in ripening. After peaking immediately after veraison, the levels of castasterone declined (similar to the profile for ABA).

The enhancement of grape berry ripening (as measured by colour increase) through the application of epi-brassinolide and its delay by the application of brassinazole, an inhibitor of BR biosynthesis, is further evidence for BR involvement in berry ripening (Symons *et al.*, 2006).

BRs have been implicated in the control of ripening of both climacteric and non-climacteric fruits, which is indicative of a similarity in ripening between the two fruit types. As can been seen from the above discussion, although there is evidence supporting a role for BRs in fruit ripening, particularly in grapes, there is still a great deal of work to be done to convincingly establish this as a general phenomenon.

12.2.3 Auxins

Auxins are involved in a multitude of processes during fruit development, but their role in the control of fruit ripening has received relatively little attention. Various patterns of accumulation in fruit have been reported for indole-3-acetic acid (IAA), the most abundant auxin found in plants, but the consensus is that levels are usually high early in development and decrease to be low at, or before, the initiation of ripening (e.g. Buta and Spaulding, 1994; Abbas *et al.*, 2000; Böttcher *et al.*, 2010). This pattern of accumulation is consistent with a possible role in ripening inhibition.

Many of the hormones described in this chapter are modified as part of normal metabolism – for example ABA is glycosylated and then further modified (for example by oxidation) – and such pathways are an important part of maintaining hormone levels. The sequestration of auxins through conjugation to amino acids seems to be important for ripening, and changes in the levels of conjugated forms of IAA occur during fruit development (Dunlap *et al.*, 1996; Purgatto *et al.*, 2002; Böttcher *et al.*, 2010). A detailed study in grape berries showed that IAA-amido synthetases appear to reduce the levels of free IAA prior to the initiation of ripening (and perhaps during ripening; Böttcher *et al.*, 2010, 2011a). This decrease may be essential for the initiation of ripening due to the proposed role of auxins as inhibitors. There is also some evidence that auxin oxidation products may accelerate fruit ripening under some circumstances, which may suggest an even more complex control of ripening by IAA and its products (Frenkel, 1975; Frenkel *et al.*, 1975; Guttridge *et al.*, 1977).

Auxin application to pre-ripening fruits produces varying effects. In some fruits, auxin treatments have resulted in a delay in ripening, while in others auxin treatment advanced or enhanced ripening. The enhancement of ripening by auxin analogues occurs most frequently in climacteric fruits, as reported in apple, loquat, nectarine, peach, pear, persimmon and plum (Agustí *et al.*, 1999, 2003, 2004; Ohmiya, 2000; Kondo *et al.*, 2004; Yuan and Carbaugh, 2007; Li and Yuan, 2008). This effect may be due to increased ethylene evolution resulting from a stimulation of ethylene biosynthesis (Kondo *et al.*, 2004; Trainotti *et al.*, 2007; Li and Yuan, 2008). There are numerous reports where the application of auxins during the pre-ripening stage delayed ripening in both climacteric and non-climacteric fruits; examples include avocado (Tingwa and Young, 1975), banana (Purgatto *et al.*, 2002), grape (Böttcher *et al.*, 2011a,b), kiwifruit (Fabbroni *et al.*, 2006), strawberry (Villarreal *et al.*, 2009), tomato (Cohen, 1996), loquat (Amorós *et al.*, 2004), peach (Ohmiya and Haji, 2002) and pear (Frenkel and Haard, 1973). These reports detail some of the specific effects of ripening-delaying auxin treatments such as delayed reduction of chlorophyll levels, delayed changes to cell-wall components that normally result in fruit softening, delayed accumulation of sugars (or other forms of carbon storage) and delayed accumulation of anthocyanins. These changes are also reflected by changes in gene transcription, and in some cases the corresponding enzyme activities. For example, in grapes, auxin treatment of pre-ripening berries maintained the expression of genes normally expressed during the pre-ripening stage and delayed the expression of genes normally associated with ripening and the ripening-associated increase in ABA levels (Davies *et al.*, 1997). In strawberry, 1-naphthaleneacetic acid (NAA) treatment reduced the expression (and in some cases enzyme activity) of ripening-associated proteins (Manning, 1994; Villarreal *et al.*, 2009). The transcript levels of many ripening-related genes, including some thought to be involved in ripening-related cell-wall changes, were downregulated by NAA treatments of immature fruit (Harpster *et al.*, 1998; Aharoni *et al.*, 2002; Bustamante *et al.*, 2009; Villarreal *et al.*, 2009), and the expression of two auxin (*AUX*)/*IAA* genes possibly involved in pre-ripening fruit development was upregulated (Liu *et al.*, 2011).

The results of high-throughput transcript analyses also suggest the importance of IAA in the control of fruit development, including ripening. In grapes, expression of the majority of auxin-related genes (including those encoding auxin-responsive factors (ARFs)) were downregulated at the initiation of ripening in line with the proposal that auxins inhibit ripening (Deluc *et al.*, 2007; Fortes *et al.*, 2011). Some *AUX/IAA* genes, which are repressors of auxin-regulated transcription, were down-regulated during ripening but others were upregulated. An upregulation of *AUX/IAA* genes at ripening was also observed in apricot (Manganaris *et al.*, 2011) and peaches (Trainotti *et al.*, 2007). In tomato, the expression of certain *AUX/IAA* genes has been correlated with the levels of particular metabolites (Rohrmann *et al.*, 2011).

The profound effect that changes in auxin signalling can have on fruit development and fruit ripening is exemplified by experiments in tomato. The downregulation, through transgenesis, of the expression of an *ARF* gene normally expressed throughout fruit development, and most highly in early red fruit, led to a pleiotropic phenotype (Jones *et al.*, 2002). The fruit contained elevated levels of chlorophyll, had a blotchy appearance during ripening and were firmer, but other measures of ripening such as total soluble solids and acid levels were similar to those of wild-type fruit.

The above discussion demonstrates the importance of IAA during fruit development and ripening but also shows that there is still much to be understood about all aspects of auxin action including interactions with other hormone signalling pathways.

12.2.4 Cytokinins

Cytokinins are involved in a range of plant processes including fruit set (NeSmith, 2002) and early growth (Werner and Schmülling, 2009). Higher levels are often found in the flesh of immature fruits, which then decrease in maturing and mature fruits (Desai and Chism, 1978; Chen, 1983; Zhang *et al.*, 2003). This pattern is consistent with a role in early fruit development (cell division) and perhaps a role in inhibiting ripening rather than promoting it.

The application of a range of cytokinins has been shown to increase fruit size, especially when applied early in development (Itai *et al.*, 1995; Famiani *et al.*, 1999; NeSmith, 2002; Peppi and Fidelibus, 2008). Effects on ripening have also been reported and, like auxins, which have a similar pattern of accumulation, exogenous cytokinins tend to inhibit ripening. Delays in ripening due to cytokinin application have been reported for a number of fruits, both climacteric and non-climacteric (Itai *et al.*, 1995; NeSmith, 2002; Peppi and Fidelibus, 2008). The delay observed in ripening may be related to the increase in fruit size caused by cytokinin application. However, as with other growth regulators, the reported effects of the application of cytokinins on fruit ripening are varied and are difficult to generalize. For example, in contrast to many reports, a synthetic cytokinin applied to kiwifruit increased fruit size but also increased sugar levels, decreased acidity and decreased flesh firmness (Famiani *et al.*, 1999).

As with the other hormones discussed, the data regarding cytokinins in fruit is far from complete. Cytokinins appear to play a role in fruit set and early fruit development, and, although there may be exceptions, it seems that cytokinins are likely to inhibit ripening.

12.2.5 Gibberellins (GAs)

GA levels generally seem to be high in the early stages of fruit development, and this appears to be linked to the phase of rapid cell division and growth (Fraser *et al.*, 1995; Symons *et al.*, 2006; Zhang *et al.*, 2007b). From their initial high levels, GA levels decline to be low throughout ripening (Fraser *et al.*, 1995; Symons *et al.*, 2006; Zhang *et al.*, 2007b), and this accumulation profile makes them likely to be involved in promoting early fruit development and possibly inhibiting ripening. Part of the role of GAs during the early stages of development may involve changes to the cell wall. Fruit firmness was increased/maintained by the application of GA to a wide range of fruit (Ju *et al.*, 1999; Southwick *et al.*, 2000; Kappel and MacDonald, 2002; Jiang *et al.*, 2004). In sweet cherry, GA application caused a delay in the conversion of pectins to the water-soluble form (Kondo and Danjo, 2001), and in tomato this delay was accompanied by reduced polygalacturonase activity (Mignani *et al.*, 1995). Zhang *et al.* (2007a) reported increased total pectins, cellulose and hemicelluloses when pears were treated with GA.

GA applications (sometimes even applications quite early in development) often lead to a delay in ripening, or at least some aspects of it (Ju *et al.*, 1999; Southwick *et al.*, 2000; Kappel and MacDonald, 2002; Jiang *et al.*, 2004; Zhang and Whiting, 2011). This delay may be an indirect effect of these treatments, as GA application is often associated with an increase in fruit size (Ju *et al.*, 1999; Kappel and MacDonald, 2002; Zhang and Whiting, 2011), probably through the ability of GAs to increase fruit sink strength (Zhang *et al.*, 2012). Larger fruit would be expected to require more photosynthate to achieve the same concentrations of storage product (e.g. sugars) compared with smaller fruit.

A microarray analysis of developing grape berries indicated that the expression of putative GA biosynthesis and response genes was generally low with little modulation (Grimplet *et al.*, 2007). The decrease in levels prior to the initiation of fruit ripening and their ability to delay some aspects of ripening when applied later in development may indicate a possible role in delaying ripening, but there is no direct evidence that they perform this function during 'normal' development.

12.2.6 Jasmonates

Jasmonates are involved in a number of processes but are most commonly associated with plant defence responses (Browse, 2009), and the evidence for their being involved in the control of fruit ripening is not extensive. Endogenous levels of jasmonates have been measured only in the fruits of a small number of species and in some reports with a lack of labelled standards. In experiments involving apples and tomatoes, jasmonic acid levels were higher than methyl jasmonate (MeJA) levels and both peaked in concentration just before the climacteric ethylene burst, which suggests a possible involvement in ripening in these fruits (Fan *et al.*, 1998). In other reports where the levels of various forms of jasmonates were measured in non-climacteric fruits such as grape skins (Kondo and Fukuda, 2001), and in the flesh of sweet cherry (Kondo *et al.*, 2000) and strawberry (Gansser *et al.*, 1997), jasmonate levels were higher early in development and decreased to be low at ripening with little or no increase during ripening. The reported effects of jasmonate application to fruits are mixed. In both climacteric and non-climacteric fruits, there is evidence of a jasmonate-induced ethylene production (Fan *et al.*, 1998; Mukkun and Singh, 2009), but reduced ethylene production has also been reported (Ziosi *et al.*, 2008). In peach, exogenous jasmonates have been reported to enhance the rate of ripening (Janoudi and Flore, 2003) or delay it (Ziosi *et al.*, 2008). When ripening was delayed in jasmonate-treated fruit, ethylene biosynthesis was repressed and overall polyamine levels were increased (Ziosi *et al.*, 2009). Curiously, in mature green tomato discs, ethylene production was increased by MeJA treatment but lycopene accumulation was inhibited (Fan *et al.*, 1998). Ripening of blackberries, considered to be non-climacteric fruit, was enhanced by MeJA treatment as measured by anthocyanin content, soluble solids content and acid levels (Wang *et al.*, 2008). As well as effects on ethylene and polyamine synthesis, jasmonate application to fruit on the plant, detached fruit or culture cells can influence the synthesis of secondary metabolites, such as anthocyanins, carotenoids, chlorophyll, sesquiterpines, tannins and stilbenes (Pérez *et al.*, 1993; Kondo *et al.*, 2000; Wang *et al.*, 2008; D'Onofrio *et al.*, 2009). From the above, it can be seen that, although jasmonates influence ripening in some cases, clear proof of a role for endogenous jasmonates in the control of fleshy fruit ripening is still lacking.

12.2.7 Polyamines (PAs)

The biosynthesis of PAs, such as putrescine, spermidine and spermine, is well understood in plants (Liu *et al.*, 2006). In developing fruits, putrescine is the most abundant species, spermine is usually the least abundant (e.g. Kushad *et al.*, 1988; Shiozaki *et al.*, 2000; Mehta *et al.*, 2002). In a range of species, the levels of free PAs are high early in fruit development, at around the time of rapid cell division and expansion (Kushad *et al.*, 1988: Shiozaki *et al.*, 2000). In a number of fruits, the levels of free PAs decline after this early stage to be low during ripening (Mehta *et al.*, 2002; Valero *et al.*, 2002; Liu *et al.*, 2006), suggestive of a possible role as a ripening inhibitor. There are some exceptions to this pattern of PA accumulation, however, as increases at around the time of ripening have been observed, although their significance has not been explored fully (Martínez-Madrid *et al.*, 1996; Valero *et al.*, 2002). Microarray analysis in grape berries has indicated that PA biosynthetic gene expression increases at around the time of ripening initiation (Deluc *et al.*, 2007; Fortes *et al.*, 2011).

A number of reports indicate that PA application (or higher endogenous levels) resulted in a delay of one or more ripening-related parameters (Dibble *et al.*, 1988; Liu *et al.*, 2006). This may be due partly to the apparently antagonistic interaction between PAs and ethylene (Harpaz-Saad *et al.*, 2012). As with some other hormones, such as jasmonates, more work is required to firmly establish a role for PAs in the control of ripening.

12.2.8 Salicylic acid

Salicylic acid and methyl salicylic acid are involved in signalling in plants, particularly in the induction of defence and stress responses (Bari and Jones, 2009). There is a good deal of information regarding the effects of salicylic acid application on fruit postharvest, where treatment generally delays senescence (Asghari and Aghdam, 2010). A limited number of reports in grapes and tomatoes suggest that salicylic acid application may delay, or slow, ripening-associated changes (Li *et al.*, 1992; Kraeva *et al.*, 1998; Wang *et al.*, 2011). However, there is very little data regarding the accumulation of salicylic acid during fruit development and so a role for salicylic acid during ripening is still unproven.

12.3 Cross-talk during Fruit Ripening

There is an increasing amount of information about cross-talk between different signalling pathways during plant development, and all of the hormones and some sugars seem to be involved (Finkelstein and Gibson, 2002; Rolland *et al.*, 2002; Rashotte *et al.*, 2005; Weiss and Ori, 2007; Teale *et al.*, 2008; Zhang *et al.*, 2009b). Evidence for direct interactions in fruit is limited, but it is increasingly obvious that cross-talk occurs in fruit and is involved in fruit ripening. At the simplest level, a number of studies have shown that the application of one hormone to a fruit induces the synthesis of another; for example, ethylene evolution in pre-ripening grapes and pears was increased by auxin application (Coombe and Hale, 1973; Kondo *et al.*, 2004), and there are numerous other examples to be found in the literature (some have been discussed above).

New technical processes, such as microarray analyses, have been useful in studying the wide-ranging changes in gene transcript level during fruit development (e.g. apple, Costa *et al.*, 2010; grape, Deluc *et al.*, 2007; Fortes *et al.*, 2011; tomato, Jones *et al.*, 2002; watermelon, Wechter *et al.*, 2008). Using this type of approach a number of examples of cross-talk between signalling pathways (e.g. between ABA and auxin, and ethylene and auxin) have also been revealed following hormone application (e.g. grape, Koyama *et al.*, 2010; peach, Trainotti *et al.*, 2007). The significant levels of cross-talk likely to occur between hormone pathways during fruit ripening has important implications for the interpretation of experiments where a hormone, or an inhibitor of its action, is applied, or where the levels of a hormone or the ability to perceive a hormone is altered through transgenesis.

In addition to the cross-talk between hormones during fruit development, it should also be noted that there is cross-talk between seeds and flesh, and that their maturation process appears to be linked. An example of this and the involvement of various hormones in this process can be found in peach, where auxin, cytokinins and GAs are proposed to be important earlier in development, while ABA and ethylene seem to be involved later (Bonghi *et al.*, 2011).

12.4 Summary

Despite the economic and social importance of fruit and fruit products and their health benefits, there are still large gaps in our understanding of fruit ripening. Clearly, hormones other than ethylene are involved in the ripening of both climacteric and non-climacteric fruits. Much of the current evidence regarding the action of various hormones is indirect, unlike the case regarding ethylene in climacteric fruits, which has been studied more intensively (see Grierson, Chapter 10, and Kumar and Sharma, Chapter 11, this volume). More work is needed to better define the roles and relative importance of these 'other' hormones. Particular areas of interest include detailed knowledge of hormone perception and signalling, the resulting transcriptional changes and the resultant changes in the protein profile and hence metabolism. Although we have discussed some examples where cross-talk between hormones seems to be occurring, there is relatively little direct evidence for this in fruits, and much is inferred from what we know from vegetative tissues. It should be remembered that effects caused by hormone application, for example the delay or enhancement of ripening, do not necessarily confirm the action of a hormone during 'normal' fruit development. The use of inhibitors and accurate measurement of endogenous hormone levels add weight to these arguments, but such studies are lacking for a number of hormones in a range of fruits.

While the role of ethylene has been established in many fruits, its role in others is still uncertain, and it seems likely that there will be a spectrum of response with the classic definitions of climacteric and non-climacteric at either end. Indeed, as our knowledge of the action of hormones other than ethylene during ripening increases, there may be a number of different models required to explain the control of ripening in a broad range of fruit species. Much of the focus is on those hormones that promote ripening, but it is understood that there are some that appear to delay ripening. This inhibition seems to occur in both climacteric and non-climacteric fruits, and, as the release of this inhibition is required for ripening, it could be argued that the true control over ripening initiation, but not the conduct of the ripening process itself, is vested in those hormones whose levels are low at the initiation of ripening, such as auxins. The difficulty in gaining direct evidence of a role for these hormones in inhibiting ripening is that they are essential for fruit development and so downregulating their levels to test function is difficult. This is especially challenging as more than one hormone may be involved in inhibition.

Powerful new techniques, such as deep RNA sequencing and microarray analysis, are well suited for studying the hormonal control of fruit development, as hormonal control often involves changes on a large scale with the transcriptional response being rapid and wide-reaching. However, it is clear that detailed functional analysis is still essential in determining the role of the hormone in plant development.

The number of hormones that may influence ripening and the complexity of the interactions between them make it difficult to be categorical about their individual roles. In many cases, more than one hormone may simply be moderating the effects of others through subtle influences on synthesis, transport, perception and signalling. This intricate network may be required to provide flexibility in controlling the complexities of ripening where the fruit must adapt to change and, in the case of fleshy fruit, store large amounts of precious carbon/energy while interacting with the developing seed.

References

Abbas, M., Abbas, M. and Abdel-Basit, O. (2000) Indole-3-acetic acid concentration during fruit development in date palm (*Phoenix dactylifera* L. cv. Hillawi). *Fruit* 55, 115–118.

Agustí, M., Almela, V., Andreu, I., Juan, M. and Zacarias, L. (1999) Synthetic auxin 3,5,6-TPA promotes fruit development and climacteric in *Prunus persica* L. Batsch. *Journal of Horticultural Science and Biotechnology* 74, 556–560.

Agustí, M., Gariglio, N., Castillo, A., Juan, M., Almela, V., Martínez-Fuentes, A. and Mesejo, C. (2003) Effect of the synthetic auxin 2,4-DP on fruit development of loquat. *Plant Growth Regulation* 41, 129–132.

Agustí, M., Juan, M., Mesejo, C., Martinez-Fuentes, A. and Almela, V. (2004) Persimmon fruit size and climacteric encouraged by 3,5,6-trichloro-2-pyridyloxyacetic acid. *Journal of Horticultural Science and Biotechnology* 79, 171–174.

Aharoni, A., Keizer, L.C.P., Van Den Broeck, H.C., Blanco-Portales, R., Muñoz-Blanco, J., Bois, G., Smit, P., De Vos, R.C.H. and O'Connell, A.P. (2002) Novel insight into vascular, stress, and auxin-dependent and -independent gene expression programs in strawberry, a non-climacteric fruit. *Plant Physiology* 129, 1019–1031.

Amorós, A., Zapata, P., Pretel, M.T., Botella, M.A., Almansa, M.S. and Serrano, M. (2004) Role of naphthalene acetic acid and phenothiol treatments on increasing fruit size and advancing fruit maturity in loquat. *Scientia Horticulturae* 101, 387–398.

Asghari, M. and Aghdam, M.S. (2010) Impact of salicylic acid on post-harvest physiology of horticultural crops. *Trends in Food Science and Technology* 21, 502–509.

Ban, T., Ishimaru, M., Kobayashi, S., Shiozaki, S., Goto-Yamamoto, N. and Horiuchi, S. (2003) Abscisic acid and 2,4-dichlorophenoxyacetic acid affect the expression of anthocyanin biosynthetic pathway genes in 'Kyoho' grape berries. *Journal of Horticultural Science and Biotechnology* 78, 586–589.

Bari, R. and Jones, J. (2009) Role of plant hormones in plant defence responses. *Plant Molecular Biology* 69, 473–488.

Bastías, A., López-Climent, M., Valcárcel, M., Rosello, S., Gómez-Cadenas, A. and Casaretto, J.A. (2011) Modulation of organic acids and sugar content in tomato fruits by an abscisic acid-regulated transcription factor. *Physiologia Plantarum* 141, 215–226.

Bombarely, A., Merchante, C., Csukasi, F., Cruz-Rus, E., Caballero, J., Medina-Escobar, N., Blanco-Portales, R., Botella, M., Munoz-Blanco, J., Sanchez-Sevilla, J. and Valpuesta, V. (2010) Generation and analysis of ESTs from strawberry (*Fragaria* × *ananassa*) fruits and evaluation of their utility in genetic and molecular studies. *BMC Genomics* 11, 503.

Bonghi, C., Trainotti, L., Botton, A., Tadiello, A., Rasori, A., Ziliotto, F., Zaffalon, V., Casadoro, G. and Ramina, A. (2011) A microarray approach to identify genes involved in seed-pericarp cross-talk and development in peach. *BMC Plant Biology* 11, 107.

Böttcher, C., Keyzers, R.A., Boss, P.K. and Davies, C. (2010) Sequestration of auxin by the indole-3-acetic acid-amido synthetase GH3-1 in grape berry (*Vitis vinifera* L.) and the proposed role of auxin conjugation during ripening. *Journal of Experimental Botany* 61, 3615–3625.

Böttcher, C., Boss, P.K. and Davies, C. (2011a) Acyl substrate preferences of an IAA-amido synthetase account for variations in grape (*Vitis vinifera* L.) berry ripening caused by different auxinic compounds indicating the importance of auxin conjugation in plant development. *Journal of Experimental Botany* 62, 4267–4280.

Böttcher, C., Harvey, K., Forde, C.G., Boss, P.K. and Davies, C. (2011b) Auxin treatment of pre-veraison grape (*Vitis vinifera* L.) berries both delays ripening and increases the synchronicity of sugar accumulation. *Australian Journal of Grape and Wine Research* 17, 1–8.

Browse, J. (2009) Jasmonate passes muster: a receptor and targets for the defense hormone. *Annual Review of Plant Biology* 60, 183–205.

Bustamante, C.A., Civello, P.M. and Martínez, G.A. (2009) Cloning of the promoter region of β-xylosidase (*FaXyl1*) gene and effect of plant growth regulators on the expression of *FaXyl1* in strawberry fruit. *Plant Science* 177, 49–56.

Buta, J. and Spaulding, D. (1994) Changes in indole-3-acetic acid and abscisic acid levels during tomato (*Lycopersicon esculentum* Mill.) fruit development and ripening. *Journal of Plant Growth Regulation* 13, 163–166.

Chai, Y.-M., Jia, H.-F., Li, C.-L., Dong, Q.-H. and Shen, Y.-Y. (2011) FaPYR1 is involved in strawberry fruit ripening. *Journal of Experimental Botany* 62, 5079–5089.

Chen, J.-Y., Liu, D.-J., Jiang, Y.-M., Zhao, M.-L., Shan, W., Kuang, J.-F. and Lu, W.-J. (2011) Molecular characterization of a strawberry *FaASR* gene in relation to fruit ripening. *PLoS One* 6, e24649.

Chen, W.-S. (1983) Cytokinins of the developing mango fruit: isolation, identification, and changes in levels during maturation. *Plant Physiology* 71, 356–361.

Chernys, J.T. and Zeevaart, J.A.D. (2000) Characterization of the 9-*cis*-epoxycarotenoid dioxygenase gene family and the regulation of abscisic acid biosynthesis in avocado. *Plant Physiology* 124, 343–354.

Cohen, J.D. (1996) *In vitro* tomato fruit cultures demonstrate a role for indole-3-acetic acid in regulating fruit ripening. *Journal of the American Society for Horticultural Science* 121, 520–524.

Coombe, B.G. and Hale, C.R. (1973) The hormone content of ripening grape berries and the effects of growth substance treatments. *Plant Physiology* 51, 629–634.

Costa, F., Alba, R., Schouten, H., Soglio, V., Gianfranceschi, L., Serra, S., Musacchi, S., Sansavini, S., Costa, G., Fei, Z. and Giovannoni, J. (2010) Use of homologous and heterologous gene expression profiling tools to characterize transcription dynamics during apple fruit maturation and ripening. *BMC Plant Biology* 10, 229.

D'Onofrio, C., Cox, A., Davies, C. and Boss, P.K. (2009) Induction of secondary metabolism in grape cell cultures by jasmonates. *Functional Plant Biology* 36, 323–338.

Davies, C., Boss, P.K. and Robinson, S.P. (1997) Treatment of grape berries, a nonclimacteric fruit with a synthetic auxin, retards ripening and alters the expression of developmentally regulated genes. *Plant Physiology* 115, 1155–1161.

Deluc, L., Grimplet, J., Wheatley, M., Tillett, R., Quilici, D., Osborne, C., Schooley, D., Schlauch, K., Cushman, J. and Cramer, G. (2007) Transcriptomic and metabolite analyses of Cabernet Sauvignon grape berry development. *BMC Genomics* 8, 429.

Deluc, L., Quilici, D., Decendit, A., Grimplet, J., Wheatley, M., Schlauch, K., Merillon, J.-M., Cushman, J. and Cramer, G. (2009) Water deficit alters differentially metabolic pathways affecting important flavor and quality traits in grape berries of Cabernet Sauvignon and Chardonnay. *BMC Genomics* 10, 212.

Desai, N. and Chism, G.W. (1978) Changes in cytokinin activity in the ripening tomato fruit. *Journal of Food Science* 43, 1324–1326.

Dibble, A.R.G., Davies, P.J. and Mutschler, M.A. (1988) Polyamine content of long-keeping alcobaca tomato fruit. *Plant Physiology* 86, 338–340.

Dunlap, J.R., Slovin, J.P. and Cohen, J.D. (1996) Indole-3-acetic acid, ethylene, and abscisic acid metabolism in developing muskmelon (*Cucumis melo* L.) fruit. *Plant Growth Regulation* 19, 45–54.

Fabbroni, C., Costa, F., Bregoli, A.M. and Costa, G. (2006). Effect of auxin on fruit morphogenesis: physiological and molecular aspects in kiwifruit ripening. In: *Proceedings of the 6th International Symposium on Kiwifruit*, Rotorua, New Zealand (International Society of Horticultural Science), pp. 541–547.

Famiani, F., Battistelli, A., Moscatello, S., Boco, M. and Antognozzi, E. (1999) Thidiazuron affects fruit growth, ripening and quality of *Actinidia deliciosa*. *Journal of Horticultural Science and Biotechnology* 74, 375–380.

Fan, X., Mattheis, J.P. and Fellman, J.K. (1998) A role for jasmonates in climacteric fruit ripening. *Planta* 204, 444–449.

Finkelstein, R.R. and Gibson, S.I. (2002) ABA and sugar interactions regulating development: cross-talk or voices in a crowd? *Current Opinion in Plant Biology* 5, 26–32.

Fortes, A., Agudelo-Romero, P., Silva, M., Ali, K., Sousa, L., Maltese, F., Choi, Y., Grimplet, J., Martinez- Zapater, J., Verpoorte, R. and Pais, M. (2011) Transcript and metabolite analysis in Trincadeira cultivar reveals novel information regarding the dynamics of grape ripening. *BMC Plant Biology* 11, 149.

Fraser, P.D., Hedden, P., Cooke, D.T., Bird, C.R., Schuch, W. and Bramley, P.M. (1995) The effect of reduced activity of phytoene synthase on isoprenoid levels in tomato pericarp during fruit development and ripening. *Planta* 196, 321–326.

Frenkel, C. (1975) Oxidative turnover of auxins in relation to the onset of ripening in Bartlett pear. *Plant Physiology* 55, 480–484.

Frenkel, C. and Haard, N.F. (1973) Initiation of ripening in Bartlett pear with an antiauxin α(*p*-chlorophenoxy)isobutyric acid. *Plant Physiology* 52, 380–384.

Frenkel, C., Haddon, V.R. and Smallheer, J.M. (1975) Promotion of softening and ethylene synthesis in Bartlett pears by 3-methylene oxindole. *Plant Physiology* 56, 647–649.

Gansser, D., Latza, S. and Berger, R.G. (1997) Methyl jasmonates in developing strawberry fruit (*Fragaria ananassa* Duch. Cv. Kent). *Journal of Agricultural and Food Chemistry* 45, 2477–2480.

Giribaldi, M., Gény, L., Delrot, S. and Schubert, A. (2010) Proteomic analysis of the effects of ABA treatments on ripening *Vitis vinifera* berries. *Journal of Experimental Botany* 61, 2447–2458.

Gomes, M.D.M.A., Campostrini, E., Leal, N.R., Viana, A.P., Ferraz, T.M., Siqueira, L.d.N., Rosa, R.C.C., Netto, A.T., Nunez-Vazquez, M. and Zullo, M.A.T. (2006) Brassinosteroid analogue effects on the yield of yellow passion fruit plants (*Passiflora edulis* f. flavicarpa). *Scientia Horticulturae* 110, 235–240.

Grimplet, J., Deluc, L., Tillett, R., Wheatley, M., Schlauch, K., Cramer, G. and Cushman, J. (2007) Tissue-specific mRNA expression profiling in grape berry tissues. *BMC Genomics* 8, 187.

Gudesblat, G.E. and Russinova, E. (2011) Plants grow on brassinosteroids. *Current Opinion in Plant Biology* 14, 530–537.

Guttridge, C.G., Jarrett, J.M., Stinchcombe, G.R. and Curtis, P.J. (1977) Chemical induction of local reddening in strawberry fruits. *Journal of the Science of Food and Agriculture* 28, 243–246.

Han, S., Lee, H., Heo, S., Sung, S.-K., Kim, W. and Kim, D. (2011) Isolation and characterization of genes expressed differently in mature fruits of 'redfield' and 'greensleeves' apples. *Horticulture, Environment, and Biotechnology* 52, 413–421.

Harpaz-Saad, S., Yoon, G.M., Mattoo, A. and Kieber, J.J. (2012) The formation of ACC and competition between polyamines and ethylene for SAM. In: McManus, M.T. (ed.) *Annual Plant Reviews*, Vol. 44, *The Plant Hormone Ethylene*. Wiley Blackwell.

Harpster, M.H., Brummell, D.A. and Dunsmuir, P. (1998) Expression analysis of a ripening-specific, auxin-repressed endo-1,4-β-glucanase gene in strawberry. *Plant Physiology* 118, 1307–1316.

Itai, A., Tanabe, K., Tamura, F., Susaki, S., Yonemori, K. and Sugiura, A. (1995) Synthetic cytokinins control persimmon fruit shape, size and quality. *Journal of Horticultural Science* 70, 867–873.

Janoudi, A. and Flore, J.A. (2003) Effects of multiple applications of methyl jasmonate on fruit ripening, leaf gas exchange and vegetative growth in fruit trees. *Journal of Horticultural Science and Biotechnology* 78, 793–797.

Jeong, S.T., Goto-Yamamoto, N., Kobayashi, S. and Esaka, M. (2004) Effects of plant hormones and shading on the accumulation of anthocyanins and the expression of anthocyanin biosynthetic genes in grape berry skins. *Plant Science* 167, 247–252.

Jia, H.-F., Chai, Y.-M., Li, C.-L., Lu, D., Luo, J.-J., Qin, L. and Shen, Y.-Y. (2011) Abscisic acid plays an important role in the regulation of strawberry fruit ripening. *Plant Physiology* 157, 188–199.

Jiang, W., Sheng, Q., Jiang, Y. and Zhou, X. (2004) Effects of 1-methylcyclopropene and gibberellic acid on ripening of Chinese jujube (*Zizyphus jujuba* M) in relation to quality. *Journal of the Science of Food and Agriculture* 84, 31–35.

Jones, B., Frasse, P., Olmos, E., Zegzouti, H., Li, Z.G., Latché, A., Pech, J.C. and Bouzayen, M. (2002) Down-regulation of DR12, an auxin-response-factor homolog, in the tomato results in a pleiotropic phenotype including dark green and blotchy ripening fruit. *Plant Journal* 32, 603–613.

Ju, Z., Duan, Y. and Ju, Z. (1999) Combinations of GA3 and AVG delay fruit maturation, increase fruit size and improve storage life of 'Feicheng' peaches. *Journal of Horticultural Science and Biotechnology* 74, 579–583.

Kappel, F. and MacDonald, R.A. (2002) Gibberellic acid increases fruit firmness, fruit size, and delays maturity of 'Sweetheart' sweet cherry. *Journal of the American Pomological Society* 56, 219–222.

Kondo, S. and Danjo, C. (2001) Cell wall polysaccharide metabolism during fruit development in sweet cherry 'Satohnishiki' as affected by gibberellic acid. *Journal of the Japanese Society for Horticultural Science* 70, 178–184.

Kondo, S. and Fukuda, K. (2001) Changes of jasmonates in grape berries and their possible roles in fruit development, *Scientia Horticulturae* 91, 275–288.

Kondo, S. and Gemma, H. (1993) Relationship between abscisic acid (ABA) content and maturation of the sweet cherry, *Journal of the Japanese Society for Horticultural Science* 62, 63–68.

Kondo, S., Tomiyama, A. and Seto, H. (2000) Changes of endogenous jasmonic acid and methyl jasmonate in apples and sweet cherries during fruit development. *Journal of the American Society for Horticultural Science* 125, 282–287.

Kondo, S., Settsu, K. and Jitratham, A. (2004) How application times of 2,4-DP influence the ripening capacity of 'La France' pears. *HortScience* 39, 101–104.

Koyama, K., Sadamatsu, K. and Goto-Yamamoto, N. (2010) Abscisic acid stimulated ripening and gene expression in berry skins of the Cabernet Sauvignon grape. *Functional and Integrative Genomics* 10, 367–381.

Kraeva, E., Andary, C., Carbonneau, A. and Deloire, A. (1998) Salicylic acid treatment of grape berries retards ripening. *Vitis* 37, 143–144.

Kushad, M.M., Yelenosky, G. and Knight, R. (1988) Interrelationship of polyamine and ethylene biosynthesis during avocado fruit development and ripening. *Plant Physiology* 87, 463–467.

Li, J. and Yuan, R. (2008) NAA and ethylene regulate expression of genes related to ethylene biosynthesis, perception, and cell wall degradation during fruit abscission and ripening in 'Delicious' apples. *Journal of Plant Growth Regulation* 27, 283–295.

Li, N., Parsons, B.L., Liu, D. and Mattoo, A.K. (1992) Accumulation of wound-inducible ACC-synthase in tomato fruit is inhibited by salicylic acid and polyamines. *Plant Molecular Biology* 18, 477–487.

Liu, D.-J., Chen, J.-Y. and Lu, W.-J. (2011) Expression and regulation of the early auxin-responsive Aux/IAA genes during strawberry fruit development. *Molecular Biology Reports* 38, 1187–1193.

Liu, J.-H., Honda, C. and Moriguchi, T. (2006) Involvement of polyamine in floral and fruit development. *Japan Agricultural Research Quarterly* 40, 51–58.

Manganaris, G., Rasori, A., Bassi, D., Geuna, F., Ramina, A., Tonutti, P. and Bonghi, C. (2011) Comparative transcript profiling of apricot (*Prunus armeniaca* L.) fruit development and on-tree ripening. *Tree Genetics and Genomes* 7, 609–616.

Manning, K. (1994) Changes in gene expression during strawberry fruit ripening and their regulation by auxin. *Planta* 194, 62–68.

Martínez-Madrid, M.C., Serrano, M., Riquelme, F. and Romojaro, F. (1996) Polyamines, abscisic acid and ethylene production in tomato fruit. *Phytochemistry* 43, 323–326.

Mehta, R.A., Cassol, T., Li, N., Ali, N., Handa, A.K. and Mattoo, A.K. (2002) Engineered polyamine accumulation in tomato enhances phytonutrient content, juice quality and vine life. *Nature Biotechnology* 20, 613–618.

Mignani, I., Greve, L.C., Ben-Arie, R., Stotz, H.U., Li, C., Shackel, K.A. and Labavitch, J.M. (1995) The effects of GA3 and divalent cations on aspects of pectin metabolism and tissue softening in ripening tomato pericarp. *Physiologia Plantarum* 93, 108–115.

Montoya, T., Nomura, T., Yokota, T., Farrar, K., Harrison, K., Jones, J.G.D., Kaneta, T., Kamiya, Y., Szekeres, M. and Bishop, G.J. (2005) Patterns of *Dwarf* expression and brassinosteroid accumulation in tomato reveal the importance of brassinosteroid synthesis during fruit development. *Plant Journal* 42, 262–269.

Mukkun, L. and Singh, Z. (2009) Methyl jasmonate plays a role in fruit ripening of 'Pajaro' strawberry through stimulation of ethylene biosynthesis. *Scientia Horticulturae* 123, 5–10.

Nambara, E. and Marion-Poll, A. (2005) Abscisic acid biosynthesis and catabolism. *Annual Review of Plant Biology* 56, 165–185.

NeSmith, D.S. (2002) Response of rabbiteye blueberry (*Vaccinium ashei* Reade) to the growth regulators CPPU and gibberellic acid. *HortScience* 37, 666–668.

Ohmiya, A. (2000) Effects of auxin on growth and ripening of mesocarp discs of peach fruit. *Scientia Horticulturae* 84, 309–319.

Ohmiya, A. and Haji, T. (2002) Promotion of ethylene biosynthesis in peach mesocarp discs by auxin. *Plant Growth Regulation* 36, 209–214.

Pan, Q.-H., Li, M.-J., Peng, C.-C., Zhang, N., Zou, X., Zou, K.-Q., Wang, X.-L., Yu, X.-C., Wang, X.-F. and Zhang, D.-P. (2005) Abscisic acid activates acid invertases in developing grape berry. *Physiologia Plantarum* 125, 157–170.

Peppi, M.C. and Fidelibus, M.W. (2008) Effects of forchlorfenuron and abscisic acid on the quality of 'Flame Seedless' grapes. *HortScience* 43, 173–176.

Pérez, A., Sanz, C., Richardson, D. and Olías, J. (1993) Methyl jasmonate vapor promotes β-carotene synthesis and chlorophyll degradation in Golden Delicious apple peel. *Journal of Plant Growth Regulation* 12, 163–167.

Pilati, S., Perazzolli, M., Malossini, A., Cestaro, A., Dematte, L., Fontana, P., Dal Ri, A., Viola, R., Velasco, R. and Moser, C. (2007) Genome-wide transcriptional analysis of grapevine berry ripening reveals a set of genes similarly modulated during three seasons and the occurrence of an oxidative burst at veraison. *BMC Genomics* 8, 428.

Purgatto, E., Oliveira do Nascimento, J.R., Lajolo, F.M. and Cordenunsi, B.R. (2002) The onset of starch degradation during banana ripening is concomitant to changes in the content of free and conjugated forms of indole-3-acetic acid. *Journal of Plant Physiology* 159, 1105–1111.

Rashotte, A.M., Chae, H.S., Maxwell, B.B. and Kieber, J.J. (2005) The interaction of cytokinin with other signals. *Physiologia Plantarum* 123, 184–194.

Ren, J., Sun, L., Wu, J., Zhao, S., Wang, C., Wang, Y., Ji, K. and Leng, P. (2010) Cloning and expression analysis of cDNAs for ABA 8′-hydroxylase during sweet cherry fruit maturation and under stress conditions. *Journal of Plant Physiology* 167, 1486–1493.

Rodrigo, M.-J., Alquezar, B. and Zacarías, L. (2006) Cloning and characterization of two 9-*cis*-epoxycarotenoid dioxygenase genes, differentially regulated during fruit maturation and under stress conditions, from orange (*Citrus sinensis* L. Osbeck). *Journal of Experimental Botany* 57, 633–643.

Rohrmann, J., Tohge, T., Alba, R., Osorio, S., Caldana, C., McQuinn, R., Arvidsson, S., van der Merwe, M.J., Riaño-Pachón, D.M., Mueller-Roeber, B., Fei, Z., Nesi, A.N., Giovannoni, J.J. and Fernie, A.R. (2011) Combined transcription factor profiling, microarray analysis and metabolite profiling reveals the transcriptional control of metabolic shifts occurring during tomato fruit development. *Plant Journal* 68, 999–1013.

Rohwer, C.L. and Erwin, J.E. (2008) Horticultural applications of jasmonates: a review. *Journal of Horticultural Science and Biotechnology* 83, 283–304.

Rolland, F., Moore, B. and Sheen, J. (2002) Sugar sensing and signaling in plants. *Plant Cell Online* 14, S185–S205.

Rook, F., Hadingham, S.A., Li, Y. and Bevan, M.W. (2006) Sugar and ABA response pathways and the control of gene expression. *Plant, Cell and Environment* 29, 426–434.

Shiozaki, S., Ogata, T. and Horiuchi, S. (2000) Endogenous polyamines in the pericarp and seed of the grape berry during development and ripening. *Scientia Horticulturae* 83, 33–41.

Southwick, S.M., Moran, R.E., Yeager, J.T. and Glozer, K. (2000) Use of gibberellin to delay maturity and improve fruit quality of 'French' prune. *Journal of Horticultural Science and Biotechnology* 75, 591–597.

Sun, L., Sun, Y., Zhang, M., Wang, L., Ren, J., Cui, M., Wang, Y., Ji, K., Li, P., Li, Q., Chen, P., Dai, S., Duan, C., Wu, Y. and Leng, P. (2012) Suppression of 9-*cis*-epoxycarotenoid dioxygenase, which encodes a key enzyme in abscisic acid biosynthesis, alters fruit texture in transgenic tomato. *Plant Physiology* 158, 283–298.

Symons, G.M., Davies, C., Shavrukov, Y., Dry, I.B., Reid, J.B. and Thomas, M.R. (2006) Grapes on steroids. Brassinosteroids are involved in grape berry ripening. *Plant Physiology* 140, 150–158.

Teale, W.D., Ditengou, F.A., Dovzhenko, A.D., Li, X., Molendijk, A.M., Ruperti, B., Paponov, I. and Palme, K. (2008) Auxin as a model for the integration of hormonal signal processing and transduction. *Molecular Plant* 1, 229–237.

Tingwa, P.O. and Young, R.E. (1975) The effect of indole-3-acetic acid and other growth regulators on the ripening of avocado fruits. *Plant Physiology* 55, 937–940.

Trainotti, L., Tadiello, A. and Casadoro, G. (2007) The involvement of auxin in the ripening of climacteric fruits comes of age: the hormone plays a role of its own and has an intense interplay with ethylene in ripening peaches. *Journal of Experimental Botany* 58, 3299–3308.

Valero, D., Martínez-Romero, D. and Serrano, M. (2002) The role of polyamines in the improvement of the shelf life of fruit. *Trends in Food Science and Technology* 13, 228–234.

Vardhini, B.V. and Rao, S.S.R. (2002) Acceleration of ripening of tomato pericarp discs by brassinosteroids. *Phytochemistry* 61, 843–847.

Villarreal, N.M., Martínez, G.A. and Civello, P.M. (2009) Influence of plant growth regulators on polygalacturonase expression in strawberry fruit. *Plant Science* 176, 749–757.

Wang, S.Y., Bowman, L. and Ding, M. (2008) Methyl jasmonate enhances antioxidant activity and flavonoid content in blackberries (*Rubus* sp.) and promotes antiproliferation of human cancer cells. *Food Chemistry* 107, 1261–1269.

Wang, Y.-Y., Li, B.-Q., Qin, G.-Z., Li, L. and Tian, S.-P. (2011) Defense response of tomato fruit at different maturity stages to salicylic acid and ethephon. *Scientia Horticulturae* 129, 183–188.

Wechter, W.P., Levi, A., Harris, K., Davis, A., Fei, Z., Katzir, N., Giovannoni, J., Salman-Minkov, A., Hernandez, A., Thimmapuram, J., Tadmor, Y., Portnoy, V. and Trebitsh, T. (2008) Gene expression in developing watermelon fruit. *BMC Genomics* 9, 275.

Weiss, D. and Ori, N. (2007) Mechanisms of cross talk between gibberellin and other hormones. *Plant Physiology* 144, 1240–1246.

Werner, T. and Schmülling, T. (2009) Cytokinin action in plant development. *Current Opinion in Plant Biology* 12, 527–538.

Yuan, R. and Carbaugh, D.H. (2007) Effects of NAA, AVG, and 1-MCP on ethylene biosynthesis, preharvest fruit drop, fruit maturity, and quality of Golden Supreme and 'Golden Delicious' apples. *HortScience* 42, 101–105.

Zhang, C. and Whiting, M.D. (2011) Improving 'Bing' sweet cherry fruit quality with plant growth regulators. *Scientia Horticulturae* 127, 341–346.

Zhang, C., Tanabe, K., Tamura, F., Itai, A. and Yoshida, M. (2007a) Roles of gibberellins in increasing sink demand in Japanese pear fruit during rapid fruit growth. *Plant Growth Regulation* 52, 161–172.

Zhang, C., Tanabe, K., Tani, H., Nakajima, H., Mori, M. and Sakuno, E. (2007b) Biologically active gibberellins and abscisic acid in fruit of two late-maturing Japanese pear cultivars with contrasting fruit size. *Journal of the American Society for Horticultural Science* 132, 452–458.

Zhang, L., Ma, G., Kato, M., Yamawaki, K., Takagi, T., Kiriiwa, Y., Ikoma, Y., Matsumoto, H., Yoshioka, T. and Nesumi, H. (2012) Regulation of carotenoid accumulation and the expression of carotenoid metabolic genes in citrus juice sacs *in vitro*. *Journal of Experimental Botany* 63, 871–886.

Zhang, M., Yuan, B. and Leng, P. (2009a) The role of ABA in triggering ethylene biosynthesis and ripening of tomato fruit. *Journal of Experimental Botany* 60, 1579–1588.

Zhang, S., Wei, Y., Lu, Y. and Wang, X. (2009b) Mechanisms of brassinosteroids interacting with multiple hormones. *Plant Signaling and Behavior* 4, 1117–1120.

Zhang, X.R., Luo, G.G., Wang, R.H., Wang, J. and Himelrick, D.G. (2003) Growth and developmental responses of seeded and seedless grape berries to shoot girdling. *Journal of the American Society for Horticultural Science* 128, 316–323.

Ziosi, V., Bonghi, C., Bregoli, A.M., Trainotti, L., Biondi, S., Sutthiwal, S., Kondo, S., Costa, G. and Torrigiani, P. (2008) Jasmonate-induced transcriptional changes suggest a negative interference with the ripening syndrome in peach fruit. *Journal of Experimental Botany* 59, 563–573.

Ziosi, V., Bregoli, A.M., Fregola, F., Costa, G. and Torrigiani, P. (2009) Jasmonate-induced ripening delay is associated with up-regulation of polyamine levels in peach fruit. *Journal of Plant Physiology* 166, 938–946.

13 Genetic Diversity of Tropical Fruit

Surendra Kumar Malik, Susheel Kumar and Kailash C. Bansal*
National Bureau of Plant Genetic Resources, Pusa Campus, New Delhi, India

13.1 Introduction

Botanically, a true fruit is a mature ovary; however, other flower and inflorescence parts also form a part of the fruit in some taxa. There is a vast diversity of fruits in angiosperms. Fruits are classified based on their morphology and development: simple (fruit from a single ovary); accessory (fruit from inferior ovary); aggregate (fruit from several separate ovaries); and multiple (fruit from several independent flowers). Simple fruits can be dry or fleshy and result from the ripening of a simple or compound ovary with only one pistil. Dry fruits may be dehiscent (opening to discharge seeds) or indehiscent (not opening to discharge seeds). Ecological parameters and the habitat of a plant play an important part in the fate of a fruit, facilitating the reproductive mechanism, dissemination of seeds and eventually propagation of the species.

Ripening is the final stage of fruit development and makes the fruit commercially important and brings about changes in cell-wall structure, ultrastructure and texture, the conversion of starch to sugars, alterations in pigment biosynthesis and accumulation, and heightened levels of flavour and aromatic volatiles (Giovannoni, 2001; White, 2002).

13.2 Diversity in Climacteric and Non-climacteric Fruit Ripening

Fruit are classified as climacteric or non-climacteric on the basis of respiration and ethylene evolution patterns (Table 13.1) as described by Hiwasa-Tanase and Ezura (Chapter 1, this volume). The molecular distinctions underlying climacteric versus non-climacteric ripening are poorly understood. Nevertheless, it seems likely that, at least in instances of the same or closely related species with examples of both climacteric and non-climacteric types, non-climacteric phenotypes may represent mutations in ethylene synthesis or signalling as opposed to more complex distinctions. Crossing of a non-climacteric melon with a climacteric one indicated that the climacteric character is genetically dominant and conferred by two duplicated loci. However, other experiments made by crossing two non-climacteric melons have generated climacteric fruit, indicating that different and complex genetic regulation exists for the climacteric character.

13.3 Fruit Diversity in Tropical Fruits

There is a vast diversity of tropical fruits in Asia, America and Africa. Tropical fruit

* kailashbansal@hotmail.com

Table 13.1. Fruit diversity in terms of climacteric (CL) and non-climacteric (NC) ripening. From http://www.quisqualis.com/climacteric.html.

Species	Common name	Fruit type[a]
Actinidia delicious	Kiwi	CL
Anacardium occidentale	Cashew	NC
Ananas comosus	Pineapple	NC
Annona squamosa	Sugarapple	CL
Artocarpus altilis	Breadfruit	CL
Artocarpus heterophyllus	Jack Fruit	CL
Averrhoa carambola	Carambola	NC
Carica papaya	Papaya	CL
Citrus aurantiifolia	Lime	NC
Citrus grandis	Pummelo	NC
Citrus paradisi	Grapefruit	NC
Citrus reticulata	Mandarin	NC
Citrus sinensis	Orange	NC
Dimocarpus longan	Longan	NC
Diospyros digyna	Black Sapote	CL
Diospyros kaki	Oriental persimmon	CL
Durio zibethinus	Durian	CL
Eriobotrya japonica	Loquat	CL
Eugenia brasiliensis	Grumichama	NC
Eugenia uniflora	Surinam cherry	NC
Feijoa sellowiana	Feijoa	CL
Ficus carica	Fig	CL
Fragaria sp.	Strawberry	NC
Garcinia mangostana	Mangosteen	CL
Litchi chinensis	Lychee	NC
Lycopersicon esculentum	Tomato	CL
Malus domestica	Apple	CL
Mangifera indica	Mango	CL
Manilkara zapota	Sapodilla	CL
Monstera deliciosa	Monstera	CL
Musa sp.	Banana	CL
Nephelium lappaceum	Rambutan	NC
Passiflora edulis	Purple passion fruit	CL
Passiflora edulis f. flavicarpa	Yellow passion fruit	CL
Passiflora quadrangularis	Giant granadilla	CL
Persea sp.	Avocado	CL
Poraqueiba sericea	Umari roxo	NC
Pouteria caimito	Caimito	CL
Pouteria campechiana	Canistel	CL
Pouteria sapota	Mamey sapote	CL
Prunus sp.	Peach	CL
Prunus sp.	Apricot	CL
Prunus sp.	Plum	CL
Prunus sp.	Cherry	NC
Psidium guajava	Guava	CL
Punica granatum	Pomegranate	NC
Pyrus i	Pear	CL
Syzygium malaccense	Malay apple	NC
Syzygium samarangense	Wax JAMBU	NC
Tamarindus indica	Tamarind	NC
Vitis sp.	Grape	NC
Ziziphus mauritiana	Indian Jujube	CL

[a]CL, climacteric; NC, non-climacteric.

tree genetic resources in the Asian region alone include more than 500 species of edible tropical fruits (Ramanatha Rao and Mal, 2002). Fruits under cultivation include banana, citrus, mango, pineapple, papaya, durian, rambutan, jackfruit, litchi, longan, tamarind, chempedak, carambola, langsat, guava, sour soup, custard apple and salak (Verheij and Coronel, 1991; Singh, 1993; Arora and Ramanatha Rao, 1995), with the predominant fruits being banana, pineapple, citrus, mango and papaya. Other than these cultivated species, many are considered rare or underutilized. These rare fruits are generally neglected because they have not been exploited commercially and therefore lack improved varieties. Fruit species, both cultivated and wild, contribute substantially towards sustainability of the ecosystems (Ramanatha Rao and Mal, 2002). They are also a source of useful genes for the genetic improvement of related crop species. Some of the genera with rare species of fruits are *Garcinia*, *Lansium*, *Baccaurea*, *Artocarpus* and *Nephelium*. Very little is known about the ripening behaviour and regulation thereof of several underutilized tropical fruits.

13.3.1 Banana (*Musa* spp., family Musaceae)

The banana is one of the most ancient food plants, having been used, and perhaps cultivated, at the dawn of recorded history. It is a very common plant in the world and is more highly commercialized than any other fruit. All of the edible bananas and plantains are indigenous to the warm, moist regions of tropical Asia: India (Assam), Burma, Thailand and Indo-China. The Island of Honduras and Jamaica are among the chief banana-exporting countries today. The banana crop is cultivated extensively in India, Jamaica, Costa Rica, Cuba, Honduras, the northern shores of Columbia, Central America, the Canary Islands and the West Indies.

Banana (*Musa* spp.), including plantain, is the third most important fruit crop of India in terms of the area under cultivation (0.77 million ha). However, with respect to production, it tops the list with an annual production of 26.47 million t (Anon., 2010). The term banana is given to varieties that are used for fruit and eaten raw, whereas plantain refers to varieties that are edible only when cooked. Banana fruit is rich in phosphorus, calcium and potassium. Banana fruit is a seedless, parthenocarpic berry that develops without pollination and fertilization. The edible pulp develops mainly from the ovary wall under an autonomous stimulus of complex nature in which growth substances are involved. Sterileness, i.e. seedlessness, is partially independent of parthenocarpy; many edible bananas are somewhat fertile if pollinated. Wild bananas have seedy fruits that develop only if pollinated effectively. Banana fruit in general are negatively geotropic and the shape of the mature fruit reflects the posture of the bunch and the position of the fruit upon it.

13.3.2 Mango (Mangifera indica L., family Anacardiaceae)

Mango is the most important fruit of India. It occupies the largest area (2.3 million ha) among fruit crops, and production-wise mango ranks second (15.0 million t) with nearly 21% contribution to the total fruit production of India. The nearly 1000 varieties of mango cultivated in India provide an unusual diversity of flavours and tastes. It is titled the 'King of fruits' because of its richness in variety, delicious taste, excellent flavour, attractive appearance and popularity among the masses. India is the largest producer of mangoes in the world and accounts for more than 55% of world production.

Mangoes are large drupes. The large, flattened, kidney-shaped central stone contains one or more large starchy embryos and can constitute up to 20% of the fruit weight. The skin has a yellow or green background colour, with a red/orange blush in many cultivars, and is thicker than usual for drupaceous fruit. The mango skin contains irritating oils, particularly in the unripe fruit. The flesh is yellow/orange in colour, sometimes astringent (turpentine-like) and can have fibres extending from the endocarp (stone). Several mango varieties have been evaluated for fruit morphology, colour, early and late maturity, shelf-life and several other economic characters (Table 13.2). Studies of the surface morphology and anatomy of mango fruit indicated that the fruit surface for varieties like Langra and Ram Kela is more resistant to the diffusion of gases and moisture, mainly due to a low density of lenticels, while varieties like Amrapali and Dusheheri show a higher density of lenticels on the surface of fruits and thereby a lesser degree of diffusion resistance (Paul *et al.*, 2007). The predominant role of lenticels in controlling the respiration and transpiration by determining the exchange of gases and loss of moisture, respectively, during the development of fruit as well as ripening is well documented (Khader *et al.*, 1992).

13.3.3 *Citrus* spp. (family Rutaceae)

Two systems of classification exist for the genus *Citrus*. One is that given by Walter T. Swingle, a USDA scientist who did much of his work in Florida, and the second by Tyosaburo Tanaka of Japan. The centre of diversity for citrus fruits ranges from north-eastern India to eastwards through the Malay Archipelago and south to Australia. Sweet oranges probably arose in India, the trifoliate orange and mandarin in China, and acid citrus types in Malaysia. Oranges and pummelos were mentioned in Chinese literature in 2400 BC, and later in Sanskrit writings (800 BC) lemons were mentioned. Citrus is the third most important fruit crop of India, with an estimated production of 9.63 million t from an area of 0.98 million ha. Mandarin (*Citrus reticulata* Blanco), sweet orange (*Citrus sinensis* Osbeck), acid lime (*Citrus aurantiifolia* (Christm.) Swingle) and lemon (*Citrus limon* Burn. f.) are the major cultivated species of the country. Other species that are cultivated to a lesser extent include seedless lime (*Citrus latifolia* Tan.), pummelo (*Citrus grandis* Osbeck), grapefruit (*Citrus paradisi* Macf.) and belladikithuli (*Citrus maderaspatana* Tan.). Citrus fruits have been put to numerous uses although primarily they are eaten fresh or prepared into juice concentrate. The pulp and seed are used for cattle feed and molasses and also for flavouring and for the production of pharmaceuticals, soaps and perfumes. Fermented orange juice produces vinegar and alcohol. Some species are known to have properties of curing fever and colic (Solley, 1997). A number of species are grown for ornamental fruits and flowers.

Table 13.2. Sources of desirable characteristics in mango. Modified from Karihaloo *et al.* (2005).

Characters	Cultivars
Dwarfness	'Creeping', 'Latra', 'Kerala Dwarf', 'Rumani', 'Neelum'
Precocious, dwarf	'MDCH-1' and 'Manipur Dwarf'
Big fruit size	'Baganpalli', 'Kensington', 'Fazli'
Red peel colour	'Janardan Pasand', 'Suvarnarekha', 'Vanraj', 'Tommy Atkins', 'Sensation'
High pulp content	'Langra', 'Dashehari', 'Pairi', 'Mallika', 'Fazli'
High total soluble solids content	'Sahib Pasand', 'Dashehari', 'Langra', 'Mallika', 'Lazzat Baksh', 'Amarpalli', 'Sakkar China'
Long shelf life	'Alphonso', 'Totapuri', 'Sunderja', 'Bangapalli', 'Kesar', 'Rataul'
Regular bearing	'Neelum', 'Kalapadi', 'Totapuri'
Early maturity	'Gaurjeet', 'Zardalu', 'Lazzat Baksh', 'Guruvarm', 'Chandrakaran', 'Panakalu'
Late maturity	'Kaitki', 'Bathui', 'Fazli', 'Neelum', 'Chausa', 'Sora'
Processing (pulp/juice/nectar)	'Alphonso', 'Totapari', 'Rani Pasand', 'Sahib Pasand', 'Padri', 'Banganpalli'

Citrus fruits are designated as non-climacteric, and the fruit type is termed a hesperidium, basically a leathery rinded berry. The endocarp is the edible portion, divided into 10–14 segments separated by thin septa, each containing up to eight seeds but usually only one. Placentation is axile. The central axis may be open as in tangerines. Each segment is composed of juice vesicles ('pulp'), with long stalks attached to the outer wall, containing juice. The mesocarp is the white tissue usually adherent to the outer surface of the endocarp, except for tangerines; it is also called the albedo. The exocarp, or flavedo, is the thin, pigmented outer portion of the rind, with numerous oil glands.

In citrus fruits, vast diversity in each group, namely mandarin, sweet orange, lime and lemon, and pumellos, exists, and in each group various genotypes with early to late ripening type, diverse shelf-life, maturity and diverse fruit characters are available.

13.4 Genetic Studies on Fruit Ripening

Fruit ripening is a complex process of molecular, physiological and biochemical mechanisms leading to changes in colour, sugar, acidity, aroma volatiles, phytochemicals and texture (Handa *et al.*, 2011). Thus, fruit ripening has been considered a step of a programmed cell death (Mattoo and Handa, 2004; Bouzayen *et al.*, 2010). These changes during ripening are driven by the coordinated expression of several ripening-related genes. Ripening in climacteric fruits is induced by the action of ethylene and results in the activation of several cell-wall hydrolases. The action of these hydrolases on cell walls results in wall disassembly leading to softening of fruit. Several studies have been carried out to investigate fruit-ripening mechanisms in tomato and have inferred that ethylene plays a crucial role in ripening in climacteric fruits (Yang, 1985; Tucker and Brady, 1987). The main role of ethylene in climacteric fruit ripening was first observed in tomato at the molecular level by analysis of ethylene-inducible and ripening-related gene expression (Lincoln *et al.*, 1987; Maunder *et al.*, 1987). Several genetic studies have been carried out to study the molecular mechanisms of fruit ripening in, for example, tomato, melon, apple, pear, grape, strawberry, mango and papaya in different laboratories (Giovannoni, 2007; Seymour *et al.*, 2007; Bouzayen *et al.*, 2010; Li *et al.*, 2010). Among these crops, tomato and melon have been commonly used as model systems for studying biosynthetic pathways and molecular mechanisms underlying fruit ripening.

Numerous fruit development and ripening-related genes have been isolated and characterized using differential gene expression patterns and biochemical functions (Gray *et al.*, 1992; Bouzayen *et al.*, 2010; Li *et al.*, 2010; Handa *et al.*, 2011), including the regulatory functions of genes involved in ethylene biosynthesis in climacteric (tomato, melon, apple) and non-climacteric (strawberry, melon) fruits (Giovannoni, 2007; Seymour *et al.*, 2007; Bouzayen *et al.*, 2010). Two genes, *ripening inhibitor* (*rin*) and *Colourless non-ripening* (*Cnr*) have been identified in tomato (*Lycopersicon esculentum*) mutants, encoding transcription factors, which play important roles in the regulation of the ethylene biosynthesis pathway (Vrebalov *et al.*, 2002; Manning *et al.*, 2006; Giovannoni, 2007). Similarly, the *Gr* mutant (*Green-Ripe*) contains a mutation in a gene encoding a conserved protein that disrupts ethylene signalling in tomato (Barry and Giovannoni, 2006). Strawberry (a non-climacteric fruit) also has a fruit-specific *LeMADS-RIN* orthologue, suggesting the presence of common ethylene-independent regulatory pathway involving MADS-box genes in both climacteric and non-climacteric types of fruit ripening (Vrebalov *et al.*, 2002; Giovannoni, 2007).

13.4.1 Molecular mechanism of fruit softening

Genetic and environmental factors simultaneously affect texture changes in ripening fruit. During fruit development and ripening, fruit softening, primarily due to cell-wall degradation and reduction of intercellular adhesion, is a prerequisite for a desirable eating quality of a fruit. Ethylene plays a main role in promoting fruit ripening at the level of flesh softening and lignification (Yin *et al.*, 2008; Johnston *et al.*, 2009; Wang *et al.*, 2010). In tomato, ethylene is required for normal fruit softening. Its fruit firmness can be maintained through transgenic plants carrying an antisense RNA to the 1-aminocyclopropane-1-carboxylic acid synthase gene and subsequent reversal of such an inhibitory effect following exogenous ethylene treatment (Oeller *et al.*, 1991). Although ripening of non-climacteric fruit may be ethylene independent, exogenous ethylene treatment increases fruit firmness in non-climacteric fruits like loquat; an inhibitor of ethylene responses, 1-methylcyclopropene, significantly delays postharvest fruit firmness, thus suggesting that ethylene plays an important role in regulating flesh lignification (Cai *et al.*, 2006; Shan *et al.*, 2008). Ethylene response factors are a family of transcription factors (potential regulators of ethylene response), and further support the important role of ethylene signalling in regulating fruit softening. In addition to studies in tomato, ethylene signalling components have also been isolated and characterized from several other fruits (Yin *et al.*, 2008; Wang *et al.*, 2010). Besides ethylene, expansins are implicated in fruit ripening/softening, which were first investigated using the *LeEXP1* gene in tomato (Rose *et al.*, 1997). *MiExpA1*, an α-expansin gene, has been isolated and characterized from a variety of mango (*Mangifera indica* cv. 'Dashehari') and is correlated with softening in mango (Sane *et al.*, 2005).

13.4.2 Epigenetic mechanisms

Epigenetic factors are important genetic determinants for plant improvement, affecting the regulation of gene expression in several biosynthesis pathways. These epigenetic variations do not affect the primary DNA sequence but consist of DNA methylation or histone modifications that affect gene expression generally at the level of chromatin organization. In fruit, the *Cnr* mutation is the only well-characterized, natural and stably inherited epigenetic mutation (Seymour *et al.*, 2007). In this mutation, a region of the *LeSPLCNR* promoter is highly methylated and the gene expression responsible for ethylene production is suppressed (Manning *et al.*, 2006). A study of epigenetic variation in *Arabidopsis* using tiling microarrays showed that at least one-third of expressed genes were methylated in parts of their coding regions, while about 5% of genes were methylated within promoter regions (Zhang *et al.*, 2006; Vaughn *et al.*, 2007). However, the promoter-methylated genes had a higher degree of tissue-specific expression (Zhang *et al.*, 2006; Zilberman *et al.*, 2007), suggesting these as preferential sites for selection of subtle *cis*-regulation during fruit development and ripening. Moreover, methylation differences among ecotypes have also been reported, which are common, heritable and stable (Vaughn *et al.*, 2007).

13.5 Quantitative Trait Loci (QTLs) Mapping of Fruit-ripening Traits

The advent of advanced molecular markers like QTLs opened up new prospects for genetic improvement of agronomic traits. Indeed, most fruit quality traits, including fruit development and ripening, are controlled by multigenic families. Thus, the QTL approach is more useful for localization of loci on genetic maps responsible for, at least, part of the phenotypic variation, and enables quantification of their individual effects. The molecular markers found in the proximity of these QTLs are now being used in marker-assisted selection to create parent lines with increased potential, or to avoid certain unfavourable traits (Fulton *et al.*, 2002). A QTL for fruit weight has been precisely localized and then cloned by chromosome walking (Frary *et al.*, 2000). Another QTL controlling sugar concentration has also been cloned and characterized (Fridman *et al.*, 2000). It has been possible to study the inheritance of the climacteric character due to the presence of both genetically compatible climacteric and non-climacteric types in melon. A segregating population resulting from a cross between a typical climacteric-type Charentais melon (*Cucumis melo* var. *cantalupensis* cv. 'Védrantais') and a non-climacteric melon, Songwhan Charmi PI 161375 (*Cucumis melo* var. *chinensis*) has been generated and used to study the segregation of abscission layer (Al) formation in the peduncle and ethylene production. It was found that the climacteric character was controlled by two duplicated independent loci (*Al-3* and *Al-4*), and the intensity of ethylene production was controlled by at least four QTLs localized in other genomic regions. None of the QTLs matched known genes of the ethylene biosynthesis or transduction pathways.

To identify major genetic components impacting the organoleptic quality of fruits, a QTL approach has also been applied to mapping a population of 144 recombinant inbred lines derived from an intraspecific cross between a cherry tomato ('Cervil') of high organoleptic quality and an inbred line ('Levovil') with larger fruits (Causse *et al.*, 2002). These authors studied 38 traits using a variety of physical and biochemical assays, plus a panel of trained tasters. The presence of QTLs for organoleptic qualities was restricted to 14% of the genome, lying on chromosomes 1, 2, 3, 4, 8, 9, 11 and 12. The latter observation confirmed and extended previous studies using other mapping populations. As small regions of the genome influenced several traits, Causse *et al.* (2002) used both QTL analysis and local genetic mapping techniques to determine the nature of specific genes conferring pleiotropic phenotypes. Such studies show promise for developing molecular markers for breeding programmes and also help in the identification of candidate genes to improve the organoleptic quality of fruits (White, 2002). Genetic analysis of traits linked with fruit texture have identified QTLs associated with fruit firmness in apple (King *et al.*, 2000, 2001), tomato (Lecomte *et al.*, 2004; Causse *et al.*, 2007), melon (Moreno *et al.*, 2007) and peach (Ogundiwin *et al.*, 2009).

13.6 Conclusions

The genetics of fruit ripening is a complex phenomenon controlled by several genetic and epigenetic factors. Ripening phenomena vary among species, and specific biochemical changes result in modification of surface colour through the alteration of chlorophyll, carotenoid and/or flavonoid accumulation, texture by alteration of cell turgor and cell-wall structure and/or metabolism, and modification of sugars, acids and volatile profiles that affect nutritional quality, flavour and aroma. Although fruit species are classically defined physiologically on the basis of the presence (climacteric) or absence (non-climacteric) of increased respiration and synthesis of the gaseous hormone ethylene at the onset of ripening, fruit displaying both ripening programmes, such as watermelon, typically follow the same general developmental changes. Tomato (*L. esculentum* Mill.) has emerged as the primary model system for climacteric fruit ripening due to its simple diploid genetics, small genome size, short generation time, routine transformation technology, availability of genetic and genomic resources including mapping populations, mapped DNA markers, extensive expressed sequence tag collections and publicly available microarrays (Tanksley *et al.*, 1992; Van der Hoeven *et al.*, 2002; Fatima *et al.*, 2008). In addition, numerous single-gene mutations that regulate fruit size, shape, development and ripening combined with dramatic and readily quantifiable ripening phenotypes (ethylene, colour index, carotenoids, softening) have enhanced the use of tomato as a model system for climacteric ripening (Giovannoni, 2007). Melon (*Cucumis melo* L.) is another important model species used to study climacteric ripening (Ayub *et al.*, 1996). Additionally, within this species, both climacteric and non-climacteric varieties are reported. For example, Cantalupensis melon types are climacteric, whereas Inodorus melon types are non-climacteric. The coexistence of both types of ripening variety make melon also a suitable model system to study the genetic control of fruit ripening (Ayub *et al.*, 1996; Nishiyama *et al.*, 2007). Both climacteric and non-climacteric regulation coexist during climacteric fruit ripening, as presented in Fig. 13.1.

The QTL approach will be of immense use for localization of loci on genetic maps responsible for at least part of the phenotypic variation, which enables the quantification of their individual effects. The molecular mechanism of fruit ripening is a complex phenomenon and is controlled by multiple genes and epigenetic factors leading to limited studies in fruits of economic importance. Tropical fruit species with a fleshy mesocarp and of a highly perishable nature need urgent attention for such studies to improve breeding strategies for enhanced shelf-life of their fruits.

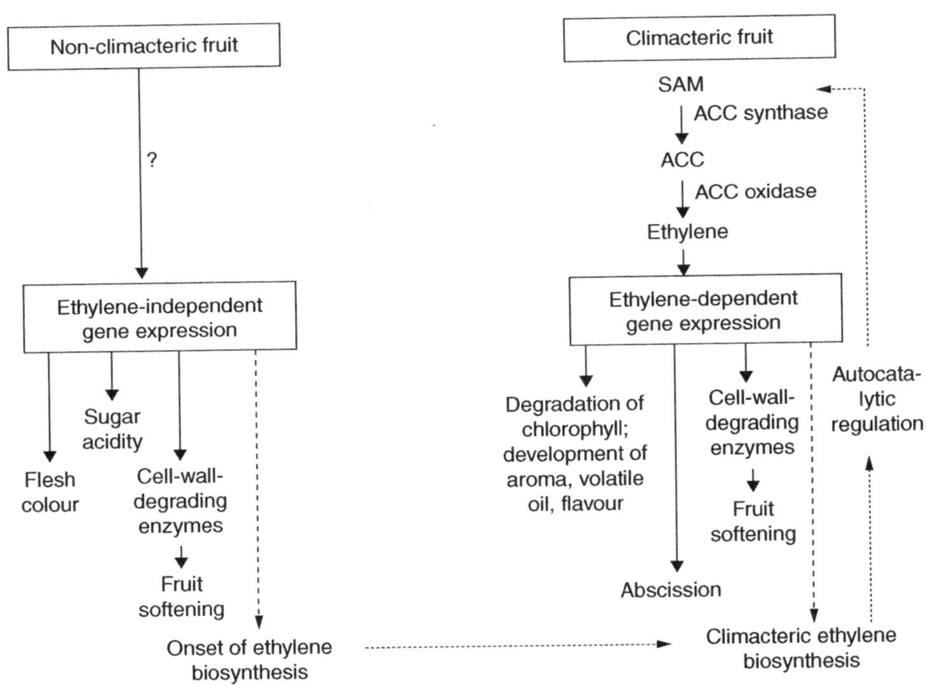

Fig. 13.1. General scheme showing the presence of ethylene-dependent and -independent processes in ripening melon fruit. Modified from Pech *et al.* (2008).

References

Anon. (2010) Indian Horticulture Database, National Horticulture Board, Ministry of Agriculture, Gurgaon, Haryana.

Arora, R.K. and Ramanatha Rao, V. (eds) (1995) *Proceedings of Expert Consultation on Tropical Fruit Species of Asia*, 17–19 May 1994, Malaysian Agricultural Research and Development Institute, Serdang, Kuala Lumpur, Malaysia. IPGRI Office for South Asia, New Delhi, India.

Ayub, R., Guis, M., Ben, A.M., Gillot, L., Roustan, J.P., Latché, A., Bouzayen, M. and Pech, J.C. (1996) Expression of ACC oxidase antisense gene inhibits ripening of cantaloupe melon fruits. *Nature Biotechnology* 14, 862–866.

Barry, C.S. and Giovannoni, J.J. (2006) Ripening in the tomato *Green-ripe* mutant is inhibited by ectopic expression of a protein that disrupts ethylene signaling. *Proceedings of the National Academy of Sciences USA* 103, 7923–7928.

Bouzayen, M., Latché, A., Nath, P. and Pech, J.C. (2010) Mechanism of fruit ripening. In: Pua, E.C. and Davey, M.R. (eds) *Plant Developmental Biology – Biotechnological Perspectives*, Vol. I. Springer, Berlin/Heidelberg, pp. 319–334.

Cai, C., Chen, K.S., Xu, W.P., Zhang, W.S., Li, X. and Ferguson, I.B. (2006) Effect of 1-MCP on postharvest quality of loquat fruit. *Postharvest Biology and Technology* 40, 155–162.

Causse, M., Sabina-Colombani, V., Lecomte, L., Duffé, P., Rouselle, P. and Buret, M. (2002) QTL analysis of fruit quality in fresh market tomato: a few chromosome regions control the variation of sensory and instrumental traits. *Journal of Experimental Botany* 53, 2089–2098.

Causse, M., Chaïb, J., Lecomte, L., Buret, M. and Hospital, F. (2007) Both additivity and epistasis control the genetic variation for fruit quality traits in tomato. *Theoretical and Applied Genetics* 115, 429–442.

Fatima, T., Rivera-Dominguez, M., Troncoso-Rojas, R., Tiznado-Hernandez, M.-E., Handa, A.K., and Mattoo, A.K. (2008) Tomato. In: Kole, C. and Hall, T.C. (eds) *Compendium of Transgenic Crop Plants: Transgenic Vegetable Crops*, Vol. 6. Blackwell Publishing, Oxford, pp. 1–46.

Frary, A., Nesbitt, T.C., Frary, A., Grandillo, E., Vander Knaap, E., Cong, B., Liu, J., Meller, J., Elber, R. and Alpert, K.B. (2000) *fw2.2*: a quantitative trait locus key to the evolution of tomato fruit size. *Science* 289, 85–88.

Fridman, E., Pleban, T. and Zamir, D. (2000) A recombination hotspot delimits a wild species QTL for tomato sugar content to 484-bp within an invertase gene. *Proceedings of the National Academy of Sciences USA* 97, 4718–4723.

Fulton, T.M., Bucheli, P., Voirol, E., López, J., Pétiard, V. and Tanksley, S.D. (2002) Quantitative trait loci (QTL) affecting sugars, organic acids and other biochemical properties possibly contributing to flavor, identified in four advanced backcross populations of tomato. *Euphytica* 127, 163–177.

Giovannoni, J.J. (2001) Molecular biology of fruit maturation and ripening. *Annual Review of Plant Physiology and Plant Molecular Biology* 52, 725–749.

Giovannoni, J.J. (2007) Fruit ripening mutants yield insights into ripening control. *Current Opinion in Plant Biology* 10, 283–289.

Gray, J.E., Picton, S., Shabbeer, J., Schuch, W. and Grierson, D. (1992) Molecular biology of fruit ripening and its manipulation with antisense genes. *Plant Molecular Biology* 19, 69–87.

Handa, A.K., Tiznado-Hernandez, M.-E. and Mattoo, A.K. (2011) Fruit development and ripening: a molecular perspective. In: Altman, A. and Hasegawa, P.M. (eds) *Plant Biotechnology and Agriculture. Prospects for the 21st Century*. Elsevier, Waltham, MA and London, pp. 405–424.

Johnston, J.W., Gunaseelan, K., Pidakala, P., Wang, M. and Schaffer, R.J. (2009) Co-ordination of early and late ripening events in apples is regulated through differential sensitivities to ethylene. *Journal of Experimental Botany* 60, 2689–2699.

Karihaloo, J.L., Malik, S.K., Rajan, S., Pathak, R.K. and Gangopadhyay, K.K. (2005) Tropical fruits. In: Dhillon, B.S., Tyagi, R.K., Saxena, S. and Randhawa, G.J. (eds) *Plant Genetic Resources: Indian Perspective*. NBPGR, New Delhi, India, pp. 121–145.

Khader, S.E., Tandon, S.A., Kalra, D.K. and Kalra, S.K. (1992) Epicuticular wax and lenticels changes during development of mango fruit cv. Dushehari. *Indian Journal of Horticulture* 49, 165–168.

King, G.J., Maliepaard, C., Lynn, J.R., Alston, F.H., Durel, C.E., Evans, K.M., Griffon, B., Laurens, F., Manganaris, A.G., Schrevens, E., Tartarini, S. and Verhaegh, J. (2000) Quantitative genetic analysis and comparison of physical and sensory descriptors relating to fruit flesh firmness in apple (*Malus pumila* Mill). *Theoretical and Applied Genetics* 100, 1074–1084.

King, G.J., Lynn, J.R., Dover, C.J., Evans, K.M. and Seymour, G.B. (2001) Resolution of quantitative trait loci for mechanical measures accounting for genetic variation in fruit texture of apple (*Malus pumila* Mill.). *Theoretical and Applied Genetics* 102, 1227–1235.

Lecomte, L., Duffe, P., Buret, M., Servin, B., Hospital, F. and Causse, M. (2004) Marker-assisted introgression of five QTLs controlling fruit quality traits into three tomato lines revealed interactions between QTLs and genetic backgrounds. *Theoretical and Applied Genetics* 109, 658–668.

Li, X., Xu, C., Korban, S.S. and Chen, K. (2010) Regulatory mechanisms of textural changes in ripening fruits. *Critical Reviews in Plant Sciences* 29, 222–243.

Lincoln, J.E., Cordes, S., Read, E. and Fischer, R.L. (1987) Regulation of gene expression by ethylene during *Lycopersicon esculentum* (tomato) fruit development. *Proceedings of the National Academy of Sciences USA* 84, 2793–2797.

Manning, K., Tor, M., Poole, M., Hong, Y., Thompson, A.J., King, G.J., Giovannoni, J.J. and Seymour, G.B. (2006) A naturally occurring epigenetic mutation in a gene encoding an SBP-box transcription factor inhibits tomato fruit ripening. *Nature Genetics* 38, 948–952.

Mattoo, A.K., and Handa, A.K. (2004) Ethylene signaling in plant cell death. In: Nooden, L. (ed.) *Plant Cell Death Processes*, Academic Press, New York, pp. 125–142.

Maunder, M., Holdsworth, M., Slater, A., Knapp, J., Bird, C., Schuch, W. and Grierson, D. (1987) Ethylene stimulates the accumulation of ripening-related mRNAs in tomatoes. *Plant, Cell and Environment* 10, 177–184.

Moreno, E., Obando, J.M., Dos-Santos, N., Fernandez-Trujillo, J.P., Monforte, A.J. and Garcia-Mas, J. (2007) Candidate genes and QTLs for fruit ripening and softening in melon. *Theoretical and Applied Genetics* 116, 589–602.

Nishiyama, K., Guis, M., Rose, J.K.C., Kubo, Y., Bennett, K.A., Wangjin, L., Kato, K., Ushijima, K., Nakano, R., Inaba, A., Bouzayen, M., Latché, A., Pech, J.C. and Bennett, A. (2007) Ethylene regulation of fruit softening and cell wall disassembly in Charentais melon. *Journal of Experimental Botany* 58, 1281–1290.

Oeller, D.C., Min-Wong, L., Taylor, L.P., Pike, D.A. and Theologis, A. (1991) Reversible inhibition of tomato fruit senescence by antisense RNA. *Science* 254, 437–439.

Ogundiwin, E.A., Peace, C.P., Gradziel, T.M., Parfitt, D.E., Bliss, F.A. and Crisosto, C.H. (2009) A fruit quality gene map of *Prunus*. *BMC Genomics* 10, 587.

Paul, V., Malik, S.K. and Srivastava, G.C. (2007) Intervarietal differences in the surface morphology and anatomy of mango (*Mangifera indica* L.) fruit. *Phytomorphology* 57, 211–220.

Ramanatha Rao, V. and Mal, B. (2002) Tropical fruit species in Asia: diversity and conservation strategies. In: Drew, R. (ed.) *Proceedings of the International Symposium on Tropical and Subtropical Fruits*. International Society for Horticulture Science, p. 405.

Rose, J.K.C., Lee, H.H. and Bennett, A.B. (1997) Expression of a divergent expansin gene is fruit-specific and ripening-regulated. *Proceedings of National Academy of Sciences USA* 94, 5955–5960.

Sane, V.A., Chourasia, A. and Nath, P. (2005) Softening in mango (*Mangifera indica* cv. Dashehari) is correlated with the expression of an early ethylene responsive, ripening related expansin gene, MiExpA1. *Postharvest Biology and Technology* 38, 223–230.

Seymour, G., Poole, M., Manning, K. and King, G. (2007) Genetics and epigenetics of fruit development and ripening. *Current Opinion in Plant Biology* 10, 1–6.

Shan, L.L., Li, X., Wang, P., Cai, C., Zhang, B., Sun, C.D., Zhang, W.S., Xu, C.J., Ferguson, I.B. and Chen, K.S. (2008) Characterization of cDNAs associated with lignification and their expression profiles in loquat fruit with different lignin accumulation. *Planta* 227, 1243–1254.

Singh, R.B. (1993) Fruit production in Asia-Pacific region. In: Singh, R.B. (ed.) *Research and Development of Fruits in Asia-Pacific Region*. RAPA/FAO, Bangkok, p. 207.

Solley, P.G. (1997) Citrus – lemons, limes and oranges, tangerines, grapefruits and pummelos <http://www.sonprong.com/citrus.html>.

Tanksley, S., Ganal, M.W., Prince, J.P., *et al.* (1992) High density molecular maps of the tomato and potato genomes. *Genetics* 132, 1141–1160.

Tucker, G.A. and Brady, C.J. (1987) Silver ions interrupt tomato fruit ripening. *Journal of Plant Physiology* 127, 165–169.

Van der Hoeven, R., Ronning, C., Martin, G., Giovannoni, J. and Tanksley, S. (2002) Deductions about the number, organization, and evolution of genes in the tomato genome based on analysis of a large expressed sequence tag collection and selective genomic sequencing. *Plant Cell* 14, 1441–1456.

Vaughn, M.W., Tanurdzic, M., Lippman, Z., Jiang, H., Carrasquillo, R., Rabinowicz, P.D., Dedhia, N., McCombie, W.R., Agier, N., Bulski, A., Colot, V., Doerge, R.W. and Martienssen, R.A. (2007) Epigenetic natural variation in *Arabidopsis thaliana*. *PLoS Biology* 5, 1617–1729.

Verheij, E.W.M. and Coronel, R.E. (1991) *Plant Resources of South-East Asia No. 2. Edible Fruits and Nuts.* Pudoc, Wageningen, the Netherlands.

Vrebalov, J., Ruezinsky, D., Padmanabahn, V., White, R., Medrano, D., Drake, R., Schuch, W. and Giovannoni, J. (2002) A MADS-box gene necessary for fruit ripening at the tomato *ripening inhibitor* (*rin*) locus. *Science* 296, 343–346.

Wang, P., Zhang, B., Li, X., Xu, C.J., Yin, X.R., Shan, L.L., Ferguson, I.B. and Chen, K.S. (2010) Ethylene signal transduction elements involved in chilling injury in non-climacteric loquat fruit. *Journal of Experimental Botany* 61, 179–190.

White, P.J. (2002) Recent advances in fruit development and ripening: an overview. *Journal of Experimental Botany* 53, 1995–2000.

Yang, S.F. (1985) Biosynthesis and action of ethylene. *HortScience* 20, 41–45.

Yin, X.R., Chen, K.S., Allan, A.C., Wu, R.M., Zhang, B., Lallu, N. and Ferguson, I.B. (2008) Ethylene-induced modulation of genes associated with the ethylene signaling pathway in ripening kiwifruit. *Journal of Experimental Botany* 59, 2097–2108.

Zhang, X., Yazaki, J., Sundaresan, A., Cokus, S., Chan, S., Chen, H., Henderson, I., Shinn, P., Pellegrini, M., Jacobsen, S. and Ecker, J.R. (2006) Genome-wide high-resolution mapping and functional analysis of DNA methylation in *Arabidopsis*. *Cell* 126, 1189–1201.

Zilberman, D., Gehring, M., Tran, R.K., Ballinger, T. and Henikoff, S. (2007) Genome-wide analysis of *Arabidopsis thaliana* DNA methylation uncovers an interdependence between methylation and transcription. *Nature Genetics* 39, 61–69.

14 Natural Diversity and Genetic Control of Fruit Sensory Quality

Bénédicte Quilot-Turion and Mathilde Causse*
INRA, Unité de Génétique et Amélioration des Fruits et Légumes, Domaine Saint-Maurice, Montfavet Cedex, France

14.1 Introduction

Fruit sensory quality has only recently become a target for breeders. Due to consumer dissatisfaction relating especially to fruit flavour, genetic improvement of this quality is now required (Ulrich and Olbricht, 2011). Fruit sensory quality is a complex trait that contributes a combination of flavour and texture components, together with general fruit appearance attributes. Most sensory traits are difficult to measure by methods other than sensory analysis. However, some of the major components of flavour and texture such as sweetness, sourness or fruit firmness can be assessed by physical or chemical measurements (Baldwin *et al.*, 1998). The complexity of fruit quality (due to the number of parameters to take into account, their polygenic inheritance and their multiple interactions) and generation length for fruit trees has limited genetic progress. Today, molecular markers enable dissection of the genetic basis of complex traits, and our increasing knowledge about the genomes offer new and efficient tools to breeders.

This chapter first briefly describes the quality components considered and then presents their genetic diversity and inheritance. The genetic basis of fruit quality traits has been studied in several progenies and we will summarize the information provided by quantitative trait loci (QTLs) studies. The major genes/QTLs identified as being involved in fruit quality will be described before presentation of the future prospects offered by the new high-throughput genomic approaches available today. Many studies concern tomato fruit quality, which is both a model species for fruit quality studies and an important crop. Nevertheless, the number of studies on fruit quality concerning other species is increasing rapidly.

14.2 Quality Components

Flavour can be divided into two categories: the first being taste. Taste is often described by the sweetness and sourness of fruit, which are related to the amount of sugars and organic acids (Stevens *et al.*, 1977; Janse and Schols, 1995; Malundo *et al.*, 1995), and are also related to their balance (Stevens *et al.*, 1979; Bucheli *et al.*, 1999). Sugars and acids represent about 60 and 70% of the dry matter weight,

* mcausse@avignon.inra.fr

  Fruit Ripening: Physiology, Signalling and Genomics (eds P. Nath *et al.*)

respectively, in tomato (Davies and Hobson, 1981) and peach fruit. In mature tomato, glucose and fructose constitute the major sugars, whereas in peach, sucrose is predominant at maturity, followed by reducing sugars (glucose and fructose) and, in smaller amounts, sorbitol (Moriguchi *et al.*, 1990; Robertson *et al.*, 1990). Two organic acids, citrate and malate, are dominant in most fruit species (Tucker, 1993), as in tomato and peach. Acidity has been related either to the fruit pH or to titratable acidity (Baldwin *et al.*, 1998; Auerswald *et al.*, 1999). Acidity affects not only the sour taste of the fruit (Sweeney *et al.*, 1970) but also sweetness, by masking the taste of sugars. Thus, both sugars and acids contribute to the sweetness and to the overall taste intensity (Baldwin *et al.*, 1998). Sweetness seems more influenced by the content of fructose than of glucose, while acidity is mostly due to citric acid, present in higher amounts than malic acid in mature fruits (Stevens *et al.*, 1977).

The other main component of flavour is aroma, due to volatile compounds. In tomato, several sensory attributes have been proposed to characterize aroma, such as fruity, green, grassy, earthy, musty, floral, candy, citrus, grapefruit and pharmaceutical aromas (Bucheli *et al.*, 1999; Causse *et al.*, 2001; Baldwin *et al.*, 2004; Sinesio *et al.*, 2010). More than 400 aroma volatiles have been identified in tomato fruit (reviewed by Petro-Turza 1987), among which about 30 seem to be important for tomato aroma (Baldwin *et al.*, 2000, 2004). In apple and strawberry, more than 300 volatile compounds have been identified (Dixon and Hewett, 2000) for which 20 volatiles are considered the aroma fingerprint of strawberry (Ulrich *et al.*, 1997). Volatiles can be divided into two groups: those produced in intact tissues, important for the flavour of the intact fresh fruit; and those formed when plant tissues are disrupted, essential for the perception of flavour when the fruit is being chewed (Yahia, 1994). The combination of volatiles may be more important than their total amount for sensory sensations. For example, the perception of aldehydes may be enhanced by the presence of esters (Fellman *et al.*, 2000).

In apple, texture is the primary limiting factor for acceptability by consumers. Firmness, mealiness, juiciness, meltiness or tough skin are the major texture attributes. They are quite difficult to correlate with instrumental measurements. Firmness in the mouth is related partly to the instrumental measure of fruit firmness in tomato (Causse *et al.*, 2002), and mealiness could be found to be related to the texture parameters of the pericarp (Verkeke *et al.*, 1998). Texture is related to several processes, such as fruit morphology, cell size and shape, cell adhesion and cell-wall properties (Lee *et al.*, 1999; Devaux *et al.*, 2005; Chaïb *et al.*, 2007).

Consumer preferences facing natural diversity indicate the most important traits for breeders but also the diversity of consumer appreciation. High tomato-like aroma intensity and sweetness but intermediate acidity are the most important characteristics for consumer preferences (Jones 1986; Baldwin *et al.*, 1998; Lê and Ledauphin, 2006). Malundo *et al.* (1995) showed that given levels of sweetness correspond to optimal acid concentrations, beyond which acceptability decreases. Baldwin *et al.* (1998) related the overall acceptability to the ratio of sugars to titratable acidity and to the concentration of several aroma compounds. Verkeke *et al.* (1998) pinpointed the major role of texture traits in the preference of consumers. The sensory evaluation of apricots revealed that overall quality was linked mainly to flavour, sweetness and juiciness (Valentini *et al.*, 2006). For melon, the most critical quality traits depend on the melon type: retronasal aroma for climacteric types and texture for non-climateric ones (Ferrer *et al.*, 2007). Each cultivar may be characterized by a particular sensory profile and a specific potential of taste and texture (Causse *et al.*, 2010; Sinesio *et al.*, 2010; Fig. 14.1). Genetic variation is the major source of fruit quality variation (Stevens 1986; Causse *et al.*, 2003), but fruit organoleptic quality is also influenced by external factors such as the environment

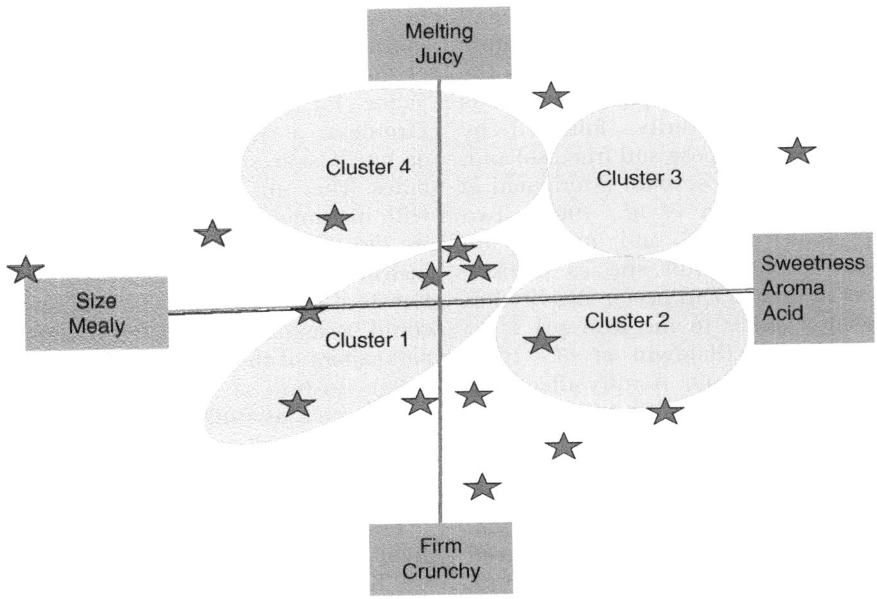

Fig. 14.1. Schematic representation of a preference mapping experiment in tomato. Trained panels and 800 consumers tasted 16 tomato cultivars. The cultivars (stars) are projected on the first multifactorial plan defined by sensory traits. The first axis is defined by sweet and acid flavours and aroma intensity (positive values) and fruit size and mealiness (negative values). The second axis is defined by texture traits. Four clusters of consumers were identified: consumers of cluster 1 liked many different tomato types; in cluster 2 they preferred the small varieties and appreciated sourness and aroma intensity; and in cluster 3 they liked small, tasty tomatoes; while in cluster 4 they were mainly looking for melting fruits. Adapted from Sinesio *et al.* (2010) and Causse *et al.* (2010).

during fruit growth (Dorais *et al.*, 2001), the maturity stage at harvest (Kader, 2008) and the conditions during fruit storage (Stern *et al.*, 1994).

14.3 Genetic Variability and Inheritance of Quality Traits

For a successful breeding programme, breeders need efficient selection criteria and must know the potential for improvement, i.e. the range of genetic variability available, the mode of inheritance and the respective influence of cultivar and environmental conditions on the traits to improve. Breeders tend to use only a few high-quality parents and thus work with a reduced genetic diversity. For example, in apple, a study of the pedigrees of 50 modern cultivars showed that five genitors were frequently used and revealed a high level of co-ancestry among many modern cultivars (Noiton and Alspach, 1996).

Wild species, in spite of their unfavourable characteristics in comparison with cultivars, can carry alleles that may contribute to the improvement of most agronomic traits (de Vicente and Tanksley, 1993; Bernacchi *et al.*, 1998). Gur and Zamir (2004) made progress by pyramiding independent yield-promoting regions introduced from the wild species *Solanum pennellii*. Wild species may provide original aromas, either favourable to tomato quality, as found in a *Solanum peruvianum* accession (Kamal *et al.*, 2001) or unfavourable as the *Malodorous* locus found in an *S. pennellii* accession (Tadmor *et al.*, 2002).

Genetic variability for quality traits has been reviewed in tomato by Davies and

Hobson (1981), Stevens (1986), Dorais *et al.* (2001), Causse *et al.* (2007) and Causse (2008). Cherry tomatoes have been identified as having the best flavour (Hobson and Bedford 1989), with fruits richer in acids and sugars than large-fruited lines. In contrast, long-shelf-life cultivars have been described as generally less tasty than traditional ones (Jones, 1986), with a lower volatile content (Baldwin *et al.*, 1991). Several studies have concerned the qualitative and quantitative composition of aroma volatiles in tomato varieties (Buttery *et al.*, 1987; Baldwin *et al.*, 1991; Krumbein and Auerswald 1998; Krumbein *et al.*, 2004). Tikunov *et al.* (2005) characterized 94 genotypes for their content in 322 different compounds. A study of the inheritance of tomato quality traits revealed that consumers seemed to particularly appreciate hybrids between old and modern lines with intermediate firmness (Causse *et al.*, 2003).

In other species, several studies have recently examined the variability within genetic resources, focusing on common fruit quality traits (Chessa and Nieddu, 2005; Nunez-Palenius *et al.*, 2008; Ruiz and Egea, 2008; Ledbetter, 2009). Concerning flavour, diversity within germplasm collections has rarely been investigated except in strawberry. Flavour compounds have been compared between wild and cultivated species (Aharoni *et al.*, 2004) and the diversity of aroma was studied by Ulrich *et al.* (2009) in wild and cultivated *Fragaria* accessions by gas chromatography/mass spectrometry and sensory assays. In apple, Sugimoto *et al.* (2007) revealed fruit aroma diversity in a core collection.

14.4 QTL Mapping for Fruit Quality Components

Identifying 'robust' quality QTLs is a prerequisite for molecular breeding or for their molecular characterization. Molecular characterization of QTLs has been performed to date by positional cloning, but many Mendelian mutants have also been identified. Tomato was among the first crop for which molecular markers were used to dissect the genetic basis of quantitative traits into QTLs (Tanksley, 1993). Since then, many QTLs controlling yield and fruit quality-related traits have been mapped (reviewed by Labate *et al.*, 2007). These studies were all performed on progenies derived from interspecific crosses between wild tomato species and processing tomato inbred lines. Some quality traits of interest for processing tomato are common to fresh market tomato (e.g. sugar content, soluble solid content, pH, acidity and firmness), and QTL locations can be compared among the progenies. In most of the studies, QTLs were detected, sometimes with strong effects. A few QTLs explaining a large fraction (20–50%) of the phenotypic variation, acting in concert with minor QTLs, are usually detected. Most of the QTLs act in an additive manner, but dominant and overdominant QTLs have been detected (Paterson *et al.*, 1988, 1991; De Vicente and Tanksley, 1993; Semel *et al.*, 2006). Epistasis (interaction among QTLs) is rarely detected unless a specific experimental design is used (Eshed and Zamir, 1996; Causse *et al.*, 2007).

Several fruit quality QTL studies have been carried out in peach using both intraspecific (Dirlewanger *et al.*, 1996, 1999; Abbott *et al.*, 1998; Etienne *et al.*, 2002) and interspecific (Quarta *et al.*, 2000; Quilot *et al.*, 2004) populations. Recently, 16 important QTLs controlling major fruit quality components were mapped, including acidity, sucrose content, fruit weight and pH (Dirlewanger *et al.*, 2007). In apple, Liebhard *et al.* (2003) detected QTLs for fruit quality traits, including fruit size and weight, fruit flesh firmness, sugar content and fruit acidity, and compared their location to previously mapped QTLs in apple. In apple again, a more recent study identified 74 QTLs major fruit physiological traits including fruit height, diameter, weight and stiffness, flesh firmness, rate of flesh browning, acidity, the soluble solid content and harvest date (Kenis *et al.*, 2008).

14.4.1 QTLs for fruit weight and shape

Grandillo *et al.* (1999) summarized the results of QTL mapping for fruit weight obtained in 17 studies on tomato based on progenies of various types and involving seven wild species. Six QTLs explained more than 20% of the phenotypic variation. A common set of 28 QTLs could be identified that frequently segregated in at least two populations. Nevertheless, only QTL cloning and complementation permits determination of whether each consensus QTL location corresponds to a single gene. For fruit shape, Grandillo *et al.* (1999) identified a common set of 11 QTLs from the six studies. Three major QTLs were identified, *ovate* on chromosome 2, *sun* on chromosome 7 and *fs8.1* on chromosome 8 (van der Knaap *et al.*, 2002).

In apple, different studies have led to the mapping of QTLs for fruit weight (King *et al.*, 2001; Liebhard *et al.*, 2003; Kenis *et al.*, 2008) with different results in terms of number of QTLs detected and the percentages of explained variability. In melon, numerous works have described QTLs for fruit size and shape (Périn *et al.*, 2002; Monforte *et al.*, 2004; Zalapa *et al.*, 2007; Paris *et al.*, 2008), which both appeared to be under complex polygenic control.

14.4.2 QTLs for fruit firmness and texture

Labate *et al.* (2007) presented a summary of QTLs controlling fruit firmness in nine populations of tomato. Forty-six QTLs controlling firmness were mapped using seven different populations. In apple, QTLs for fruit firmness evaluated by penetrometer measurements have been detected in different studies (King *et al.*, 2000; Maliepaard *et al.*, 2001; Liebhard *et al.*, 2003; Kenis *et al.*, 2008). In addition, QTLs accounting for stiffness determined by acoustic resonance and for sensory attributes of texture assessed by a trained panel were detected by King *et al.* (2000). Other components of texture were assessed through mechanical measures such as wedge fracture tests, distance at maximum force, specific gravity and measures of cell size and shape, and allowed the detection of other QTLs for specific attributes of texture (King *et al.*, 2001).

14.4.3 QTLs for sugar and acid content

The review of Labate *et al.* (2007) in tomato also summarized the chromosome regions carrying QTLs for sugar content or related traits (soluble solids; fructose, glucose or sucrose content), based on 14 populations involving eight different species. From three to 19 QTLs were detected per progeny, with a total of 95 QTLs gathered in 56 chromosomal regions. For the majority of QTLs, the wild species alleles increased the trait value. The large number of regions involved suggested that many mechanisms are responsible for increasing fruit sugar content. The same results were obtained for acid content (Causse *et al.*, 2002, 2004; Fulton *et al.*, 2002), with only a few regions common to acid and sugar content. In contrast, frequent colocations between QTLs for sugar content and fruit weight (Grandillo *et al.*, 1999) with opposite allelic effects could be detected, suggesting pleiotropic effects of some common QTLs.

In melon, a large number of QTLs have been detected for sugar accumulation (Monforte *et al.*, 2004; Sinclair *et al.*, 2006; Eduardo *et al.*, 2007; Obando-Ulloa *et al.*, 2008; Paris *et al.*, 2008; Park *et al.*, 2009), whereas for individual organic acids more extended research is needed (Obando *et al.*, 2008).

14.4.4 QTLs for volatile compounds and secondary metabolites

QTLs for volatile compounds have been mapped in two populations of tomato. Saliba-Colombani *et al.* (2001) detected QTLs for 12 volatile compounds among 18 that were quantified in the progeny of a cross involving a cherry tomato. Tieman *et al.* (2006a) identified QTLs for 23 volatiles in a population of introgression lines derived from *S. pennellii*. Twenty-five loci showing alterations in the concentration of one or more volatiles were identified. Although ten volatiles were analysed in both studies, only three QTLs were detected in the same regions. In both studies, QTLs for several volatiles were frequently in clusters. In a few cases, these clusters corresponded to volatiles derived from the same metabolic pathway (related to fatty acid, carotenoid or amino acid degradation), suggesting the action of a gene within a single pathway. More frequently, colocalizations of QTLs for volatiles derived from various metabolic pathways were shown, suggesting the presence of a regulatory gene acting on several pathways. In *Solanum habrochaites* introgression lines, 30 QTLs affecting the emission of one or more volatiles were mapped (Mathieu *et al.*, 2009). In apple, Dunemann *et al.* (2009) identified 50 putative QTLs for a total of 27 different fruit volatiles. QTLs for volatile compounds putatively involved in apple aroma were found on 12 of the 17 apple chromosomes, but they were clustered mainly on linkage groups 2, 3 and 9.

Few studies have focused on QTL mapping of secondary metabolites so far. Cuevas *et al.* (2008) mapped QTL regulating β-carotene composition in melon. Recently, Huang *et al.* (2012) studied the genetic basis of proanthocyanidin composition in grape via a QTL analysis on a 191-individual pseudo-F1 progeny. Proanthocyanidins are flavonoid polymers determinant in food quality. The study revealed a complex genetic control for proanthocyanidin traits and different genetic architectures for grape proanthocyanidin composition between berry skin and seeds.

14.5 Impact of Environment

Environmental conditions, including cultural practices, are well-known determinants of fruit growth and quality (Heuvelink 1997; Bertin *et al.*, 2003; Gautier *et al.*, 2008). In addition to macro-environment variations (due to year, location and growing conditions), micro-climatic gradients (Corelli-Grappadelli and Coston, 1991; Marini *et al.*, 1991) may cause within-plant variations of quality. Consequently, the stability of QTLs involved in fruit quality partly depends on the environment, as shown for sugar concentration and firmness in tomato by Chaïb *et al.* (2006). The importance of the environment in the stability of QTLs for sugar concentration has also been reported in a QTL analysis performed under two fruit load conditions (Prudent *et al.*, 2009). In another progeny, the genotype × environment (G × E) interaction appeared strong for fruit weight and aroma intensity but not very significant for firmness and fruit composition (Causse *et al.*, 2003). QTL detection under different saline conditions revealed a QTL specific to saline conditions (Villalta *et al.*, 2007), indicating strong G × E interactions. In melon, 27 near-isogenic lines have been evaluated in four different locations, for different fruit quality traits (fruit weight, soluble solids content, maximum fruit diameter, fruit length, fruit shape index, ovary shape index, external colour and flesh colour) (Eduardo *et al.*, 2007). Among these traits, soluble solids content showed the highest G × E interaction, whereas G × E interactions for fruit shape and fruit weight were low. Kenis *et al.* (2008) conducted a comparison of newly detected QTLs in apple with published QTL results obtained using other populations (King *et al.*, 2001; Liebhard *et al.*, 2003) and revealed that, for the six fruit quality traits that were measured in all populations, only nine of a total of 45 QTLs were common or stable across all population × environment combinations. Few studies have analysed the effect of environment on other quality traits such as flavour, but some clues already come from Tieman *et al.* (2006a) who showed that content in some volatile compounds of tomato is strongly variable over years or environments.

Although such QTL studies in different environments allow the suggestion of clues as to the instability of some chromosome regions towards environmental variations, the results cannot be used to predict the behaviour of a genotype in a climatic scenario that differs from that in which QTLs were detected. On the contrary, an ecophysiological model predicts the behaviour of one genotype in many environments. It decomposes the development of a trait into various processes subjected to environmental factors, with model parameters independent of the environment. Therefore, combining ecophysiological model and QTL analysis, carried out on parameters of the ecophysiological model, further allow prediction of the behaviour of different genotypes in different environments (Tardieu, 2003). Such an approach has been applied to study fruit quality in peach (Quilot *et al.*, 2005a,b).

14.6 Major Genes and Mutations Involved in Fruit Quality

In tomato, many mutations have been used to improve fruit quality, and the molecular basis of some of these mutations and of the major QTLs have been characterized over the last 20 years, as reviewed by Barry, Chapter 15, Handa *et al.*, Chapter 16, and Giovannoni, Chapter 17 (this volume). In peach, two recent studies allowed saturation of the genome with candidate genes for quality involved in metabolic pathways affecting fruit growth and maturity, texture, sugar and organic acid content, aroma and colour (Ogundiwin *et al.*, 2009; Illa *et al.*, 2011). Colocations between candidate genes and published QTLs responsible for natural variability of fruit quality characters in *Prunus* were identified.

14.6.1 Molecular basis for fruit size and shape

The first fruit-size QTL to be cloned was *fw2.2* (Frary *et al.*, 2000). This QTL controls up to 30% of the fruit size variation and was identified by a map-based cloning approach. It corresponds to a gene of unknown function, *ORFX*, which acts on cell number before anthesis. The two main QTLs controlling locule number, *fasciated* (*fas*) and *locule number* (*lc*), were cloned by positional cloning (Cong *et al.*, 2008; Muños *et al.*, 2011). Two other QTLs, *Ovate* and *SUN*, both responsible for elongated fruit shape, have been cloned (Liu *et al.*, 2002; Xiao *et al.*, 2008).

14.6.2 Genes involved in sugar content

The first QTL cloned controlling sugar content in tomato is a region encompassing *Lin5* (Fridman *et al.*, 2000), a gene encoding an apoplastic invertase expressed exclusively in fruits and flowers (Godt and Roitsch, 1997; Fridman and Zamir, 2003). Starch accumulates at the early stages of tomato fruit development, contributing approximately 20% of the dry weight of the fruit tissue at peak concentration, prior to the mature green stage. This starch is completely degraded in the ripe fruit, serving as a reservoir contributing to the soluble solids content (Dinar and Stevens, 1981; Ho, 1996). ADP-glucose pyrophosphorylase (AGPase) catalyses the synthesis of ADP- glucose and is considered the first committed step in starch synthesis. Tomato plants harbouring the allele for the AGPase large subunit derived from the wild species *S. habrochaites* are characterized by higher AGPase activity and increased starch content in the immature fruit, as well as higher soluble solids in the mature fruit following the breakdown of the transient starch, compared with fruits from plants harbouring the cultivated tomato allele (Schaffer *et al.*, 2000).

14.6.3 Genes involved in acidity

Two organic acids, citrate and malate, are dominant in most fruit species (Tucker, 1993). Genetic studies of different species (tomato, apple, peach, pomelo and grape) have shown that malic and citric acids are each governed by a major gene (Yoshida, 1970; Stevens, 1972; Cameron and Soost,

1977; Fang *et al.*, 1997; Maliepaard *et al.*, 1998). In tomato, fruit acidity was shown to be polygenic, but malic and citric acids are each governed by a major gene (Fulton *et al.*, 2000; Saliba-Colombani *et al.*, 2001). Peach fruits can be grouped into two types according to their pH value: the normal-acid (pH below 4.0) and the low-acid phenotypes (pH above 4.0) (Yoshida 1970; Dirlewanger *et al.*, 1998). A major dominant allele, *D*, is responsible for low acidity (Monet, 1979). Boudehri *et al.* (2009) generated a high-resolution genetic map of the *D* locus, currently delimited at a genetic interval of 0.4 cM. In contrast, the low-acid characteristic was found to be recessive in apple and citrus (Visser and Verhaegh, 1978; Maliepaard *et al.*, 1998). In apple, the position of the *Ma* gene, controlling malic acid accumulation, was determined by genetic mapping (Maliepaard *et al.*, 1998). More recently, Yao *et al.* (2009) isolated three genes encoding key enzymes involved in malic acid metabolism and transportation. Their expression pattern and the corresponding enzyme activity equated to the difference in fruit acidity between low- and high-acid genotypes. In contrast, in peach, Etienne *et al.* (2002) did not find any clear-cut difference between normal-acid and low-acid fruits for gene expression of six genes implicated in organic acid metabolism (mitochondrial citrate synthase, cytosolic NAD-dependent malate dehydrogenase and cytosolic NADP-dependent isocitrate dehydrogenase) and storage (vacuolar proton translocating pumps: one vacuolar H^+-ATPase and two vacuolar H^+-pyrophosphatases).

14.6.4 Genes involved in volatile compounds

Volatiles are derived from the degradation of amino acids, fatty acids, carotenoids or phenolic compounds (Klee, 2010). Due to the diversity of compounds, little is known about the genes controlling their accumulation, but a few genes have been identified as responsible for their accumulation. The *ADH* gene encoding an alcohol dehydrogenase is involved in the ratio of hexanal to hexanol in the tomato fruit (Speirs *et al.*, 1998). *TomloxC*, a gene encoding a fruit-specific lipoxygenase, has been shown to be related to the generation of volatile C6 aldehyde and alcohol compounds including hexanal, hexenal and hexenol (Chen *et al.*, 2004). Two genes, *LeAADC1* and *LeAADC2*, are responsible for the decarboxylation of phenylalanine and subsequent synthesis of phenylethanol and related compounds in tomato (Tieman *et al.*, 2006b). The gene coding for the carotenoid cleavage dioxygenase 1 enzyme (CCD1) is involved in the synthesis of several aroma volatiles derived from carotenoid cleavage (Vogel *et al.*, 2008). Tieman *et al.* (2007) showed that phenylacetaldehyde reductases catalyse the last step in the synthesis of the aroma volatile 2-phenylethanol. A salycylic acid methyl transferase has been shown to be involved in the synthesis of methyl salicylate (Tieman *et al.*, 2010).

14.6.5 Genes involved in fruit texture

Peach cultivars can be grouped into two main types according to the flesh texture: melting (M) or non-melting types. This trait, known to be Mendelian (Monet, 1989), is highly linked to the *Freestone* locus (*F*) mapped in linkage group 4 (Dettori *et al.*, 2001). Peace *et al.* (2005) showed evidence of a single locus containing at least one gene for endo-polygalacturonase, and controlling both *F* and M with at least three effective alleles. It is thus possible to differentiate by a PCR test the three major phenotypes of peach: freestone melting flesh, clingstone melting flesh, and clingstone non-melting flesh.

14.6.6 Genes involved in fruit ripening

Our current understanding of ripening mechanisms and the molecular basis of fruit texture evolution in fleshy fruits is largely due to tomato and relies mainly on transgenic or mutant plant analysis

(Seymour *et al.*, 2002; Giovannoni, 2007; Vicente *et al.*, 2007). Fruit ripening is the last step of a developmental programme made up of a complex string of physiological and metabolic processes including modification of pigment composition, enhancement of sugars, acids, aroma volatiles and fruit softening. Internal hormonal stimulation as well as environmental factors such as light, temperature, water and nutrient supply regulate ripening. Several mutations affecting fruit ripening and shelf-life are known. The most widely used in tomato breeding is *ripening inhibitor* (*rin*), which, in the heterozygous state, enables fruits to be kept for several weeks (Davies and Hobson, 1981). Long-shelf-life cultivars have invaded the tomato market, but in the 1990s their quality, particularly their colour and flavour, was criticized by consumers (Jones, 1986; McGlasson *et al.*, 1987).

14.7 Marker-assisted Selection for Fruit Sensory Quality

Very few experiments using marker-assisted selection for fruit sensory quality are available. A marker-assisted backcross scheme was set up in tomato for introducing favourable alleles of five major QTLs regions into three tomato lines with an ordinary taste (Lecomte *et al.*, 2004). In all three genetic backgrounds, the introduced regions had a favourable effect on the traits controlled by QTLs from the donor line (a cherry tomato), with the exception of fruit weight. Consumer tests revealed that the prototypes developed were significantly preferred over their corresponding recurrent parents. The QTLs appeared to be mainly specific to the genetic background (Chaïb *et al.*, 2006). About 50% of the QTLs were stable over different years in the recombinant inbred line population and in the genetic background used for QTL detection, but a lower number of QTLs was detected in the two other genetic backgrounds.

Regarding fruit traits, major genes or QTLs associated for example with firmness, melting flesh, fruit skin colour, freestone, fruit shape, acidity and sugar content, non-acid fruit and fruit skin pubescence have been mapped in apple and/or peach. Despite the identification of molecular markers associated with these many traits, their use in marker-assisted selection remains limited. One of the reasons for the lack of the extensive application of marker-assisted selection has been the relatively high costs of infrastructure and consumables required to run it, although these are expected to reduce as the rapidly evolving molecular industries become able to deliver cheaper technologies. In peach, experiments using marker-assisted selection have been launched within the FruitBreedomics European project (http://www.fruitbreedomics.com) for major genes controlling fruit traits: fruit low acidity (*D* locus), fruit shape (flat/round) (*S* locus), glabrous fruit epidermis (peach/nectarine) (*G* locus), fruit flesh colour (white/yellow) (locus *Y*) and non-melting (*F* locus).

14.8 Conclusions and Prospects

Fruit quality is a complex character due to the number of components involved and because it is dependent on environmental conditions throughout the entire process of plant and fruit development. Genetic variation for fruit quality is extensive, especially if one considers the possibilities offered by natural variation and wild species. A few mutations have been shown to be involved in fruit quality, particularly in ripening, and QTL studies have revealed a number of genomic regions involved in the variation of quality traits. Several genomic hot-spots for fruit quality QTLs have been identified. These clusters of QTLs for several quality traits may permit simultaneous marker-assisted selection for multiple traits. Today, very few QTLs have been identified at the molecular level, but one can expect a rapid increase in the number of genes identified in the near future, due to systems biology approaches combining transcriptomics, proteomics and metabolomics studies and information from genome sequences. These discoveries will greatly facilitate breeding for improved fruit quality. To date, tomato plants have been transformed with many genes whose function is more or less related to fruit quality, leading to variations in fruit quality profiles, but none of these studies has been used in cultivar development since the modification of the polygalacturonase gene for the modification of fruit texture (Kramer and Redenbaugh, 1994). Consumer concern about genetically modified crops, particularly in Europe, hampers the use of such plants and highlights the need for natural variation.

With genome sequences available in several species (tomato, peach, apple, grape and melon), many additional candidate genes will be identified. High-throughput technologies of genomics, proteomics and metabolomics enable the simultaneous quantification of the products of these genes during development (Fei *et al.*, 2004; Alba *et al.*, 2005; Baxter *et al.*, 2005; Carrari *et al.*, 2006; Faurobert *et al.*, 2007), in different plant tissues (Lemaire-Chamley *et al.*, 2005) or in different genotypes (Schauer *et al.*, 2006; Prudent *et al.*, 2009; Zanor *et al.*, 2009). These studies will provide necessary data for hypothesis testing to rapidly elucidate the genes underlying favourable QTLs or mutations and subsequently incorporate them into cultivars.

Thousands of SNPs are being discovered thanks to the new sequencing technologies that can be used for genome mapping, MAS, and positional cloning. Once a target gene has been characterized, it will be important to find new allelic variants within the large germplasm collections. An increased knowledge of gene function and regulation, as well as the development of more precise and efficient marker-assisted selection, will help to avoid introgression of large segments and undesirable loci into elite lines. Mutational and transgenic tools, such as mutation libraries in a uniform genetic background (Menda *et al.*, 2004), and techniques to screen for genetic lesions in specific genes (Comai and Henikoff, 2006; Rothan and Causse, 2007) will aid in the description of desirable alleles. Thanks to a wealth of novel tools and techniques, to its vast natural polymorphism at intra- and interspecific levels, and to consumer demands for nutritious foods, tomato will retain its status as a valuable model crop for fruit development, but major progress is also rapidly needed in other fruit crops for the satisfaction of consumers.

References

Abbott, A.G., Rajapakse, S., Sosinski, B., Lu, Z.X., Sossey-Alaoui, K., Scorza, R. and Callahan, A. (1998) Construction of saturated linkage maps of peach crosses segregating for characters controlling fruit quality, tree architecture and pest disease. *Acta Horticulturae* 465, 41–50.

Aharoni, A., Giri, A.P., Verstappen, F.W., Bertea, C.M., Sevenier, R., Sun, Z.K., Jongsma, M.A., Schwab, W. and Bouwmeester, H.J. (2004) Gain and loss of fruit flavor compounds produced by wild and cultivated strawberry species. *Plant Cell* 16, 3110–3131.

Alba, R., Payton, P., Fei, Z.J., McQuinn, R., Debbie, P., Martin, G.B., Tanksley, S.D. and Giovannoni, J.J. (2005) Transcriptome and selected metabolite analyses reveal multiple points of ethylene control during tomato fruit development. *Plant Cell* 17, 2954–2965.

Auerswald, H., Peters, P., Bruckner, B., Krumbein, A. and Kuchenbuch, R. (1999) Sensory analysis and instrumental measurements of short-term stored tomatoes (*Lycopersicon esculentum* Mill.). *Postharvest Biology and Technology* 15, 323–334.

Baldwin, E.A., Nisperos-Carriedo, M.O., Baker, R. and Scott, J.W. (1991) Quantitative analysis of flavor parameters in six Florida tomato cultivars. *Journal of Agriculture and Food Chemistry* 39, 1135–1140.

Baldwin, E.A., Scott, J.W., Einstein, M.A., Malundo, T.M.M., Carr, B.T., Shewfelt, R.L. and Tandon, K.S. (1998) Relationship between sensory and instrumental analysis for tomato flavor. *Journal of the American Society of Horticulture Science* 123, 906–915.

Baldwin, E.A., Scott, J.W., Shewmaker, C.K. and Schuw, W. (2000) Flavor trivia and tomato aroma: biochemistry and possible mechanisms for control of important aroma components. *Hortscience* 35, 1013–1022.

Baldwin, E.A., Goodner, K., Plotto, A., Pritcett, K. and Einstein, M. (2004) Effect of volatiles and their concentration on perception of tomato descriptors. *Journal of Food Science* 69, 310–318.

Baxter, C.J., Sabar, M., Quick, W.P. and Sweetlov, L.J. (2005) Comparison of changes in fruit gene expression in tomato introgression lines provides evidence of genome-wide transcriptional changes and reveals links to mapped QTLs and described traits. *Journal of Experimental Botany* 56, 1591–1604.

Bernacchi, D., Beck-Bunn, T., Emmatty, D., Eshed, Y., Inai, S., Lopez, J., Petiard, V., Sayama, H., Uhlig, J., Zamir, D. and Tansley, S. (1998) Advanced backcross QTL analysis in tomato. I.I. Evaluation of near-isogenic lines carrying single-donor introgressions for desirable wild QTL-alleles derived from *Lycopersicon hirsutum* and *L. pimpinellifolium*. *Theoretical and Applied Genetics* 97, 170–180.

Bertin, N., Borel, C., Brunel, B., Cheniclet, C. and Causse, M. (2003) Do genetic make-up and growth manipulation affect tomato fruit size by cell number, or cell size and DNA endoreduplication. *Annals of Botany* 92, 415–424.

Boudehri, K., Bendahmane, A., Cardinet, G., Troadec, C., Moing, A. and Dirlewanger, E. (2009). Phenotypic and fine genetic characterization of the D locus controlling fruit acidity in peach. *BMC Plant Biology* 9, 1–14.

Bucheli, P., Voirol, E., de la Torre, R., López, J., Rytz, A., Tanksley, S.D. and Pétiard, V. (1999) Definition of nonvolatile markers for flavor of tomato (*Lycopersicon esculentum* Mill.) as tools in selection and breeding. *Journal and Agriculture and Food Chemistry* 47, 659–664.

Buttery, R.G., Teranishi, R. and Ling, L. (1987) Fresh tomato aroma volatiles – a quantitative study. *Journal of Agricultural and Food Chemistry* 35, 540–544.

Cameron, J.W. and Soost, R.K. (1977) Acidity and total soluble solids in Citrus hybrids and advanced crosses involving acidless orange and acidless pummel. *Journal of the American Society for Horticultural Science* 120, 510–514.

Carrari, F., Baxter, C., Usadel, B., Urbanczyk-Wochniak, E., Zanor, M.I., Nunes-Nesi, A., Nikiforova, V., Centero, D., Ratzka, A., Pauly, M., Sweetlove, L.J. and Fernie, A.R. (2006) Integrated analysis of metabolite and transcript levels reveals the metabolic shifts that underlie tomato fruit development and highlight regulatory aspects of metabolic network behaviour. *Plant Physiology* 142, 1380–1396.

Causse, M. (2008) Genetic background of flavour: the case of the tomato. In: Brückner, B. and Wyllie, S.G. (eds) *Fruit and Vegetable Flavour. Recent Advances and Future Prospects*. CRC Press, Boca Raton, FL., pp. 229–253.

Causse, M., Saliba-Colombani, V., Lesschaeve, I., Duffé, P., Rousselle, P. and Buret, M. (2001) Genetic analysis of organoleptic quality in fresh market tomato. 2. Mapping QTLs for sensory attributes. *Theoretical and Applied Genetics* 102, 273–283.

Causse, M., Saliba-Colombani, V., Lesschaeve, I., Duffé, P., Rousselle, P. and Buret, M. (2002) QTL analysis of fruit quality in fresh market tomato: a few chromosome regions control the variation of sensory and instrumental traits. *Journal of Experimental Botany* 53, 2089–2098.

Causse, M., Buret, M., Robini, K. and Verschave, P. (2003) Inheritance of nutritional and sensory quality traits in fresh market tomato and relation to consumer preferences. *Journal of Food Science* 68, 2342–2350.

Causse, M., Duffe, P., Gomez, M.C., Buret, M., Damidaux, R., Zamir, D., Gur, A., Chevalier, C., Lemaire-Chamley, M. and Rothan, C. (2004) A genetic map of candidate genes and QTLs involved in tomato fruit size and composition. *Journal of Experimental Botany* 55, 1671–85.

Causse, M., Damidaux, R. and Rousselle, P. (2007) Traditional and enhanced breeding for fruit quality traits in tomato. In: Razdan, M.K. and Mattoo, A.K. (eds) *Genetic Improvement of Solanaceous Crops*, Vol. 2: *Tomato*. Science Publishers, Enfield, C.T., pp. 153–192.

Causse, M., Friguet, C., Coiret, C., Lépicier, M., Navez, B., Lee, M., Holthuysen, N., Sinesio, F., Moneta, E. and Grandillo, S. (2010) Consumer preferences for fresh tomato at the European scale: a common segmentation on taste and firmness. *Journal of Food Science* 75, S531–S541.

Chaïb, J., Lecomte, L., Buret, M. and Causse, M. (2006) Stability over genetic backgrounds, generations and years of quantitative trait locus (QTLs) for organoleptic quality in tomato. *Theoretical and Applied Genetics* 112, 934–944.

Chaïb, J., Devaux, M.F., Grotte, M.G., Robini, K., Causse, M., Lahaye, M. and Marty, I. (2007) Physiological relationships among physical, sensory, and morphological attributes of texture in tomato fruits. *Journal of Experimental Botany* 58, 1915–1925.

Chen, G.P., Hackett, R., Walker, D., Taylor, A., Lin, Z. and Grierson, D. (2004) Identification of a specific isoform of tomato lipoxygenase (TomloxC) involved in the generation of fatty acid-derived flavor compounds. *Plant Physiology* 136, 2641–2651.

Chessa, I. and Nieddu, G. (2005) Analysis of diversity in the fruit tree genetic resources from a Mediterranean island. *Genetic Resources and Crop Evolution* 52, 267–276.

Comai, L. and Henikoff, S. (2006) TILLING: practical single-nucleotide mutation discovery. *Plant Journal* 45, 684–694.

Cong, B., Barrero, L.S. and Tanksley, S.D. (2008) Regulatory change in YABBY-like transcription factor led to evolution of extreme fruit size during tomato domestication. *Nature Genetics* 40, 800–804.

Corelli-Grappadelli, L. and Coston, D.C. (1991) Thinning pattern and light environment in peach tree canopies influence fruit quality. *HortScience* 26, 1464–1466.

Cuevas, H.E., Staub, J.E., Simon, P.W., Zalapa, J.E. and McCreight, J.D. (2008). Mappng of genetic loci that regulate quantity of beta-carotene in fruit of U.S. Western Shipping melon (*Cucumis melo* L.). *Theoretical and Applied Genetics* 33, 301–310.

Davies, J.N. and Hobson, G.E. (1981) The constituents of tomato fruit – the influence of environment, nutrition and genotype. *Critical Review of Food Science and Nutrition* 15, 205–280.

De Vicente, M.C. and Tanksley, S.D. (1993) QTL analysis of transgressive segregation in an interspecific tomato cross. *Genetics* 134, 585–596.

Dettori, M.T., Quarta, R. and Verde, I. (2001) A peach linkage map integrating RFLPs, SSRs, RAPDs, and morphological markers. *Genome* 44, 783–790.

Devaux, M.F., Barakat, A., Robert, P., Bouchet, B., Guillon, F., Navez, B. and Lahaye, M. (2005) Mechanical breakdown and cell wall structure of mealy tomato pericarp tissue. *Postharvest Biology and Technology* 37, 209–221.

Dinar, M. and Stevens, M.A. (1981) The relationship between starch accumulation and soluble solids content of tomato fruits. *Journal of the American Society for Horticultural Science* 106, 415–418.

Dirlewanger, E., Pascual, M., Zuger, C. and Kervella, J. (1996) Analysis of molecular markers associated with powdery mildew resistance genes in peach (*Prunus persica* (L.) Batsch) × *Prunus davidiana* hybrids. *Theoretical and Applied Genetics* 93, 909–919.

Dirlewanger, E., Moing, A., Pronier, V., Svanella, L., Guye, A., Monet, R. and Rothan, C. (1998) Detection of QTLs controlling peach fruit acidity and sweetness. *Acta Horticulturae* 465, 89–98.

Dirlewanger, E., Moing, A., Rothan, C., Svanella, L., Pronier, V., Guye, A., Plomion, C. and Monet, R. (1999) Mapping QTLs controlling fruit quality in peach (*Prunus persica* (L.) Batsch). *Theoretical and Applied Genetics* 98, 18–31.

Dirlewanger, E., Cosson, P., Boudehri, K., Renaud, C., Capdeville, G., Tauzin, Y., Laigret, F. and Moing, A. (2007) Development of a second-generation genetic linkage map for peach *Prunus persica* (L.) Batsch and characterization of morphological traits affecting flower and fruit. *Tree Genetics and Genomes* 3, 1–13.

Dixon, J. and Hewett, E.W. (2000) Factors affecting apple aroma/flavour volatile concentration: a review. *New Zealand Journal of Crop and Horticultural Science* 28, 155–173.

Dorais, M., Papadopoulos, A.P. and Gosselin, A. (2001) Greenhouse tomato fruit quality. *Horticulture Review* 26, 239–319.

Dunemann, F., Ulrich, D., Boudichevskaia, A., Grafe, C. and Weber, W.E. (2009). QTL Mapping of aroma compounds analysed by headspace solid-phase microextraction gas chromatography in the apple progeny 'Discovery' × 'Prima'. *Molecular Breeding* 23, 501–521.

Eduardo, I., Arus, P., Monforte, A.J., Obando, J., Fernandez-Trujillo, J.P., Martinez, J.A., Alarcon, A.L., Alvarez, J.M. and van der Knaap, E. (2007) Estimating the genetic architecture of fruit quality traits in melon using a genomic library of near isogenic lines. *Journal of the American Society for Horticultural Science* 132, 80–89.

Eshed, Y. and Zamir, D. (1996) Less-than-additive epistatic interactions of quantitative trait loci in tomato. *Genetics* 143, 1807–1817.

Etienne, C., Rothan, C., Moing, A., Plomion, C., Bodénès, C., Svanella-Dumas, L., Cosson, P., Pronier, V., Monet, R. and Dirlewanger, E. (2002) Candidate genes and QTLs for sugar and organic acid content in peach *Prunus persica* (L.) Batsch. *Theoretical and Applied Genetics* 105, 145–159.

Fang, D.Q., Federici, C.T. and Roose, M.L. (1997) Development of molecular markers linked to a gene controlling fruit acidity in citrus. *Genome* 40, 841–849.

Faurobert, M., Mihr, C., Bertin, N., Pawlowski, T., Negroni, L., Sommerer, N. and Causse, M. (2007) Major proteome variations associated with cherry tomato pericarp development and ripening. *Plant Physiology* 143, 1327–1346.

Fei, Z.J., Tang, X., Alba, R.M., White, J.A., Ronning, C.M., Martin, G.B., Tanksley, S.D. and Giovannoni, J.J. (2004) Comprehensive EST analysis of tomato and comparative genomics of fruit ripening. *Plant Journal* 40, 47–59.

Fellman, J.K., Miller, T.W., Mattinson, D.S. and Mattheis, J.P. (2000) Factors that influence biosynthesis of volatile flavor compounds in apple fruits. *HortScience* 35, 1026–1033.

Ferrer, M.J., Obando, J., Moreno, E. *et al.* (2007). Differences in sensory quality associated with climacteric or non-climacteric behavior by using near-isogenic lines. In: Artes, A. (ed.) *V Congreso Iberoamericano de Tecnologia Postcosecha y Agroexportaciones*. Cartagena, Spain, pp. 504–514.

Frary, A., Nesbitt, T.C., Grandillo, S., Knaap, E., Cong, B., Liu, J., Meller, J., Elber, R., Alpert, K.B. and Tanksley, S.D. (2000) *fw2.2*: A quantitative trait locus key to the evolution of tomato fruit size. *Science* 289, 85–88.

Fridman, E. and Zamir, D. (2003) Functional divergence of a syntenic invertase gene family in tomato, potato, and Arabidopsis. *Plant Physiology* 131, 603–609.

Fridman, E., Pleban, T. and Zamir, D. (2000) A recombination hotspot delimits a wild-species quantitative trait locus for tomato sugar content to 484 bp within an invertase gene. *Proceedings of the National Academy of Sciences USA* 97, 4718–4723.

Fulton, T.M., Grandillo, S., BeckBunn, T., Fridman, E., Frampton, A., Lopez, J., Petiard, V., Uhlig, J., Zamir, D. and Tanksley, S.D. (2000) Advanced backcross QTL analysis of a *Lycopersicon esculentum* × *Lycopersicon parviflorum* cross. *Theoretical and Applied Genetics* 100, 1025–1042.

Fulton, T.M., Bucheli, P., Voirol, E., López, J., Pétiard, V. and Tanksley, S.D. (2002) Quantitative trait loci (QTL) affecting sugars, organic acids and other biochemical properties possibly contributing to flavor, identified in four advanced backcross populations of tomato. *Euphytica* 127, 163–177.

Gautier, H., Diakou-Verdin, V., Benard, C., Reich, M., Buret, M., Bourgaud, F., Poessel, J.L., Caris-Veyrat, C. and Genard, M. (2008) How does tomato quality (sugar, acid, and nutritional quality) vary with ripening stage, temperature, and irradiance? *Journal of Agricultural and Food Chemistry* 56, 1241–1250.

Giovannoni, J.J. (2007) Fruit ripening mutants yield insights into ripening control. *Current Opinion in Plant Biology* 10, 283–289.

Godt, D.E. and Roitsch, T. (1997) Regulation and tissue-specific distribution of mRNAs for three extracellular invertase isoenzymes of tomato suggests an important function in establishing and maintaining sink metabolism. *Plant Physiology* 115, 273–282.

Grandillo, S., Ku, H.M. and Tanksley, S.D. (1999) Identifying the loci responsible for natural variation in fruit size and shape in tomato. *Theoretical and Applied Genetics* 99, 978–987.

Gur, A. and Zamir, D. (2004) Unused natural variation can lift yield barriers in plant breeding. *PLOS Biology* 2, 1610–1615.

Heuvelink, E. (1997) Effect of fruit load on dry matter partitioning in tomato. *Scientia Horticulturae* 69, 51–59.

Ho, L.C. (1996) The mechanism of assimilate partitioning and carbohydrate compartmentation in fruit in relation to the quality and yield of tomato. *Journal of Experimental Botany* 47, 1239–1243.

Hobson, G.E. and Bedford, L. (1989) The composition of cherry tomatoes and its relation to consumer acceptability. *Journal of Horticulture Science* 64, 321–329.

Huang, Y.F., Doligez, A., Fournier-Level, A., Le Cunff, L., Bertrand, Y., Canaguier, A., Morel, C., Miralles, V., Veran, F., Souquet, J.M., Cheynier, V., Terrier, N. and This, P. (2012) Dissecting genetic architecture of grape proanthocyanidin composition through quantitative trait locus mapping. *BMC Plant Biology* 12, 30.

Illa, E., Eduardo, I., Audergon, J.M., Barale, F., Dirlewanger, E., Li, X., Moing, A., Lambert, P., Le Dantec, L., Gao, Z., Poessel, J.-L., Pozzi, C., Rossini, L., Vecchietti, A., Arus, P. and Howad, W. (2011) Saturating the *Prunus* (stone fruits) genome with candidate genes for fruit quality. *Molecular Breeding* 28, 667–682.

Janse, J. and Schols, M. (1995) Une préférence pour un goût sucré et non farineux. *Groenten and Fruit* 26, 16–17.

Jones, R.A. (1986) Breeding for improved post-harvest tomato quality: genetical aspects. *Acta Horticulturae* 190, 77–87.

Kader, A.A. (2008) Flavor quality of fruits and vegetables. *Journal of the Science of Food and Agriculture* 88, 1863–1868.

Kamal, A.H.M., Takashina, T., Egashira, H., Satoh, H. and Imanishi, S. (2001) Introduction of aromatic fragrance into cultivated tomato from the 'peruvianum complex'. *Plant Breeding* 120, 179–181.

Kenis, K., Keulemans, J. and Davey, M.W. (2008) Identification and stability of QTLs for fruit quality traits in apple. *Tree Genetics and Genomes* 4, 647–661.

King, G.J., Maliepaard, C., Lynn, J.R., Alston, F.H., Durel, C.E., Evans, K.M., Griffon, B., Laurens, F., Manganaris, A.G., Schrevens, T., Tartarini, S. and Verhaegh, J. (2000) Quantitative genetic analysis and comparison of physical and sensory descriptors relating to fruit flesh firmness in apple (*Malus pumila* Mill.). *Theoretical and Applied Genetics* 100, 1074–1084.

King, G.J., Lynn, J.R., Dover, C.J., Evans, K.M. and Seymour, G.B. (2001) Resolution of quantitative trait loci for mechanical measures accounting for genetic variation in fruit texture of apple (*Malus pumila* Mill.). *Theoretical and Applied Genetics* 102, 1227–1235.

Klee, H.J. (2010) Improving the flavor of fresh fruits: genomics, biochemistry, and biotechnology. *New Phytologist* 187, 44–56.

Kramer, M.G. and Redenbaugh, K. (1994) Commercialization of a tomato with an antisense polygalacturonase gene: the FLAVR SAVR™ tomato story. *Euphytica* 79, 293–297.

Krumbein, A. and Auerswald, H. (1998) Characterization of aroma volatiles in tomatoes by sensory analyses. *Food/Nahrung* 42, 395–399.

Krumbein, A., Peters, P. and Bruckner, B. (2004), Flavour compounds and a quantitative descriptive analysis of tomatoes (*Lycopersicon esculentum* Mill.) of different cultivars in short-term storage. *Postharvest Biology and Technology* 32, 15–28.

Labate, J.A., Grandillo, S., Fulton, T., Muños, S., Caicedo, A.L., Peralta, I., Ji, Y., Chetelat, R.T., Scott, J.W., Gonzalo, M.J., Francis, D., Yang, W., Knaap, E.V.D., Baldo, A.M., Smith-White, B., Mueller, L.A., Prince, J.P., Blanchard, N.E., Storey, D.B., Stevens, M.R., Robbins, M.D., Wang, J.-F., Liedl, B.E., O'Connell, M.A., Stommel, J.R., Aoki, K., Iijima, Y., Slade, A.J., Hurst, S.R., Loeffler, D., Steine, M.N., Vafeados, D., McGuire, C., Freeman, C., Amen, A., Goodstal, J., Facciotti, D., Eck, J.V. and Causse, M. (2007) Tomato. In: Kole, C. (ed.) *Genome Mapping and Molecular Breeding in Plants*, Vol. 5, *Vegetables*. Springer-Verlag, Berlin/Heidelberg, pp. 1–125.

Lê, S. and Ledauphin, S. (2006) You like tomato, I like tomato: segmentation of consumers with missing values. *Food Quality and Preference* 17, 228–233.

Lecomte, L., Duffé, P., Buret, M., Servin, B., Hospital, F. and Causse, M. (2004) Marker-assisted introgression of five QTLs controlling fruit quality traits into three tomato lines revealed interactions between QTLs and genetic backgrounds. *Theoretical and Applied Genetics* 109, 658–668.

Ledbetter, C.A. (2009) Using Central Asian germplasm to improve fruit quality and enhance diversity in California adapted apricots. *Acta Horticulturae* 814, 77–80.

Lee, S.Y., Luna-Guzman, I., Chang, S., Barrett, D.M. and Guinard, J.-X. (1999) Relating descriptive analysis and instrumental texture data of processed diced tomatoes. *Food Quality and Preference* 10, 447–455.

Lemaire-Chamley, M., Petit, J., Garcia, V., Just, D., Baldet, P., Germain, V., Fagard, M., Mouassite, M., Cheniclet, C. and Rothan, C. (2005) Changes in transcriptional profiles are associated with early fruit tissue specialization in tomato. *Plant Physiology* 139, 750–769.

Liebhard, R., Kellerhals, M., Pfammatter, W., Jertmini, M. and Gessler, C. (2003). Mapping quantitative physiological traits in apple (*Malus × domestica* Borkh.). *Plant Molecular Biology* 52, 511–526.

Liu, J.P., Van Eck, J., Cong, B. and Tanksley, S.D. (2002) A new class of regulatory genes underlying the cause of pear-shaped tomato fruit. *Proceedings of the National Academy of Sciences USA* 99, 13302–13306.

Maliepaard, C., Alston, F.H., Arkel, G.V., Brown, L.M., Chevreau, E., Dunemann, F., Evans, K.M., Gardiner, S., Guilford, P., Heusden, A.W.V., Janse, J., Laurens, F., Lynn, J.R., Manganaris, A.G., den Nijs, A.P.M., Periam, N., Rikkerink, E., Roche, P., Ryder, C., Sansavini, S., Schmidt, H.,

Tartarini, S., Verhaegh, J.J., Vrielink-van Ginkel, M. and King, G.J. (1998) Aligning male and female linkage maps of apple (*Malus pumila* Mill.) using multi-allelic markers. *Theoretical and Applied Genetics* 97, 60–73.

Maliepaard, C., Sillanpaa, M.J., van Ooijen, J.W., Jansen, R.C. and Arjas, E. (2001) Bayesian versus frequentist analysis of multiple quantitative trait loci with an application to an outbred apple cross. *Theoretical and Applied Genetics* 103, 1243–1253.

Malundo, T.M.M., Shewfelt, R.L. and Scott, J.W. (1995) Flavor quality of fresh market tomato (*Lycopersicon esculentum* Mill.) as affected by sugar and acid levels. *Postharvest Biology and Technology* 6, 103–110.

Marini, R.P., Sowers, D.L. and Marini, M.C. (1991) Peach fruit quality is affected by shade during final swell of fruit growth. *Journal of the American Society for Horticultural Science* 116, 383–389.

Mathieu, S., Cin, V.D., Fei, Z.J., Li, H., Bliss, P., Taylor, M.G., Klee, H.J. and Tieman, D.M. (2009) Flavour compounds in tomato fruits: identification of loci and potential pathways affecting volatile composition. *Journal of Experimental Botany* 60, 325–337.

McGlasson, W.B., Last, J.H., Shaw, K.J. and Meldrum, S.K. (1987) Influence of the non-ripening mutants rin and nor on the aroma of tomato fruit. *Hortscience* 22, 632–634.

Menda, N., Semel, Y., Peled, D., Eshed, Y. and Zamir, D. (2004) In silico screening of a saturated mutation library of tomato. *Plant Journal* 38, 861–872.

Monet, R. (1979) *Transmission Génétique du Caractère 'Fruit Doux' Chez le Pêcher. Incidence sur la Sélection pour la Qualité. Eucarpia Fruit Section.* INRA, Angers, France.

Monet, R. (1989) Peach genetics: past, present and future. *Acta Horticulturae* 254, 49–53.

Monforte, A.J., Oliver, M., Gonzalo, M.J., Alvarez, J.M., Dolcet-Sanjuan, R. and Arus, P. (2004) Identification of quantitative trait loci involved in fruit quality traits in melon (*Cucumis melo* L.). *Theoretical and Applied Genetics* 108, 750–758.

Moriguchi, T., Ishizawa, Y. and Sanada, T. (1990). Differences in sugar composition in *Prunus persica* fruit and the classification by the principal component analysis. *Journal of the Japanese Society for Horticultural Science* 59, 307–312.

Muños, S., Ranc, N., Botton, E., Bérard, A., Rolland, S., Duffé, P., Carretero, Y., Le Paslier, M.C., Delalande, C., Bouzayen, M., Brunel, D. and Causse, M. (2011) Increase in tomato locule number is controlled by two key SNP located near Wuschel. *Plant Physiology* 4, 2244–2254.

Noiton, D.A.M. and Alspach, P.A. (1996) Founding clones, inbreeding, coancestry, and status number of modern apple cultivars. *Journal of the American Society for Horticultural Science* 121, 773–782.

Nunez-Palenius, H.G., Gomez-Lim, M., Ochoa-Alejo, N., Grumet, R., Lester, G. and Cantliffe, D.J. (2008) Melon fruits: genetic diversity, physiology, and biotechnology features. *Critical Reviews in Biotechnology* 28, 13–55.

Obando, J., Fernandez-Trujillo, J.P., Martinez, J.A., Alarcon, A.L., Eduardo, I., Arus, P. and Monforte, A.J. (2008) Identification of melon fruit quality quantitative trait loci using near-isogenic lines. *Journal of the American Society for Horticultural Science* 133, 139–151.

Obando-Ulloa, J.M., Moreno, E., Garcia-Mas, J., Nicolai, B., Lammertyn, J., Monforte, A.J. and Fernandez-Trujillo, J.P. (2008) Climacteric or non-climacteric behavior in melon fruit – 1. Aroma volatiles. *Postharvest Biology and Technology* 49, 27–37.

Ogundiwin, E.A., Peace, C.P., Gradziel, T.M., Parfitt, D.E., Bliss, F.A. and Crisosto, C.H. (2009) A fruit quality gene map of *Prunus*. *BMC Genomics* 10, 587.

Paris, M.K., Zalapa, J.E., McCreight, J.D. and Staub, J.E. (2008) Genetic dissection of fruit quality components in melon (*Cucumis melo* L.) using a RIL population derived from exotic × elite US Western Shipping germplasm. *Molecular Breeding* 22, 405–419.

Park, S.O., Hwang, H.Y. and Crosby, K.M. (2009) A genetic linkage map including loci for male sterility, sugars, and ascorbic acid in melon. *Journal of the American Society for Horticultural Science* 13, 67–76.

Paterson, A.H., Lander, E.S., Hewitt, J.D., Peterson, S., Lincoln, S.E. and Tanksley, S.D. (1988) Resolution of quantitative traits into Mendelian factors by using a complete linkage map of restriction fragment length polymorphisms. *Nature* 335, 721–726.

Paterson, A.H., Damon, S., Hewitt, J.D., Zamir, D., Rabinowitch, H.D., Lincoln, S.E., Lander, E.S. and Tanksley, S.D. (1991) Mendelian factors underlying quantitative traits in tomato: comparison across species, generations, and environments. *Genetics* 127, 181–197.

Peace, C.P., Crisosto, C.H. and Gradziel, T.M. (2005) Endopolygalacturonase: a candidate gene for Freestone and Melting flesh in peach. *Molecular Breeding* 16, 21–31.

Périn, C., Hagen, L.S., Giovinazzo, N., Besombes, D., Dogimont, C. and Pitrat, M. (2002) Genetic control of fruit shape acts prior to anthesis in melon (*Cucumis melo* L.). *Molecular Genetics and Genomics* 266, 933–941.

Petro-Turza, M. (1987) Flavor of tomato and tomato products. *Food Review International* 2, 309–351.

Prudent, M., Causse, M., Génard, M., Tripodi, P., Grandillo, S. and Bertin, N. (2009) Genetic and physiological analysis of tomato fruit weight and composition: influence of carbon availability on QTL detection. *Journal of Experimental Botany* 60, 923–937.

Quarta, R., Dettori, M.T., Sartori, A. and Verde, I. (2000) Genetic linkage map and QTL analysis in peach. In: VanderPlas L.H.W. (ed.) *Proceedings of the XXV International Horticultural Congress, Part 11: Applications of Biotechnology and Molecular Biology and Breeding Gene Expression and Molecular Breeding Genome Analysis*, pp. 233–241.

Quilot, B., Wu, B.H., Kervella, J., Genard, M., Foulongne, M. and Moreau, K. (2004) QTL analysis of quality traits in an advanced backcross between *Prunus persica* cultivars and the wild relative species *P. davidiana*. *Theoretical and Applied Genetics* 109, 884–897.

Quilot, B., Genard, M., Lescourret, F. and Kervella, J. (2005a) Simulating genotypic variations of fruit quality in an advanced peach × *Prunus davidiana* cross. *Journal of Experimental Botany* 56, 3071–3081.

Quilot, B., Kervella, J., Genard, M. and Lescourret, F. (2005b). Analysing the genetic control of peach fruit quality through an ecophysiological model combined with a quantitative trait loci approach. *Journal of Experimental Botany* 56, 3083–3092.

Robertson, J.A., Horvat, J.A., Lyon, B.G., Meredith, F.I., Senter, S.D. and Okie, W.R. (1990) Comparison of quality characteristics of selected yellow- and white-fleshed peach cultivar. *Journal of Food Science* 55, 1308–1311.

Rothan, C. and Causse, M. (2007) Natural and artificially induced genetic variability in crop and model plant species for plant systems biology. *Experientia Supplementum* 97, 21–53.

Ruiz, D. and Egea, J. (2008). Phenotypic diversity and relationships of fruit quality traits in apricot (*Prunus armeniaca* L.) germplasm. *Euphytica* 163, 143–158.

Saliba-Colombani, V., Causse, M., Langlois, D., Philouze, J. and Buret, M. (2001) Genetic analysis of organoleptic quality in fresh market tomato. 1. Mapping QTLs for physical and chemical traits. *Theoretical and Applied Genetics* 102, 259–272.

Schaffer, A.A., Levin, I., Oguz, I., Petreikova, M., Cincarevskyb, F., Yeselsona, Y., Shena, S., Gilboab, N. and Bar, M. (2000) ADPglucose pyrophosphorylase activity and starch accumulation in immature tomato fruit: the effect of a *Lycopersicon hirsutum*-derived introgression encoding for the large subunit. *Plant Science* 152, 135–144.

Schauer, N., Semel, Y., Roessner, U., Gur, A., Balbo, I., Carrari, F., Pleban, T., Perez-Melis, A., Bruedigam, C., Kopka, J., Willmitzer, L., Zamir, D. and Fernie, A.R. (2006) Comprehensive metabolic profiling and phenotyping of interspecific introgression lines for tomato improvement. *Nature Biotechnology* 24, 447–454.

Semel, Y., Nissenbaum, J., Menda, N., Zinder, M., Krieger, U., Issman, N., Pleban, T., Lippman, Z., Gur, A. and Zamir, D. (2006) Overdominant quantitative trait loci for yield and fitness in tomato. *Proceedings of the National Academy of Sciences USA* 103, 12981–12986.

Seymour, G.B., Manning, K., Eriksson, E.M., Popovich, A.H. and King, G.J. (2002) Genetic identification and genomic organization of factors affecting fruit texture. *Journal of Experimental Botany* 53, 2065–2071.

Sinclair, J.W., Park, S.O., Lester, G.E., Yoo, K.S. and Crosby, K.M. (2006) Identification and confirmation of RAPD markers and andromonoecious associated with quantitative trait loci for sugars in melon. *Journal of the American Society for Horticultural Science* 131, 360–371.

Sinesio, F., Cammareri, M., Moneta, E., Navez, B., Peparaio, M., Causse, M. and Grandillo, S. (2010) Sensory quality of fresh French and Dutch market tomatoes: a preference mapping study with Italian consumers. *Journal of Food Science* 75, 55–67.

Speirs, J., Lee, E., Holt, K., Yong-Duk, K., Steele Scott, N., Loveys, B. and Schuch, W. (1998) Genetic manipulation of alcohol dehydrogenase levels in ripening tomato fruit affects the balance of some flavor aldehydes and alcohols. *Plant Physiology* 117, 1047–1058.

Stern, D.J., Buttery, R.G., Teranishi, R. and Ling, L. (1994) Effect of storage and ripening on fresh tomato quality. *Food Chemistry* 49, 225–231.

Stevens, M.A. (1972) Citrate and malate concentration in tomato fruits: genetic control and maturational effects. *Journal of the American Society for Horticultural Science* 97, 655–658.

Stevens, M.A. (1986) Inheritance of tomato fruit quality components. *Plant Breeding Reviews* 4, 273–311.

Stevens, M.A., Kader, A.A., Albright-Holton, M. and Algazi, M. (1977) Genotypic variation for flavor and composition in fresh market tomatoes. *Journal of the American Society for Horticulture Science* 102, 680–689.

Stevens, M.A., Kader, A.A. and Albright, M. (1979) Potential for increasing tomato flavor via increased sugar and acid content. *Journal of the American Society of Horticulture Science* 104, 40–52.

Sugimoto, N., Forsline, P., Srmack, W. and Beaudry, R. (2007) Evaluation of apple fruit aroma character from the Geneva *Malus* core collection. *HortScience* 42, 869–870.

Sweeney, J.P., Chapman, V.J. and Hepner, P.A. (1970) Sugar, acid and flavour in flesh fruits. *Journal of the American Dietetic Association* 57, 432–435.

Tadmor, Y., Fridman, E., Gur, A., Larkov, O., Lastochkin, E., Ravid, U., Zamir, D. and Lewinsohn, E. (2002) Identification of malodorous, a wild species allele affecting tomato aroma that was selected against during domestication. *Journal of Agricultural and Food Chemistry* 50, 2005–2009.

Tanksley, S.D. (1993) Mapping polygenes. *Annual Review of Genetics* 27, 205–233.

Tardieu, F. (2003) Virtual plants: modelling as a tool for the genomics of tolerance to water deficit. *Trends in Plant Science* 8, 9–14.

Tieman, D., Taylor, M., Schauer, N., Fernie, A.R., Hanson, A.D. and Klee, H.J. (2006b) Tomato aromatic amino acid decarboxylases participate in synthesis of the flavor volatiles 2-phenylethanol and 2-phenylacetaldehyde. *Proceedings of the National Academy of Sciences USA* 103, 8287–8292.

Tieman, D., Zeigler, M., Schmelz, E., Taylor, M.G., Rushing, S., Jones, J.B. and Klee, H.J. (2010) Functional analysis of a tomato salicylic acid methyl transferase and its role in synthesis of the flavor volatile methyl salicylate. *Plant Journal* 62, 113–123.

Tieman, D.M., Zeigler, M., Schmelz, E.A., Taylor, M.G., Bliss, P., Kirst, M. and Klee, H.J. (2006a) Identification of loci affecting flavour volatile emissions in tomato fruits. *Journal of Experimental Botany* 57, 887–896.

Tieman, D.M ., Loucas, H.M., Kim, J.Y., Clark, D.G. and Klee, H.J. (2007) Tomato phenylacetaldehyde reductases catalyze the last step in the synthesis of the aroma volatile 2-phenylethanol. *Phytochemistry* 68, 2660–2669.

Tikunov, Y., Lommen, A., de Vos, C.H.R., Verhoeven, H.A., Bino, R.J., Hall, R.D. and Bovy, A.G. (2005) A novel approach for nontargeted data analysis for metabolomics. Large-scale profiling of tomato fruit volatiles. *Plant Physiology* 139, 1125–1137.

Tucker, G.A. (1993) Introduction: respiration and energy. In: Seymour, G.B., Taylor, J.E. and Tucker, G.A. (eds) *Biochemistry of Fruit Ripening*. Chapman & Hall, London, pp. 1–51.

Ulrich, D. and Olbricht, K. (2011) Fruit organoleptic properties and potential for their genetic improvement. In: Jenks, M.A. and Bebelli, P.J. (eds) *Breeding for Fruit Quality*. John Wiley & Sons, Hoboken, NJ, pp. 41–60.

Ulrich, D., Hoberg, E., Rapp, A. and Kecke, S. (1997) Analysis of strawberry flavour – discrimination of aroma types by quantification of volatile compounds. *Zeitschrift fur Lebensmittel-Untersuchung und Forschung A* 205, 218–223.

Ulrich, D., Hoberg, E. and Fischer, C. (2009). Diversity and dynamic of sensory related traits in different apple cultivars. *Journal of Applied Botany and Food Quality* 83, 70–75.

Valentini, N., Mellano, M.G., Antonioni, I. and Botta, R. (2006) Chemical, physical and sensory analysis for evaluating quality of apricot cultivars. In: Audergon (ed.) *Proceedings of the XIIth ISHS Symposium on Apricot Culture and Decline*, Vols 1 and 2. Acta Horticulturae, Brugge, Belgium, pp. 559–563.

Van der Knaap, E., Lippman, Z.B. and Tanksley, S.D. (2002) Extremely elongated tomato fruit controlled by four quantitative trait loci with epistatic interactions. *Theoretical and Applied Genetics* 104, 241–247.

Verkeke, W., Janse, J. and Kersten, M. (1998) Instrumental measurement and modelling of tomato fruit taste. *Acta Horticulturae* 456, 199–205.

Vicente, A.R., Saladie, M., Rose, J.K.C. and Labavitch, J.M. (2007) The linkage between cell wall metabolism and fruit softening: looking to the future. *Journal of the Science of Food and Agriculture* 87, 1435–1448.

Villalta, I., Bernet, G.P., Carbonell, E.A. and Asins, M.J. (2007) Comparative QTL analysis of salinity tolerance in terms of fruit yield using two solanum populations of F-7 lines. *Theoretical and Applied Genetics* 114, 1001–1017.

Visser, T. and Verhaegh, J.J. (1978) Inheritance and selection of some fruit characters of apple. I. Inheritance of low and high acidity. *Euphytica* 27, 753–760.

Vogel, J.T., Tan, B.C., McCarty, D.R. and Klee, H.J. (2008) The carotenoid cleavage dioxygenase 1 enzyme has broad substrate specificity, cleaving multiple carotenoids at two different bond positions. *Journal of Biological Chemistry* 283, 11364–11373.

Xiao, H., Jiang, N., Schaffner, E., Stockinger, E.J. and van der Knaap, E. (2008) A retrotransposon-mediated gene duplication underlies morphological variation of tomato fruit. *Science* 319, 1527–1530.

Yahia, E.M. (1994) Apple flavor. *Horticultural Reviews* 16, 197–234.

Yao, Y.X., Li, M., Liu, Z., You, C.X., Wang, D.M., Zhai, H. and Hao, Y.J. (2009) Molecular cloning of three malic acid related genes *MdPEPC, MdVHA-A, MdcyME* and their expression analysis in apple fruits. *Scientia Horticulturae* 122, 404–408.

Yoshida, M. (1970) Genetical studies on the fruit quality of peach varieties, I. Acidity. *Bulletin of the Horticultural Research Station Japan A* 9, 1–15.

Zalapa, J.E., Staub, J.E., McCreight, J.D., Chung, S.M. and Cuevas, H. (2007) Detection of QTL for yield-related traits using recombinant inbred lines derived from exotic and elite, U.S. Western Shipping melon germplasm. *Theoretical and Applied Genetics* 114, 1185–1201.

Zanor, M.I., Rambla, J.L., Chaïb, J., Steppa, A., Medina, A., Granell, A., Fernie, A.R. and Causse, M. (2009) Metabolic characterization of loci affecting sensory attributes in tomato allows an assessment of the influence of the levels of primary metabolites and volatile organic contents. *Journal of Experimental Botany* 60, 2139–2154.

15 Ripening Mutants

Cornelius S. Barry*
Department of Horticulture, Michigan State University, MI, USA

15.1 Introduction

Fleshy fruits are botanically and chemically diverse, yet ripening processes are surprisingly conserved and often include changes in colour and cell-wall dissolution, together with subsequent fruit softening, the synthesis of aroma compounds and conversion of starch to sugars. These changes increase palatability and help to signal seed maturity, facilitating dispersal by frugivores. Due to the importance of fleshy fruits in providing sources of sugars, vitamins, minerals, antioxidants and fibre to the human diet, considerable research effort has focused on identifying the processes, enzymes and regulatory proteins that contribute to the development and ripening of fleshy fruits yet limit their postharvest deterioration. In recent years, the development of genomics resources for fleshy fruit-bearing species, including available genome sequences, large expressed sequence tag collections and publically available gene expression data, have greatly increased our understanding of the genes correlated with events that occur during the ripening process. However, despite the development of these resources, functional analyses of putative ripening-related genes is the main factor limiting understanding of the ripening process. While gene-silencing approaches are useful in some fruit crop species, some of the most significant insights into the genetic factors that control fruit ripening and quality have come from the characterization of mutants that inhibit or modify the ripening process, together with the identification of the underlying genes. This chapter highlights recent research on the characterization of ripening mutants and their impact on our current understanding of the ripening process. In compiling topics, a broad perspective on what defines a ripening mutant has been adopted so as to encompass mutants that influence different aspects of ripening and fruit quality. In addition, while the majority of the chapter is focused on advances using tomato as a model system, where appropriate, research describing fruit-related mutants in other species is presented.

15.2 Ripening Mutants Define Master Transcriptional Regulators of the Ripening Process

Mutations at the *ripening inhibitor* (*rin*), *non-ripening* (*nor*) and *Colorless non-ripening* (*Cnr*) loci result in dramatic inhibition of the characteristic phenotypes associated with ripening, including respiratory climacteric, ripening-related ethylene synthesis, chlorophyll degradation, carotenoid biosynthesis, softening and aroma formation (Robinson and Tomes 1968; Tigchelaar *et al.*, 1973, 1978;

* barrycs@msu.edu

Thompson *et al.*, 1999; Kovacs *et al.*, 2009). Furthermore, although the expression of ethylene-regulated genes can be restored by ethylene treatment, the mutant phenotypes are not reversed, suggesting that these loci act upstream of ethylene to control the ripening process (Tigchelaar *et al.*, 1978; Lincoln and Fischer 1988; Thompson *et al.*, 1999). The *rin*, *nor* and *Cnr* loci have been isolated using positional cloning strategies and each encodes a transcription factor (Vrebalov *et al.*, 2002; Manning *et al.*, 2006; Klee and Giovannoni, 2011).

RIN encodes a MADS-box gene designated MADS-RIN that is a member of the *SEPALLATA* subfamily (Vrebalov *et al.*, 2002). *RIN* influences the expression of many ripening-related genes, and several studies have identified direct targets of the RIN protein including the promoters of the transcription factors *RIN*, *NOR*, *CNR*, *TDR4* and *HB1*, together with the promoters of genes involved in ethylene biosynthesis, aroma formation and cell-wall degradation (Ito *et al.*, 2008; Fujisawa *et al.*, 2011; Martel *et al.*, 2011; Osorio *et al.*, 2011). RNA interference-mediated suppression of a second *MADS* gene, designated *TOMATO AGAMOUS-LIKE 1* (*TAGL1*), also results in pleiotropic phenotypes including inhibition of fruit ripening and reduced carpel expansion in tomato (Itkin *et al.*, 2009; Vrebalov *et al.*, 2009). Furthermore, suppression of *MADS*-box genes in strawberry and bilberry also inhibit ripening, suggesting a broad role for these transcriptional regulators in the ripening process (Jaakola *et al.*, 2010; Seymour *et al.*, 2011). Overexpression of *TAGL1* in tomato causes a range of altered floral and fruit phenotypes and most strikingly leads to the development of fleshy sepals that undergo a form of ripening to accumulate lycopene (Itkin *et al.*, 2009; Vrebalov *et al.*, 2009). A T-DNA insertion mutant within the 5′ untranslated region of *TAGL1*, designated *Arlequin* (*Alq*), causes ectopic expression of *TAGL1*, also leading to the development of fleshy sepals (Gimenez *et al.*, 2010). Interestingly, ectopic expression of *TAGL1* in the *rin* and *nor* mutant backgrounds leads to partial ripening, suggesting that *TAGL1* acts downstream of these regulatory loci (Gimenez *et al.*, 2010). However, *TAGL1* expression is not dramatically reduced in the *rin* mutant background (Vrebalov *et al.*, 2009), implying a more complex relationship between these two transcriptional regulators.

The *Cnr* mutant is caused by an epigenetic mutation that increases methylation in the promoter region of a SQUAMOSA PROMOTER BINDING PROTEIN homologue leading to reduced expression of *CNR* and inhibition of ripening (Manning *et al.*, 2006). The expression of *CNR* is reduced in *rin* and slightly elevated in *TAGL1*-silenced lines (Vrebalov *et al.*, 2009; Martel *et al.*, 2011). RIN binds the *CNR* promoter, but chromatin immunoprecipitation experiments in the *Cnr* mutant background indicate that *CNR* or a protein regulated by CNR is required for the promoter binding activity of RIN (Martel *et al.*, 2011). Overall, these data highlight the importance of transcriptional regulators in regulating ripening but also illustrate that the transcriptional cascade probably involves a complex network of interactions rather than a linear chain of transcriptional events.

15.3 Mutants that Disrupt Hormone Biosynthesis and Signalling Impact on Ripening

Plant hormones control many aspects of plant growth and development together with responses to diverse environmental stimuli. Several of the major plant hormones have been implicated in the development and ripening of fleshy fruit and, for selection of these, their role has been defined through the use of mutants that disrupt either hormone biosynthesis or perception (Srivastava and Handa, 2005; Barry, 2010). Notably, the plant hormone ethylene acts downstream of the ripening regulators *RIN*, *NOR* and *CNR* to regulate ripening in many fleshy fruit species (Klee and Giovannoni, 2011). The importance of ethylene in regulating fruit ripening is well documented and has been reviewed in detail elsewhere (Barry and Giovannoni, 2007; Pech *et al.*, 2008). Briefly, the role of ethylene in the ripening process was defined using a range of approaches including chemical inhibitors that disrupt ethylene biosynthesis or action (Hobson *et al.*, 1984; Yang and Hoffman, 1984; Sisler, 2006), the use of gene-silencing approaches to reduce the activity of 1-aminocyclopropane-1-carboxylate (ACC) synthase or ACC oxidase (Oeller *et al.*, 1991; Picton *et al.*, 1993; Ayub *et al.*, 1996) and the heterologous expression of mutant forms of ethylene receptors to create dominant ethylene insensitivity (Wilkinson *et al.*, 1997). Furthermore, the classical ripening mutants of tomato, *Never-ripe* (*Nr*) and *Green-ripe* (*Gr*), confer dominant ethylene insensitivity leading to inhibition of fruit ripening, and identification of the underlying genes revealed that they are mutated in components of the ethylene signalling pathway (Lanahan *et al.*, 1994; Wilkinson *et al.*, 1995; Barry *et al.*, 2005; Barry and Giovannoni, 2006). In addition, natural variation exists within the germplasm of several fruit crop species as a result of mutations that directly or indirectly influence ethylene biosynthesis or responses leading to slow ripening varieties with improved shelf-life characteristics (Sunako *et al.*, 1999; Harada *et al.*, 2000; Perin *et al.*, 2002; Tatsuki *et al.*, 2006; Wang *et al.*, 2009).

Several studies have implicated abscisic acid (ABA) as playing a regulatory role in fruit ripening. In tomato, ABA levels peak prior to the increase in ethylene synthesis, and ABA treatments stimulate ethylene synthesis and ripening whereas treatment with ABA biosynthesis inhibitors slow ripening (Zhang *et al.*, 2009). Furthermore, silencing of ABA biosynthesis and signalling components led to ripening inhibition in strawberry and reduced rates of cell-wall dissolution in tomato leading to improved shelf-life (Chai *et al.*, 2011; Jia *et al.*, 2011; Sun *et al.*, 2012). ABA-deficient mutants of tomato also exhibit a range of altered fruit phenotypes including increased cuticular permeability, altered pectin composition, increased susceptibility to *Botrytis cinerea*, an increase in plastid compartment size and plastid size, and increased fruit pigmentation (Galpaz *et al.*, 2008; Curvers *et al.*, 2010). These data highlight broad roles for ABA in determining fruit ripening and quality.

15.4 Mutants that Alter Plastid Phenotypes Influence Fruit Ripening and Quality

The chloroplast is a defining organelle in plants that is not only the site of photosynthesis but also functions in the biosynthesis of starch, fatty acids, amino acids, vitamins and diverse isoprenoids, including carotenoids that determine the ripe colour of many fleshy fruits (Armbruster *et al.*, 2011). As mutants that influence components of the chloroplast or chromoplast typically lead to alterations in fruit colour, they represent some of the most easily recognizable mutations that influence fruit ripening. Consequently, colour variants of many fleshy fruit crop species have been selected by breeders and integrated into commercially grown varieties. For example, mutants that disrupt carotenoid biosynthesis in tomato and pepper result in several widely used colour variants, and identification and characterization of the underlying genes have helped to define the carotenoid biosynthesis pathway (Paran and van der Knaap, 2007; Tanaka *et al.*, 2008).

Similarly, mutations that disrupt chlorophyll degradation have been identified in tomato, pepper and orange, and typically lead to fruits that ripen to a brown colour. The *green-flesh* (*gf*) mutant of tomato and the *chlorophyll retainer* (*cl*) mutant of pepper each carry point mutations in a homologue of the *STAY-GREEN* (*SGR*) gene of rice that encodes a protein that interacts with chlorophyll catabolic enzymes at the light-harvesting complex II (Park *et al.*, 2007; Barry *et al.*, 2008; Borovsky and

Paran, 2008; Sakuraba *et al.*, 2012). Interestingly, a survey of heirloom tomatoes that exhibit the *gf* mutant phenotype identified four additional alleles, indicating that this colour trait has been selected on at least five separate occasions, illustrating the novelty value of such colour variants to gardeners, breeders and growers (Barry and Pandey, 2009). The *navel negra* (*nan*) mutant of *Citrus sinensis* also exhibits a stay-green phenotype and the expression of an *SGR* homologue was reduced in *nan* fruit compared with the wild type. However, sequencing of *SGR* from the *nan* mutant failed to reveal any nucleotide changes, suggesting that *nan* may be the result of a mutation in a regulatory factor rather than in a catalytic step of the chlorophyll degradation pathway (Alos *et al.*, 2008).

In addition to simply influencing colour, mutants that directly or indirectly influence fruit plastids have additional pleiotropic effects. For example, carotenoid-deficient mutants of tomato impact fruit flavour characteristics, and both the *gf* and *cl* mutants influence the rate of carotenoid biosynthesis (Ramirez and Tomes, 1964; Roca *et al.*, 2006; Vogel *et al.*, 2010). Similarly, the *high-pigment 1* and *2* mutants that encode tomato homologues of DNA DAMAGE BINDING PROTEIN 1 and DETIOLATED 1, which are involved in ubiquitin-mediated protein turnover, have pleiotropic effects on fruit quality resulting in an increase in chloroplast number and altered plastid ultrastructure, leading to higher levels of chlorophyll, carotenoids and flavonoids, together with altered patterns of aroma volatile production (Yen *et al.*, 1997; Mustilli *et al.*, 1999; Liu *et al.*, 2004; Bino *et al.*, 2005; Kolotilin *et al.*, 2007; Galpaz *et al.*, 2008; Kovacs *et al.*, 2009).

Recent research has shown that mutations disrupting plastid components in tomato can also inhibit or slow the rate of ripening. The *Orange ripening* (*Orr^{DS}*) mutant encodes the M subunit of the plastidial NADH dehydrogenase complex and exhibits inhibition of fruit ripening, including a delay in the onset of ethylene biosynthesis, together with a range of altered ripening-related phenotypes (Nashilevitz *et al.*, 2010). Similarly, the *lutescent 1* and *2* (*l1* and *l2*) mutants display a range of phenotypes indicative of chloroplast defects and rapidly lose chlorophyll, leading to mature fruits that can be almost white prior to the onset of ripening. Fruit of the *l1* and *l2* mutants also show a delay in the onset of ripening and ripening-related ethylene biosynthesis, but once ripening is initiated, the fruit ripen at the same rate as those of wild-type plants (Barry *et al.*, 2012). Positional cloning of the *l2* locus revealed that it encodes a chloroplast-targeted zinc metalloprotease related to ETHYLENE-DEPENDENT GRAVITROPISM DEFICIENT AND YELLOW GREEN 1 (EGY1) of *Arabidopsis* (Barry *et al.*, 2012).

15.5 Mutants With Altered Flavonoid Biosynthesis: The Role of MYB Transcription Factors

Flavonoids, particularly anthocyanins, contribute to the nutritional and visual quality of ripe fruits, and engineering of anthocyanin biosynthesis in fleshy fruits to improve nutritional content has been achieved (Bovy *et al.*, 2002; Butelli *et al.*, 2008). Mutations that influence fruit flavonoid or anthocyanin accumulation have been identified in several species. For example, the *colorless epidermis* (*y*) locus of tomato (*Solanum lycopersicum*) encodes a MYB transcription factor, designated *SlMYB12*, that is required for synthesis of the flavonoid naringenin chalcone in fruit peels (Adato *et al.*, 2009; Ballester *et al.*, 2010). Mutants, genetic variants and transgenic lines in which *SlMYB12* is inactivated have a characteristic pink colour, whereas varieties that contain a functional *Y* gene possess a more vibrant red colour due to the presence of the orange pigmentation associated with naringenin chalcone. In addition, mutants at the *y* locus have reduced cuticle thickness, a lower cutin monomer content, an altered cuticular wax profile and altered elastic properties, although these changes did not result in altered cuticular permeability (Adato *et al.*, 2009). Mutant alleles at the *y* locus do not express *SlMYB12* and, although *SlMYB12* complements the *y* mutation, the exact molecular basis for the mutation remains unknown (Adato *et al.*, 2009; Ballester *et al.*, 2010).

Mutant loci that influence anthocyanin accumulation in diverse fruit crop species have been identified, and several of these loci have been attributed to altered activity of *MYB* transcription factors that directly regulate anthocyanin biosynthesis. In pepper, the dominant *A* locus controls anthocyanin accumulation in pepper fruit through increased expression of a *MYB* transcription factor homologous to *ANTHOCYANIN 2* of petunia (Borovsky *et al.*, 2004). Recently, the *Anthocyanin fruit* (*Aft*) locus of tomato, which is derived from introgression of a chromosomal segment from the wild tomato species *Solanum chilense* into cultivated tomato, was attributed to increased activity of the wild species allele of a second *MYB* transcription factor known as *ANTHOCYANIN 1* (Schreiber *et al.*, 2012). Similarly, red-fleshed apple varieties can be attributed to ectopic expression of *MdMYB10*, and differential methylation within the promoter region of *MdMYB10* correlates with its altered expression and the presence of red stripes versus green stripes in some apple varieties (Chagne *et al.*, 2007; Espley *et al.*, 2007; Telias *et al.*, 2011). The presence of transposable element insertions in *MYB* transcription factors also alters anthocyanin pigmentation in fleshy fruit. A retrotransposon insertion into *VvmybA1* in grape (*Vitis vinifera*) disrupts function leading to loss of anthocyanin pigmentation in white-fruited varieties (Kobayashi *et al.*, 2004). Conversely, distinct retrotransposon insertions within the promoter regions of a *MYB* gene from blood orange confer its cold-dependent fruit-specific expression, leading to an increase in anthocyanin formation (Butelli *et al.*, 2012). Together, these data illustrate that different mechanisms have evolved to control *MYB* gene expression and anthocyanin biosynthesis in diverse fruit crop species.

15.6 Cuticle and Wax Biosynthesis Mutants Alter the Physical Properties of Fruit Surfaces

The aerial surfaces of terrestrial plants are covered with a cuticle, a heterogeneous lipid-based polymer comprised of cutin, together with intra- and epicuticular waxes and polysaccharides (Pollard *et al.*, 2008). The cuticle forms a structural framework, acts as the principal barrier to restrict non-stomatal water loss, possesses reflective properties that reduce heat load and limit the effects of UV radiation, and forms an anti-adhesive layer that helps protect against insect feeding and pathogen infection (Bargel *et al.*, 2006). Many of the genes involved in the synthesis and transport of cutin and waxes have been identified and characterized in *Arabidopsis* (Li-Beisson *et al.*, 2010). However, the cuticle is of central importance to the quality of fleshy fruits, and mechanical failure of the cuticle can result in either pre- or postharvest cracking leading to crop losses (Dominguez *et al.*, 2011). Furthermore, significant cuticle lipid deposition can occur in some fleshy fruits, particularly tomato, where the cuticle can be up to 10 μm thick, yielding more than 1 mg cm^{-2} (Isaacson *et al.*, 2009; Nadakuduti *et al.*, 2012). The cuticular lipid deposition, coupled with the importance of tomato as a model crop species has driven interest in examining the physical and chemical properties of tomato fruit cuticles, and several studies have utilized either genomics- or proteomics-based approaches to identify genes and proteins preferentially associated with epidermal cells and cuticle deposition (Mintz-Oron *et al.*, 2008; Samuels *et al.*, 2008; Yeats *et al.*, 2010; Dominguez *et al.*, 2011; Matas *et al.*, 2011).

Tomato mutants have been characterized that alter the properties of the fruit cuticle. Fruits of the *cutin deficient 1, 2* and *3* (*cd1*, *cd2* and *cd3*) mutants exhibit a 95–98% reduction in cutin content compared with the wild type, and this is accompanied by altered wax composition (Isaacson *et al.*, 2009). The altered chemical properties of the *cd* mutant cuticles influence their physical properties, leading to increased glossiness and altered elasticity. The reduced cutin load of the *cd1* mutant correlated with an increase in the rate of fruit water loss, leading to rapid shrivelling of fruits following harvest. However, the relationship between reduced cutin load and increased rates of water loss is not absolute, as the *cd2* and *cd3* mutants do not exhibit increased rates of postharvest water loss. In contrast, the importance of the cuticle in reducing the incidence of pathogen infection in tomato fruit was highlighted by the observations that *cd1*, *cd2* and *cd3* each displayed an increased susceptibility to postharvest pathogens (Isaacson *et al.*, 2009). Positional cloning revealed that *CD2* encodes a member of the class IV homeodomain-leucine zipper (HD-Zip IV) gene family, a class of transcription factors that have roles in epidermal cell development, including those reported to influence cutin and wax biosynthesis in *Arabidopsis* and maize (Javelle *et al.*, 2010; Wu *et al.*, 2011; Nadakuduti *et al.*, 2012). Recent characterization of the *sticky peel* (*pe*) mutant of tomato indicated that it encodes an allele of *CD2* and possesses an altered fruit phenotype identical to that of the *cd2* mutant. However, a more in-depth characterization of *pe* mutant phenotypes also revealed cutin and wax deficiencies in leaves together with an associated increase in cuticular permeability. In addition, the *pe* mutant has a pale phenotype due to reduced anthocyanin content, and its leaves possess fewer type VI glandular trichomes and consequently have reduced volatile terpene emissions (Nadakuduti *et al.*, 2012). These data reveal that, although *CD2* dramatically alters fruit cuticle formation, it has a broader role in regulating epidermal cell development and specialized metabolism.

The wax composition, and in particular the long-chain alkane (C28–C32) content of cuticles, is typically associated with cuticular permeability (Samuels *et al.*, 2008). Mutants with reduced long-chain alkane content often display enhanced rates of water loss, and the importance of alkanes in maintaining water balance is further highlighted by the findings that the alkane content of cuticular waxes can increase in response to water stress (Riederer and Schreiber, 2001; Kosma *et al.*, 2009; Buschhaus and Jetter, 2011; Wang *et al.*, 2011). Tomato mutants deficient in alkane content in fruit cuticles have increased cuticular permeability, which can lead to enhanced rates of shrivelling on or off the vine. This was observed in a mutant of a very-long-chain fatty acid β-ketoacyl-CoA synthase, designated *LeCER6*, and the *positional sterile* mutant (Vogg *et al.*, 2004; Leide *et al.*, 2011).

15.7 TILLING and the Potential to Identify Additional Mutations in Ripening Genes

The majority of ripening mutants identified to date have strong visible phenotypes that affect pleiotropic ripening processes, fruit colour or cuticular properties. Furthermore, with the exception of the large mutant collection developed in the M82 background (Menda *et al.*, 2004), ripening mutants have not been identified through systematic screens, with the majority of the classical ripening mutants occurring spontaneously and identified in grower fields or during large breeding programmes. Consequently, mutants that alter more subtle phenotypes associated with the ripening process such as cell-wall breakdown or aroma formation are generally not available, and progress in elucidating these ripening pathways have generally been made through gene-silencing approaches such as RNA interference. However, while gene-silencing approaches can be powerful for determining gene function, they typically lead to 'knockdown' phenotypes rather than 'knockout' phenotypes. In contrast, an allelic series of mutations within a gene can provide a range of activities from no phenotypic effect through to complete loss of function and, in the case of substitution mutations, can provide valuable information on the contribution of individual amino acid residues to protein function.

Targeting-induced local lesions in genomes (TILLING) has been utilized for identifying allelic series of mutations within genes of interest in ethyl methanesulfonate-generated mutant populations (McCallum *et al.*, 2000a,b; Colbert *et al.*, 2001). TILLING platforms have been established for several model species, including the fleshy fruit-bearing species of tomato and melon (Gady *et al.*, 2009; Dahmani-Mardas *et al.*, 2010; Piron *et al.*, 2010; Okabe *et al.*, 2011; Gady *et al.*, 2012). A TILLING population generated in the 'Micro-Tom' cultivar was utilized for identifying mutants in several ripening-related genes including the family of six tomato ethylene receptors (Okabe *et al.*, 2011). Characterization of point mutations within the N-terminal domain of the ETR1 ethylene receptor identified two mutants designated *Sletr1-1* and *Sletr1-2* carrying the amino acid substitutions P51L and V69D, respectively, which resulted in dominant ethylene insensitivity. The utility of TILLING for recovering mutant alleles of varying strengths was demonstrated, as the *Sletr1-1* mutant allele consistently resulted in stronger ethylene-insensitive phenotypes than observed with the *Sletr1-2* allele (Okabe *et al.*, 2011). Similarly, two TILLING-derived alleles of varying severity were identified in *PHYTOENE SYNTHASE 1* (*PSY1*) of tomato, which controls substrate flow into the carotenoid biosynthesis pathway during fruit ripening (Gady *et al.*, 2012). A TILLING platform has also been established in melon and mutants in several genes influencing fruit quality traits recovered, including mutants within the ACC oxidase (*ACO1*) gene, with a G194D substitution resulting in a delayed fruit-ripening and extended shelf-life phenotype (Dahmani-Mardas *et al.*, 2010). Together, these studies demonstrate the utility of TILLING as an approach to identify new alleles in known ripening-related genes, but the technology should also be applicable for investigating the function of candidate genes identified through expression analyses, protein–protein interaction studies and other functional approaches.

15.8 Conclusions and Perspectives

Mutants that disrupt aspects of fruit development and ripening have yielded important insights into multiple aspects of the ripening process and have played a pivotal role in identifying genes involved in ripening. Notably, the transcriptional cascade involving RIN, NOR, CNR and TAGL1 was defined in part through analysis of the corresponding ripening mutants. Chromatin immunoprecipitation approaches are being utilized to identify the targets of these transcription factors, and as genome-wide approaches for determination of transcription factor binding sites become more routine, it is likely that a complete transcriptional network required for ripening will become available. This will provide insights into the unique and overlapping roles of each of these transcription factors during ripening. Furthermore, as genomics-based methodologies continue to develop, the ability to generate molecular and metabolic phenotypes increases, creating the potential to establish a more complete understanding of altered phenotypes associated with each ripening mutant. These technologies will help to create a systems-level understanding of ripening-related networks that are likely to have the genes identified through characterizing ripening mutants as the principal nodes.

The underlying genes for the majority of the known ripening mutants of tomato have been cloned and characterized. Furthermore, given the logistics of performing large-scale mutant screens to uncover fruit ripening genes, it seems unlikely that additional large genetic screens for ripening mutants will be conducted, and it seems more likely that reverse genetics-based methods for determining gene function will become more widespread. In addition to gene-silencing approaches, TILLING currently would be the method of choice for functional analysis of ripening-related genes and could be applied to a range of fleshy fruit-bearing species. The attractiveness of TILLING lies in the potential to generate an allelic series of mutants in a gene of interest, enabling a range of substitution alleles to be recovered and their effects on the phenotype established. However, in general, current population sizes and mutagenesis rates are insufficient to generate on average more than one or two mutant alleles per gene in tomato and melon. Therefore, additional efforts will be needed to increase population sizes to increase the likelihood of recovering multiple alleles and for the TILLING platforms to become more widely accessible.

This review has focused on research using tomato as a model system, where it is comparatively easier than in other fleshy fruit-bearing species to identify the underlying genes responsible for mutant phenotypes. However, as advances in next-generation sequencing technology continue to progress, it will be possible to directly sequence multiple cultivars or individuals from populations that have altered ripening behaviours. Such approaches will allow natural allelic variation in ripening and other fruit quality traits to be explored in a range of fruit crop species, thereby uncovering more subtle phenotypes associated with ripening genes, including those that could be directly incorporated into elite breeding germplasm to facilitate crop improvement.

References

Adato, A., Mandel, T., Mintz-Oron, S., Venger, I., Levy, D., Yativ, M., Dominguez, E., Wang, Z.H., De Vos, R.C.H., Jetter, R., Schreiber, L., Heredia, A., Rogachev, I. and Aharoni, A. (2009) Fruit-surface flavonoid accumulation in tomato is controlled by a *SlMYB12*-regulated transcriptional network. *PLOS Genetics* 5, e1000777.

Alos, E., Roca, M., Iglesias, D.J., Minguez-Mosquera, M.I., Damasceno, C.M.B., Thannhauser, T.W., Rose, J.K.C., Talon, M. and Cercos, M. (2008) An evaluation of the basis and consequences of a stay-green mutation in the *navel negra* citrus mutant using transcriptomic and proteomic profiling and metabolite analysis. *Plant Physiology* 147, 1300–1315.

Armbruster, U., Pesaresi, P., Pribil, M., Hertle, A. and Leister, D. (2011) Update on chloroplast research: new tools, new topics, and new trends. *Molecular Plant* 4, 1–16.

Ayub, R., Guis, M., BenAmor, M., Gillot, L., Roustan, J.P., Latche, A., Bouzayen, M. and Pech, J.C. (1996) Expression of ACC oxidase antisense gene inhibits ripening of cantaloupe melon fruits. *Nature Biotechnology* 14, 862–866.

Ballester, A.R., Molthoff, J., de Vos, R., Hekkert, B.T.L., Orzaez, D., Fernandez-Moreno, J.P., Tripodi, P., Grandillo, S., Martin, C., Heldens, J., Ykema, M., Granell, A. and Bovy, A. (2010) Biochemical and molecular analysis of pink tomatoes: deregulated expression of the gene encoding transcription factor SlMYB12 leads to pink tomato fruit color. *Plant Physiology* 152, 71–84.

Bargel, H., Koch, K., Cerman, Z. and Neinhuis, C. (2006) Structure–function relationships of the plant cuticle and cuticular waxes – a smart material? *Functional Plant Biology* 33, 893–910.

Barry, C.S. (2010). Factors influencing the ripening and quality of fleshy fruits. *Annual Plant Reviews* 38, 296–326.

Barry, C.S. and Giovannoni, J.J. (2006) Ripening in the tomato *Green-ripe* mutant is inhibited by ectopic expression of a protein that disrupts ethylene signaling. *Proceedings of the National Academy of Sciences USA* 103, 7923–7928.

Barry, C.S. and Giovannoni, J.J. (2007) Ethylene and fruit ripening. *Journal of Plant Growth Regulation* 26, 143–159.

Barry, C.S. and Pandey, P. (2009) A survey of cultivated heirloom tomato varieties identifies four new mutant alleles at the *green-flesh* locus. *Molecular Breeding* 24, 269–276.

Barry, C.S., McQuinn, R.P., Thompson, A.J., Seymour, G.B., Grierson, D. and Giovannoni, J.J. (2005) Ethylene insensitivity conferred by the *Green-ripe* and *Never-ripe 2* ripening mutants of tomato. *Plant Physiology* 138, 267–275.

Barry, C.S., McQuinn, R.P., Chung, M.-Y., Besuden, A. and Giovannoni, J.J. (2008) Amino acid substitutions in homologs of the STAY-GREEN protein are responsible for the *green-flesh* and *chlorophyll retainer* mutations of tomato and pepper. *Plant Physiology* 147, 179–187.

Barry, C.S., Aldridge, G.M., Herzog, G., Ma, Q., Hirschberg, J. and Giovannoni, J.J. (2012) Altered chloroplast development and delayed fruit ripening caused by mutations in a zinc metalloprotease at the *lutescent 2* locus of tomato. *Plant Physiology* 159, 1086–1098.

Bino, R.J., de Vos, C.H.R., Lieberman, M., Hall, R.D., Bovy, A., Jonker, H.H., Tikunov, Y., Lommen, A., Moco, S. and Levin, I. (2005) The light-hyperresponsive *high pigment-2* (*dg*) mutation of tomato: alterations in the fruit metabolome. *New Phytologist* 166, 427–438.

Borovsky, Y. and Paran, I. (2008) Chlorophyll breakdown during pepper fruit ripening in the *chlorophyll retainer* mutation is impaired at the homolog of the senescence-inducible *stay-green* gene. *Theoretical and Applied Genetics* 117, 235–240.

Borovsky, Y., Oren-Shamir, M., Ovadia, R., De Jong, W. and Paran, I. (2004) The *A* locus that controls anthocyanin accumulation in pepper encodes a MYB transcription factor homologous to Anthocyanin2 of Petunia. *Theoretical and Applied Genetics* 109, 23–29.

Bovy, A., de Vos, R., Kemper, M., Schijlen, E., Pertejo, M.A., Muir, S., Collins, G., Robinson, S., Verhoeyen, M., Hughes, S., Santos-Buelga, C. and van Tunen, A. (2002) High-flavonol tomatoes resulting from the heterologous expression of the maize transcription factor genes *LC* and *C1*. *Plant Cell* 14, 2509–2526.

Buschhaus, C. and Jetter, R. (2011) Composition differences between epicuticular and intracuticular wax substructures: how do plants seal their epidermal surfaces? *Journal of Experimental Botany* 62, 841–853.

Butelli, E., Titta, L., Giorgio, M., Mock, H.-P., Matros, A., Peterek, S., Schijlen, E.G.W.M., Hall, R.D., Bovy, A.G., Luo, J. and Martin, C. (2008) Enrichment of tomato fruit with health-promoting anthocyanins by expression of select transcription factors. *Nature Biotechnology* 26, 1301–1308.

Butelli, E., Licciardello, C., Zhang, Y., Liu, J., Mackay, S., Bailey, P., Reforgiato-Recupero, G. and Martin, C. (2012) Retrotransposons control fruit-specific, cold-dependent accumulation of anthocyanins in blood oranges. *Plant Cell* 24, 1242–1255.

Chagne, D., Carlisle, C.M., Blond, C., Volz, R.K., Whitworth, C.J., Oraguzie, N.C., Crowhurst, R.N., Allan, A.C., Espley, R.V., Hellens, R.P. and Gardiner, S.E. (2007) Mapping a candidate gene (*MdMYB10*) for red flesh and foliage colour in apple. *BMC Genomics* 8, 212.

Chai, Y.M., Jia, H.F., Li, C.L., Dong, Q.H. and Shen, Y.Y. (2011) *FaPYR1* is involved in strawberry fruit ripening. *Journal of Experimental Botany* 62, 5079–5089.

Colbert, T., Till, B.J., Tompa, R., Reynolds, S., Steine, M.N., Yeung, A.T., McCallum, C.M., Comai, L. and Henikoff, S. (2001) High-throughput screening for induced point mutations. *Plant Physiology* 126, 480–484.

Curvers, K., Seifi, H., Mouille, G., de Rycke, R., Asselbergh, B., Van Hecke, A., Vanderschaeghe, D., Hofte, H., Callewaert, N., Van Breusegem, F. and Hofte, M. (2010) Abscisic acid deficiency causes changes in cuticle permeability and pectin composition that influence tomato resistance to *Botrytis cinerea*. *Plant Physiology* 154, 847–860.

Dahmani-Mardas, F., Troadec, C., Boualem, A., Leveque, S., Alsadon, A.A., Aldoss, A.A., Dogimont, C. and Bendahmane, A. (2010) Engineering melon plants with improved fruit shelf life using the TILLING approach. *PLOS One* 5, e15776.

Dominguez, E., Cuartero, J. and Heredia, A. (2011) An overview on plant cuticle biomechanics. *Plant Science* 181, 77–84.

Espley, R.V., Hellens, R.P., Putterill, J., Stevenson, D.E., Kutty-Amma, S. and Allan, A.C. (2007) Red colouration in apple fruit is due to the activity of the MYB transcription factor, MdMYB10. *Plant Journal* 49, 414–427.

Fujisawa, M., Nakano, T. and Ito, Y. (2011) Identification of potential target genes for the tomato fruit-ripening regulator RIN by chromatin immunoprecipitation. *BMC Plant Biology* 11, 26.

Gady, A.L.F., Hermans, F.W.K., Van de Wal, M.H.B.J., van Loo, E.N., Visser, R.G.F. and Bachem, C.W.B. (2009) Implementation of two high through-put techniques in a novel application: detecting point mutations in large EMS mutated plant populations. *Plant Methods* 5, 13.

Gady, A.L.F., Vriezen, W.H., Van de Wal, M., Huang, P.P., Bovy, A.G., Visser, R.G.F. and Bachem, C.W.B. (2012) Induced point mutations in the phytoene synthase 1 gene cause differences in carotenoid content during tomato fruit ripening. *Molecular Breeding* 29, 801–812.

Galpaz, N., Wang, Q., Menda, N., Zamir, D. and Hirschberg, J. (2008) Abscisic acid deficiency in the tomato mutant *high-pigment 3* leading to increased plastid number and higher fruit lycopene content. *Plant Journal* 53, 717–730.

Gimenez, E., Pineda, B., Capel, J., Teresa Anton, M., Atares, A., Perez-Martin, F., Garcia-Sogo, B., Angosto, T., Moreno, V. and Lozano, R. (2010) Functional analysis of the *Arlequin* mutant corroborates the essential role of the *ARLEQUIN/TAGL1* gene during reproductive development of tomato. *PLOS One* 5, e14427.

Harada, T., Sunako, T., Wakasa, Y., Soejima, J., Satoh, T. and Niizeki, M. (2000) An allele of the 1-aminocyclopropane-1-carboxylate synthase gene (*Md-ACS1*) accounts for the low level of ethylene production in climacteric fruits of some apple cultivars. *Theoretical and Applied Genetics* 101, 742–746.

Hobson, G.E., Nichols, R., Davies, J.N. and Atkey, P.T. (1984) The inhibition of tomato *Lycopersicon esculentum* fruit ripening by silver. *Journal of Plant Physiology* 116, 21–30.

Isaacson, T., Kosma, D.K., Matas, A.J., Buda, G.J., He, Y.H., Yu, B.W., Pravitasari, A., Batteas, J.D., Stark, R.E., Jenks, M.A. and Rose, J.K.C. (2009) Cutin deficiency in the tomato fruit cuticle consistently affects resistance to microbial infection and biomechanical properties, but not transpirational water loss. *Plant Journal* 60, 363–377.

Itkin, M., Seybold, H., Breitel, D., Rogachev, I., Meir, S. and Aharoni, A. (2009) *TOMATO AGAMOUS-LIKE 1* is a component of the fruit ripening regulatory network. *Plant Journal* 60, 1081–1095.

Ito, Y., Kitagawa, M., Ihashi, N., Yabe, K., Kimbara, J., Yasuda, J., Ito, H., Inakuma, T., Hiroi, S. and Kasumi, T. (2008) DNA-binding specificity, transcriptional activation potential, and the rin mutation effect for the tomato fruit-ripening regulator RIN. *Plant Journal* 55, 212–223.

Jaakola, L., Poole, M., Jones, M.O., Kamarainen-Karppinen, T., Koskimaki, J.J., Hohtola, A., Haggman, H., Fraser, P.D., Manning, K., King, G.J., Thomson, H. and Seymour, G.B. (2010) A SQUAMOSA MADS box gene involved in the regulation of anthocyanin accumulation in bilberry fruits. *Plant Physiology* 153, 1619–1629.

Javelle, M., Vernoud, V., Depege-Fargeix, N., Arnould, C., Oursel, D., Domergue, F., Sarda, X. and Rogowsky, P.M. (2010) Overexpression of the epidermis-specific homeodomain-leucine zipper IV transcription factor *OUTER CELL LAYER1* in maize identifies target genes involved in lipid metabolism and cuticle biosynthesis. *Plant Physiology* 154, 273–286.

Jia, H.F., Chai, Y.M., Li, C.L., Lu, D., Luo, J.J., Qin, L. and Shen, Y.Y. (2011) Abscisic acid plays an important role in the regulation of strawberry fruit ripening. *Plant Physiology* 157, 188–199.

Klee, H.J. and Giovannoni, J.J. (2011) Genetics and control of tomato fruit ripening and quality attributes. *Annual Review of Genetics* 45, 41–59.

Kobayashi, S., Goto-Yamamoto, N. and Hirochika, H. (2004) Retrotransposon-induced mutations in grape skin color. *Science* 304, 982.

Kolotilin, I., Koltai, H., Tadmor, Y., Bar-Or, C., Reuveni, M., Meir, A., Nahon, S., Shlomo, H., Chen, L. and Levin, I. (2007) Transcriptional profiling of *high pigment-2* (*dg*) tomato mutant links early fruit plastid biogenesis with its overproduction of phytonutrients. *Plant Physiology* 145, 389–401.

Kosma, D.K., Bourdenx, B., Bernard, A., Parsons, E.P., Lu, S., Joubes, J. and Jenks, M.A. (2009) The impact of water deficiency on leaf cuticle lipids of Arabidopsis. *Plant Physiology* 151, 1918–1929.

Kovacs, K., Fray, R.G., Tikunov, Y., Graham, N., Bradley, G., Seymour, G.B., Bovy, A.G. and Grierson, D. (2009) Effect of tomato pleiotropic ripening mutations on flavour volatile biosynthesis. *Phytochemistry* 70, 1003–1008.

Lanahan, M.B., Yen, H.-C., Giovannoni, J.J. and Klee, H.J. (1994) The *Never ripe* mutation blocks ethylene perception in tomato. *Plant Cell* 6, 521–530.

Leide, J., Hildebrandt, U., Vogg, G. and Riederer, M. (2011) The *positional sterile* (*ps*) mutation affects cuticular transpiration and wax biosynthesis of tomato fruits. *Journal of Plant Physiology* 168, 871–877.

Li-Beisson, Y., Shorrosh, B., Beisson, F., Andersson, M.X., Arondel, V., Bates, P.D., Baud, S.B., Bird, D., DeBono, A., Durrett, T.P., Franke, R.B., Graham, I.A., Katayama, K., Kelly, A.L.A., Larson, T., Markham, J.E., Miquel, M., Molina, I., Nishida, I., Rowland, O., Samuels, L., Schmid, K.M., Wada, H., Welti, R., Xu, C., Zallot, R.M. and Ohlrogge, J. (2010) Acyl-lipid metabolism. *The Arabidopsis Book* 8, e0133.

Lincoln, J.E. and Fischer, R.L. (1988) Regulation of gene-expression by ethylene in wild-type and *rin* tomato (*Lycopersicon esculentum*) fruit. *Plant Physiology* 88, 370–374.

Liu, Y.S., Roof, S., Ye, Z.B., Barry, C., van Tuinen, A., Vrebalov, J., Bowler, C. and Giovannoni, J. (2004) Manipulation of light signal transduction as a means of modifying fruit nutritional quality in tomato. *Proceedings of the National Academy of Sciences USA* 101, 9897–9902.

Manning, K., Tor, M., Poole, M., Hong, Y., Thompson, A.J., King, G.J., Giovannoni, J.J. and Seymour, G.B. (2006) A naturally occurring epigenetic mutation in a gene encoding an SBP-box transcription factor inhibits tomato fruit ripening. *Nature Genetics* 38, 948–952.

Martel, C., Vrebalov, J., Tafelmeyer, P. and Giovannoni, J.J. (2011) The tomato MADS-box transcription factor RIPENING INHIBITOR interacts with promoters involved in numerous ripening processes in a COLORLESS NONRIPENING-dependent manner. *Plant Physiology* 157, 1568–1579.

Matas, A.J., Yeats, T.H., Buda, G.J., Zheng, Y., Chatterjee, S., Tohge, T., Ponnala, L., Adato, A., Aharoni, A., Stark, R., Fernie, A.R., Fei, Z.J., Giovannoni, J.J. and Rose, J.K.C. (2011) Tissue- and cell-type specific transcriptome profiling of expanding tomato fruit provides insights into metabolic and regulatory specialization and cuticle formation. *Plant Cell* 23, 3893–3910.

McCallum, C.M., Comai, L., Greene, E.A. and Henikoff, S. (2000a) Targeted screening for induced mutations. *Nature Biotechnology* 18, 455–457.

McCallum, C.M., Comai, L., Greene, E.A. and Henikoff, S. (2000b) Targeting induced local lesions in genomes (TILLING) for plant functional genomics. *Plant Physiology* 123, 439–442.

Menda, N., Semel, Y., Peled, D., Eshed, Y. and Zamir, D. (2004) *In silico* screening of a saturated mutation library of tomato. *Plant Journal* 38, 861–872.

Mintz-Oron, S., Mandel, T., Rogachev, I., Feldberg, L., Lotan, O., Yativ, M., Wang, Z., Jetter, R., Venger, I., Adato, A. and Aharoni, A. (2008) Gene expression and metabolism in tomato fruit surface tissues. *Plant Physiology* 147, 823–851.

Mustilli, A.C., Fenzi, F., Ciliento, R., Alfano, F. and Bowler, C. (1999) Phenotype of the tomato *high pigment-2* mutant is caused by a mutation in the tomato homolog of *DEETIOLATED1*. *Plant Cell* 11, 145–157.

Nadakuduti, S.S., Pollard, M., Kosma, D.K., Allen, C.J., Ohlrogge, J.B. and Barry, C.S. (2012) Pleiotropic phenotypes of the *sticky peel* (*pe*) mutant provide new insight into the role of *CUTIN DEFICIENT 2* in epidermal cell function in tomato. *Plant Physiology* 159, 945–960.

Nashilevitz, S., Melamed-Bessudo, C., Izkovich, Y., Rogachev, I., Osorio, S., Itkin, M., Adato, A., Pankratov, I., Hirschberg, J., Fernie, A.R., Wolf, S., Usadel, B., Levy, A.A., Rumeau, D. and Aharoni, A. (2010) An *Orange Ripening* mutant links plastid NAD (P)H dehydrogenase complex activity to central and specialized metabolism during tomato fruit maturation. *Plant Cell* 22, 1977–1997.

Oeller, P.W., Min Wong, L., Taylor, L.P., Pike, D.A. and Theologis, A. (1991) Reversible inhibition of tomato fruit senescence by antisense RNA. *Science* 254, 437–439.

Okabe, Y., Asamizu, E., Saito, T., Matsukura, C., Ariizumi, T., Bres, C., Rothan, C., Mizoguchi, T. and Ezura, H. (2011) Tomato TILLING technology: development of a reverse genetics tool for the efficient isolation of mutants from Micro-Tom mutant libraries. *Plant and Cell Physiology* 52, 1994–2005.

Osorio, S., Alba, R., Damasceno, C.M.B., Lopez-Casado, G., Lohse, M., Zanor, M.I., Tohge, T., Usadel, B., Rose, J.K.C., Fei, Z.J., Giovannoni, J.J. and Fernie, A.R. (2011) Systems biology of tomato fruit development: combined transcript, protein, and metabolite analysis of tomato transcription factor (*nor*, *rin*) and ethylene receptor (*Nr*) mutants reveals novel regulatory interactions. *Plant Physiology* 157, 405–425.

Paran, I. and van der Knaap, E. (2007) Genetic and molecular regulation of fruit and plant domestication traits in tomato and pepper. *Journal of Experimental Botany* 58, 3841–3852.

Park, S.Y., Yu, J.W., Park, J.S., Li, J., Yoo, S.C., Lee, N.Y., Lee, S.K., Jeong, S.W., Seo, H.S., Koh, H.J., Jeon, J.S., Park, Y.I. and Paek, N.C. (2007) The senescence-induced staygreen protein regulates chlorophyll degradation. *Plant Cell* 19, 1649–1664.

Pech, J.C., Bouzayen, M. and Latche, A. (2008) Climacteric fruit ripening: ethylene-dependent and independent regulation of ripening pathways in melon fruit. *Plant Science* 175, 114–120.

Perin, C., Gomez-Jimenez, M., Hagen, L., Dogimont, C., Pech, J.-C., Latche, A., Pitrat, M. and Lelievre, J.-M. (2002) Molecular and genetic characterization of a non-climacteric phenotype in melon reveals two loci conferring altered ethylene response in fruit. *Plant Physiology* 129, 300–309.

Picton, S., Barton, S.L., Bouzayen, M., Hamilton, A.J. and Grierson, D. (1993) Altered fruit ripening and leaf senescence in tomatoes expressing an antisense ethylene-forming enzyme transgene. *Plant Journal* 3, 469–481.

Piron, F., Nicolai, M., Minoia, S., Piednoir, E., Moretti, A., Salgues, A., Zamir, D., Caranta, C. and Bendahmane, A. (2010) An induced mutation in tomato eIF4E leads to immunity to two potyviruses. *PLOS One* 5, e11313.

Pollard, M., Beisson, F., Li, Y.H. and Ohlrogge, J.B. (2008) Building lipid barriers: biosynthesis of cutin and suberin. *Trends in Plant Science* 13, 236–246.

Ramirez, D.A. and Tomes, M.L. (1964) Relationship between chlorophyll and carotenoid biosynthesis in dirty-red (*green-flesh*) mutant in tomato. *Botanical Gazette* 125, 221–226.

Riederer, M. and Schreiber, L. (2001) Protecting against water loss: analysis of the barrier properties of plant cuticles. *Journal of Experimental Botany* 52, 2023–2032.

Robinson, R.W. and Tomes, M.L. (1968) Ripening inhibitor: a gene with multiple effects on ripening. *Report of the Tomato Genetics Cooperative* 18, 36–37.

Roca, M., Hornero-Mendez, D., Gandul-Rojas, B. and Minguez-Mosquera, M.I. (2006) Stay-green phenotype slows the carotenogenic process in *Capsicum annuum* (L.) fruits. *Journal of Agricultural and Food Chemistry* 54, 8782–8787.

Sakuraba, Y., Schelbert, S., Park, S.-Y., Han, S.-H., Lee, B.-D., Besagni Andres, C., Kessler, F., Hortensteiner, S. and Paek, N.-C. (2012) STAY-GREEN and chlorophyll catabolic enzymes interact at light-harvesting complex II for chlorophyll detoxification during leaf senescence in *Arabidopsis*. *Plant Cell* 24, 507–518.

Samuels, L., Kunst, L. and Jetter, R. (2008) Sealing plant surfaces: cuticular wax formation by epidermal cells. *Annual Review of Plant Biology* 59, 683–707.

Schreiber, G., Reuveni, M., Evenor, D., Oren-Shamir, M., Ovadia, R., Sapir-Mir, M., Bootbool-Man, A., Nahon, S., Shlomo, H., Chen, L. and Levin, I. (2012) *ANTHOCYANIN1* from *Solanum chilense* is more efficient in accumulating anthocyanin metabolites than its *Solanum lycopersicum* counterpart in association with the ANTHOCYANIN FRUIT phenotype of tomato. *Theoretical and Applied Genetics* 124, 295–307.

Seymour, G.B., Ryder, C.D., Cevik, V., Hammond, J.P., Popovich, A., King, G.J., Vrebalov, J., Giovannoni, J.J. and Manning, K. (2011) A *SEPALLATA* gene is involved in the development and ripening of strawberry (*Fragaria × ananassa* Duch.) fruit, a non-climacteric tissue. *Journal of Experimental Botany* 62, 1179–1188.

Sisler, E.C. (2006) The discovery and development of compounds counteracting ethylene at the receptor level. *Biotechnology Advances* 24, 357–367.

Srivastava, A. and Handa, A.K. (2005) Hormonal regulation of tomato fruit development: a molecular perspective. *Journal of Plant Growth Regulation* 24, 67–82.

Sun, L., Sun, Y.F., Zhang, M., Wang, L., Ren, J., Cui, M.M., Wang, Y.P., Ji, K., Li, P., Li, Q., Chen, P., Dai, S.J., Duan, C.R., Wu, Y. and Leng, P. (2012) Suppression of 9-*cis*-epoxycarotenoid dioxygenase, which encodes a key enzyme in abscisic acid biosynthesis, alters fruit texture in transgenic tomato. *Plant Physiology* 158, 283–298.

Sunako, R., Sakuraba, W., Senda, M., Akada, S., Ishikawa, R., Niizeki, M. and Harada, T. (1999) An allele of the ripening-specific 1-aminocyclopropane-1-carboxylic acid synthase gene (*ACS1*) in apple fruit with a long storage life. *Plant Physiology* 119, 1297–1303.

Tanaka, Y., Sasaki, N. and Ohmiya, A. (2008) Biosynthesis of plant pigments: anthocyanins, betalains and carotenoids. *Plant Journal* 54, 733–749.

Tatsuki, M., Haji, T. and Yamaguchi, M. (2006) The involvement of 1-aminocyclopropane-1-carboxylic acid synthase isogene, *Pp-ACS1*, in peach fruit softening. *Journal of Experimental Botany* 57, 1281–1289.

Telias, A., Kui, L.W., Stevenson, D.E., Cooney, J.M., Hellens, R.P., Allan, A.C., Hoover, E.E. and Bradeen, J.M. (2011) Apple skin patterning is associated with differential expression of *MYB10*. *BMC Plant Biology* 11, 93.

Thompson, A.J., Tor, M., Barry, C.S., Vrebalov, J., Orfila, C., Jarvis, M.C., Giovannoni, J.J., Grierson, D. and Seymour, G.B. (1999) Molecular and genetic characterization of a novel pleiotropic tomato-ripening mutant. *Plant Physiology* 120, 383–389.

Tigchelaar, E.C., Tomes, M.L., Kerr, E.A. and Barman, R.J. (1973) A new fruit ripening mutant, *non-ripening* (*nor*). *Report of the Tomato Genetics Cooperative* 23, 33.

Tigchelaar, E.C., McGlasson, W.B. and Buescher, R.W. (1978) Genetic regulation of tomato fruit ripening. *HortScience* 13, 508–513.

Vogel, J.T., Tieman, D.M., Sims, C.A., Odabasi, A.Z., Clark, D.G. and Klee, H.J. (2010) Carotenoid content impacts flavor acceptability in tomato (*Solanum lycopersicum*). *Journal of the Science of Food and Agriculture* 90, 2233–2240.

Vogg, G., Fischer, S., Leide, J., Emmanuel, E., Jetter, R., Levy, A.A. and Riederer, M. (2004) Tomato fruit cuticular waxes and their effects on transpiration barrier properties: functional characterization of a mutant deficient in a very-long-chain fatty acid β-ketoacyl-CoA synthase. *Journal of Experimental Botany* 55, 1401–1410.

Vrebalov, J., Ruezinsky, D., Padmanabhan, V., White, R., Medrano, D., Drake, R., Schuch, W. and Giovannoni, J. (2002) A *MADS*-box gene necessary for fruit ripening at the tomato *ripening-inhibitor* (*rin*) locus. *Science* 296, 343–346.

Vrebalov, J., Pan, I.L., Arroyo, A.J.M., McQuinn, R., Chung, M., Poole, M., Rose, J.K.C., Seymour, G., Grandillo, S., Giovannoni, J. and Irish, V.F. (2009) Fleshy fruit expansion and ripening are regulated by the tomato *SHATTERPROOF* gene *TAGL1*. *Plant Cell* 21, 3041–3062.

Wang, A., Yamakake, J., Kudo, H., Wakasa, Y., Hatsuyama, Y., Igarashi, M., Kasai, A., Li, T.Z. and Harada, T. (2009) Null mutation of the *MdACS3* gene, coding for a ripening-specific 1-aminocyclopropane-1-carboxylate synthase, leads to long shelf life in apple fruit. *Plant Physiology* 151, 391–399.

Wang, Z.Y., Xiong, L.M., Li, W.B., Zhu, J.K. and Zhu, J.H. (2011) The plant cuticle is required for osmotic stress regulation of abscisic acid biosynthesis and osmotic stress tolerance in *Arabidopsis*. *Plant Cell* 23, 1971–1984.

Wilkinson, J.Q., Lanahan, M.B., Yen, H.-C., Giovannoni, J.J. and Klee, H.J. (1995) An ethylene-inducible component of signal transduction encoded by *Never-ripe*. *Science* 270, 1807–1809.

Wilkinson, J.Q., Lanahan, M.B., Clark, D.G., Bleecker, A.B., Chang, C., Meyerowitz, E.M. and Klee, H.J. (1997) A dominant mutant receptor from Arabidopsis confers ethylene insensitivity in heterologous plants. *Nature Biotechnology* 15, 444–447.

Wu, R.H., Li, S.B., He, S., Wassmann, F., Yu, C.H., Qin, G.J., Schreiber, L., Qu, L.J. and Gu, H.Y. (2011) CFL1, a WW domain protein, regulates cuticle development by modulating the function of HDG1, a class IV homeodomain transcription factor, in rice and *Arabidopsis*. *Plant Cell* 23, 3392–3411.

Yang, S.F. and Hoffman, N.E. (1984) Ethylene biosynthesis and its regulation in higher plants. *Annual Review of Plant Physiology and Plant Molecular Biology* 35, 155–190.

Yeats, T.H., Howe, K.J., Matas, A.J., Buda, G.J., Thannhauser, T.W. and Rose, J.K.C. (2010) Mining the surface proteome of tomato (*Solanum lycopersicum*) fruit for proteins associated with cuticle biogenesis. *Journal of Experimental Botany* 61, 3759–3771.

Yen, H.C., Shelton, B.A., Howard, L.R., Lee, S., Vrebalov, J. and Giovannoni, J.J. (1997) The tomato *high-pigment* (*hp*) locus maps to chromosome 2 and influences plastome copy number and fruit quality. *Theoretical and Applied Genetics* 95, 1069–1079.

Zhang, M., Yuan, B. and Leng, P. (2009) The role of ABA in triggering ethylene biosynthesis and ripening of tomato fruit. *Journal of Experimental Botany* 60, 1579–1588.

16 Biotechnology of Fruit Quality

Avtar K. Handa,[1]* Raheel Anwar[1,2] and Autar K. Mattoo[3]
[1]Department of Horticulture and Landscape Architecture, Purdue University, West Lafayette, IN, USA; [2]Institute of Horticultural Sciences, University of Agriculture, Faisalabad, Punjab, Pakistan; [3]Sustainable Agricultural Systems Laboratory, USDA-ARS, Beltsville Agricultural Research Center, Beltsville, MD, USA

16.1 Introduction

Fruit and vegetable crops are the major dietary source of vitamins, antioxidants and minerals and have the potential not only to ameliorate physiological disorders but also to decrease the incidence of human diseases such as cancer. Consequently, consumption of fruits and vegetables has increased in recent years, further increasing their global demand. Consumers expect good-quality fruit to be flavourful, succulent, juicy and nutritional, in addition to being attractive in size and appearance. Other consumer-desirable characteristics of fruits include crispness, chewiness and oiliness. However, for the fruit handler, shipper and retailer, the desirable fruit quality attributes include being less prone to handling and shipping damages, slow softening during storage and longer shelf-life, without affecting consumer appeal. Fruit processors consider better-quality fruit to have a higher proportion of solids, appropriate rheological properties, tolerance to mechanical processing including during peeling or crushing, and prolonged maintenance of the processed products during marketing. A recent trend towards organic farming adds another set of desirable parameters to fruit quality (Lind *et al.*, 2003; Reich, 2012). Enhanced phytonutrient levels add to the overall quality of fruit crops (Mattoo *et al.*, 2010), although consumers expect fruits at the same time to be free of unfavourable chemicals such as cyanogenic glucosides, oxalates, heavy metals, dioxane and pesticides, and contaminations due to microbes.

Following domestication of crop plants, the traditional breeding approaches have extensively improved certain qualities of horticultural crops. In the last three decades, several new tools, especially quantitative trait locus (QTL) mapping, have allowed identification of regions of the genome associated with particular phenotypic traits (Grandillo *et al.*, 1999; Seymour *et al.*, 2002; Causse *et al.*, 2007). Genomic tools such as chromosome walking, DNA sequencing and bioinformatics have further facilitated isolation, identification and characterization of the genomic regions controlling fruit quality parameters. In addition, an understanding of the molecular basis of impaired ripening in different tomato (*Solanum*

* ahanda@purdue.edu

lycopersicum) mutants has added to our knowledge on the regulatory mechanisms underlying the fruit ripening process (Giovannoni, 2004). Molecular genetics has provided many additional tool kits that have enhanced the molecular engineering of crop plants. These include plant transformation methods that have made the candidate gene approach a reality to test the phenotypic role of a particular gene by altering its expression during plant growth and development (Fatima *et al.*, 2009). The gain-of-function (ectopic overexpression) or loss-of-function (repression by antisense RNA interference (RNAi)) approaches have made it possible to characterize the phenotypes associated with a single gene and its potential to regulate desirable phenotype in crop plants. This chapter summarizes some of the progress made using these tools to enhance fruit quality attributes.

Fruits are derived from different parts of a flower, including the inflorescence, and have enormous diversity in their structure and physiological functions (Handa *et al.*, 2012). As development, maturation and ripening of diverse classes of fruits differ significantly, it is a challenge to improve quality attributes of a chosen fruit by biotechnology. None the less, many biochemical and regulatory mechanisms impacting the quality of fruits during ripening are similar, and therefore it is possible to genetically alter ripening and/or slow down deterioration to enhance fruit quality. Ethylene is a gaseous plant hormone decidedly integral to fruit ripening, especially in fruit types that have a burst of respiration during ripening, classified as climacteric fruits (Mattoo and Suttle, 1991; Abeles *et al.*, 1992). The elucidation of its biosynthesis and perception has eased biotechnological strategies to regulate ripening and senescence processes in plants. Thus, regulation of both production and perception of ethylene in fruit crops via molecular engineering has led to remarkable effects on various aspects of fruit quality (Lin *et al.*, 2009; Klee and Giovannoni, 2011), a topic discussed in Part III of this book. Molecular engineering of a number of fruit crops including apple, banana, berries, citrus, cucumber, grape, melon, potato, aubergine and tomato is the subject of research in many laboratories worldwide. Tomato has become a model fruit crop of choice to elucidate the role of various genes in fruit quality (Giovannoni, 2007; Fatima *et al.*, 2009; Klee and Giovannoni, 2011). In this chapter, we have focused primarily on the molecular engineering of shape, size, texture, phytonutrient levels and volatiles in tomato fruit and also make reference to genetic engineering studies in other fruit crops.

16.2 Molecular Engineering of Fruit Appearance

Fruit size and shape are attributes that are quantitatively inherited and determine yield and consumer appeal in most fruit crops. These attributes were given considerable attention during the domestication and selection of new fruit cultivars (Rodríguez *et al.*, 2011a). During domestication, small fruited wild-type *Solanum pimpinellifolium* was developed to larger fruit varieties such as Giant Heirloom. In the process, the fruit mass of 1–2 g per fruit was increased to over 1000 g per fruit and locule numbers from two to more than ten (Lippman and Tanksley, 2001). Other fruit species were also bred for similar increases in size during domestication of their wild progenitors (Smartt and Simmonds, 1995). The application of molecular marker and high-resolution fine-mapping approaches made it possible to identify QTLs and genes encoded within these loci affecting fruit size and shape. In tomato alone, over 30 QTLs have been identified, although only ten of them contribute to most of the observed phenotypic variation (Grandillo *et al.*, 1999; Doganlar *et al.*, 2002; van der Knaap *et al.*, 2002; Tanksley, 2004; van der Knaap *et al.*, 2004). Among them, *fruit weight* (*fw2.2*) controls fruit size without affecting fruit shape or seed production (Frary *et al.*, 2000; Cong *et al.*, 2002; Liu *et al.*, 2003); *sun*, *ovate* and *fruit shapechromosome* (*fs8.1*) regulate fruit shape with minimum effect on fruit size; and *fasciated* (*fas*) and *locule number* (*lc*) determine carpel number and affect both fruit size and shape (Ku *et al.*, 2000; Rodríguez *et al.*, 2011a). *fw2.2*, cloned by high-resolution positional mapping, has been reported to share homology with the cell-membrane-localized Ras-like G-protein (Frary *et al.*, 2000) and negatively regulates fruit size. A mutation in its 2.7 kb upstream promoter region resulted in null expression and a large tomato fruit phenotype (Nesbitt and Tanksley, 2002). *fw2.2* has been further shown to suppress the anticlinal, but not periclinal, cell division in the placenta and pericarp, causing a reduction in the fruit length-to-perimeter ratio but not the pericarp thickness (Liu *et al.*, 2003). In pepper (*Capsicum chinense* and *Capsicum frutescens*), *fw2.1*, but not *fw2.2*, is the single major fruit-weight QTL responsible for 62% of the trait variation (Zygier *et al.*, 2005; Ben Chaim *et al.*, 2006).

Fruit size and weight are a function of the number of cells within the ovary prior to fertilization and cell expansion (Bohner and Bangerth, 1988). Additionally, endoreduplication, which increases cell expansion, contributes to the final fruit size (Cheniclet *et al.*, 2005). Cyclins and cyclin-dependent kinase (CDK) complexes regulate the progression of cell division, while CDK inhibitors such as *WEE1* induce endoreduplication (Sun *et al.*, 1999). Expression of antisense *Slwee1* under the control of the cauliflower mosaic virus (CaMV) 35S promoter reduced ploidy levels, fruit mass, plant growth and seed size (Gonzalez *et al.*, 2007). Another gene that promotes endoreduplication is *cell cycle switch* (*CCS52A*), arresting cell division (Cebolla *et al.*, 1999). Overexpression of *CCS52A*, which activates anaphase-promoting complex E3 ubiquitin ligase, led to increased tomato fruit size (Mathieu-Rivet *et al.*, 2010).

A retrotransposon-mediated gene duplication at the *sun* locus resulted in morphological variation of tomato fruit (Xiao *et al.*, 2008). Overexpression of *IQD12*, one of the five genes at the *sun* locus, significantly increased fruit elongation, while impairing its expression by RNAi significantly decreased fruit elongation (Xiao *et al.*, 2008). The molecular function of IQD12 is not yet known, but it exhibits homology with a member of the IQ67 protein family containing the calmodulin-binding domain and probably changes the fruit shape by affecting the pattern along the apical–basal axis (Xiao *et al.*, 2008). The *ovate* locus, another important QTL responsible for the development of a pear-shaped instead of an oval-shaped tomato fruit, encodes a transcription repressor regulating *GA20ox1*, a gibberellic acid (GA) biosynthesis enzyme (Wang *et al.*, 2007). Overexpression of the ovate family protein 1 gene, *OFP1*, reduced fruit elongation in tomato (Ku *et al.*, 1999) and pepper (Tsaballa *et al.*, 2011). Complementation of pear-shaped fruit phenotype TA503 by either native *OVATE* or ectopic overexpression of *OVATE* under the control of the CaMV 35S promoter (*35S:OVATE*) produced round-shaped fruit (Liu *et al.*, 2002). Silencing of *OVATE* in round-fruited pepper cv. 'Mytilini' resulted in increased expression of *GA20ox1* and an oblong-shaped fruit (Tsaballa *et al.*, 2011). The molecular identity of genes present at other QTLs determining fruit shape and size including *fs8.1*, *fs10.1*, *fs3.1*, *fas*, and *lc* remains to be determined. Similarly, biochemical signals regulating fruit size and shape genes are also still largely unknown (Handa *et al.*, 2012). It will be interesting to explore downstream and upstream regulators of these QTLs through which these loci impart their effect on fruit quality attributes such as size and shape. Although the fruit shape and size genes can be used to alter fruit architecture by molecular genetics approaches, they have not yet been used to develop fruit with a novel architecture for commercial purposes. However, all emerging evidence indicates that these genes would provide a rich resource to develop desirable fruit phenotypes.

16.3 Molecular Engineering of Fruit Texture

Changes in fruit texture are essential for fruit softening and for making a fruit edible and desirable for human consumption. Fruit softening is associated with several attributes including crispness, mealiness, grittiness, chewiness, succulence and juiciness, fibrousness, toughness and oiliness. Furthermore, fruit textural changes are connected with the development of organoleptic characteristics, such as sweetness, sourness, astringency, bitterness and the production of volatile compounds that provide the aroma. However, excessive fruit softening can cause some undesirable attributes including the development of off flavours and susceptibility to phytopathogens. The fact that excessive fruit softening makes most fruit unacceptable, leading to large economic losses, has generated considerable interest among plant biologists to understand the molecular basis of fruit softening and modify this process using recombinant technology (Negi and Handa, 2008). As softening and cell-wall metabolism are intertwined during fruit ripening, this subject is the focus of another chapter (see Tucker, Chapter 4, this volume). We limit this chapter to biotechnological approaches for enhancing the textural qualities of fruits.

Based on observed modifications of the polysaccharides in the primary cell wall and dissolution of the middle lamella during fruit softening, it had been hypothesized that cell-wall depolymerizing enzymes play important roles in fruit textural changes (Brady, 1987). This hypothesis gained further credence when it was shown that expression of several cell-wall-degrading enzymes was severely reduced in tomato mutants impaired in fruit ripening (Tigchelaar *et al.*, 1978; Biggs and Handa, 1989; DellaPenna *et al.*, 1989). A test of this hypothesis led to the development of the first genetically engineered tomato cultivar designated 'FlavrSavr'. In 'FlavrSavr' fruit, the polygalacturonase (PG) gene (*SlPG2*) was silenced by antisense RNA technology (Kramer and Redenbaugh, 1994). The impaired *SlPG2* expression resulted in enhanced juice viscosity, but fruit softening was not significantly affected, and thus it failed to meet the market expectation of an extended-shelf-life fruit (Giovannoni *et al.*, 1989; Thakur *et al.*, 1997). The ectopic expression of *SlPG2* also failed to enhance softening of the ripening mutant, *rin*, suggesting a limited role of *SlPG2* in tomato fruit softening (Giovannoni *et al.*, 1989). In contrast, antisense inhibition of *FaPG1* expression in strawberry (*Fragaria* × *ananassa*) resulted in reduced fruit softening (Quesada *et al.*, 2009). Reduced softening of *FaPG1*-antisense fruit occurred in spite of only a slight reduction in total PG activity, as most of the PG activity was contributed by another isozyme, *FaPG2*, whose expression was not impaired by the *FaPG1* antisense gene (Quesada *et al.*, 2009).

Multiple isozymes of pectin methylesterase (PME), an enzyme that demethoxylates pectin, are expressed during fruit development, but their roles in fruit texture are not as yet understood (Harriman *et al.*, 1991; Gaffe *et al.*, 1994; Tieman and Handa, 1994; Phan *et al.*, 2007). Over a 95% reduction in PME transcripts, protein and enzymatic activity by antisense expression of *SlPME3* under the CaMV 35S promoter did not affect fruit softening but greatly enhanced juice viscosity and increased the total soluble solids (Tieman *et al.*, 1992; Tieman and Handa, 1994; Thakur *et al.*, 1996a,b). The fruit integrity, however, was compromised if low-PME fruits were stored for extended periods (Tieman and Handa, 1994). In another study, silencing of *SlPMU1*, a ubiquitously expressed PME isozyme, enhanced softening of transgenic fruit even when the reduction in PME activity was only 25% compared with wild-type fruit (Gaffe *et al.*, 1997; Phan *et al.*, 2007). These studies on PG and PME further emphasize that only specific isozymes among cell-wall-modifying isozymes contribute to fruit textural changes. Interestingly, tomato fruit with reduced *PME* expression also exhibited a reduction in fruit blossom end rot, a calcium-associated fruit disorder (de Freitas *et al.*, 2012). These authors showed that apoplastic calcium levels increased due to reduced calcium binding to high methoxyl pectin, a consequence of low *PME* activity, and influenced the development of blossom end rot symptoms in tomato fruit (de Freitas *et al.*, 2012).

Preferential loss of galactose and/or arabinose from cell walls during early fruit ripening has led to the suggestion that β-galactosidase plays an important role in fruit textural changes (Gross and Sams, 1984). Among the seven β-galactosidase genes (*SlTBG1–7*) expressed in developing fruit, only silencing of *SlTGB4* (about 90% reduction in extractable exogalactanase activity) led to about 40% increase in fruit firmness compared with the wild-type fruits at comparable stages of ripening (Smith and Gross, 2000; Smith *et al.*, 2002). Total exogalactanase activity, cell-wall galactose content and fruit softening were not affected in transgenic fruit exhibiting an approximate 90% reduction in *SlTBG1* transcripts obtained by homology-dependent gene silencing (Carey *et al.*, 2001). The role of endo-β-mannanase, which hydrolyses mannose in hemicellulose polymers to mannobiose and mannotriose, was tested by developing transgenic plants expressing its antisense RNA or a gene-specific hairpin RNAi gene. These transgenic fruits exhibited reduced endo-β-mannanase activity, but a clear correlation between fruit firmness and endo-β-mannanase activity was not found (Bewley *et al.*, 2000).

The role of xyloglucan xyloglucosyl transferase/endohydrolase (*XTH*) in fruit textural changes was examined by over-expressing *SlXTH1*, a tomato homologue of the *Nicotiana tabacum NtXET-1* gene, under the control of CaMV 35S promoter (Miedes *et al.*, 2010). XTHs have been suggested to play dual roles in cell-wall chemistry by integrating newly secreted xyloglucan chains into an existing cell-wall-bound xyloglucan and by catalysing trans-glucosylation during restructuring of existing cell-wall-bound xyloglucan molecules. The transgenic fruits had a more than fourfold increase in *XET* activity associated with reduced xyloglucan depolymerization and reduced fruit softening, suggesting its role in maintaining the structural integrity of cell walls (Miedes *et al.*, 2010). Most fruit species contain multiple genes for pectate lyase (PL), an enzyme that hydrolyses the unesterified galacturonosyl linkages by a β-elimination reaction. Although expression of several *PL* isozymes increases during fruit ripening, understanding their role in pectinolysis and fruit texture changes is still in its early stages. Introduction of an antisense gene of a strawberry *PL* (*njjs25*) under the control of the CaMV 35S promoter inhibited the expression of *PL*, and the transgenic strawberry fruit registered a decrease in ripening-associated firmness. These transgenic fruit showed an extended postharvest shelf-life, a reduction in pectin solubility, decreased depolymerization of bound polyuronides and loss of cell–cell adhesion in the transgenic fruits (Jiménez-Bermúdez *et al.*, 2002; Santiago-Doménech *et al.*, 2008). Transgenic inhibition of *Cel1* and *Cel2*, two endo-β-1,4-glucanases (EGases, cellulases) present in many fruits, had little effect on fruit softening (Lashbrook *et al.*, 1998; Brummell *et al.*, 1999a). Downregulation of *Cel1* and *Cel2* in strawberry fruits yielded similar results with little influence on fruit softening, but a slightly higher abundance of the larger hemicellulosic polymers was present in the fruit (Mercado *et al.*, 2010; Pang *et al.*, 2010).

Expansins are a family of proteins that induce extension in isolated plant cell walls, expressed during fruit development and ripening, and their roles in fruit textural changes have been examined using molecular genetic techniques (Choi *et al.*, 2006). The antisense RNA inhibition of a ripening-specific expansin, *SlExp1*, caused a reduction in polyuronide depolymerization without affecting the breakdown of other structurally important hemicelluloses, and the transgenic fruit retained a firmer texture than the wild-type fruit (Rose *et al.*, 1997). The constitutive expression of *SlExp1* caused an opposite phenotype, and the transgenic fruit was softer and associated with precocious and extensive depolymerization of structural hemicelluloses without altering polyuronide depolymerization (Brummell *et al.*, 1999b). It was proposed that *Exp1* modulates relaxation of the cell walls and regulates polyuronide depolymerization by controlling access of a pectinase to its substrate, whereas the depolymerization of hemicellulose occurs independently or requires only very small amounts of Exp1 protein (Brummell *et al.*, 1999b). A firmer fruit texture and higher cellular integrity during longer storage was observed in the fruit in which *Exp1* and *PG* were simultaneously downregulated (Powell *et al.*, 2003).

After the initial demonstration that a protein glycosylation inhibitor, tunicamycin, impaired fruit ripening (Handa *et al.*, 1985), the role of protein glycosylation in fruit ripening and textural changes has begun to emerge using transgenic technologies. Tunicamycin inhibits the UDP-HexNAc:polyprenol-P HexNAc-1-P family of enzymes and blocks the synthesis of all *N*-linked glycoproteins (*N*-glycans). Suppression by antisense RNA technology of two *N*-glycosylating enzymes, β-mannosidase and β-D-*N*-acetylhexosaminidase, led to reduced ripening-associated softening and improved fruit shelf-life (Meli *et al.*, 2010), whereas their ectopic expression caused excessive softening of the transgenic fruit. These studies provided a novel way to alter fruit ripening and extend their shelf-life.

16.4 Molecular Engineering of Carotenoids

Fruits are naturally rich in carotenoids, one of the most abundant groups of plant pigments. Over 600 carotenoids have been structurally identified, and this list continues to increase as new compounds are added. Carotenoids play several roles in plants including photosystem assembly, light harvesting, free-radical detoxification, photomorphogenesis, non-photochemical quenching, lipid peroxidation and as a substrate for the phytohormone abscisic acid (ABA) (Namitha and Negi, 2010). However, it is the human health benefits of carotenoids that have attracted significant attention in recent years (Dixon, 2005; Mattoo *et al.*, 2010). The role of vitamin A (retinal) in preserving eyesight, especially in preventing night blindness, is one of the best-known functions of carotenoids in human health (Cook, 2010). Due to their high antioxidant activity, carotenoids are implicated in protection against cataract and macular degeneration of the eye, and against cervical, lung, prostate, colorectal, stomach, pancreatic and oesophagus cancers. Carotenoids may also reduce low-density lipoproteins, implicated in cardiovascular disease, and boost the immune system to provide protection against many other diseases such as osteoporosis, hypertension and neurodegenerative diseases like Alzheimer's, Parkinson's and vascular dementia (Mattoo *et al.*, 2010; Namitha and Negi, 2010). The emerging consensus in favour of the beneficial role of carotenoids has led to significant research activity to raise their cellular levels in fruit and vegetable crops using novel approaches.

Genes encoding carotenoid biosynthetic pathway enzymes have been identified and cloned from several species, but regulation of their accumulation in plants is complicated and poorly understood (Klee and Giovannoni, 2011). A detailed carotenoid biosynthesis pathway is illustrated in Plate 7. A series of addition and condensation reactions convert isopentenyl diphosphate to form geranylgeranyl diphosphate. Two different pathways, mevalonate-dependent (cytosolic) and mevalonate-independent (plastid), generate isopentenyl diphosphate (Rodríguez-Concepción, 2010). Phytoene synthase (PSY) is the first committed step in carotenoid biosynthesis and catalyses the condensation of two geranylgeranyl diphosphates to form phytoene, which is converted into ζ-carotene by phytoene desaturase (PDS). ζ-Carotene desaturase (ZDS) converts ζ-carotene to lycopene, which in turn is converted into either β-carotene by lycopene β-cyclase (CRTL-B), a precursor of vitamin A, or α-carotene by lycopene α-cyclase (CRTL_E). Lutein, a major xanthophyll involved in light harvesting and preventing macular degeneration of the eyes in older people, is synthesized from α-carotene (Ronen *et al.*, 1999).

Transgenic approaches have been widely used to enhance levels of carotenoids in many crop species by expressing various genes of the carotenoid pathways (Table 16.1). Ectopic expression of a bacterial PSY (*crtB*) under the control of a fruit-specific promoter increased phytoene (2.4-fold), lycopene (1.8-fold), β-carotene (2.2-fold) and lutein levels in tomato fruit (Fraser *et al.*, 2002). The constitutive expression of citrus lycopene β-cyclase (*CRTL-B*) increased β-carotene levels 4.1-fold with a 30% increase in total carotenoids while suppressing fluxes downstream into the β-carotene pathway and the concomitant increase in α-carotene (Guo *et al.*, 2012). A mutation in lycopene ε-cyclase (*CRTL-E*) caused accumulation of δ-carotene at the expense of lycopene in *Delta* (*Del*), a fruit-colour mutant (Ronen *et al.*, 1999). Two other genes, *CYP97A29* and *CYP97C11*, have been functionally characterized by expressing them in tomato under the control of the CaMV 35S promoter. *CYP97A29* and *CYP97C11* encode P450 carotenoid β-hydroxylase (*CRTR-B*) and carotenoid ε-hydroxylase (*CRTR-E*), respectively. *CRTR-E* converts α-carotene into lutein, and *CRTR-B* converts β-carotene into zeaxanthin and α-carotene into lutein (Stigliani *et al.*, 2011). Zeaxanthin is further converted to violaxanthin by zeaxanthin epoxidase (*ZEP1*). A mutation in *zep1* caused ABA-deficiency in tomato plants with a concomitant accumulation of 30% more carotenoids in mature red tomato fruit (Galpaz *et al.*, 2008). RNAi-mediated fruit-specific suppression of 9-*cis*-epoxycarotenoid dioxygenase 3 (*NCED3*), an enzyme that catalyses the first step of ABA biosynthesis converting 9-*cis*-violaxanthin to 2-*cis*,4-*trans*-xanthoxin, not only suppressed ABA synthesis but also stimulated accumulation of upstream compounds such as β-carotene and lycopene in transgenic tomato fruits (Sun *et al.*, 2012).

Characterization of several tomato mutants that accumulate higher levels of carotenoids than wild-type fruits has helped to discover factors regulating flux through the carotenoid pathways. *UV-DAMAGED DNA BINDING PROTEIN 1* (*DDB1*) and *DE-ETIOLATED 1* (*DET1*) are transcription factors that negatively regulate photomorphogenic responses (Azari *et al.*, 2010a). Mutations in *DDB1* and *DET1* exhibit recessive *high-pigment 1* (*hp-1*) and *hp-2* phenotypes with severe developmental defects (Mustilli *et al.*, 1999; Levin *et al.*, 2003; Davuluri *et al.*, 2004; Azari *et al.*, 2010b). However, organ-specific silencing of *DET1* by RNAi under a fruit-specific promoter resulted in a twofold increase in lycopene, a fourfold increase in β-carotene and up to a 3.5-fold increase in flavonoids without significant changes in fruit weight and total soluble solids in red-ripe fruit (Davuluri *et al.*, 2005). The transgenic overexpression of cryptochrome 2 (*35S:CRY2*) resulted in a 1.7-fold increase in carotenoids and a 2.9-fold increase in flavonoids (Giliberto *et al.*, 2005). RNAi-mediated repression of *ELONGATED HYPOCOTYL 5* (*HY5*) reduced carotenoid accumulation and repression of *CONSTITUTIVELY PHOTOMORPHOGENIC 1* (*COP1*)-like showed an elevation in tomato fruit carotenoids suggesting involvement of light-signalling factors in carotenoid biosynthesis (Liu *et al.*, 2004).

The biogenic amines spermidine and spermine, which belong to the group of ubiquitous polycations called polyamines, have also been implicated in delaying fruit ripening and increasing carotenoid content in tomato (Mehta *et al.*, 2002; Nambeesan *et al.*, 2010). Constitutive overexpression of *Saccharomyces cerevisiae* spermidine synthase or fruit-specific overexpression of yeast *S*-adenosylmethionine decarboxylase (*E8:ySAMdc*) in tomato led to a 40% or 200–300% increase in lycopene content,

Table 16.1. Engineering of carotenoid pathways in tomato fruit to alter carotenoid levels.

Transgene	Metabolic reaction or function	Promoter expression	Metabolic phenotype	Reference
3-Hydroxy-3-methyl-glutaryl CoA reductase (*HMGR-1*) (*Arabidopsis thaliana*)	3-Hydroxy-3-methylglutaryl CoA → mevalonic acid	CaMV *35S* (OE)	2.4-fold ↑total phytosterol; no change in lycopene or β-carotene	Enfissi *et al.* (2005)
1-Deoxy-D-xyluose-5-phosphate synthase (*DXS*) (*Escherichia coli*)	Pyruvate and D-glyceraldehyde-3-phosphate → 1-deoxy-D-xylulose-5-phosphate synthase	CaMV *35S* or fibrillin (OE)	1.6-fold ↑ carotenoids; 2.4-fold ↑ phytoene; 2.2-fold ↑ β-carotene	Enfissi *et al.* (2005)
Phytoene synthase (*crtB*) (*Erwinia uredovora*)	Geranyldiphosphate → phytoene	*SlPG* (OE)	2.4-fold ↑phytoene; 1.8-fold ↑ lycopene; 2.2-fold ↑ β-carotene	Fraser *et al.* (2002)
Phytoene synthase (*psy-1*) (*Solanum lycopersicum*)	As above	CaMV *35S* (OE)	1.2-fold ↑ total carotenoids; 1.3-fold ↑ β-carotene; 2.3-fold ↑ phytoene; 1.8-fold phytofluene	Fraser *et al.* (2007)
Phytoene synthase (*psy-1*) (*S. lycopersicum*)	As above	CaMV *35S* (OE)	↑ lycopene (386 μg g^{-1} dry weight); ↓ plant height; 30-fold ↓ GA_1	Fray *et al.* (1995)
Phytoenedesaturase (*crtI*) (*E. uredovora*)	Phytoene → ζ-carotene	CaMV *35S* (OE)	3-fold ↑ β-carotene; no effect on total carotenoids: reduction in lycopene and phytoene; no effect on plant growth and development	Römer *et al.* (2000)
Lycopene β-cyclase (*SpB*) (*Solanum pennellii*)	Lycopene → β-carotene, δ-carotene → α-carotene	CaMV *35S* (OE)	>6-fold ↑ β-carotene; 1.8-fold ↓ lycopene	Ronen *et al.* (2000)
Lycopene β-cyclase (*SpB*) (*S. pennellii*)	As above	CaMV *35S* (AS)	>6-fold ↓ β-carotene; slight ↑ lycopene	Ronen *et al.* (2000)
Lycopene β-cyclase (*Lyc-b*) (*S. lycopersicum*)	As above	CaMV 35iS (OE)	31.7-fold ↑ β-carotene at the expense of lycopene; no morphological and developmental defects	D'Ambrosio *et al.* (2004)
Lycopene β-cyclase (*β-Lcy*) (*A. thaliana*)	As above	*SlPds* (OE)	>6-fold ↑ β-carotene; no change in lycopene	Rosati *et al.* (2000)
Lycopene β-cyclase (*β-Lcy*) (*S. lycopersicum*)	As above	*SlPds* (AS)	1.3-fold ↑ lycopene; 1.7-fold ↑ lutein; 50% ↓ β-lycopene expression	Rosati *et al.* (2000)
Lycopene β-cyclase (*Lycb-1*) (*Citrus*)	As above	CaMV *35S* (OE)	4.1-fold ↑ β-carotene; 30% ↑ total carotenoids	Guo *et al.* (2012)
Lycopene β-cyclase (*crtY*) (*E. herbicola*) or (*carRA*) (*Phycomyces blakesleeanus*)	As above	*atpI* (OE)	4-fold ↑ β-carotene; slight ↓lycopene and total carotenoids	Wurbs *et al.* (2007)

Lycopene β-cyclase (*b-Lcy*) (*A. thaliana*) + carotene β-hydroxylase (*b-Chy*) (*Capsicum annuum*)	Lycopene → β-carotene, δ-carotene → α-carotene + β-carotene → zeaxanthin, α-carotene → lutein	*SlPds* (OE)	12-fold ↑ β-carotene; 10-fold ↑ total xanthophyll	Dharmapuri *et al.* (2002)
9-*cis*-Epoxycarotenoid dioxygenase 3 (*NCED1*) (*S. lycopersicum*)	9-*cis*-Violaxanthin → xanthoxin, 9-*cis*-neoxanthin → xanthoxin	*E8* (RNAi)	↑ β-carotene and lycopene; 20–50% ↓ abscisic acid	Sun *et al.* (2012)
DE-ETIOLATED (*DET1*) (*S. lycopersicum*)	TFs negatively regulate photomorphogenic responses	*P119, 2A11, TFM7* (RNAi)	2-fold ↑ lycopene; 4-fold ↑ β-carotene; 3.5-fold ↑ flavonoids; no change in fruit weight and total soluble solids in red-ripe fruit	Davuluri *et al.* (2005)
Cryptochrome 2 (*CRY2*) (*S. lycopersicum*)	Blue light photoreceptor	CaMV *35S* (OE)	1.5-fold ↑ lutein; 1.7-fold ↑ carotenoids; 2.9-fold ↑ flavonoids	Giliberto *et al.* (2005)
ELONGATED HYPOCOTYL 5 (*HY5*) (*S. lycopersicum*)	Positive regulator of phytochrome signal transduction	CaMV *35S* (RNAi)	↓ carotenoid	Liu *et al.* (2004)
CONSTITUTIVELY PHOTOMORPHOGENIC 1 (*COP1-like*) (*S. lycopersicum*)	TFs negatively regulate photomorphogenic responses	CaMV *35S* (RNAi)	2-fold ↑ carotenoids	Liu *et al.* (2004)
Spermidine synthase (*SPE3*) (*Saccharomyces cerevisiae*)	Putrescine → spermidine	CaMV *35S* (OE)	40% ↑ lycopene	Nambeesan *et al.* (2010)
Spermidine synthase (*Md-SPDS1*) (*Malus* × *domestica*)	As above	CaMV *35S* (OE)	↑ *PSY* and *PDS* and ↓ *CRTL-B* and *CRTL-E* transcripts; 1.3–2.2-fold ↑ lycopene	Neily *et al.* (2011)
SAM decarboxylase (*SPE2*) (*S. cerevisiae*)	SAM → decarboxylated SAM	*E8* (OE)	2–3-fold ↑ lycopene	Mehta *et al.* (2002)

CaMV *35S*, cauliflower mosaic virus *35S* promoter; *E8*, tomato fruit-specific *E8* promoter; *Pds*, fruit-specific phytoene desaturase promoter; *PG*, fruit ripening-specific polygalacturonase 2 promoter; *atpl* (ATPase IV subunit), tobacco plastid-specific promoter; *P119, 2A11, TFM7*, fruit-specific promoters; fibrillin, ripening-enhanced promoter of fibrillin; RNAi, RNAi-mediated repression of target gene; OE, overexpression of the introduced gene; AS, antisense-mediated downregulation; TF, transcription factor; SAM, S-adenosylmethionine; ↑, increased levels; ↓, decreased levels; →, substrate to product conversion.

respectively (Mehta *et al.*, 2002; Nambeesan *et al.*, 2010). Transgenic overexpression of apple spermidine synthase in tomato fruit not only upregulated PSY and PDS but also downregulated the catabolic enzymes lycopene β- and ε-cyclases, resulting in an overall 1.3–2.2-fold increase in lycopene content (Neily *et al.*, 2011). These findings indicate a positive correlation between polyamines with carotenoids levels during fruit ripening and microarray-based transcriptional profiling of *E8:ySAMdc* tomato fruit (Kolotilin *et al.*, 2011).

Deficiency of vitamin A is a major issue affecting child health, especially in developing nations. To increase its synthesis in staple foods, transgenic technologies have successfully been used to develop rice varieties (Golden Rice) engineered to accumulate higher levels of the provitamin A 'β-carotene'. Introduction of maize *PSY* in combination with carotene desaturase from *Erwinia uredovora* resulted in 23-fold increase in total carotenoids in rice (Paine *et al.*, 2005). During rice processing, an aleurone layer is removed to avoid rancidity of rice grains during storage, while rice endosperm lacks β-carotene. Using DNA recombination technology, three transgene constructs were co-transformed and transformants containing all three transgenes were selected and characterized. The three transgenes introduced were daffodil *PSY* under the control of an endosperm-specific gluten promoter, *E. uredovora* phytoene desaturase under the control of the CaMV 35S promoter and *Narcissus pseudonarcissus* lycopene β-cyclase under the control of a rice gluten promoter. Expression of these transgenes in rice endosperm led to higher accumulation of β-carotene (Ye *et al.*, 2000).

Transgenic approaches have also been used to enhance levels of carotenoids in flaxseed (Fujisawa and Misawa, 2010), maize (Naqvi *et al.*, 2011), kumquat citrus (Zhang *et al.*, 2009), wheat (Cong *et al.*, 2009), *Brassica* (Yu *et al.*, 2008; Fujisawa *et al.*, 2009; Wei *et al.*, 2009), rice (Burkhardt *et al.*, 1997; Rai *et al.*, 2007), tobacco (Qin and Zeevaart, 2002; Frey *et al.*, 2006) and canola (Ravanello *et al.*, 2003).

16.5 Molecular Engineering of Flavonoids

Flavonoids are aromatic, low-molecular-weight secondary metabolites classified as plant phenolics (Robards and Antolovich, 1997). Their hydrophilic properties (Rice-Evans *et al.*, 1997) complement the hydrophobic nature of carotenoids. More than 6000 naturally occurring flavonoids have been identified (Harborne and Herbert, 1999) and classified based on the degree of unsaturation and oxidation of a three-carbon bridge in the flavone skeleton between their phenyl groups.

Antioxidant and free-radical scavenging properties of flavonoids have been associated with reducing the risks of heart and age-related diseases and cancers (Ross and Kasum, 2002). Fruit juice is a major source of flavonoids in the human diet, and total fruit juice consumption seems to account for 20–30% of the dietary intake of flavonoids (Robards and Antolovich, 1997). As well as their emerging therapeutic role in alternative medicinal science, flavonoids are also known to provide protection to plants against UV-B light and microbial interactions (Harborne and Williams, 2000). This attribute is important for fruits to maintain their resistance against fungi during storage. Flavonoids also contribute towards various fruit quality attributes including colour (red, violet, blue), flavour and texture. In contrast, undesirable brown pigmentation (bruises) on the fruit surface has been attributed to oxidation of phenols to quinones, which then polymerize into brown pigments, for example, flavan-3-ols in apples (Amiot *et al.*, 1992; Goupy *et al.*, 1995; Robards and Antolovich, 1997). Different classes of flavonoids also combine with proteins and cause sedimentation in fruit juices and wines (Amiot *et al.*, 1992). The various flavonoid compounds found in different fruits and vegetables have been summarized elsewhere (Robards and Antolovich, 1997; Nicoletti *et al.*, 2007; USDA, 2007; Slimestad and Verheul, 2009).

The genetic regulation of the flavonoid biosynthesis pathway was initially investigated primarily by the inheritance pattern of flower colour and radiolabelling. However, genetic engineering technology has added a new dimension to our understanding of flavonoid biosynthetic enzymes and substrates and their diversity among various plant species (Table 16.2). Mutants and transgenic plants have provided direct evidence for the function of various genes involved in flavonoid biosynthesis pathways (reviewed by Ververidis *et al.*, 2007a). Flavonoids are mainly synthesized from phenylalanine via the phenylpropanoid pathway. Following cinnamate hydroxylation by the cinnamate 4-hydroxylase and 4-coumarate:CoA ligase step in phenylpropanoid pathway, the flavonoid biosynthesis pathway branches out into phenolics (chlorogenic acid) and flavonols (naringenin, quercetin and their derivatives) (Anterola and Lewis, 2002; Ververidis *et al.*, 2007b).

Tomato fruits synthesize significant amounts of carotenoids but are poor in the production of flavonoids in fruit flesh. Flavonoid production in fruits is restricted mainly to the peel with accumulation of naringenin chalcone, the flavonol rutin and kaempferol 3-*O*-rutinoside (Crozier *et al.*, 1997; Muir *et al.*, 2001; Bovy *et al.*, 2002). The major focus of flavonoid biotechnology research is to increase flavonoid accumulation in fruit flesh and determine the potential to induce the production of new flavonoids (Table 16.2). Tomato fruit do not have stilbene synthase gene (*StSy*) (Giovinazzo *et al.*, 2005) and cannot normally produce resveratrol, a stilbenoid flavonoid. However, expression of a grape *StSy* not only induced the production of resveratrol in transgenic fruits but also led to the accumulation of *trans*-resveratrol and *trans*-resveratrol glucopyranosides (piceid), further elevating the antioxidant capacity in tomato fruit (D'Introno *et al.*, 2009). Transgenic tomato lines were developed that constitutively expressed flavonoid genes from different plant species (Schijlen *et al.*, 2006). It was shown that the expression of grape *StSy* produced a higher amount of resveratrol and piceid (a stilbenoid glucoside), while combined expression of petunia chalcone synthase and lucerne chalcone reductase induced higher levels of butein and isoliquiritigenin (deoxychalcones). Combined expression of petunia chalcone isomerase and gerbera flavone synthase resulted in elevated production of luteolin-7-glucoside, luteolinaglycon (flavone) and quercetin glycosides (flavonol). Although the constitutive overexpression of *StSy* produced up to tenfold higher levels of resveratrol, it resulted in complete male sterility, probably due to a lack of coumaric and ferulic acid production (Ingrosso *et al.*, 2011). The seedless parthenocarpic fruit phenotype resulting from male sterility is of much interest because of its desirability by both the consumer and food industry (Rotino *et al.*, 1997; Ficcadenti *et al.*, 1999; Pandolfini *et al.*, 2002).

The ectopic expression of petunia chalcone isomersae in tomato resulted in a 78-fold increase in peel flavonols, which was mainly due to the accumulation of rutin, a quercetin glycoside. After processing the tomato fruit, the paste still retained 65% of the total flavonols present in the fresh fruit (Muir *et al.*, 2001). Although isoflavones are legume-specific flavonoids, tomato plants engineered to constitutively overexpress soybean isoflavone synthase (*35S:GmIFS2*) showed significant accumulation of genistin (a major isoflavone metabolite) in leaves with a marginal increase in fruit peel. Naringenin chalcone biosynthesis was also upregulated in these transgenic fruit, indicating naringenin as a limiting factor (substrate) for isoflavone biosynthesis in fruit peel (Shih *et al.*, 2008).

In addition to the candidate gene approach to enhance flavonoid content, transcription factors have also been tested to achieve similar objectives. Coordinated expression of maize MYB-type C1 and MYC-type LC, transcription factors implicated in anthocyanin production, in tomato induced flavonoid biosynthesis in fruit flesh the tissues where flavonoids are poorly synthesized (Bovy *et al.*, 2002). Overall, a tenfold increase in total flavonoids and a 20-fold increase in total

Table 16.2. Studies on tomato engineered to alter fruit flavonoids.

Transgene	Metabolic reaction or function	Promoter expression	Metabolic phenotype	Reference(s)
Stilbene synthase (*StSy*) (*Vitis vinifera*)	Malonyl-CoA and *p*-coumaroyl-CoA → resveratrol	CaMV *35S* (OE)	↑ *Trans*-resveratrol (48.48 mg kg^{-1} fresh weight); ↑ *trans*-piceid (126.58 mg kg^{-1} fresh weight); 2-fold ↓ rutin, 2.4-fold ↓ naringenin; seedless fruit	Giovinazzo *et al.* (2005); Nicoletti *et al.* (2007)
Stilbene synthase (*StSy*) (*V. vinifera*)	As above	*TomLoxB* (OE)	↑ Resveratrol, *trans*-resveratrol and piceid	D'Introno *et al.* (2009)
Stilbene synthase (*STS*) (*V. vinifera*)	As above	CaMV *d35S* (OE)	↑ Stilbenes (resveratrol and piceid); ↑ naringenin chalcone and rutin	Schijlen *et al.* (2006)
Chalcone synthase (*Chs1*) (*Solanum lycopersicum*)	Phenylpropanoids → chalcones	CaMV *d35S* (RNAi)	↓ Total flavonoid; parthenocarpic fruit	Schijlen *et al.* (2007)
Chalcone isomersae (*Chi-A*) (*Petunia hybrida*)	Chalcones → flavanones	CaMV *d35S* (OE)	78-fold ↑ peel flavonols, mainly due to ↑ rutin	Muir *et al.* (2001)
Chalcone synthase (*Chs1*) (*P. hybrida*) + chalcone reductase (*CHR*) (*Medicago sativa*)	Phenylpropanoids → chalcones + Phenylpropanoids → deoxychalcones	CaMV *35S* (OE)	↑ Butein and isoliquiritigenin; ↑ naringenin chalcone and rutin	Schijlen *et al.* (2006)
Chalcone isomerase (*CHI*) (*P. hybrida*) + flavone synthase (*CYP93B2*) (*Gerbera hybrida*)	Chalcones → flavanones + flavanones → flavones	CaMV *35S* (OE)	16-fold ↑ rutinflavonol ↑ luteolin-7-glucoside, luteolinaglycon, quercetin glycosides, naringenin chalcone and rutin	Schijlen *et al.* (2006)
Isoflavone synthase (*IFS2*) (*Glycine max*)	Naringenin → genistein	CaMV *35S* (OE)	↑ Genistin in leaves; only marginal increase in fruit peel; ↑ naringenin chalcone in fruit peel	Shih *et al.* (2008)
LC (*LC*) (*Zea mays*)	LC, a member of maize R gene family of MYC-type TFs, determines the tissue-specific expression of anthocyanin in maize	CaMV *35S* (OE)	↑ Anthocyanins in all vegetative tissues but to a lesser extent in green fruit	Goldsbrough *et al.* (1996)
C1 (*C1*)+ LC (*LC*) (*Z. mays*)	MYB-type C1 and MYC-type LC are TFs required for production of anthocyanin in plants	*E8* or CaMV *d35S* (OE)	Induced flavonoid synthesis in fruit flesh; 10-fold ↑ total flavonoids; 20-fold ↑ total flavonol, mainly due to ↑ kaempferol	Bovy *et al.* (2002)
RP (*Myc-rp*) (*Perilla frutescens*)	Myc-like TF regulating anthocyanin biosynthesis	CaMV *35S* (OE)	↑ Anthocyanin in vegetative tissues and flowers	Gong *et al.* (1999)
Delila (*Del*) (*Antirrhinum majus*)	Myc TFs that activate biosynthesis of anthocyanin	CaMV *35S* (OE)	↑ Anthocyanins in mature leaves (23-fold), corolla (40-fold) and stamen (50-fold) but none in fruit	Mooney *et al.* (1995)

Rosea1 (*AmRos1*) + Delila (*Del*) (*A. majus*)	TFs that activate biosynthesis of anthocyanin	*E8* (OE)	↑ Anthocyanin in pericarp comparable to blackberries and blueberries	Butelli *et al.* (2008)
MYB12 (*MYB12*) (*Arabidopsis thaliana*)	R2R3-MYB TF that mediates the accumulation of flavonoids in tomato peel	CaMV *35S* (OE)	27-fold ↑ chlorogenic acid; 26-fold ↑ dicaffeoylquinic acid; 42-fold ↑ tricaffeoylquinic acid; 67-fold ↑ quercetinrutinoside; 593-fold ↑ kaempferolrutinoside	Luo *et al.* (2008)
MYB12 (*MYB12*) (*S. lycopersicum*)	As above	CaMV *35S* (RNAi)	↓ Flavonoid pigment naringenin chalcone; exhibited a *y*-like phenotype	Adato *et al.* (2009)
MYB12 (*MYB12*) (*S. lycopersicum*)	As above	CaMV *35S* (OE)	Rescued colourless-peel *y* tomato mutant phenotype	Adato *et al.* (2009)
ANTHOCYANIN 1 (*ANT1*) (*S. lycopersicum*)	Flavonoid-related R2R3-MYB TF	*CVM* (OE)	500-fold ↑ anthocyanin	Mathews *et al.* (2003)
ANTHOCYANIN 1 (*ANT1*) (*Solanum chilense*)	As above	CaMV *35S* (OE)	↑ Anthocyanadins (petunidin, malvidin, delphinidin) in tomato (*S. lycopersicum*) fruit	Schreiber *et al.* (2012)
TOMATO AGAMOUS-LIKE 1 (*TAGL1*) (*S. lycopersicum*)	MADS-box TF	*E8* (*TAGL1-SRDX*)	↓ Lycopene and isoprenoids	Itkin *et al.* (2009)
TOMATO AGAMOUS-LIKE 1 (*TAGL1*) (*S. lycopersicum*)	As above	CaMV *35S* (OE)	↑ Lycopene and naringenin chalcone	Itkin *et al.* (2009)
Cullin 4 (*CUL4*) (*S. lycopersicum*)	DDB1a and DET1 form a complex with CUL4, a ubiquitin-conjugating E3 ligase, and target proteins for proteolysis	CaMV *35S* (RNAi)	Pleiotropic phenotype; ↑anthocyanins and carotenoids; 2-fold ↑ lycopene	Wang *et al.* (2008)
UV-DAMAGED DNA BINDING PROTEIN 1 (*DDB1*) (*S. lycopersicum*)	TF that negatively regulates photomorphogenic responses	*E8* (RNAi)	↑ Pigment accumulation due to ↑ plastid compartment space	Wang *et al.* (2008)
SAM decarboxylase (*SPE2*) (*Saccharomyces cerevisiae*)	As above	*E8* (OE)	↑ Transcripts related to flavonoid biosynthesis genes	Mehta *et al.* (2002); Mattoo *et al.* (2007)
Phytoene synthase (*Psy-1*) (*S. lycopersicum*)	As above	CaMV *35S* (OE)	↓ Phenylpropanoids and flavonoids	Fraser *et al.* (2007)

CaMV *d35S*, cauliflower mosaic virus double *35S* promoter; *TomLoxB*, tomato fruit-specific promoter; *CVM*, constitutive cassava vein mosaic promoter; TAGL1-SRDX, chimeric TAGL1 fused to EAR (ERF-associated amphiphilic repression) motif that functions as a dominant repressor; SAM, *S*-adenosylmethionine. See Table 16.1 for other abbreviations.

flavonols in ripe tomato fruit was achieved, mainly due to increased production of kaempferolin transgenic fruit (Bovy *et al.*, 2002; Le Gall *et al.*, 2003). Expression of *Rosea1* and *Delila*, transcription factors that activate the biosynthesis of anthocyanin, driven by an *E8* promoter resulted in enhanced anthocyanin production in tomato pericarp at concentrations comparable to that in blackberries and blueberries (Butelli *et al.*, 2008). Transgenic tomato fruit constitutively expressing *ANTHOCYANIN1* (*ANT1*), a flavonoid-related R2R3-MYB transcription factor, had higher levels of anthocyanadins including petunidin, malvidin and delphinidin (Schreiber *et al.*, 2012). Downregulation of *TOMATO AGAMOUS-LIKE 1* (*TAGL1*), a MADS-box transcription factor, resulted in lowering the levels of lycopene and isoprenoids whereas its overexpression caused higher accumulation of lycopene and naringenin chalcone (Itkin *et al.*, 2009).

Altering the expression of transcriptional regulators of photomorphogenic responses enhanced the production of flavonoids. Fruit-specific RNAi-mediated silencing of *DE-ETIOLATED 1* (*DET1*), transcriptional repressor of photomorphogenic responses, not only increased carotenoid levels but also increased flavonoids by 3.5-fold (Davuluri *et al.*, 2005). Constitutive overexpression of cryptochrome 2 resulted in about a threefold increase in flavonoids (Giliberto *et al.*, 2005). The fruit-colour tomato mutant, *high-pigment-1* (*hp-1*), carrying a mutation in *UV-DAMAGED DNA BINDING PROTEIN 1* (*DDB1*) increased levels of both carotenoids and flavonoids (chlorogenic acid and rutin) (Long *et al.*, 2006). Likewise, RNAi-mediated repression of the DDB1-interacting protein CUL4 in tomato lines (*35S:CUL4-RNAi*) resulted in elevated levels of anthocyanins and carotenoids (Wang *et al.*, 2008).

Transcriptome analysis of high polyamine-accumulating *E8:ySAMdc* tomato fruits showed upregulation of the transcription profiles related to the carotenoid and flavonoid biosynthesis pathways (Mattoo *et al.*, 2007). A mutation in *PSY-1* (tomato mutant *rr*) did not increase the levels of phenylpropanoids and flavonoids (chlorogenic acid, caffeic acid, *p*-coumaric acid and ferulic acid) in pericarp tissues (Long *et al.*, 2006), but constitutive overexpression of *PSY-1* showed an increase in phenylpropanoids and flavonoids including 3-caffeoylquinic acid, naringenin chalcone and quercetin derivatives in red-ripe tissues (Fraser *et al.*, 2007).

Other studies have highlighted the importance of the aforementioned strategies in either enhancing the biological activity of endogenous flavonoids or achieving fruit quality attributes, rather than just enhancement of flavonoids. For example, prenylated flavonoids, derived from the addition of hydrophobic molecules to flavonoids, are biologically more active than their native forms, possibly because of the lipophilicity of the prenyl moiety, which makes flavonoids more membrane permeable (Maitrejean *et al.*, 2000; Murakami *et al.*, 2000). Fruit-specific overexpression of *Streptomyces* prenyltransferase *HypSc* in tomato fruit resulted in the accumulation of 3′-dimethylallyl naringenin, a prenylated form of the native naringenin flavonoid (Koeduka *et al.*, 2011). The other example of achieving an industry-driven objective is to induce parthenocarpy (Rotino *et al.*, 1997; Ficcadenti *et al.*, 1999; Pandolfini *et al.*, 2002).

Together, these results provide strong evidence in favour of biotechnological interventions not only for enhancing the levels and composition of health-related polyphenols in fruits but also to produce novel compounds by the engineering of flavonoid and other pathways (Schijlen *et al.*, 2006). Studies on the tomato model system described above clearly support the significance of transgenic approaches in enhancing sensory fruit quality attributes. Similar approaches have also been adopted to manipulate flavonoid biosynthetic pathways in strawberry (Lunkenbein *et al.*, 2006), maize (Sidorenko *et al.*, 2000; Li *et al.*, 2007), grape (Boss *et al.*, 1996; Bogs *et al.*, 2007), rice (Shin *et al.*, 2006; Furukawa *et al.*, 2007), *Medicago truncatula* (Pang *et al.*, 2007), citrus (Moriguchi *et al.*, 2001; Frydman *et al.*, 2004; Koca *et al.*, 2009), *Brassica* (Hüsken *et al.*, 2005; Auger *et al.*, 2009; Nesi *et al.*, 2009; Wei *et al.*, 2009), flax seed (Lorenc-Kukula *et al.*, 2005; Zuk *et al.*, 2011), apple (Rühmann *et al.*, 2006; Ban *et al.*, 2007; Flachowsky *et al.*, 2010; Flachowsky *et al.*, 2012; Han *et al.*, 2012), soybean (Nagamatsu *et al.*, 2007) and tobacco (Aharoni *et al.*, 2001).

16.6 Molecular Engineering of Flavour Volatiles

Flavour, an important quality attribute of a fruit, is the sum of specific interactions of fruit constituents among which sugars, acids and a number of volatile molecules are significant components (Mathieu *et al.*, 2009). Preference for a specific flavour (sugar:acid ratio) and perception of volatiles by olfactory receptors in the human nose are partly a social/cultural science that vary with diversity in ethnicity, age, and personal likes and dislikes. In general, components concentration and odour threshold are important variables in determining the contribution of various volatiles to fruit flavour (Baldwin *et al.*, 2000). Most fruits and vegetables produce aromatic volatiles, as has been revealed by studies on mango (MacLeod and Snyder, 1985; MacLeod *et al.*, 1988; Andrade *et al.*, 2000), guava (Wilson *et al.*, 1982; Porat *et al.*, 2011), watermelon (Lewinsohn *et al.*, 2005), apple (Dixon and Hewett, 2000), strawberry (Song *et al.*, 1998) and tomato (Buttery *et al.*, 1988; Buttery and Ling, 1993; Maul *et al.*, 1997; Krumbein and Auerswald, 1998; Marković *et al.*, 2007; Mayer *et al.*, 2008; Christiansen *et al.*, 2011). Over 400 aroma volatiles have been detected in tomato, but fewer than 30 have been proposed to impact on organoleptic properties (Baldwin *et al.*, 2000; Tieman *et al.*, 2006a). These aroma and flavour volatiles include *cis*-3-hexenal, β-ionone, hexanal, β-damascenone, 1-penten-3-one, 2-methylbutanal, 3-methylbutanal, *trans*-2-hexenal, isobutylthiozole and *trans*-2-heptenal (Goff and Klee, 2006; Zeigler, 2007; Mathieu *et al.*, 2009; Klee, 2010).

Fruit breeding programmes that have focused on developing larger and firmer fruits with an extended shelf-life have largely ignored organoleptic attributes with the unintended consequence of loss of flavour components (Mathieu *et al.*, 2009). The manipulation of flavour components in fruits via biotechnology has been limited, particularly because biosynthetic pathways are complex and are known only for a limited number of volatile compounds. Thus, the nature and biosynthetic pathways of many volatile compounds remain to be discovered (Tieman *et al.*, 2006a). The availability of new molecular genetics tools has begun to change this inactivity, and efforts to improve fruit flavour components by genetic engineering have a good future. QTLs regulating the production and accumulation of several volatiles compounds in tomato have been identified, and functional characterization of genes present at these loci has begun (Tieman *et al.*, 2006a; Mathieu *et al.*, 2009). Transgenic studies on tomato engineered to alter the fruit flavour volatiles are listed in Table 16.3.

Most of the flavour volatiles are synthesized during fruit ripening, reaching a maximum at or before full ripening (Klee and Giovannoni, 2011). This temporal regulation of volatile compounds is maintained through the production of their precursors including lipids, carotenoids, amino acids and keto acids (Iijima *et al.*, 2004; Kochevenko *et al.*, 2012). Aromatic volatiles, 2-phenylacetaldehdye and 2-phenylethanol are derived from phenylalanine and contribute significantly to tomato fruit flavour. A family of aromatic amino acid decarboxylases (*SlAADC1A*, *SlAADC1B* and *SlAADC2*) has been characterized (Tieman *et al.*, 2006b). Constitutive overexpression of either *SlAADC1A* or *SlAADC2* increased the emission of 2-phenylacetaldehyde, 2-phenylethanol and 1-nitro-2-phenylethane more than tenfold in transgenic tomato

Table 16.3. Studies on tomato engineered to alter fruit flavour volatiles.

Transgene	Metabolic reaction or function	Promoter expression	Metabolic phenotype	Reference
Amino acid aromatic decarboxylase (*AADC1A*) (*Solanum lycopersicum*)	Phenylalanine → phenylethylamine	FMV *35S* (OE)	10-fold ↑ 1-nitro-2-phenylethane, 2-phenylethanol and 2-phenylacetaldehyde	Tieman *et al.* (2006b)
Salicylic acid methyltransferase (*SAMT*) (*S. lycopersicum*)	Salicylic acid → methyl salicylate	FMV *35S* (OE)	123-fold ↑ methyl salicylate	Tieman *et al.* (2010)
ω-3 fatty acid desaturase (*FAD3*) (*Brassica napus*) or/and (*FAD7*) (*S. tuberosum*)	Linoleic acid (18:2) → linolenic acid (18:3)	CaMV *35S* (OE)	↑ 18:3/18:2 ratio	Domínguez *et al.* (2010)
α-Zingiberene synthase (*ZIS*) (*Ocimum basilicum*)	Farnesyldiphosphate → α-zingiberene	*PG* (OE)	↑ α-zingiberene, other sesquiterpenes and monoterpenes	Davidovich-Rikanati *et al.* (2008)
Geraniol synthase (*GES*) (*O. basilicum*)	Geranyldiphosphate → geraniol	*PG* (OE)	↑ carotenoid-derived aroma volatiles; ↓ phytoene, lycopene and β-carotene	Davidovich-Rikanati *et al.* (2007)
Carotenoid cleavage dioxygenase (*CCDIB*) (*S. lycopersicum*)	Carotenoids → volatileterpenoid compounds	FMV *35S* (AS)	50% ↓ β-ionone (50%); ≥60% ↓ geranylacetone; no morphological alterations or changes in carotenoids	Simkin *et al.* (2004)
Lipoxygenase (*TomLoxC*) (*S. lycopersicum*)	Chloroplast-targeted lipoxygenase isoform, C6 volatiles made at the expense of linoleic and linolenic acids	CaMV *35S* (AS)	1.5% ↓ hexanal, hexenal and hexanol	Chen *et al.* (2004)
S-linalool synthase (*LIS*) (*Clarkia breweri*)	Geranyldiphosphate → *S*-linalool	*E8* (OE)	↑ *S*-Linalool and 8-hydroxylinalool; no change in phenotype or in terpenoids	Lewinsohn *et al.* (2001)
Fibrillin (*FIB1*, *FIB2*) (*Capsicum annuum*)	Involved in synthesis of lipoproteins in certain chromoplast types	Native (OE)	2-fold ↑ carotenoids, i.e. 118% ↑ lycopene; 64% ↑ β-carotene, 36% ↑ β-ionone, 74% ↑ β-cyclocitral, 50% ↑ citral, 122% ↑ 6-methyl-5-hepten-2-one and 223% ↑ geranylacetone	Simkin *et al.* (2007)
ODORANT 1 (*ODO1*) (*Petunia hybrida*)	R2R3-type MYB transcription factor that positively regulates volatile benzoid levels, synthesizing precursors from the shikimate pathway	*E8* (OE)	No increase in phenylalanine-derived volatile compounds	Dal Cin *et al.* (2011)

FMV, figwort mosaic virus. See Tables 16.1 and 16.2 for other abbreviations.

fruit compared with levels in the wild-type fruit. Antisense inhibition of *SlAADC2* also resulted in reduced emission of these volatiles. Expression of tomato phenylaldehyde reductase (*SlPAR1* and *SlPAR2*) in transgenic petunia accelerated the emission of 2-phenylethanol at the expense of 2-phenylacetaldehyde (Tieman *et al.*, 2007). However, how expression of this gene affects the quality and quantity of volatiles in fruits has not yet been evaluated.

Hexanals and (*Z*)-hex-3-enal are derived from lipoxygenase pathway, and a higher (*Z*)-hex-3-enal/hexanal ratio correlates with a higher consumer appreciation of tomato varieties (Carbonell-Barrachina *et al.*, 2006). The enzyme ω-3 fatty acid desaturase converts linoleic acid (18:2) to linolenic acid (18:3), the precursor of hexanal and its derivatives. Expression of ω-3 fatty acid desaturase (*BnFAD3*) from *Brassica napus* in transgenic tomato increased the ratios of 18:3/18:2 and (*Z*)-hex-3-enal/hexanal (Domínguez *et al.*, 2010). The constitutive expression of an antisense gene of chloroplast-targeted lipoxygenase, *TomloxC*, greatly reduced the production of hexanal, hexenal and hexanol compared with wild-type levels (Chen *et al.*, 2004).

Monoterpenes and sesquiterpenes are other important contributors to fruit aroma and volatile components, and are connected with the early steps of carotenoid biosynthesis pathway (see Plate 7). α-Zingiberene synthase catalyses the formation of α-zingiberene and other sesquiterpenes from farnesyl diphosphate, while geraniol synthase catalyses the conversion of geranyl diphosphate to geraniol (Iijima *et al.*, 2004). Geraniol is an acyclic monoterpene and the precursor of geranial, nerol, citronellol, and geraniol and citronellol acetate esters (Davidovich-Rikanati *et al.*, 2007), compounds that are produced in minute amounts in ripe tomato fruit (Baldwin *et al.*, 2000). Overexpression of lemon basil geraniol synthase under the control of the tomato *PG* promoter resulted in multiple-fold enrichment of endogenous carotenoid-derived aroma volatiles at the expense of phytoene, lycopene and β-carotene, and induced the biosynthesis of geraniol and its derivatives and monoterpenes, which were not detected in wild-type fruit (Davidovich-Rikanati *et al.*, 2007). Linalool, another monoterpene, is synthesized directly from geranyl diphosphate by linalool synthase. Tomato fruit does not contain any linalool synthase activity. However, tomato transformed with the *Clarkia breweri* *S*-linalool synthase gene under control of the *E8* promoter exhibited greatly induced production of *S*-linalool and 8-hydroxylinalool with no alteration in the phenotype or in the level of terpenoids (Lewinsohn *et al.*, 2001). Transgenic tomato fruit overexpressing lemon basil α-zingiberene synthase under the fruit ripening-specific *PG* promoter produced higher levels of α-zingiberene and other sesquiterpenes and monoterpenes, which were otherwise undetectable in the wild-type fruit (Davidovich-Rikanati *et al.*, 2008).

Apocarotenoid volatiles, 6-methyl-5-hepten-2-one, 6-methyl-5-hepten-2-ol, β-ionone, β-cyclocitral and geranylacetone, are derived from carotenoid degradation. Thus, production of apocarotenoid volatiles depends on the type and amount of carotenoids being synthesized and the stage of ripening. Constitutive overexpression of the native carotenoid cleavage dioxygenase antisense gene did not alter plant morphology or carotenoid accumulation in fruit tissues, but reduced β-ionone levels by 50% and geranylacetone by ≥60% (Simkin *et al.*, 2004). As carotenoids are synthesized in plastoglobules – lipid bodies within plastids – and none of the carotenoid cleavage dioxygenase genes is upregulated during ripening, the ripening-associated increase in these volatiles was attributed to physical changes in plastids, such as chromoplast differentiation (Klee and Giovannoni, 2011). Several studies have suggested that carotenoid-derived synthesis of aroma volatiles is ethylene dependent. However, the pericarp discs of ACC synthase-suppressed transgenic tomato fruit deficient in ethylene production converted the exogenously applied lycopene into carotenoid-related volatiles. These results suggest that carotenoid biosynthesis is ethylene dependent, but degradation into volatile compounds is ethylene independent (Gao *et al.*, 2008). More *in vivo* experiments are needed to separate out the role of ethylene in carotenoid production and their catabolism. Carotenoids are also synthesized from chloroplast-derived isoprenoids, and their levels increase with total chromoplast area per cell in ripe fruit pericarp of the *high-pigment-1* (*hp-1*) tomato mutant (Cookson *et al.*, 2003; Wang *et al.*, 2008). This led to testing of the hypothesis that elevating biosynthesis of structural chromoplast proteins would increase the emission of carotenoid-derived volatile compounds. Transgenic tomato lines overexpressing pepper fibrillin, a protein involved in the synthesis of lipoprotein in the chromoplast, exhibited increased lycopene (118%) and β-carotene (64%) (Simkin *et al.*, 2007). Elevations in the emission of β-ionone (36%), β-cyclocitral (74%), citral (50%), 6-methyl-5-hepten-2-one (122%) and geranylacetone (223%) were also recorded in these transgenic fruit as a consequence of increases in the availability of carotenoids for cleavage activity (Simkin *et al.*, 2007).

In addition to the production of volatile compounds from phenylalanine, carotenoid or lipoxygenase-mediated pathways, other sources of important volatile compounds are now known and include guaiacol, synthesized by methylation of catechol, which contributes smoky aroma to tomato flavour. Tomato lines silenced for or overexpressing catechol-*O*-methyltransferase (*CTOMT1*) provided evidence that this gene is responsible for the production of guaiacol in tomato (Mageroy *et al.*, 2012). Transgenic tomato lines constitutively over- or underexpressing salicylic acid methyl transferase (*SlSAMT*) due to a sense or antisense chimeric gene construct confirmed the functional role of *SlSAMT* in the production and emission of methyl salicylate (Tieman *et al.*, 2010).

Transgenic approaches applied to apple (Dandekar *et al.*, 2004; Defilippi *et al.*, 2004, 2005a,b; Schaffer *et al.*, 2007; Brown, 2009), cucumber (Zawirska-Wojtasiak *et al.*, 2009), grape (Battilana *et al.*, 2011), berries (Malowicki *et al.*, 2008), strawberry and banana (Beekwilder *et al.*, 2004), potato (Di, 2009), basil (Dudai and Belanger, 2009), melon (Flores *et al.*, 2002) and oranges (Rodríguez *et al.*, 2011b,c) have identified various enzymes involved in the volatile biosynthesis pathway and their interaction with genetic and environmental factors such as ethylene and pathogen responsiveness.

16.7 Future Perspectives

The first edible transgenic crop, 'FlavrSavr' tomato, was released for human consumption in 1992, some 20 years ago (USDA-APHIS, 1991, 1992; Kramer and Redenbaugh, 1994). 'FlavrSavr' was produced by antisense RNA technology to have reduced *PG* expression and a promise to maintain texture of the ripened tomato fruit after harvest and during long-distance transportation (Kramer and Redenbaugh, 1994). This was a large leap but was not sufficient to meet market expectation (Giovannoni *et al.*, 1989; Thakur *et al.*, 1997). However, it provided the impetus and a path to genetically modified crops for enhancing various desirable traits, some of which have been discussed in this chapter. Most of the first-generation genetically engineered agronomical crops were developed based on manipulation of simple monogenic traits such as herbicide or insect resistance. Examples of successful genetic engineering of fruit crops, discussed in this chapter, are a testament to an approach that is robust and powerful. Thus, rational strategies have resulted in enhancing several desirable quality attributes in fruit crops and have produced novel phenotypes by using the gain or loss of function of a candidate gene.

Many desirable crop traits are, however, multigenic in nature, the final outcome being a function of a group of genes.

Therefore, enhancing a multigenic trait became a focus of the second-generation genetically engineered crops. These basic strategies were used to accomplish this objective. Because transcription factors could control a number of downstream genes, using them to engineer crops to introduce complex polygenic traits such as tolerance against abiotic stresses and enhancing production of secondary metabolites was another means of co-expressing multiple genes. Some success using such an approach has been achieved. However, most metabolic pathways have rate limiting step(s), and the simultaneous expression of a number of genes may not always help boost the intended trait(s). Also, such a strategy may introduce a negative override of metabolism and result in lowering the desired attribute(s). None the less, simultaneous introduction of several genes helped develop high-level β-carotene rice, designated Golden Rice (Paine *et al.*, 2005). This study also demonstrated that the source of gene(s) plays a significant role in increasing a preferred molecule/nutrient; for instance, the use of carotene desaturase from *E. uredovora* resulted in a 23-fold increase in total carotenoids in rice (Paine *et al.*, 2005). It is implicit from such studies that a clear understanding of the complex gene expression and the process of production of a desired metabolite is needed to enable targeted expression of a transgene at a desired stage of development of a specific tissue. In this context and to solve such complex hurdles, significant new knowledge is desired to develop chimeric promoters and accomplish targeted expression of the introduced genes at a specific stage of fruit development.

In the near future, we see a need for complementary interaction of practitioners of biotechnology and conventional breeding methods to accelerate the development of novel fruit varieties with enhanced and much desired attributes. Molecular genetic tools such as QTL mapping, chromosome walking, genome sequencing and bioinformatics are powerful catalysts whose use can help bring these approaches together. Whereas the recombinant DNA approach via transformation provides a direct path to introducing new traits in elite germplasm, more work and effort are required to get rid of undesirable traits introduced by a QTL-based approach. The availability of molecular markers associated with known traits should, however, facilitate the use of this approach to introduce desirable attributes in fruit crops. We see a bright and exciting future for precision-based engineering of quality attributes in fruit crops.

Acknowledgements

A.K.H. and A.K.M. are supported by USDA/NIFA 2010-65115-20374 and USDA/NIFA 2012-67017-30159. R.A. is supported by the Higher Education Commission of Pakistan. Trade names or commercial products mentioned in this publication are only to provide specific information and do not imply any recommendation or endorsement by the authors.

References

Abeles, F.B., Morgan, P.W. and Saltveit, M.E.J. (1992) *Ethylene in Plant Biology*. Academic Press, San Diego.

Adato, A., Mandel, T., Mintz-Oron, S., Venger, I., Levy, D., Yativ, M., Domínguez, E., Wang, Z., De Vos, R.C.H., Jetter, R., Schreiber, L., Heredia, A., Rogachev, I. and Aharoni, A. (2009) Fruit-surface flavonoid accumulation in tomato is controlled by a SlMYB12-regulated transcriptional network. *PLOS Genetics* 5, e1000777.

Aharoni, A., De Vos, C.H.R., Wein, M., Sun, Z., Greco, R., Kroon, A., Mol, J.N.M. and O'Connell, A.P. (2001) The strawberry FaMYB1 transcription factor suppresses anthocyanin and flavonol accumulation in transgenic tobacco. *Plant Journal* 28, 319–332.

Amiot, M.J., Tacchini, M., Aubert, S. and Nicolas, J. (1992) Phenolic composition and browning susceptibility of various apple cultivars at maturity. *Journal of Food Science* 57, 958–962.

Andrade, E.H.A., Maia, J.G.S. and Zoghbi, M.d.G.B. (2000) Aroma volatile constituents of Brazilian varieties of mango fruit. *Journal of Food Composition and Analysis* 13, 27–33.

Anterola, A.M. and Lewis, N.G. (2002) Trends in lignin modification: a comprehensive analysis of the effects of genetic manipulations/mutations on lignification and vascular integrity. *Phytochemistry* 61, 221–294.

Auger, B., Baron, C., Lucas, M.-O., Vautrin, S., Bergès, H., Chalhoub, B., Fautrel, A., Renard, M. and Nesi, N. (2009) Brassica orthologs from *BANYULS* belong to a small multigene family, which is involved in procyanidin accumulation in the seed. *Planta* 230, 1167–1183.

Azari, R., Tadmor, Y., Meir, A., Reuveni, M., Evenor, D., Nahon, S., Shlomo, H., Chen, L. and Levin, I. (2010a) Light signaling genes and their manipulation towards modulation of phytonutrient content in tomato fruits. *Biotechnology Advances* 28, 108–118.

Azari, R., Reuveni, M., Evenor, D., Nahon, S., Shlomo, H., Chen, L. and Levin, I. (2010b) Overexpression of *UV-DAMAGED DNA BINDING PROTEIN 1* links plant development and phytonutrient accumulation in *high pigment-1* tomato. *Journal of Experimental Botany* 61, 3627–3637.

Baldwin, E.A., Scott, J.W., Shewmaker, C.K. and Schuch, W. (2000) Flavor trivia and tomato aroma: biochemistry and possible mechanisms for control of important aroma components. *Hortscience* 35, 1013–1022.

Ban, Y., Honda, C., Hatsuyama, Y., Igarashi, M., Bessho, H. and Moriguchi, T. (2007) Isolation and functional analysis of a MYB transcription factor gene that is a key regulator for the development of red coloration in apple skin. *Plant and Cell Physiology* 48, 958–970.

Battilana, J., Emanuelli, F., Gambino, G., Gribaudo, I., Gasperi, F., Boss, P.K. and Grando, M.S. (2011) Functional effect of grapevine 1-deoxy-D-xylulose 5-phosphate synthase substitution K284N on Muscat flavour formation. *Journal of Experimental Botany* 62, 5497–5508.

Beekwilder, J., Alvarez-Huerta, M., Neef, E., Verstappen, F.W.A., Bouwmeester, H.J. and Aharoni, A. (2004) Functional characterization of enzymes forming volatile esters from strawberry and banana. *Plant Physiology* 135, 1865–1878.

Ben Chaim, A., Borovsky, Y., Rao, G., Gur, A., Zamir, D. and Paran, I. (2006) Comparative QTL mapping of fruit size and shape in tomato and pepper. *Israel Journal of Plant Sciences* 54, 191–203.

Bewley, J.D., Banik, M., Bourgault, R., Feurtado, J.A., Toorop, P. and Hilhorst, H.W.M. (2000) Endo-β-mannanase activity increases in the skin and outer pericarp of tomato fruits during ripening. *Journal of Experimental Botany* 51, 529–538.

Biggs, M.S. and Handa, A.K. (1989) Temporal regulation of polygalacturonase gene-expression in fruits of normal, mutant, and heterozygous tomato genotypes. *Plant Physiology* 89, 117–125.

Bogs, J., Jaffé, F.W., Takos, A.M., Walker, A.R. and Robinson, S.P. (2007) The grapevine transcription factor VvMYBPA1 regulates proanthocyanidin synthesis during fruit development. *Plant Physiology* 143, 1347–1361.

Bohner, J. and Bangerth, F. (1988) Cell number, cell size and hormone levels in semi-isogenic mutants of *Lycopersicon pimpinellifolium* differing in fruit size. *Physiologia Plantarum* 72, 316–320.

Boss, P.K., Davies, C. and Robinson, S.P. (1996) Analysis of the expression of anthocyanin pathway genes in developing *Vitis vinifera* L. cv Shiraz grape berries and the implications for pathway regulation. *Plant Physiology* 111, 1059–1066.

Bovy, A., de Vos, R., Kemper, M., Schijlen, E., Almenar Pertejo, M., Muir, S., Collins, G., Robinson, S., Verhoeyen, M., Hughes, S., Santos-Buelga, C. and van Tunen, A. (2002) High-flavonol tomatoes resulting from the heterologous expression of the maize transcription factor genes *LC* and *C1*. *Plant Cell Online* 14, 2509–2526.

Brady, C.J. (1987) Fruit ripening. *Annual Review of Plant Physiology* 38, 155–178.

Brown, S.K. (2009) Breeding and biotechnology for flavor development in apple (*Malus × Domestica* Borkh.). In: Havkin-Frenkel, D. and Belanger, F. (eds) *Biotechnology in Flavor Production.* Blackwell Publishing, Ames, IA, pp. 147–160.

Brummell, D., Hall, B. and Bennett, A. (1999a) Antisense suppression of tomato endo-1,4-β-glucanase *Cel2* mRNA accumulation increases the force required to break fruit abscission zones but does not affect fruit softening. *Plant Molecular Biology* 40, 615–622.

Brummell, D.A., Harpster, M.H., Civello, P.M., Palys, J.M., Bennett, A.B. and Dunsmuir, P. (1999b) Modification of expansin protein abundance in tomato fruit alters softening and cell wall polymer metabolism during ripening. *Plant Cell Online* 11, 2203–2216.

Burkhardt, P.K., Beyer, P., Wünn, J., Klöti, A., Armstrong, G.A., Schledz, M., von Lintig, J. and Potrykus, I. (1997) Transgenic rice (*Oryza sativa*) endosperm expressing daffodil (*Narcissus pseudonarcissus*) phytoene synthase accumulates phytoene, a key intermediate of provitamin A biosynthesis. *Plant Journal* 11, 1071–1078.

Butelli, E., Titta, L., Giorgio, M., Mock, H.-P., Matros, A., Peterek, S., Schijlen, E.G.W.M., Hall, R.D., Bovy, A.G., Luo, J. and Martin, C. (2008) Enrichment of tomato fruit with health-promoting anthocyanins by expression of select transcription factors. *Nature Biotechnology* 26, 1301–1308.

Buttery, R.G. and Ling, L.C. (1993) Volatile components of tomato fruit and plant parts. In: *Bioactive Volatile Compounds from Plants*. American Chemical Society, Washington, DC, pp. 23–34.

Buttery, R.G., Teranishi, R., Ling, L.C., Flath, R.A. and Stern, D.J. (1988) Quantitative studies on origins of fresh tomato aroma volatiles. *Journal of Agricultural and Food Chemistry* 36, 1247–1250.

Carbonell-Barrachina, A., Agustí, A. and Ruiz, J. (2006) Analysis of flavor volatile compounds by dynamic headspace in traditional and hybrid cultivars of Spanish tomatoes. *European Food Research and Technology* 222, 536–542.

Carey, A.T., Smith, D.L., Harrison, E., Bird, C.R., Gross, K.C., Seymour, G.B. and Tucker, G.A. (2001) Down-regulation of a ripening-related β-galactosidase gene (TBG1) in transgenic tomato fruits. *Journal of Experimental Botany* 52, 663–668.

Causse, M., Chaïb, J., Lecomte, L., Buret, M. and Hospital, F. (2007) Both additivity and epistasis control the genetic variation for fruit quality traits in tomato. *Theoretical and Applied Genetics* 115, 429–442.

Cebolla, A., Maria Vinardell, J., Kiss, E., Olah, B., Roudier, F., Kondorosi, A. and Kondorosi, E. (1999) The mitotic inhibitor *ccs52* is required for endoreduplication and ploidy-dependent cell enlargement in plants. *EMBO Journal* 18, 4476–4484.

Chen, G., Hackett, R., Walker, D., Taylor, A., Lin, Z. and Grierson, D. (2004) Identification of a specific isoform of tomato lipoxygenase (TomloxC) involved in the generation of fatty acid-derived flavor compounds. *Plant Physiology* 136, 2641–2651.

Cheniclet, C., Rong, W.Y., Causse, M., Frangne, N., Bolling, L., Carde, J.-P. and Renaudin, J.-P. (2005) Cell expansion and endoreduplication show a large genetic variability in pericarp and contribute strongly to tomato fruit growth. *Plant Physiology* 139, 1984–1994.

Choi, D., Cho, H.-T. and Lee, Y. (2006) Expansins: expanding importance in plant growth and development. *Physiologia Plantarum* 126, 511–518.

Christiansen, K.F., Olsen, E., Vegarud, G., Langsrud, T., Lea, P., Haugen, J.E. and Egelandsdal, B. (2011) Flavor release of the tomato flavor enhancer, 2-isobutylthiazole, from whey protein stabilized model dressings. *Food Science and Technology International* 17, 143–154.

Cong, B., Liu, J. and Tanksley, S.D. (2002) Natural alleles at a tomato fruit size quantitative trait locus differ by heterochronic regulatory mutations. *Proceedings of the National Academy of Sciences USA* 99, 13606–13611.

Cong, L., Wang, C., Chen, L., Liu, H., Yang, G. and He, G. (2009) Expression of phytoene synthase1 and carotene desaturase *crtI* genes result in an increase in the total carotenoids content in transgenic elite wheat (*Triticum aestivum* L.). *Journal of Agricultural and Food Chemistry* 57, 8652–8660.

Cook, M.S. (2010) Phyto power in the fight against disease. In: Cook, M.S. (ed.) *The Phytozyme Cure: Treat or Reverse More Than 30 Serious Health Conditions with Powerful Plant Nutrients*. Wiley, Mississauga, Ontario, pp. 11–36.

Cookson, P.J., Kiano, J.W., Shipton, C.A., Fraser, P.D., Romer, S., Schuch, W., Bramley, P.M. and Pyke, K.A. (2003) Increases in cell elongation, plastid compartment size and phytoene synthase activity underlie the phenotype of the *high pigment-1* mutant of tomato. *Planta* 217, 896–903.

Crozier, A., Lean, M.E.J., McDonald, M.S. and Black, C. (1997) quantitative analysis of the flavonoid content of commercial tomatoes, onions, lettuce, and celery. *Journal of Agricultural and Food Chemistry* 45, 590–595.

D'Ambrosio, C., Giorio, G., Marino, I., Merendino, A., Petrozza, A., Salfi, L., Stigliani, A.L. and Cellini, F. (2004) Virtually complete conversion of lycopene into β-carotene in fruits of tomato plants transformed with the tomato *lycopene β-cyclase* (*tlcy-b*) cDNA. *Plant Science* 166, 207–214.

D'Introno, A., Paradiso, A., Scoditti, E., D'Amico, L., De Paolis, A., Carluccio, M.A., Nicoletti, I., DeGara, L., Santino, A. and Giovinazzo, G. (2009) Antioxidant and anti-inflammatory properties of tomato fruits synthesizing different amounts of stilbenes. *Plant Biotechnology Journal* 7, 422–429.

Dal Cin, V., Tieman, D.M., Tohge, T., McQuinn, R., de Vos, R.C.H., Osorio, S., Schmelz, E.A., Taylor, M.G., Smits-Kroon, M.T., Schuurink, R.C., Haring, M.A., Giovannoni, J., Fernie, A.R. and Klee, H.J. (2011) Identification of genes in the phenylalanine metabolic pathway by ectopic expression of a MYB transcription factor in tomato fruit. *Plant Cell Online* 23, 2738–2753.

Dandekar, A.M., Teo, G., Defilippi, B.G., Uratsu, S.L., Passey, A.J., Kader, A.A., Stow, J.R., Colgan, R.J. and James, D.J. (2004) Effect of down-regulation of ethylene biosynthesis on fruit flavor complex in apple fruit. *Transgenic Research* 13, 373–384.

Davidovich-Rikanati, R., Sitrit, Y., Tadmor, Y., Iijima, Y., Bilenko, N., Bar, E., Carmona, B., Fallik, E., Dudai, N., Simon, J.E., Pichersky, E. and Lewinsohn, E. (2007) Enrichment of tomato flavor by diversion of the early plastidial terpenoid pathway. *Nature Biotechnology* 25, 899–901.

Davidovich-Rikanati, R., Lewinsohn, E., Bar, E., Iijima, Y., Pichersky, E. and Sitrit, Y. (2008) Overexpression of the lemon basil α-zingiberene synthase gene increases both mono- and sesquiterpene contents in tomato fruit. *Plant Journal* 56, 228–238.

Davuluri, G.R., van Tuinen, A., Mustilli, A.C., Manfredonia, A., Newman, R., Burgess, D., Brummell, D.A., King, S.R., Palys, J., Uhlig, J., Pennings, H.M.J. and Bowler, C. (2004) Manipulation of DET1 expression in tomato results in photomorphogenic phenotypes caused by post-transcriptional gene silencing. *Plant Journal* 40, 344–354.

Davuluri, G.R., van Tuinen, A., Fraser, P.D., Manfredonia, A., Newman, R., Burgess, D., Brummell, D.A., King, S.R., Palys, J., Uhlig, J., Bramley, P.M., Pennings, H.M.J. and Bowler, C. (2005) Fruit-specific RNAi-mediated suppression of DET1 enhances carotenoid and flavonoid content in tomatoes. *Nature Biotechnology* 23, 890–895.

de Freitas, S.T., Handa, A.K., Wu, Q., Park, S. and Mitcham, E.J. (2012) Role of pectin methylesterases in cellular calcium distribution and blossom-end rot development in tomato fruit. *Plant Journal*, 71, 824–835.

Defilippi, B.G., Dandekar, A.M. and Kader, A.A. (2004) Impact of suppression of ethylene action or biosynthesis on flavor metabolites in apple (*Malus domestica* Borkh) fruits. *Journal of Agricultural and Food Chemistry* 52, 5694–5701.

Defilippi, B.G., Dandekar, A.M. and Kader, A.A. (2005a) Relationship of ethylene biosynthesis to volatile production, related enzymes, and precursor availability in apple peel and flesh tissues. *Journal of Agricultural and Food Chemistry* 53, 3133–3141.

Defilippi, B.G., Kader, A.A. and Dandekar, A.M. (2005b) Apple aroma: alcohol acyltransferase, a rate limiting step for ester biosynthesis, is regulated by ethylene. *Plant Science* 168, 1199–1210.

DellaPenna, D., Lincoln, J.E., Fischer, R.L. and Bennett, A.B. (1989) Transcriptional analysis of polygalacturonase and other ripening associated genes in Rutgers, *rin*, *nor*, and *Nr* tomato fruit. *Plant Physiology* 90, 1372–1377.

Dharmapuri, S., Rosati, C., Pallara, P., Aquilani, R., Bouvier, F., Camara, B. and Giuliano, G. (2002) Metabolic engineering of xanthophyll content in tomato fruits. *FEBS Letters* 519, 30–34.

Di, R. (2009) Increasing the methional content in potato through biotechnology. In: Havkin-Frenkel, D. and Belanger, F. (eds) *Biotechnology in Flavor Production*. Blackwell Publishing, Ames, IA, pp. 185–193.

Dixon, J. and Hewett, E.W. (2000) Factors affecting apple aroma/flavour volatile concentration: a review. *New Zealand Journal of Crop and Horticultural Science* 28, 155–173.

Dixon, R.A. (2005) A two-for-one in tomato nutritional enhancement. *Nature Biotechnology* 23, 825–826.

Doganlar, S., Frary, A., Daunay, M.-C., Lester, R.N. and Tanksley, S.D. (2002) Conservation of gene function in the solanaceae as revealed by comparative mapping of domestication traits in eggplant. *Genetics* 161, 1713–1726.

Domínguez, T., Hernández, M.L., Pennycooke, J.C., Jiménez, P., Martínez-Rivas, J.M., Sanz, C., Stockinger, E.J., Sánchez-Serrano, J.J. and Sanmartín, M. (2010) Increasing ω-3 desaturase expression in tomato results in altered aroma profile and enhanced resistance to cold stress. *Plant Physiology* 153, 655–665.

Dudai, N. and Belanger, F.C. (2009) Aroma as a factor in the breeding process of fresh herbs – the case of basil. In: Havkin-Frenkel, D. and Belanger, F. (eds) *Biotechnology in Flavor Production*. Blackwell Publishing, Ames, IA pp. 161–184.

Enfissi, E.M.A., Fraser, P.D., Lois, L.-M., Boronat, A., Schuch, W. and Bramley, P.M. (2005) Metabolic engineering of the mevalonate and non-mevalonate isopentenyl diphosphate-forming pathways for the production of health-promoting isoprenoids in tomato. *Plant Biotechnology Journal* 3, 17–27.

Fatima, T., Mattoo, A.K., Rivera-Domínguez, M., Troncoso-Rojas, R., Tiznado-Hernández, M.-E. and Handa, A.K. (2009) Tomato. In: Kole, C. and Hall, T.C. (eds) *Compendium of Transgenic Crop Plants*. Blackwell Publishing, Oxford, 2008, pp. 1–46.

Ficcadenti, N., Sestili, S., Pandolfini, T., Cirillo, C., Rotino, G.L. and Spena, A. (1999) Genetic engineering of parthenocarpic fruit development in tomato. *Molecular Breeding* 5, 463–470.

Flachowsky, H., Szankowski, I., Fischer, T., Richter, K., Peil, A., Höfer, M., Dörschel, C., Schmoock, S., Gau, A., Halbwirth, H. and Hanke, M.-V. (2010) Transgenic apple plants overexpressing the Lc gene of maize show an altered growth habit and increased resistance to apple scab and fire blight. *Planta* 231, 623–635.

Flachowsky, H., Halbwirth, H., Treutter, D., Richter, K., Hanke, M.-V., Szankowski, I., Gosch, C., Stich, K. and Fischer, T.C. (2012) Silencing of flavanone-3-hydroxylase in apple (*Malus × domestica* Borkh.) leads to accumulation of flavanones, but not to reduced fire blight susceptibility. *Plant Physiology and Biochemistry* 51, 18–25.

Flores, F., El Yahyaoui, F., de Billerbeck, G., Romojaro, F., Latché, A., Bouzayen, M., Pech, J.C. and Ambid, C. (2002) Role of ethylene in the biosynthetic pathway of aliphatic ester aroma volatiles in Charentais Cantaloupe melons. *Journal of Experimental Botany* 53, 201–206.

Frary, A., Nesbitt, T.C., Frary, A., Grandillo, S., Knaap, E.v.d., Cong, B., Liu, J., Meller, J., Elber, R., Alpert, K.B. and Tanksley, S.D. (2000) *fw2.2*: a quantitative trait locus key to the evolution of tomato fruit size. *Science* 289, 85–88.

Fraser, P.D., Romer, S., Shipton, C.A., Mills, P.B., Kiano, J.W., Misawa, N., Drake, R.G., Schuch, W. and Bramley, P.M. (2002) Evaluation of transgenic tomato plants expressing an additional phytoene synthase in a fruit-specific manner. *Proceedings of the National Academy of Sciences USA* 99, 1092–1097.

Fraser, P.D., Enfissi, E.M.A., Halket, J.M., Truesdale, M.R., Yu, D., Gerrish, C. and Bramley, P.M. (2007) Manipulation of phytoene levels in tomato fruit: effects on isoprenoids, plastids, and intermediary metabolism. *Plant Cell Online* 19, 3194–3211.

Fray, R.G., Wallace, A., Fraser, P.D., Valero, D., Hedden, P., Bramley, P.M. and Grierson, D. (1995) Constitutive expression of a fruit phytoene synthase gene in transgenic tomatoes causes dwarfism by redirecting metabolites from the gibberellin pathway. *Plant Journal* 8, 693–701.

Frey, A., Boutin, J.-P., Sotta, B., Mercier, R. and Marion-Poll, A. (2006) Regulation of carotenoid and ABA accumulation during the development and germination of *Nicotiana plumbaginifolia* seeds. *Planta* 224, 622–632.

Frydman, A., Weisshaus, O., Bar-Peled, M., Huhman, D.V., Sumner, L.W., Marin, F.R., Lewinsohn, E., Fluhr, R., Gressel, J. and Eyal, Y. (2004) Citrus fruit bitter flavors: isolation and functional characterization of the gene *Cm1,2RhaT* encoding a 1,2 rhamnosyltransferase, a key enzyme in the biosynthesis of the bitter flavonoids of citrus. *Plant Journal* 40, 88–100.

Fujisawa, M. and Misawa, N. (2010) Enrichment of carotenoids in flaxseed by introducing a bacterial phytoene synthase gene. In: Fett-Neto, A.G. (ed.) *Plant Secondary Metabolism Engineering: Methods and Applications*. Springer, New York, NY, pp. 201–211.

Fujisawa, M., Takita, E., Harada, H., Sakurai, N., Suzuki, H., Ohyama, K., Shibata, D. and Misawa, N. (2009) Pathway engineering of *Brassica napus* seeds using multiple key enzyme genes involved in ketocarotenoid formation. *Journal of Experimental Botany* 60, 1319–1332.

Furukawa, T., Maekawa, M., Oki, T., Suda, I., Iida, S., Shimada, H., Takamure, I. and Kadowaki, K.-i. (2007) The *Rc* and *Rd* genes are involved in proanthocyanidin synthesis in rice pericarp. *Plant Journal* 49, 91–102.

Gaffe, J., Tieman, D.M. and Handa, A.K. (1994) Pectin methylesterase isoforms in tomato (*Lycopersicon esculentum*) tissues (effects of expression of a pectin methylesterase antisense gene). *Plant Physiology* 105, 199–203.

Gaffe, J., Tiznado, M.E. and Handa, A.K. (1997) Characterization and functional expression of a ubiquitously expressed tomato pectin methylesterase. *Plant Physiology* 114, 1547–1556.

Galpaz, N., Wang, Q., Menda, N., Zamir, D. and Hirschberg, J. (2008) Abscisic acid deficiency in the tomato mutant high-pigment 3 leading to increased plastid number and higher fruit lycopene content. *Plant Journal* 53, 717–730.

Gao, H., Zhu, H., Shao, Y., Chen, A., Lu, C., Zhu, B. and Luo, Y. (2008) Lycopene accumulation affects the biosynthesis of some carotenoid-related volatiles independent of ethylene in tomato. *Journal of Integrative Plant Biology* 50, 991–996.

Giliberto, L., Perrotta, G., Pallara, P., Weller, J.L., Fraser, P.D., Bramley, P.M., Fiore, A., Tavazza, M. and Giuliano, G. (2005) Manipulation of the blue light photoreceptor Cryptochrome 2 in tomato affects vegetative development, flowering time, and fruit antioxidant content. *Plant Physiology* 137, 199–208.

Giovannoni, J.J. (2004) Genetic regulation of fruit development and ripening. *Plant Cell* 16, S170–S180.

Giovannoni, J.J. (2007) Fruit ripening mutants yield insights into ripening control. *Current Opinion in Plant Biology* 10, 283–289.

Giovannoni, J.J., Dellapenna, D., Bennett, A.B. and Fischer, R.L. (1989) Expression of a chimeric polygalacturonase gene in transgenic *rin* (ripening inhibitor) tomato fruit results in polyuronide degradation but not fruit softening. *Plant Cell* 1, 53–63.

Giovinazzo, G., D'Amico, L., Paradiso, A., Bollini, R., Sparvoli, F. and DeGara, L. (2005) Antioxidant metabolite profiles in tomato fruit constitutively expressing the grapevine stilbene synthase gene. *Plant Biotechnology Journal* 3, 57–69.

Goff, S.A. and Klee, H.J. (2006) Plant volatile compounds: sensory cues for health and nutritional value? *Science* 311, 815–819.

Goldsbrough, A.P., Tong, Y. and Yoder, J.I. (1996) *Lc* as a non-destructive visual reporter and transposition excision marker gone for tomato. *Plant Journal* 9, 927–933.

Gong, Z.-Z., Yamagishi, E., Yamazaki, M. and Saito, K. (1999) A constitutively expressed Myc-like gene involved in anthocyanin biosynthesis from *Perilla frutescens*: molecular characterization, heterologous expression in transgenic plants and transactivation in yeast cells. *Plant Molecular Biology* 41, 33–44.

Gonzalez, N., Gévaudant, F., Hernould, M., Chevalier, C. and Mouras, A. (2007) The cell cycle-associated protein kinase WEE1 regulates cell size in relation to endoreduplication in developing tomato fruit. *Plant Journal* 51, 642–655.

Goupy, P., Amiot, M.J., Richard-Forget, F., Duprat, F., Aubert, S. and Nicolas, J. (1995) Enzymatic browning of model solutions and apple phenolic extracts by apple polyphenoloxidase. *Journal of Food Science* 60, 497–501.

Grandillo, S., Ku, H.M. and Tanksley, S.D. (1999) Identifying the loci responsible for natural variation in fruit size and shape in tomato. *Theoretical and Applied Genetics* 99, 978–987.

Gross, K.C. and Sams, C.E. (1984) Changes in cell wall neutral sugar composition during fruit ripening: a species survey. *Phytochemistry* 23, 2457–2461.

Guo, F., Zhou, W., Zhang, J., Xu, Q. and Deng, X. (2012) Effect of the citrus lycopene β-cyclase transgene on carotenoid metabolism in transgenic tomato fruits. *PLOS One* 7, e32221.

Han, Y., Vimolmangkang, S., Soria-Guerra, R.E. and Korban, S.S. (2012) Introduction of apple *ANR* genes into tobacco inhibits expression of both *CHI* and *DFR* genes in flowers, leading to loss of anthocyanin. *Journal of Experimental Botany* 63, 2437–2447.

Handa, A.K., Singh, N.K. and Biggs, M.S. (1985) Effect of tunicamycin on *in vitro* ripening of tomato pericarp tissue. *Physiologia Plantarum* 63, 417–424.

Handa, A.K., Tiznado-Hernández, M.-E. and Mattoo, A.K. (2012) Fruit development and ripening: a molecular perspective. In: Altman, A. and Hasegawa, P.M. (eds) *Plant Biotechnology and Agriculture: Prospects for 21st Century*. Elsevier, New York, NY, pp. 405–424.

Harborne, J.B. and Herbert, B. (1999) *The Handbook of Natural Flavonoids*. John Wiley, Chichester.

Harborne, J.B. and Williams, C.A. (2000) Advances in flavonoid research since 1992. *Phytochemistry* 55, 481–504.

Harriman, R.W., Tieman, D.M. and Handa, A.K. (1991) Molecular cloning of tomato pectin methylesterase gene and its expression in Rutgers, ripening inhibitor, nonripening, and never ripe tomato fruits. *Plant Physiology* 97, 80–87.

Hüsken, A., Baumert, A., Milkowski, C., Becker, H., Strack, D. and Möllers, C. (2005) Resveratrol glucoside (Piceid) synthesis in seeds of transgenic oilseed rape (*Brassica napus* L.). *Theoretical and Applied Genetics* 111, 1553–1562.

Iijima, Y., Davidovich-Rikanati, R., Fridman, E., Gang, D.R., Bar, E., Lewinsohn, E. and Pichersky, E. (2004) The biochemical and molecular basis for the divergent patterns in the biosynthesis of terpenes and phenylpropenes in the peltate glands of three cultivars of basil. *Plant Physiology* 136, 3724–3736.

Ingrosso, I., Bonsegna, S., De Domenico, S., Laddomada, B., Blando, F., Santino, A. and Giovinazzo, G. (2011) Over-expression of a grape stilbene synthase gene in tomato induces parthenocarpy and causes abnormal pollen development. *Plant Physiology and Biochemistry* 49, 1092–1099.

Itkin, M., Seybold, H., Breitel, D., Rogachev, I., Meir, S. and Aharoni, A. (2009) TOMATO AGAMOUS-LIKE 1 is a component of the fruit ripening regulatory network. *Plant Journal* 60, 1081–1095.

Jiménez-Bermúdez, S., Redondo-Nevado, J., Muñoz-Blanco, J., Caballero, J.L., López-Aranda, J.M., Valpuesta, V., Pliego-Alfaro, F., Quesada, M.A. and Mercado, J.A. (2002) Manipulation of strawberry fruit softening by antisense expression of a pectate lyase gene. *Plant Physiology* 128, 751–759.

Klee, H.J. (2010) Improving the flavor of fresh fruits: genomics, biochemistry, and biotechnology. *New Phytologist* 187, 44–56.

Klee, H.J. and Giovannoni, J.J. (2011) Genetics and control of tomato fruit ripening and quality attributes. *Annual Review of Genetics* 45, 41–59.

Koca, U., Berhow, M.A., Febres, V.J., Champ, K.I., Carrillo-Mendoza, O. and Moore, G.A. (2009) Decreasing unpalatable flavonoid components in Citrus: the effect of transformation construct. *Physiologia Plantarum* 137, 101–114.

Kochevenko, A., Araújo, W.L., Maloney, G.S., Tieman, D.M., Do, P.T., Taylor, M.G., Klee, H.J. and Fernie, A.R. (2012) Catabolism of branched chain amino acids supports respiration but not volatile synthesis in tomato fruits. *Molecular Plant* 5, 366–375.

Koeduka, T., Shitan, N., Kumano, T., Sasaki, K., Sugiyama, A., Linley, P., Kawasaki, T., Ezura, H., Kuzuyama, T. and Yazaki, K. (2011) Production of prenylated flavonoids in tomato fruits expressing a prenyltransferase gene from *Streptomyces coelicolor* A3(2). *Plant Biology* 13, 411–415.

Kolotilin, I., Koltai, H., Bar-Or, C., Chen, L., Nahon, S., Shlomo, H., Levin, I. and Reuveni, M. (2011) Expressing yeast *SAMdc* gene confers broad changes in gene expression and alters fatty acid composition in tomato fruit. *Physiologia Plantarum* 142, 211–223.

Kramer, M.G. and Redenbaugh, K. (1994) Commercialization of a tomato with an antisense polygalacturonase gene: the FLAVR SAVR™ tomato story. *Euphytica* 79, 293–297.

Krumbein, A. and Auerswald, H. (1998) Characterization of aroma volatiles in tomatoes by sensory analyses. *Nahrung* 42, 395–399.

Ku, H.M., Doganlar, S., Chen, K.Y. and Tanksley, S.D. (1999) The genetic basis of pear-shaped tomato fruit. *Theoretical and Applied Genetics* 99, 844–850.

Ku, H.M., Grandillo, S. and Tanksley, S.D. (2000) fs8.1, a major QTL, sets the pattern of tomato carpel shape well before anthesis. *Theoretical and Applied Genetics* 101, 873–878.

Lashbrook, C.C., Giovannoni, J.J., Hall, B.D., Fischer, R.L. and Bennett, A.B. (1998) Transgenic analysis of tomato endo-β-1,4-glucanase gene function. Role of cel1 in floral abscission. *Plant Journal* 13, 303–310.

Le Gall, G., DuPont, M.S., Mellon, F.A., Davis, A.L., Collins, G.J., Verhoeyen, M.E. and Colquhoun, I.J. (2003) Characterization and content of flavonoid glycosides in genetically modified tomato (*Lycopersicon esculentum*) fruits. *Journal of Agricultural and Food Chemistry* 51, 2438–2446.

Levin, I., Frankel, P., Gilboa, N., Tanny, S. and Lalazar, A. (2003) The tomato dark green mutation is a novel allele of the tomato homolog of the *DEETIOLATED1* gene. *Theoretical and Applied Genetics* 106, 454–460.

Lewinsohn, E., Schalechet, F., Wilkinson, J., Matsui, K., Tadmor, Y., Nam, K.-H., Amar, O., Lastochkin, E., Larkov, O., Ravid, U., Hiatt, W., Gepstein, S. and Pichersky, E. (2001) Enhanced levels of the aroma and flavor compound S-linalool by metabolic engineering of the terpenoid pathway in tomato fruits. *Plant Physiology* 127, 1256–1265.

Lewinsohn, E., Sitrit, Y., Bar, E., Azulay, Y., Ibdah, M., Meir, A., Yosef, E., Zamir, D. and Tadmor, Y. (2005) Not just colors – carotenoid degradation as a link between pigmentation and aroma in tomato and watermelon fruit. *Trends in Food Science and Technology* 16, 407–415.

Li, H., Flachowsky, H., Fischer, T., Hanke, M.-V., Forkmann, G., Treutter, D., Schwab, W., Hoffmann, T. and Szankowski, I. (2007) Maize Lc transcription factor enhances biosynthesis of anthocyanins, distinct proanthocyanidins and phenylpropanoids in apple (*Malus domestica* Borkh.). *Planta* 226, 1243–1254.

Lin, Z.F., Zhong, S.L. and Grierson, D. (2009) Recent advances in ethylene research. *Journal of Experimental Botany* 60, 3311–3336.

Lind, K., Lafer, G., Schloffer, K. and Innerhofer, H.M. (2003) *Organic Fruit Growing*. CABI, Wallingford, UK.

Lippman, Z. and Tanksley, S.D. (2001) Dissecting the genetic pathway to extreme fruit size in tomato using a cross between the small-fruited wild species *Lycopersicon pimpinellifolium* and *L. esculentum* var. Giant Heirloom. *Genetics* 158, 413–422.

Liu, J., Van Eck, J., Cong, B. and Tanksley, S.D. (2002) A new class of regulatory genes underlying the cause of pear-shaped tomato fruit. *Proceedings of the National Academy of Sciences USA* 99, 13302–13306.

Liu, J.P., Cong, B. and Tanksley, S.D. (2003) Generation and analysis of an artificial gene dosage series in tomato to study the mechanisms by which the cloned quantitative trait locus *fw2.2* controls fruit size. *Plant Physiology* 132, 292–299.

Liu, Y., Roof, S., Ye, Z., Barry, C., van Tuinen, A., Vrebalov, J., Bowler, C. and Giovannoni, J. (2004) Manipulation of light signal transduction as a means of modifying fruit nutritional quality in tomato. *Proceedings of the National Academy of Sciences USA* 101, 9897–9902.

Long, M., Millar, D.J., Kimura, Y., Donovan, G., Rees, J., Fraser, P.D., Bramley, P.M. and Bolwell, G.P. (2006) Metabolite profiling of carotenoid and phenolic pathways in mutant and transgenic lines of tomato: identification of a high antioxidant fruit line. *Phytochemistry* 67, 1750–1757.

Lorenc-Kukula, K., Amarowicz, R., Oszmianski, J., Doermann, P., Starzycki, M., Skala, J., Żuk, M., Kulma, A. and Szopa, J. (2005) Pleiotropic effect of phenolic compounds content increases in transgenic flax plant. *Journal of Agricultural and Food Chemistry* 53, 3685–3692.

Lunkenbein, S., Coiner, H., de Vos, C.H.R., Schaart, J.G., Boone, M.J., Krens, F.A., Schwab, W. and Salentijn, E.M.J. (2006) Molecular characterization of a stable antisense chalcone synthase phenotype in strawberry (*Fragaria × ananassa*). *Journal of Agricultural and Food Chemistry* 54, 2145–2153.

Luo, J., Butelli, E., Hill, L., Parr, A., Niggeweg, R., Bailey, P., Weisshaar, B. and Martin, C. (2008) AtMYB12 regulates caffeoyl quinic acid and flavonol synthesis in tomato: expression in fruit results in very high levels of both types of polyphenol. *Plant Journal* 56, 316–326.

MacLeod, A.J. and Snyder, C.H. (1985) Volatile components of two cultivars of mango from Florida. *Journal of Agricultural and Food Chemistry* 33, 380–384.

MacLeod, A.J., Macleod, G. and Snyder, C.H. (1988) Volatile aroma constituents of mango (cv Kensington). *Phytochemistry* 27, 2189–2193.

Mageroy, M.H., Tieman, D.M., Floystad, A., Taylor, M.G. and Klee, H.J. (2012) A *Solanum lycopersicum* catechol-*O*-methyltransferase involved in synthesis of the flavor molecule guaiacol. *Plant Journal* 69, 1043–1051.

Maitrejean, M., Comte, G., Barron, D., El Kirat, K., Conseil, G. and Di Pietro, A. (2000) The flavanolignan silybin and its hemisynthetic derivatives, a novel series of potential modulators of p-glycoprotein. *Bioorganic and Medicinal Chemistry Letters* 10, 157–160.

Malowicki, S.M.M., Martin, R. and Qian, M.C. (2008) Comparison of sugar, acids, and volatile composition in raspberry bushy dwarf virus-resistant transgenic raspberries and the wild type 'Meeker' (*Rubus Idaeus* L.). *Journal of Agricultural and Food Chemistry* 56, 6648–6655.

Marković, K., Vahčić, N., Ganić, K.K. and Banović, M. (2007) Aroma volatiles of tomatoes and tomato products evaluated by solid-phase microextraction. *Flavour and Fragrance Journal* 22, 395–400.

Mathews, H., Clendennen, S.K., Caldwell, C.G., Liu, X.L., Connors, K., Matheis, N., Schuster, D.K., Menasco, D.J., Wagoner, W., Lightner, J. and Wagner, D.R. (2003) Activation tagging in tomato identifies a transcriptional regulator of Anthocyanin biosynthesis, modification, and transport. *Plant Cell Online* 15, 1689–1703.

Mathieu, S., Cin, V.D., Fei, Z., Li, H., Bliss, P., Taylor, M.G., Klee, H.J. and Tieman, D.M. (2009) Flavour compounds in tomato fruits: identification of loci and potential pathways affecting volatile composition. *Journal of Experimental Botany* 60, 325–337.

Mathieu-Rivet, E., Gévaudant, F., Sicard, A., Salar, S., Do, P.T., Mouras, A., Fernie, A.R., Gibon, Y., Rothan, C., Chevalier, C. and Hernould, M. (2010) Functional analysis of the anaphase promoting complex activator CCS52A highlights the crucial role of endo-reduplication for fruit growth in tomato. *Plant Journal* 62, 727–741.

Mattoo, A.K. and Suttle, J.C. (1991) *The Plant Hormone Ethylene*. CRC Press, Boca Raton, FL.

Mattoo, A.K., Chung, S.H., Goyal, R.K., Fatima, T., Solomos, T., Srivastava, A. and Handa, A.K. (2007) Overaccumulation of higher polyamines in ripening transgenic tomato fruit revives metabolic memory, upregulates anabolism-related genes, and positively impacts nutritional quality. *Journal of AOAC International* 90, 1456–1464.

Mattoo, A.K., Shukla, V., Fatima, T., Handa, A.K. and Yachha, S.K. (2010) Genetic engineering to enhance crop-based phytonutrients (nutraceuticals) to alleviate diet-related diseases. In: Giardi, G.M., Rea, T. and Berra, B. (eds) *Bio-farms for Nutraceuticals: Functional Food and Safety Control by Biosensors*. Landes Bioscience and Springer, Austin, TX, pp. 122–143.

Maul, F., Sargent, S., Huber, D., Balaban, M., Luzuriaga, D. and Baldwin, E. (1997) Non-destructive quality screening of tomato fruit using "Electronic Nose" technology. *Proceedings of the Florida State Horticultural Society* 110, 188–194.

Mayer, F., Takeoka, G.R., Buttery, R.G., Whitehand, L.C., Naim, M. and Rabinowitch, H.D. (2008) Studies on the aroma of five fresh tomato cultivars and the precursors of *cis*- and *trans*-4,5-epoxy-(E)-2-decenals and methional. *Journal of Agricultural and Food Chemistry* 56, 3749–3757.

Mehta, R.A., Cassol, T., Li, N., Ali, N., Handa, A.K. and Mattoo, A.K. (2002) Engineered polyamine accumulation in tomato enhances phytonutrient content, juice quality, and vine life. *Nature Biotechnology* 20, 613–618.

Meli, V.S., Ghosh, S., Prabha, T.N., Chakraborty, N., Chakraborty, S. and Datta, A. (2010) Enhancement of fruit shelf life by suppressing *N*-glycan processing enzymes. *Proceedings of the National Academy of Sciences USA* 107, 2413–2418.

Mercado, J.A., Trainotti, L., Jiménez-Bermúdez, L., Santiago-Doménech, N., Posé, S., Donolli, R., Barceló, M., Casadoro, G., Pliego-Alfaro, F. and Quesada, M.A. (2010) Evaluation of the role of the endo-β-(1,4)-glucanase gene *FaEG3* in strawberry fruit softening. *Postharvest Biology and Technology* 55, 8–14.

Miedes, E., Herbers, K., Sonnewald, U. and Lorences, E.P. (2010) Overexpression of a cell wall enzyme reduces xyloglucan depolymerization and softening of transgenic tomato fruits. *Journal of Agricultural and Food Chemistry* 58, 5708–5713.

Mooney, M., Desnos, T., Harrison, K., Jones, J., Carpenter, R. and Coen, E. (1995) Altered regulation of tomato and tobacco pigmentation genes caused by the *delila* gene of *Antirrhinum*. *Plant Journal* 7, 333–339.

Moriguchi, T., Kita, M., Tomono, Y., Endo-Inagaki, T. and Omura, M. (2001) Gene expression in flavonoid biosynthesis: correlation with flavonoid accumulation in developing citrus fruit. *Physiologia Plantarum* 111, 66–74.

Muir, S.R., Collins, G.J., Robinson, S., Hughes, S., Bovy, A., Ric De Vos, C.H., van Tunen, A.J. and Verhoeyen, M.E. (2001) Overexpression of petunia chalcone isomerase in tomato results in fruit containing increased levels of flavonols. *Nature Biotechnology* 19, 470–474.

Murakami, A., Gao, G., Omura, M., Yano, M., Ito, C., Furukawa, H., Takahashi, D., Koshimizu, K. and Ohigashi, H. (2000) 1,1-Dimethylallylcoumarins potently supress both lipopolysaccharide- and interferon-γ-induced nitric oxide generation in mouse macrophage RAW 264.7 cells. *Bioorganic and Medicinal Chemistry Letters* 10, 59–62.

Mustilli, A.C., Fenzi, F., Ciliento, R., Alfano, F. and Bowler, C. (1999) Phenotype of the tomato *high pigment-2* mutant is caused by a mutation in the tomato homolog of *DEETIOLATED1*. *Plant Cell Online* 11, 145–158.

Nagamatsu, A., Masuta, C., Senda, M., Matsuura, H., Kasai, A., Hong, J.-S., Kitamura, K., Abe, J. and Kanazawa, A. (2007) Functional analysis of soybean genes involved in flavonoid biosynthesis by virus-induced gene silencing. *Plant Biotechnology Journal* 5, 778–790.

Nambeesan, S., Datsenka, T., Ferruzzi, M.G., Malladi, A., Mattoo, A.K. and Handa, A.K. (2010) Overexpression of yeast spermidine synthase impacts ripening, senescence and decay symptoms in tomato. *Plant Journal* 63, 836–847.

Namitha, K.K. and Negi, P.S. (2010) Chemistry and biotechnology of carotenoids. *Critical Reviews in Food Science and Nutrition* 50, 728–760.

Naqvi, S., Zhu, C., Farre, G., Sandmann, G., Capell, T. and Christou, P. (2011) Synergistic metabolism in hybrid corn indicates bottlenecks in the carotenoid pathway and leads to the accumulation of extraordinary levels of the nutritionally important carotenoid zeaxanthin. *Plant Biotechnology Journal* 9, 384–393.

Negi, P.S. and Handa, A.K. (2008) Structural deterioration of the produce – the breakdown of cell wall components. In: Paliyath, G., Murr, D.P., Handa, A.K. and Lurie, S. (eds) *Postharvest Biology and Technology of Fruits, Vegetables and Flowers*, Wiley-Blackwell, Ames, IA, pp. 162–194.

Neily, M.H., Matsukura, C., Maucourt, M., Bernillon, S., Deborde, C., Moing, A., Yin, Y.-G., Saito, T., Mori, K., Asamizu, E., Rolin, D., Moriguchi, T. and Ezura, H. (2011) Enhanced polyamine accumulation alters carotenoid metabolism at the transcriptional level in tomato fruit over-expressing spermidine synthase. *Journal of Plant Physiology* 168, 242–252.

Nesbitt, T.C. and Tanksley, S.D. (2002) Comparative sequencing in the genus *Lycopersicon*: implications for the evolution of fruit size in the domestication of cultivated tomatoes. *Genetics* 162, 365–379.

Nesi, N., Lucas, M.-O., Auger, B., Baron, C., Lécureuil, A., Guerche, P., Kronenberger, J., Lepiniec, L., Debeaujon, I. and Renard, M. (2009) The promoter of the *Arabidopsis thaliana BAN* gene is active in proanthocyanidin-accumulating cells of the *Brassica napus* seed coat. *Plant Cell Reports* 28, 601–617.

Nicoletti, I., De Rossi, A., Giovinazzo, G. and Corradini, D. (2007) Identification and quantification of stilbenes in fruits of transgenic tomato plants (*Lycopersicon esculentum* Mill.) by reversed phase HPLC with photodiode array and mass spectrometry detection. *Journal of Agricultural and Food Chemistry* 55, 3304–3311.

Paine, J.A., Shipton, C.A., Chaggar, S., Howells, R.M., Kennedy, M.J., Vernon, G., Wright, S.Y., Hinchliffe, E., Adams, J.L., Silverstone, A.L. and Drake, R. (2005) Improving the nutritional value of Golden Rice through increased pro-vitamin A content. *Nature Biotechnology* 23, 482–487.

Pandolfini, T., Rotino, G., Camerini, S., Defez, R. and Spena, A. (2002) Optimisation of transgene action at the post-transcriptional level: high quality parthenocarpic fruits in industrial tomatoes. *BMC Biotechnology* 2, 1.

Pang, X., Nada, K., Kurosawa, T., Ban, Y. and Moriguchi, T. (2010) Effect of methylglyoxal *bis*-(guanylhydrazone) on polyamine and ethylene biosynthesis of apple fruit after harvest. *Acta Physiologia Plantarum* 32, 1005–1010.

Pang, Y., Peel, G.J., Wright, E., Wang, Z. and Dixon, R.A. (2007) Early steps in proanthocyanidin biosynthesis in the model legume *Medicago truncatula*. *Plant Physiology* 145, 601–615.

Phan, T.D., Bo, W., West, G., Lycett, G.W. and Tucker, G.A. (2007) Silencing of the major salt-dependent isoform of pectinesterase in tomato alters fruit softening. *Plant Physiology* 144, 1960–1967.

Porat, R., Tietel, Z., Zippori, I. and Dag, A. (2011) Aroma volatile compositions of high- and low-aromatic guava varieties. *Journal of the Science of Food and Agriculture* 91, 2794–2798.

Powell, A.L.T., Kalamaki, M.S., Kurien, P.A., Gurrieri, S. and Bennett, A.B. (2003) Simultaneous transgenic suppression of *LePG* and *LeExp1* influences fruit texture and juice viscosity in a fresh market tomato variety. *Journal of Agricultural and Food Chemistry* 51, 7450–7455.

Qin, X. and Zeevaart, J.A.D. (2002) Overexpression of a 9-*cis*-epoxycarotenoid dioxygenase gene in *Nicotiana plumbaginifolia* increases abscisic acid and phaseic acid levels and enhances drought tolerance. *Plant Physiology* 128, 544–551.

Quesada, M.A., Blanco-Portales, R., Posé, S., García-Gago, J.A., Jiménez-Bermúdez, S., Muñoz-Serrano, A., Caballero, J.L., Pliego-Alfaro, F., Mercado, J.A. and Muñoz-Blanco, J. (2009) Antisense down-regulation of the *FaPG1* gene reveals an unexpected central role for polygalacturonase in strawberry fruit softening. *Plant Physiology* 150, 1022–1032.

Rai, M., Datta, K., Parkhi, V., Tan, J., Oliva, N., Chawla, H. and Datta, S. (2007) Variable T-DNA linkage configuration affects inheritance of carotenogenic transgenes and carotenoid accumulation in transgenic indica rice. *Plant Cell Reports* 26, 1221–1231.

Ravanello, M.P., Ke, D., Alvarez, J., Huang, B. and Shewmaker, C.K. (2003) Coordinate expression of multiple bacterial carotenoid genes in canola leading to altered carotenoid production. *Metabolic Engineering* 5, 255–263.

Reich, L. (2012) *Grow Fruit Naturally: A Hands-On Guide to Luscious, Home-Grown Fruit*. Taunton Press, Newton, CT.

Rice-Evans, C., Miller, N. and Paganga, G. (1997) Antioxidant properties of phenolic compounds. *Trends in Plant Science* 2, 152–159.

Robards, K. and Antolovich, M. (1997) Analytical chemistry of fruit bioflavonoids – a review. *Analyst* 122, 11R–34R.

Rodríguez, A., Andrés, V., Cervera, M., Redondo, A., Alquézar, B., Shimada, T., Gadea, J., Rodrigo, M., Zacarías, L. and Palou, L. (2011b) The monoterpene limonene in orange peels attracts pests and microorganisms. *Plant Signaling and Behavior* 6, 1820.

Rodríguez, A., San Andrés, V., Cervera, M., Redondo, A., Alquézar, B., Shimada, T., Gadea, J., Rodrigo, M.J., Zacarías, L., Palou, L., López, M.M., Castañera, P. and Peña, L. (2011c) Terpene down-regulation in orange reveals the role of fruit aromas in mediating interactions with insect herbivores and pathogens. *Plant Physiology* 156, 793–802.

Rodríguez, G.R., Muños, S., Anderson, C., Sim, S.-C., Michel, A., Causse, M., Gardener, B.B.M., Francis, D. and van der Knaap, E. (2011a) Distribution of SUN, OVATE, LC, and FAS in the tomato germplasm and the relationship to fruit shape diversity. *Plant Physiology* 156, 275–285.

Rodríguez-Concepción, M. (2010) Supply of precursors for carotenoid biosynthesis in plants. *Archives of Biochemistry and Biophysics* 504, 118–122.

Römer, S., Fraser, P.D., Kiano, J.W., Shipton, C.A., Misawa, N., Schuch, W. and Bramley, P.M. (2000) Elevation of the provitamin A content of transgenic tomato plants. *Nature Biotechnology* 18, 666–669.

Ronen, G., Cohen, M., Zamir, D. and Hirschberg, J. (1999) Regulation of carotenoid biosynthesis during tomato fruit development: expression of the gene for lycopene epsilon-cyclase is down-regulated during ripening and is elevated in the mutant *Delta*. *Plant Journal* 17, 341–351.

Ronen, G., Carmel-Goren, L., Zamir, D. and Hirschberg, J. (2000) An alternative pathway to β-carotene formation in plant chromoplasts discovered by map-based cloning of Beta and old-gold color mutations in tomato. *Proceedings of the National Academy of Sciences USA* 97, 11102–11107.

Rosati, C., Aquilani, R., Dharmapuri, S., Pallara, P., Marusic, C., Tavazza, R., Bouvier, F., Camara, B. and Giuliano, G. (2000) Metabolic engineering of β-carotene and lycopene content in tomato fruit. *Plant Journal* 24, 413–420.

Rose, J.K.C., Lee, H.H. and Bennett, A.B. (1997) Expression of a divergent expansin gene is fruit-specific and ripening-regulated. *Proceedings of the National Academy of Sciences USA* 94, 5955–5960.

Ross, J.A. and Kasum, C.M. (2002) Dietary flavonoids: bioavailability, metabolic effects, and safety. *Annual Review of Nutrition* 22, 19–34.

Rotino, G.L., Perri, E., Zottini, M., Sommer, H. and Spena, A. (1997) Genetic engineering of parthenocarpic plants. *Nature Biotechnology* 15, 1398–1401.

Rühmann, S., Treutter, D., Fritsche, S., Briviba, K. and Szankowski, I. (2006) Piceid (resveratrol glucoside) synthesis in stilbene synthase transgenic apple fruit. *Journal of Agricultural and Food Chemistry* 54, 4633–4640.

Santiago-Doménech, N., Jiménez-Bemúdez, S., Matas, A.J., Rose, J.K.C., Muñoz-Blanco, J., Mercado, J.A. and Quesada, M.A. (2008) Antisense inhibition of a pectate lyase gene supports a role for pectin depolymerization in strawberry fruit softening. *Journal of Experimental Botany* 59, 2769–2779.

Schaffer, R.J., Friel, E.N., Souleyre, E.J.F., Bolitho, K., Thodey, K., Ledger, S., Bowen, J.H., Ma, J.-H., Nain, B., Cohen, D., Gleave, A.P., Crowhurst, R.N., Janssen, B.J., Yao, J.-L. and Newcomb, R.D. (2007) A genomics approach reveals that aroma production in apple is controlled by ethylene predominantly at the final step in each biosynthetic pathway. *Plant Physiology* 144, 1899–1912.

Schijlen, E., Ric de Vos, C.H., Jonker, H., Van Den Broeck, H., Molthoff, J., Van Tunen, A., Martens, S. and Bovy, A. (2006) Pathway engineering for healthy phytochemicals leading to the production of novel flavonoids in tomato fruit. *Plant Biotechnology Journal* 4, 433–444.

Schijlen, E.G.W.M., de Vos, C.H.R., Martens, S., Jonker, H.H., Rosin, F.M., Molthoff, J.W., Tikunov, Y.M., Angenent, G.C., van Tunen, A.J. and Bovy, A.G. (2007) RNA interference silencing of chalcone synthase, the first step in the flavonoid biosynthesis pathway, leads to parthenocarpic tomato fruits. *Plant Physiology* 144, 1520–1530.

Schreiber, G., Reuveni, M., Evenor, D., Oren-Shamir, M., Ovadia, R., Sapir-Mir, M., Bootbool-Man, A., Nahon, S., Shlomo, H., Chen, L. and Levin, I. (2012) *ANTHOCYANIN1* from *Solanum chilense* is more efficient in accumulating anthocyanin metabolites than its *Solanum lycopersicum* counterpart in association with the ANTHOCYANIN FRUIT phenotype of tomato. *Theoretical and Applied Genetics* 124, 295–307.

Seymour, G.B., Manning, K., Eriksson, E.M., Popovich, A.H. and King, G.J. (2002) Genetic identification and genomic organization of factors affecting fruit texture. *Journal of Experimental Botany* 53, 2065–2071.

Shih, C.-H., Chen, Y., Wang, M., Chu, I.K. and Lo, C. (2008) Accumulation of isoflavone genistin in transgenic tomato plants overexpressing a soybean isoflavone synthase gene. *Journal of Agricultural and Food Chemistry* 56, 5655–5661.

Shin, Y.-M., Park, H.-J., Yim, S.-D., Baek, N.-I., Lee, C.-H., An, G. and Woo, Y.-M. (2006) Transgenic rice lines expressing maize C1 and R-S regulatory genes produce various flavonoids in the endosperm. *Plant Biotechnology Journal* 4, 303–315.

Sidorenko, L.V., Li, X., Cocciolone, S.M., Chopra, S., Tagliani, L., Bowen, B., Daniels, M. and Peterson, T. (2000) Complex structure of a maize *Myb* gene promoter: functional analysis in transgenic plants. *Plant Journal* 22, 471–482.

Simkin, A.J., Schwartz, S.H., Auldridge, M., Taylor, M.G. and Klee, H.J. (2004) The tomato carotenoid cleavage dioxygenase 1 genes contribute to the formation of the flavour volatiles β-ionone, pseudoionone, and geranylacetone. *Plant Journal* 40, 882–892.

Simkin, A.J., Gaffé, J., Alcaraz, J.-P., Carde, J.-P., Bramley, P.M., Fraser, P.D. and Kuntz, M. (2007) Fibrillin influence on plastid ultrastructure and pigment content in tomato fruit. *Phytochemistry* 68, 1545–1556.

Slimestad, R. and Verheul, M. (2009) Review of flavonoids and other phenolics from fruits of different tomato (*Lycopersicon esculentum* Mill.) cultivars. *Journal of the Science of Food and Agriculture* 89, 1255–1270.

Smartt, J. and Simmonds, N.W. (1995) *Evolution of Crop Plants*. Longman Scientific & Technical, Harlow, UK.

Smith, D.L. and Gross, K.C. (2000) A family of at least seven β-galactosidase genes is expressed during tomato fruit development. *Plant Physiology* 123, 1173–1184.

Smith, D.L., Abbott, J.A. and Gross, K.C. (2002) Down-regulation of tomato β-galactosidase 4 results in decreased fruit softening. *Plant Physiology* 129, 1755–1762.

Song, J., Fan, L. and Beaudry, R.M. (1998) Application of solid phase microextraction and gas chromatography/time-of-flight mass spectrometry for rapid analysis of flavour volatiles in tomato and strawberry fruits. *Journal of Agricultural and Food Chemistry* 46, 3721–3726.

Stigliani, A.L., Giorio, G. and D'Ambrosio, C. (2011) Characterization of P450 carotenoid β- and ε-hydroxylases of tomato and transcriptional regulation of xanthophyll biosynthesis in root, leaf, petal and fruit. *Plant and Cell Physiology* 52, 851–865.

Sun, L., Yuan, B., Zhang, M., Wang, L., Cui, M., Wang, Q. and Leng, P. (2012) Fruit-specific RNAi-mediated suppression of *SlNCED1* increases both lycopene and β-carotene contents in tomato fruit. *Journal of Experimental Botany* 63, 3097–3108.

Sun, Y., Dilkes, B.P., Zhang, C., Dante, R.A., Carneiro, N.P., Lowe, K.S., Jung, R., Gordon-Kamm, W.J. and Larkins, B.A. (1999) Characterization of maize (*Zea mays* L.) Wee1 and its activity in developing endosperm. *Proceedings of the National Academy of Sciences USA* 96, 4180–4185.

Tanksley, S.D. (2004) The genetic, developmental, and molecular bases of fruit size and shape variation in tomato. *Plant Cell* 16, S181–S189.

Thakur, B.R., Singh, R.K. and Handa, A.K. (1996a) Effect of an antisense pectin methylesterase gene on the chemistry of pectin in tomato (*Lycopersicon esculentum*) juice. *Journal of Agricultural and Food Chemistry* 44, 628–630.

Thakur, B.R., Singh, R.K., Tieman, D.M. and Handa, A.K. (1996b) Tomato product quality from transgenic fruits with reduced pectin methylesterase. *Journal of Food Science* 61, 85–87.

Thakur, B.R., Singh, R.K., Handa, A.K. and Rao, M.A. (1997) Chemistry and uses of pectin - A review. *Critical Reviews in Food Science and Nutrition* 37, 47–73.

Tieman, D., Taylor, M., Schauer, N., Fernie, A.R., Hanson, A.D. and Klee, H.J. (2006b) Tomato aromatic amino acid decarboxylases participate in synthesis of the flavour volatiles 2-phenylethanol and 2-phenylacetaldehyde. *Proceedings of the National Academy of Sciences USA* 103, 8287–8292.

Tieman, D., Zeigler, M., Schmelz, E., Taylor, M.G., Rushing, S., Jones, J.B. and Klee, H.J. (2010) Functional analysis of a tomato salicylic acid methyl transferase and its role in synthesis of the flavour volatile methyl salicylate. *Plant Journal* 62, 113–123.

Tieman, D.M. and Handa, A.K. (1994) Reduction in pectin methylesterase activity modifies tissue integrity and cation levels in ripening tomato (Lycopersicon esculentum Mill.) fruits. *Plant Physiology* 106, 429–436.

Tieman, D.M., Harriman, R.W., Ramamohan, G. and Handa, A.K. (1992) An antisense pectin methylesterase gene alters pectin chemistry and soluble solids in tomato fruit. *Plant Cell* 4, 667–679.

Tieman, D.M., Zeigler, M., Schmelz, E.A., Taylor, M.G., Bliss, P., Kirst, M. and Klee, H.J. (2006a) Identification of loci affecting flavour volatile emissions in tomato fruits. *Journal of Experimental Botany* 57, 887–896.

Tieman, D.M., Loucas, H.M., Kim, J.Y., Clark, D.G. and Klee, H.J. (2007) Tomato phenylacetaldehyde reductases catalyze the last step in the synthesis of the aroma volatile 2-phenylethanol. *Phytochemistry* 68, 2660–2669.

Tigchelaar, E., McGlasson, W. and Buescher, R. (1978) Genetic regulation of tomato fruit ripening. *Hortscience* 13, 508–513.

Tsaballa, A., Pasentsis, K., Darzentas, N. and Tsaftaris, A. (2011) Multiple evidence for the role of an *Ovate*-like gene in determining fruit shape in pepper. *BMC Plant Biology* 11, 46.

USDA (2007) *USDA Database for the Flavonoid Content of Selected Foods*. US Department of Agriculture, Beltsville, MD, pp. 128.

USDA-APHIS (1991) Environmental assessment and finding of no significant impact on tomato containing an antisense polygalacturonase gene. Permit Number 91-268-01.

USDA-APHIS (1992) Interpretive ruling on Calgene, Inc., petition for determination of regulatory status of FLAVR SAVR™ tomato (Docket no. 92-087-2). *Federal Register* 57, 47608–47616.

van der Knaap, E., Lippman, Z.B. and Tanksley, S.D. (2002) Extremely elongated tomato fruit controlled by four quantitative trait loci with epistatic interactions. *Theoretical and Applied Genetics* 104, 241–247.

van der Knaap, E., Sanyal, A., Jackson, S.A. and Tanksley, S.D. (2004) High-resolution fine mapping and fluorescence in situ hybridization analysis of *sun*, a locus controlling tomato fruit shape, reveals a region of the tomato genome prone to DNA rearrangements. *Genetics* 168, 2127–2140.

Ververidis, F., Trantas, E., Douglas, C., Vollmer, G., Kretzschmar, G. and Panopoulos, N. (2007a) Biotechnology of flavonoids and other phenylpropanoid-derived natural products. Part II: Reconstruction of multienzyme pathways in plants and microbes. *Biotechnology Journal* 2, 1235–1249.

Ververidis, F., Trantas, E., Douglas, C., Vollmer, G., Kretzschmar, G. and Panopoulos, N. (2007b) Biotechnology of flavonoids and other phenylpropanoid-derived natural products. Part I: Chemical diversity, impacts on plant biology and human health. *Biotechnology Journal* 2, 1214–1234.

Wang, S., Chang, Y., Guo, J. and Chen, J.-G. (2007) Arabidopsis Ovate Family Protein 1 is a transcriptional repressor that suppresses cell elongation. *Plant Journal* 50, 858–872.

Wang, S., Liu, J., Feng, Y., Niu, X., Giovannoni, J. and Liu, Y. (2008) Altered plastid levels and potential for improved fruit nutrient content by downregulation of the tomato DDB1-interacting protein CUL4. *Plant Journal* 55, 89–103.

Wei, S., Li, X., Gruber, M.Y., Li, R., Zhou, R., Zebarjadi, A. and Hannoufa, A. (2009) RNAi-mediated suppression of DET1 alters the levels of carotenoids and sinapate esters in seeds of *Brassica napus*. *Journal of Agricultural and Food Chemistry* 57, 5326–5333.

Wilson, C.W., Shaw, P.E. and Campbell, C.W. (1982) Determination of organic acids and sugars in guava (*Psidium guajava* L.) cultivars by high-performance liquid chromatography. *Journal of the Science of Food and Agriculture* 33, 777–780.

Wurbs, D., Ruf, S. and Bock, R. (2007) Contained metabolic engineering in tomatoes by expression of carotenoid biosynthesis genes from the plastid genome. *Plant Journal* 49, 276–288.

Xiao, H., Jiang, N., Schaffner, E., Stockinger, E.J. and van der Knaap, E. (2008) A retrotransposon-mediated gene duplication underlies morphological variation of tomato fruit. *Science* 319, 1527–1530.

Ye, X., Al-Babili, S., Klöti, A., Zhang, J., Lucca, P., Beyer, P. and Potrykus, I. (2000) Engineering the provitamin A (β-carotene) biosynthetic pathway into (carotenoid-free) rice endosperm. *Science* 287, 303–305.

Yu, B., Lydiate, D., Young, L., Schäfer, U. and Hannoufa, A. (2008) Enhancing the carotenoid content of *Brassica napus* seeds by downregulating lycopene epsilon cyclase. *Transgenic Research* 17, 573–585.

Zawirska-Wojtasiak, R., Gośliński, M., Szwacka, M., Gajc-Wolska, J. and Mildner-Szkudlarz, S. (2009) Aroma evaluation of transgenic, thaumatin II-producing cucumber fruits. *Journal of Food Science* 74, C204–C210.

Zeigler, M.L. (2007) *Role of Methylsalicylate in Tomato Flavour and Response to Bacterial Pathogen Infection*. University of Florida, FL.

Zhang, J., Tao, N., Xu, Q., Zhou, W., Cao, H., Xu, J. and Deng, X. (2009) Functional characterization of *Citrus PSY* gene in Hongkong kumquat (*Fortunella hindsii* Swingle). *Plant Cell Reports* 28, 1737–1746.

Zuk, M., Kulma, A., Dymińska, L., Szołtysek, K., Prescha, A., Hanuza, J. and Szopa, J. (2011) Flavonoid engineering of flax potentiate its biotechnological application. *BMC Biotechnology* 11, 10.

Zygier, S., Chaim, A.B., Efrati, A., Kaluzky, G., Borovsky, Y. and Paran, I. (2005) QTLs mapping for fruit size and shape in chromosomes 2 and 4 in pepper and a comparison of the pepper QTL map with that of tomato. *Theoretical and Applied Genetics* 111, 437–445.

17 Insights into Plant Epigenome Dynamics

James Giovannoni*

US Department of Agriculture and Boyce Thompson Institute for Plant Research, Cornell University, Ithaca, NY, USA

17.1 Introduction

Genetic information is housed and passed to subsequent generations via the DNA code. The epigenome provides additional information, context and regulatory constraint in addition to both transient and long-term genetic memory. The epigenome consists of information carried in the nature of chromatin packaging and organization, histone modifications (e.g. acetylation, methylation) and DNA (specifically cytosine) methylation. Specific genes involved in DNA or protein methylation, acetylation, small RNA (sRNA) processing and sRNA transcription contribute to epigenome architecture, and their mutations have provided opportunities to develop insights into the intricacies of the epigenome. Advanced sequencing and informatics capabilities permit genome-scale analyses at modest cost, resulting in a wealth of data on epigenomes, and their variation and dynamics in response to development and external stimuli. Together these tools and genetic resources have begun to shed light on what may well prove to be a repository of genetic information and context as important as the DNA code itself.

The DNA code provides information in forms beyond DNA sequence alone. The packaging, three-dimensional structure, protein interactions and specific modifications of individual nucleotides and the proteins that interact with them all provide information and represent constituents of the epigenome. Information harboured in the epigenome in some instances is stable and reliably inherited, while in others it is more transient, relevant in a single or small number of generations, and may reflect adaptation to localized and changing conditions. Recent discoveries indicate that the plant epigenome can play a myriad of functions ranging from serving to provide appropriate expression environments, silencing mobile genetic elements and facilitating a means of short- or long-term genetic 'memory', to providing regulatory oversight for developmental processes. Understanding these functions and the means through which they are attained will broaden our collective understanding of the form and function of the epigenome. To the extent that the epigenome is proved to possess conserved or flexible variation, it may yield additional targets for selection for crop improvement. Plants have proved to be important in discovery of the nature and effects of the epigenome heralding important initial findings in *Arabidopsis*, in addition to others related to important

* jjg33@cornell.edu

aspects of crop genetics and biology. These discoveries are opening new paths towards genetic and biological innovation, with implications relevant to agricultural improvement and that are certain to be heavily pursued in coming years. The present chapter briefly describes the basic mechanisms underlying epigenetic regulation with a special focus on the role of the epigenome in fruit development and ripening using the tomato as a reference species.

17.2 Components of the Epigenome

17.2.1 Histones and their modifications

Eukaryotic DNA possesses the conserved feature of DNA wrapped in 147-base intervals around nucleosomes comprised of eight histone proteins – two each of histones H2A, H2B, H3 and H4. The ordering of nucleosomes on DNA strands and the specific chemical modifications of histones vary widely and contribute to the functional characteristics of specific DNA regions along the chromatin (Filion *et al.*, 2010). These features are achieved at least in part through regulating access to DNA by both general and specific mediators of transcriptional activity (Berger, 2007). Indeed, specific histone modifications have been closely associated with distinct gene features and activities. For example, methylation of lysine 9 of H3 (H3K9) and H3K27 methylation are found predominantly in regions of the genome populated by long terminal repeat retroelements (Zhang *et al.*, 2007; Bernatavichute *et al.*, 2008) but also in the gene bodies of some expressed genes (Turck *et al.*, 2007). H3K4 and H3K36 methylation are often associated with the 5′ and 3′ ends, respectively, of transcribed genes, while H3K27 methylation is more generally associated with genes displaying reduced transcriptional activity (Oh *et al.*, 2008). The compact size and high-quality genome sequence of *Arabidopsis thaliana* has, in conjunction with efficient immunoprecipitation, tiling arrays and sequencing technologies, facilitated recent and in-depth analysis of the histone landscape and its intricacies in this model plant system (Roudier *et al.*, 2011, and references therein). Characterization of multiple histone modifications across the *Arabidopsis* genome revealed a degree of nucleosome-associated functionality beyond euchromatin/heterochromatin bifurcation, including at least four general functional domains termed CS1–CS4. CS1 is composed of active genes with enriched H3K4 and K3K27 methylation, while C2, C3 and C4 are associated with various repressed states (Roudier *et al.*, 2011).

Characterization of histone profiles in combination with cytosine methylation and sRNAs provides additional insight into the fidelity and impact of epigenome variation on chromatin activity (Luo *et al.*, 2013; see below). Few plant genomes are as complete as that of *Arabidopsis*, a necessity for genome-scale epigenome analyses. Information on the rice genome is of sufficient quality to allow similar analyses, and provides a bridge between the less repetitive *Arabidopsis* genome and the repetitive genomes of many cereals and other important crop species. Characterization of histone modifications in rice revealed the same H3K4 and H3K27 methylation association with active genes (Hu *et al.*, 2012). The ability to determine the degree to which similar chromatin states persist in additional plants and with similar effect will become feasible as high-quality ordered reference genomes become available.

17.2.2 Cytosine methylation and small interfering RNAs (siRNAs)

Eukaryotes display a complex array of sRNA molecules that mediate a range of functions related to regulation of gene expression. sRNAs classifications include microRNAs (miRNAs) and siRNAs, the former derived from DICER or DICER-like activity upon small RNA hairpins and the latter from cleavage of long dsRNAs (Brodersen and Voinnet, 2006). miRNAs target specific genes or gene families via pathways involving RNA cleavage or translational suppression (Reinhart *et al.*, 2002; Brodersen *et al.*, 2008), while siRNAs are primarily associated with cytosine methylation and gene repression, especially with regard to retroelements (Mosher *et al.*, 2008).

siRNAs are capable of mediating sequence-specific methylation in all cytosine contexts (CG, CHG and CHH, where H is A, C or T) by targeting via DOMAINS REARRANGED METHYLTRANSFERASE2 protein (DRM2; Cao and Jacobsen, 2002) to siRNA homologous sites. As a result of base pairing, CG and CHG are symmetrical, and cytosines in these contexts have their methylation maintained during DNA replication by the cytosine methyltransferases MET1 and CMT3 (Ronemus *et al.*, 1996; Lindroth *et al.*, 2001), while CHH methylation is probably maintained by siRNA-mediated DMR2 activity at each replication event. While mechanisms exist to maintain DNA methylation through subsequent generations, these epigenome modifications are not fixed and can be altered by lack of MET1, CMT3 and DRM2 maintenance activities, or more directly through the demethylation activities of the REPRESSOR OF SILENCING1 (ROS1) and the homologous DEMETER family genes (Zhu, 2009).

The complex machinery of plant cytosine methylation, maintenance and variation, capable of discretely targeting specific DNA sequences at specific loci, is an area of considerable interest and increasing understanding. A substantial proportion of RNA-mediated DNA methylation is conferred by the plant-specific RNA polymerase IV (Pol IV) and Pol V, which specifically produce non-coding sRNA precursors. Pol IV-derived siRNAs target cytosine methylation of Pol V-transcribed regions through interaction with ARGONAUTE (AGO) proteins to develop a complex with Pol V transcripts and Pol V itself (He *et al.*, 2009). This complex recruits the DMR2 methyltransferase and possibly additional chromatin remodelling proteins (Gao *et al.*, 2010, and references therein). Mutations in *Arabidopsis* Pol IV and V genes reveal that, while phenotypic consequences ensue, viable organisms can still be recovered (Haag and Pikaard, 2011), indicating their importance but also the likely redundancy of equivalent activities. Interestingly, in the cases of Pol IV, Pol V or Pol IV/V double mutants, CHH cytosine methylation was shown to occur but in different positions and with increased activity in pericentromeric loci, suggesting that these polymerases themselves also influence the specified targeting of cytosine methyltransferases (Wierzbicki *et al.*, 2012). Additional components of targeted plant cytosine methylation remain to be identified and will contribute along with those already described to our understanding of the mechanisms and control of epigenome architecture.

17.3 Epigenome Plasticity

While the epigenome harbours information that can be transferred reliably from generation to generation, the epigenome is also dynamic and changeable, demonstrating variation through aspects of plant development, in response to stress, and the potential for reprogramming during reproduction. This dynamism presents opportunities for long- and short-term genetic adaptation, regulatory fidelity during stress and developmental responses, and prospects for novel approaches to genetic improvement via the development of novel advanced crop varieties.

17.3.1 Embryo development and heterosis: insights into imprinting, memory and epigenome heritability

DNA methylation status during plant sexual reproduction can result in variation from parent to offspring (Becker *et al.*, 2011; Jullien *et al.*, 2012), furthering additional interest into how DNA methylation status is managed through plant reproduction and subsequent development. Fertilization in flowering plants is accomplished by double fertilization, resulting in an embryo and nutritive endosperm tissues. Studies on DNA methylation during this process in *Arabidopsis* have revealed several major themes regarding heredity of cytosine methylation and its transgenerational variation. First, maintenance of cytosine methylation via the MET, CMT and DRM methyltransferases is reduced in both the male and female germlines and endosperm but increases dramatically in the embryo, suggesting a process of general demethylation followed by a system that remethylates the progeny genome (Jullien *et al.*, 2012) and probably also contributes to hybrid vigour (Groszmann *et al.*, 2011). DME specifically demethylates endosperm genes throughout the genome, especially those associated with transposon repeats, resulting in maternally derived, or imprinted, patterns of endosperm gene expression facilitating the functional attributes of this tissue in embryo support (Gehring *et al.*, 2006, 2009). Endosperm genome demethylation occurs in concert with increased siRNA production from the demethylated DNA regions, indicating an RNA-mediated process (Hsieh *et al.*, 2011; Mosher *et al.*, 2011). siRNAs have been shown to be mobile between cell and tissue types (Dunoyer *et al.*, 2010; Molnar *et al.*, 2010), so that endosperm and even pollen (Borges *et al.*, 2011) may contribute information directing the methylation pattern of the embryo genome. This process of epigenome transfer, maintenance and imprinting represents one of the few examples of well-characterized epigenome dynamics in eukaryotes, and is also likely to provide opportunities for selective modification of the epigenome between generations.

Studies on transgenerational methylation patterns in *Arabidopsis*, where single seed descent lineages have been studied for both DNA sequence and methylome polymorphisms, indicate that patterns of DNA methylation from generation to generation can be quite stable (Becker *et al.*, 2011; Schmitz *et al.*, 2011). In these elegant reports, while stably inherited methylation polymorphisms were observed and infrequent, they were more common than DNA sequence polymorphisms, suggesting that the epigenome provides additional plasticity for heritable change. The ability of siRNAs to mediate heritable *cis*- and *trans*-effects on siRNA production in progeny of tomato lines harbouring wild species introgressions provides further evidence for a plastic epigenome and its possible contributions to heterosis and transgressive segregation (Shivaprasad *et al.*, 2012). The fact that all of these processes occurs within a biochemical context of sequence-specific siRNAs and methylases/demethylases also suggests that such changes have increased opportunity for reversion or modifications that may have selective advantages, as in the examples of stress epigenome changes described below. In short, the accumulation of ovule, endosperm and pollen siRNAs contribute to the process of genome and epigenome inheritance from parents to their embryos and in a manner facilitating both short- and long-term genome modifications with selective advantage.

Finally, possibly related to the underlying processes that mediate germline and embryo epigenome dynamics, cytosine methylation status appears to influence meiotic recombination rates in ways more complex than originally thought, as determined through the study of a population segregating for the *met1* hypomethylation mutation (Mirouze *et al.*, 2012). This analysis suggested that, rather than an overall repressive effect on recombination, DNA methylation actually may reduce the recombination-repressive effect of heterochromatin and repress recombination in euchromatin. Localized variation also was observed within chromatin states, indicating additional complexity.

17.3.2 Fruit development and ripening

Tomato is a long-studied model system for analysis of fleshy fruit development and ripening, in part due to the availability of molecular resources including a recently sequenced high-quality genome (The Tomato Genomics Consortium, 2012) and a suite of well-studied ripening mutations whose genes have been described (reviewed by Klee and Giovannoni, 2011; Seymour *et al.*, 2013). As in the case of *Arabidopsis* and rice, high-throughput low-cost sequencing technologies have facilitated genome-scale analyses including those related to the epigenome. Characterization of sRNA accumulation in a series of floral tissues from early floral development to fruit ripening revealed a dynamic and diverse collection of miRNAs and additional sRNAs. Specific subsets of these sRNAs associated preferentially with aspects of floral development including ripening and were derived from both genic and non-genic sequences (Mohorianu *et al.*, 2011), suggesting a possible association with epigenome architecture and dynamics. It is noteworthy that a transposon-mediated change in methylation state of a hormone synthesis gene was previously shown to confer developmental control over sex determination during melon floral development (Martin *et al.*, 2009).

Indeed, one of the few well-characterized epigenetic mutations (or so-called epi-alleles) in plants resides at the *Colourless non-ripening* (*Cnr*) locus of tomato and dramatically impacts on fruit ripening. *Cnr* encodes a ripening-related *SQUAMOSA PROMOTER BINDING PROTEIN* gene (also termed *SPB* or *SPL* gene) that is silenced in the mutant due to heritable hypermethylation of a discrete region of the *Cnr* promoter (Manning *et al.*, 2006). Together, these reports indicate that the epigenome probably influences floral development and, specifically, development of carpel tissues in fleshy fruits.

A more recent analysis of whole-genome cytosine methylation patterns during tomato fruit maturation provides even stronger evidence for regulation of fruit development and indicates that epigenome dynamics may serve roles in development beyond those reported in germline and embryo development. It has long been known that tomato fruit ripening occurs in response to the gaseous hormone ethylene but that the fruit also must achieve a certain state of maturity, or developmental competence, to respond (McMurchie *et al.*, 1973). Zhong *et al.* (2013) treated immature unripe fruit well in advance of this ripening-responsive stage (in tomato referred to as 'mature green' designating seed maturity but still several days to a week or more preceding normal ripening) with 5-azacytidine, a methyltransferase inhibitor. The result was immature fruit that ripened in sectors corresponding to pericarp regions that received the treatment. When the ripening pericarp tissues were separated from non-ripening tissues in the same fruit and analysed for cytosine methylation via bisulfite sequencing, methylation differences in the promoters of ripening regulatory genes (e.g. *CNR*, *ACC SYNTHASE 2*) were noted. Specifically, relative hypomethylation was observed in the ripe sectors compared with the unripe sectors and unripe control fruits (which were injected with water).

This intriguing observation was followed by whole-genome methylome analysis of pericarp from immature, mature green, early ripening (termed breaker) and red-ripe fruit in addition to leaves and breaker-age fruit nearly isogenic for the *Cnr* and *ripening-inhibitor* (*rin*) non-ripening mutations. The latter results from a loss-of-function mutation in a MADS-box protein-coding gene of the *SEPALLATA* clade (Vrebalov *et al.*, 2002). The RIN protein is necessary for ripening and binds numerous promoters of ripening-related genes, and its binding activity is dependent on a functional *CNR* gene (Martell *et al.*, 2011). Cytosine methylation and siRNAs were most prevalent in the paracentric heterochromatin in all tissues assayed as anticipated, but euchromatic changes in methylation and associated with ripening were most predominant in the promoters of ripening-related genes. Thus, those genes were differentially expressed during ripening and when comparing the normal and mutant (*Cnr*, *rin*) gene expression profiles. The impact of two distinct ripening transcription factors on promoter cytosine methylation of numerous ripening-related genes also suggests their participation in methylation dynamics. Previously described tomato fruit-expressing methyltransferases (Teyssier *et al.*, 2008) may contribute to these methylation patterns in addition to demethylases. Additionally, these differentially methylated regions (DMRs) generally did not co-localize with siRNAs or other sRNAs, suggesting non-sRNA-mediated methylation control of these loci.

While the localization of DMRs in promoters of ripening-related genes that became hypomethylated during fruit maturation was correlated with ripening, further evidence for a causal relationship between epigenome dynamics and ripening control resulted from analysis of binding of the RIN protein to many of the same regions populated with ripening-related DMRs, as determined by chromatin immunoprecipitation sequencing analysis using an antibody to RIN (Zhong *et al.*, 2013). Together, these results indicate that the dynamics of the epigenome provides a layer of regulatory oversight over a developmental process. Whether this represents an avenue towards increased regulatory stringency over a process that would be highly detrimental if not lethal in the wrong tissue or stage of development, or a common phenomenon that remains to be elucidated in additional species and developmental programmes, remains to be determined. In addition, the fact that inhibitor-induced hypomethylation induced premature ripening and that ripening transcription factors inhibited demethylation suggest a system of regulatory feedback between ripening transcription factors and cytosine methylation processes that remains to be fully understood, and that may provide a regulatory model for additional control systems (see Plate 8).

17.3.3 Response to stress

In addition to developmental effects, the epigenome has been shown to influence the response to both abiotic and biotic stresses. Both cytosine methylation and histone acetylation can mediate the pathogen response of *Arabidopsis* to *Pseudomonas syringae* infection (Wang *et al.*, 2013), and descendents of similarly infected *Arabidopsis* plants were more capable of mounting a defensive response to subsequent infections of *Pseudomonas* or additional pathogens (Slaughter *et al.*, 2012). Dowen *et al.* (2012) observed that *Arabidopsis* methylation mutants were more resistant to *Pseudomonas* infection and performed whole-genome surveys of cytosine methylation in *Arabidopsis* plants subject to either virulent or avirulent *Pseudomonas* inoculation or salicylic acid defence hormone treatment. All treatments resulted in changes in DNA methylation patterns, often in transposon repeats adjacent to genes. These genes frequently displayed differential gene expression profiles compared with untreated controls, and the differentially methylated transposon sequences were also shown to have corresponding differences in siRNA accumulation. Together, these results indicate that pathogen attack influences specific gene activities through epigenome modifications, specifically variation in cytosine methylation of transposons via siRNAs and probably via hormone-mediated mechanisms. The parallels to fruit sex determination in melon described above, which also includes hormone (ethylene) and DNA methylation processes, suggest common pathways for utilizing epigenome dynamics to mediate useful gene expression responses.

17.4 Challenges and Opportunities

The epigenome clearly provides additional context and information beyond that harboured within the DNA sequence code. Recent studies have provided broad insights into methods of RNA-mediated cytosine methylation and mechanisms for histone covalent modifications that change the functional attributes of associated DNA sequences. As sessile organisms, plants, in contrast to animals, may gain additional benefits from a diverse and dynamic epigenome capable of providing additional flexibility and regulatory constraint over gene activity and opportunities for conveying aspects of both short- and long-term genetic memory to their offspring. While these findings are exciting, many specific features of epigenome modification, maintenance and inheritance remain to be elucidated. Mounting and continued efforts to reveal the mysteries of the epigenome structure, dynamics and effects will provide new avenues through which to understand biology and heredity, in addition to unique opportunities for selecting and improving plant species. As plants either directly or indirectly provide the majority of human calories and nutrition in addition to feed, fuel and fibre, increased understanding of the epigenome will certainly have profound biological and practical impacts on future generations and regarding the ways in which humans both exploit and conserve the Earth's treasures.

References

Becker, C., Hagmann, J., Muller, J., Koenig, D., Stegle, O., Borgwardt, K. and Weigel, D. (2011) Spontaneous epigenetic variation in the *Arabidopsis thaliana* methylome. *Nature* 480, 245–249.

Berger, S. (2007) The complex language of chromatin regulation during transcription. *Nature* 447, 407–412.

Bernatavichute, Y., Zhang, X., Cokus, S., Pellegrini, M. and Jacobsen, S. (2008) Genome-wide association of histone H3 lysine nine methylation with CHG DNA methylation in *Arabidopsis thaliana*. *PLOS One* 3, e3156.

Borges, F., Pereira, P., Slotkin, R., Martienssen, R. and Becker, J. (2011) MicroRNA activity in the *Arabidopsis* male germline. *Journal of Experimental Botany* 62, 1611–1620.

Brodersen, P. and Voinnet, O. (2006) The diversity of RNA silencing pathways in plants. *Trends in Genetics* 22, 268–280.

Brodersen, P., Sakvarelidze-Achard, L., Bruun-Rasmussen, M., Dunoyer, P., Yamamoto, Y., Sieburth, L. and Voinnet, O. (2008) Widespread translational inhibition by plant miRNAs and siRNAs. *Science* 320, 1185–1190.

Cao, X. and Jacobsen, S. (2002) Locus-specific control of asymmetric and CpNpG methylation by the DRM and CMT3 methytransferase genes. *Proceedings of the National Academy of Sciences USA* 99, 16491–16498.

Dowen, R., Pelizzola, M., Schmitz, R., Lister, R., Dowen, J., Nery, J., Dixon, J. and Ecker, J. (2012) Widespread dynamic DNA methylation in response to biotic stress. *Proceedings of the National Academy of Sciences USA* E2183–E2191.

Dunoyer, P., Schott, G., Himber, C., Meyer, D., Takeda, A., Carrington, J. and Voinnet, O. (2010) Small RNA duplexes function as mobile silencing signals between plant cells. *Science* 328, 912–916.

Filion, G., van Bemmel, J., Braunschweig, U., Talhout, W., Kind, J., Ward, L., Brugman, W., de Castro, I., Kerkhoeven, R., Bussemaker, H. and van Steensel, B. (2010) Systematic protein location mapping reveals five principal chromatin types in *Drosophila* cells. *Cell* 143, 212–224.

Gao, Z., Liu, H., Daxinger, L., Pontes, O., He, X., Qian, W., Lin, H., Xie, M., Lorkovic, Z., Zhang, S., Miki, D., Zhan, X., Pontier, D., Lagrange, T., Jin, H., Matzke, A., Matzke, M., Pikaard, C. and Zhu, J. (2010) An RNA polymerase II- and AGO4-associated protein acts in RNA-directed DNA methylation. *Nature* 465, 106–109.

Gehring, M., Huh, J., Hsieh, T., Penterman, J., Choi, Y., Harada, J., Goldberg, R. and Fisher, R. (2006) DEMETER DNA glycosylase establishes MEDEA polycomb gene self-imprinting by allele-specific demethylation. *Cell* 124, 495–506.

Gehring, M., Bubb, K. and Henikoff, S. (2009) Extensive demethylation of repetitive elements during seed development underlies gene imprinting. *Science* 324, 1447–1451.

Groszmann, M., Greaves, I., Albertyn, Z., Scofield, G., Peacock, W. and Dennis, E. (2011) Changes in 24-nt siRNA levels in *Arabidopsis* hybrids suggest and epigenetic contribution to hybrid vigor. *Proceedings of the National Academy of Sciences USA* 108, 2617–2622.

Haag, J. and Pikaard, C. (2011) Multisubunit RNA polymerases IV and V; purveyors of non-coding RNA for plant gene silencing. *Nature Reviews Molecular Cell Biology* 12, 483–492.

He, X., Hsu, Y., Zhu, S., Wierzbicki, A., Pontes, O., Pikaard, C., Liu, H., Wang, C., Jin, H. and Zhu, J. (2009) An effector of RNA-directed DNA methylation in *Arabidopsis* is an ARGONAUTE 4 and RNA-binding protein. *Cell* 137, 498–508.

Hsieh, T., Shin, J., Uzawa, R., Silva, P., Cohen, S., Bauer, M., Hashimoto, M., Kirkbride, R., Harada, J., Zilberman, D. and Fischer, R. (2011) Regulation of imprinted gene expression in *Arabidopsis* endosperm. *Proceedings of the National Academy of Sciences USA* 108, 1755–1762.

Hu, Y., Liu, D., Zhong, X., Zhang, C. and Zhou, D. (2012) CHD3 protein recognizes and regulated methylated histone H3 lysines 4 and 27 over a subset of targets in the rice genome. *Proceedings of the National Academy of Sciences USA* 109, 5773–5778.

Jullien, P., Susaki, D., Yelagandula, R., Higashiyama, T. and Berger, F. (2012) DNA methylation dynamics during sexual reproduction in *Arabidopsis thaliana*. *Current Biology* 22, 1825–1830.

Klee, H. and Giovannoni, J. (2011) Genetics and control of tomato fruit ripening and quality attributes. *Annual Review of Genetics* 45, 41–59.

Lindroth, A., Cao, X., Jackson, J., Zilberman, D., McCallum, C., Henikoff, S. and Jacobsen, S. (2001) Requirement of CHROMOMETHYLASE3 for maintenance of CpXpG methylation. *Science* 292, 2077–2080.

Luo, C., Sidote, D., Zhang, Y., Kerstetter, R., Michael, T. and Lam, E. (2013) Integrative analysis of chromatin states in Arabidopsis identified potential regulatory mechanisms for natural antisense transcript production. *Plant Journal* 73, 77–90.

Manning, K., Tor, M., Poole, M., Hong, Y., Thompson, A., King. G., Giovannoni, J. and Seymour, G. (2006) A naturally occurring epigenetic mutation in a gene encoding an SPB-box transcription factor inhibits tomato fruit ripening. *Nature Genetics* 38, 949–952.

Martell, C., Vrebalov, J., Tafelmeyer, P. and Giovannoni, J. (2011) The tomato MADS-box transcription factor RIPENING INHIBITOR interacts with promoters involved in numerous ripening processes in a COLORLESS NONRIPENING-dependent manner. *Plant Physiology* 157, 1568–1579.

Martin, A., Troadec, C., Boualem, A., Rajab, M., Fernandez, R., Morin, H., Pitrat, M., Dogimont, C. and Bendahmane, A. (2009) A transposon-induced epigenetic change leads to sex determination in melon. *Nature* 461, 1135–1138.

McMurchie, E., McGlasson, W. and Eaks, I. (1973) Treatment of fruit with propylene gives information about biogenesis of ethylene. *Nature* 237, 235–236.

Mirouze, M., Lieberman-Lazarovich, M., Aversano, R., Bucher, E., Nicolet, J., Reinders, J. and Paskowski, J. (2012) Loss of DNA methylation affects the recombination landscape in *Arabidopsis*. *Proceedings of the National Academy of Sciences USA* 109, 5880–5885.

Mohorianu, I., Schwach, F., Jing, R., Lopez-Gomollon, S., Moxon, S., Szittya, G., Sorefan, K., Moulton, V. and Dalmay, T. (2011) Profiling short RNAs during fleshy fruit development reveals stage-specific sRNAome expression patterns. *Plant Journal* 67, 231–246.

Molnar, A., Melnyk, C., Bassett, A., Hardcastle, T., Dunn, R. and Baulcombe, D. (2010) Small silencing RNAs in plants are mobile and direct epigenetic modification in recipient cells. *Science* 328, 872–875.

Mosher, R., Schwach, F., Studhollme, D. and Baulcombe, D. (2008) PolIVb influences RNA-directed DNA-methylation independently of its role in siRNA biogenesis. *Proceedings of the National Academy of Sciences USA* 105, 3145–3150.

Mosher, R., Tan, E., Shin, J., Fischer, R., Pikaard, C. and Baulcombe, D. (2011) An atypical epigenetic mechanism affects uniparental expression of Pol IV-dependent siRNAs. *PLOS One* 6, e27756.

Oh, S., Park, S. and van Nocker, S. (2008) Genic and global functions for Paf1C in chromatin modification and gene expression in *Arabidopsis*. *PLOS Genetics* 4, e1000077.

Reinhart, B., Weinstein, E., Rhoades, M., Bartel, B. and Bartel, D. (2002) MicroRNAs in plants. *Genes and Development* 16, 1616–1626.

Ronemus, M., Galbiati, M., Ticknor, C., Chen, J. and Dellaporta, S. (1996) Demethylation-induced developmental pleiotropy in *Arabidopsis*. *Science* 273, 654–657.

Roudier, F., Ahmed, I., Berard, C., Sarazin, A., Mary-Huard, T., Cortijo, S., Bouyer, D., Caillieux, E., Duvernois-Berthet, E., Al-Shikhley, L., Giraut, L., Despres, B., Drevensek, S., Barneche, F., Derozier, S., Brunaud, V., Aubourg, S., Schnittger, A., Bowler, C., Martin-Magniette, M., Robin, S., Caboche, M. and Colot, V. (2011) Integrative epigenomic mapping defines four main chromatin states in Arabidopsis. *EMBO Journal* 30, 1928–1938.

Schmitz, R., Schultz, M., Lewsey, M., O'Malley, R., Urich, M., Libiger, O., Schork, N. and Ecker, J. (2011) Transgenerational epigenetic instability is a source of novel methylation variants. *Science* 334, 369–373.

Seymour, G., Ostergaard, L., Chapman, N., Knapp, S. and Martin, C. (2013) Fruit development and ripening. *Annual Review of Plant Biology* 64, 219–241.

Shivaprasad, P., Dunn, R., Santos, B., Bassett, A. and Baulcombe, D. (2012) Extraordinary transgressive phenotypes of hybrid tomato are influenced by epigenetics and small silencing RNAs. *EMBO Journal* 31, 257–266.

Slaughter A., Daniel, X., Flors, V., Luna, E., Hohn, B. and Mauch-Mani, B. (2012) Descendants of primed Arabidopsis plants exhibit resistance to biotic stress. *Plant Physiology* 158, 835–843.

Teyssier, E., Bernacchia, G., Maury, S., How Kit, A., Stammitti-Bert, L., Rolin, D. and Gallusci, P. (2008) Tissue dependent variations of DNA methylation and endoreduplication levels during tomato fruit development and ripening. *Planta* 228, 391–399.

The Tomato Genomics Consortium (2012) The tomato genome sequence provides insights into fleshy fruit evolution. *Nature* 485, 635–641.

Turck, F., Roudier, F., Farrona, S., Martin-Magniette, M., Guillaume, E., Buisine, N., Gagnot, S., Martienssen, R., Coupland, G. and Colot, V. (2007) *Arabidopsis* TFL2/JHP1 specifically associates with genes marked by trimethylation of histone H3 lysine 27. *PLOS Genetics* 3, 855–866.

Vrebalov, J., Ruezinsky, D., Padmanabhan, V., White, R., Medrano, D., Drake, R., Schuch, W. and Giovannoni, J. (2002) A MADS-box gene necessary for fruit ripening at the tomato ripening-inhibitor (*rin*) locus. *Science* 296, 343–346.

Wang, Y., An, C., Zhang, X., Yao, J., Zhang, Y., Sun, Y., Yu, F., Amador, D. and Mou, Z. (2013) The *Arabidopsis* Elongator complex subunit2 epigenetically regulates plant immune responses. *Plant Cell* 25, 762–776.

Wierzbicki, A., Cocklin, R., Mayampurath, A., Lister, R., Rowley, M., Gregory, B., Ecker, J., Tang, H. and Pikaard, C. (2012) Spatial and functional relationships among Pol V-associated loci, Pol IV-dependent siRNAs, and cytosine methylation in the *Arabidopsis* epigenome. *Genes and Development* 26, 1825–1836.

Zhang, X., Clarenz, O., Cokus, S., Bernatavichute, Y., Pellegrini, M., Goodrich, J. and Jacobsen, S. (2007) Whole-genome analysis of histone H3 and lysine 27 trimethylation in *Arabidopsis*. *PLOS Biology* 5, 1026–1035.

Zhong, S., Fei, Z., Chen, Y., Zheng, Y., Huang, M., Vrebalov, J., McQuinn, R., Gapper, N., Lui, B., Xiang, J., Shao, Y. and Giovannoni, J. (2013) Single-base resolution methylomes of tomato fruit development reveal epigenome modifications associated with ripening. *Nature Biotechnology* 31, 154–159.

Zhu, J. (2009) Active DNA demethylation mediated by DNA glycosylases. *Annual Review of Genetics* 43, 143–166.

18 Functional Genomics for the Study of Fruit Ripening and Quality: Towards an Integrative Approach

Federico Martinelli[1,2] and Abhaya Dandekar[3*]

[1]*Dipartimento di Scienze Agrarie e Forestali, University of Palermo, Palermo, Italy;* [2]*Istituto Euro Mediterraneo di Scienza e Tecnologia, Palermo, Italy;* [3]*Department of Plant Sciences, University of CaliforniaDavis, CA, USA*

18.1 Introduction

Fruit development is controlled by genetically programmed processes influenced by environmental factors. Different 'omics' approaches (deep sequencing, microarray analysis, suppression subtractive hybridization) have identified and characterized genes involved in this process in several fruit species. The mass of knowledge concerning transcriptional regulatory networks affecting important physiological and developmental processes has expanded in the last two decades.

Expressed sequence tag (EST) sequencing uses microarray technology and real-time PCR to generate comprehensive data for functional genomics studies. Following the pioneering work of Aharoni and co-workers (2000) on strawberry, microarrays have been used in many different fruit species. In tomato, large-scale EST sequencing projects have clarified molecular mechanisms of fruit ripening and identified important transcription factors (Moore *et al.*, 2002). In grape berry, ESTs were used to discover genes with potential roles in fruit development. In apple, EST sequencing was employed to study the molecular regulation of fruit growth and development (Park *et al.*, 2006).

Emerging genomics tools and approaches have added new candidate genes to expand the known fruit-ripening regulatory network. Ripening is influenced by internal and external factors, including developmental gene regulation, hormones, light and temperature. Until recently, studies at the molecular level were focused on the role and regulation of ethylene biosynthesis (Adams-Phillips *et al.*, 2004).

Fruits are generally categorized as fleshy or dry. Fleshy fruits typically undergo ripening, while dry fruits such as cereals and legumes mature in a process more similar to senescence, dispersing their seeds using abscission-like processes. The model plant *Arabidopsis* has provided insights into the molecular regulation of the early steps in fruit formation and development (Adams-Phillips *et al.*, 2004), although it does not produce fleshy ripe

* amdandekar@ucdavis.edu

fruit. Much of the knowledge on ethylene perception and ethylene signalling derives from research work on *Arabidopsis*. In *Arabidopsis*, ethylene is perceived by a family of five ethylene receptors (ETR1, ETR2, ERS1, ERS2 and EIN4), which are similar to bacterial two-component histidine kinase receptors. Recent experiments have linked ethylene phenotypes to the unregulated activity of EIN2 (Hall and Bleecker, 2003).

Research studies on ethylene and light signal transduction pathways, performed mainly in *Arabidopsis*, have advanced ripening research in fleshy fruit species such as tomato. Tomato has emerged as a model for an understanding of fleshy fruit development and ripening due to important features such as the availability of mutants, a rapid life cycle, routine transformation, and numerous molecular and genomics tools (http://solgenomics.net/). Characterization of gene regulatory networks during tomato fruit ripening have clarified understanding of molecular mechanisms of the ripening process. (Adams-Phillips *et al*., 2004). Several ethylene signal transduction components homologous to those identified in *Arabidopsis* have been isolated from various plant species. Although the sequences of genes involved in ethylene-related pathways are conserved, the regulation and the number of genes vary among fleshy fruit species. Six ethylene receptors have been isolated in tomato, five of which bind ethylene.

Ripe fruits come in diverse forms, colours, textures, aromas, flavours and nutrient compositions. Severe cell-wall modifications occur during fruit ripening. Among the different processes, those that are noteworthy are the conversion of starch to sugars, the modification of pigment biosynthesis/accumulation, and increased synthesis of flavour and aromatic volatiles. Several ripening features can be problematic, decreasing shelf-life and needing highly expensive harvest and post-harvest procedures. Particularly important in this respect are the changes in firmness and susceptibility to microbial and fungal infection caused by tissue breakdown associated with ripening.

Much progress over the last 15 years has allowed us to gain insight into the molecular regulation of specific ripening processes, especially those involved in cell-wall metabolism and ethylene biosynthesis (reviewed by Giovannoni, 2001). The molecular understanding of ripening allowed the development of novel biotechnological approaches to improve quanti-qualitative aspects of fruits. Ripening impacts on important components of the human diet such as fibre abundance and composition, lipid metabolism and the concentrations of vitamins and various antioxidants (Ronen *et al*., 1999). The ability to understand and manipulate, through breeding or biotechnology, key control points of ripening or to regulate the synthesis of carotenoids, flavonoids, vitamins and flavour volatiles, could improve the control of nutrition and quality changes associated with ripening. Currently unpopular genetic engineering techniques might be viewed more favourably by the public if they were used to improve the quality and nutrition of food.

18.2 Tomato as a Model Organism for Fruit Ripening

Tomato has been a key plant model for molecular fruit ripening studies over the past two decades for several reasons. Tomatoes are easily cultivated and have a short life cycle. Unlike rice or the classic molecular model organism *Arabidopsis*, tomato has fleshy fruit. Because tomato and *Arabidopsis* diverged from their common ancestor early in dicot radiation, the similarities and differences between the two model organisms are particularly informative. The tomato genome is moderately sized (950 Mb) and was recently sequenced through an international initiative entitled the 'International Solanaceae Genome Project'. Homozygous inbred lines and other well-characterized genetic and genomic resources are available.

The Tomato Genetics Resource Center (http://tgrc.ucdavis.edu/; University of California, Davis, CA, USA) maintains a large collection of wild relatives and monogenic mutants affecting many aspects of plant development and responses such as disease resistance.

The Solanaceae have adapted to diverse niches with diverse phenotypes, but their genome structure is relatively well conserved. Tomato, as member of this family, offers an opportunity to understand the diversification of these plants. The genome sequence, recently obtained, is expected to benefit breeding and genetic engineering programmes for solanaceous crops and other fleshy fruits. It is also of interest for phylogenetic studies because of its intermediate diversity between the rosid and asterid clades.

Tomato was developed for use in model DNA marker technology because line populations are easy to develop by interspecific introgression and to assess for identifying quantitative trait loci (QTLs) (Frary *et al.*, 2000). ESTs are currently used to study functional genomics and as a complement to genome sequencing (Rudd, 2003). Tomato has a large EST collection (>298,000 entries) in the public domain. This collection contains 18,051 unigenes (unique consensus sequences), of which ~70% have homologues in the *Arabidopsis* genome and the remaining 30% have unknown functions. Continuing research on *Arabidopsis* may speed up the tomato functional genomics, while EST and full-length cDNA sequencing should help to predict the functions of the remaining genes.

18.3 Functional Genomics Approaches in Tomato

ESTs are created by partially sequencing analysed transcripts that have been converted into cDNA. Recent progresses in 'omics' technologies and advanced DNA sequencing technology have allowed large-scale EST sequencing. There are now millions of ESTs in the NCBI public dbEST database (http://www.ncbi.nlm.nih.gov/dbEST/). Current and future large-scale EST sequencing projects are likely to increase the number of ESTs in the public domain, providing additional opportunities to compare intra- and interspecific genome expression and expanding opportunities for digital gene expression analysis. Progresses in bioinformatics and biostatistics make it possible to functionally analyse large-scale EST data sets in a highly efficient manner (Ronning *et al.*, 2003).

Insertional mutagenesis is a powerful tool for identifying gene function. Transposons have also been isolated for promoter trapping, and β-glucuronidase is used for enhancer trapping (Meissner *et al.*, 2000). Transcriptional enhancers can also be placed on a binary transformation vector to induce genetic mutants by expressing endogenous genes close to the T-DNA insertion site. This method was successfully used to generate 10,427 transgenic tomato lines, of which 1338 had visually observable novel traits; virus-induced gene silencing (VIGS) technology was used to determine the function of many of these unknown genes. Transcriptome profiling monitors, microarrays with 30,000 gene probes, have been used to study plant defence-related responses such as fusicoccin-induced changes in gene expression and systemic wound responses (Strassner *et al.*, 2002). A widely used tomato microarray with >150,000 ESTs and 12,000 unigenes (Tom1 Microarray, http://ted.bti.cornell.edu/) has allowed broader analysis of gene expression (Alba *et al.*, 2004). Another genome array chip with >10,000 unigenes is commercially available from Affymetrix. Another recent 'omics' approach to dissect biological systems is metabolomic analysis (Aharoni *et al.*, 2002).

18.4 Micro-Tom as a Model Tomato Line

Small organisms such as the fruit fly, *Drosophila melanogaster*, are often used as model systems in genetics studies.

Arabidopsis is still the most-used plant model, but recently a miniature tomato line called Micro-Tom has attracted the interest of researchers. This cultivar was originally developed for home gardening but has several qualities favourable for functional genomics studies. Like *Arabidopsis*, it grows well in a laboratory setting under artificial light. It has a short, 70–90-day life cycle and can grow at densities of up to 1357 plants m^{-2}.

Micro-Tom has been widely used for transposon tagging and promoter trapping, activation tagging, VIGS and cDNA libraries for EST isolation, and also as a source of mutants (Meissner *et al.*, 2000). Micro-Tom was challenged with 16 tomato pathogens to determine its susceptibility. With the release of the tomato genome sequence, this laboratory-friendly miniature cultivar will further research into plant–pathogen interactions (Meissner *et al.*, 1997) and aid 'omics' approaches such as transcriptome and metabolome profiling under strictly controlled environments.

18.5 Tools for Gene Expression Profiling

Efficient, high-throughput cloning and sequencing methods have driven the development of novel ways to analyse ever-larger genome data sets (Rounsley and Briggs, 1999). New methodologies have increased the number of available platforms for forward and reverse genetics, examining expression changes of hundreds or thousands of genes simultaneously. In tomato, a collaborative effort constructed and sequenced cDNA libraries from many tissues and conditions and created a tomato EST database (Van der Hoeven *et al.*, 2002).

This information allows parallel gene studies to detect and quantify gene expression. Parallel studies provide both static (single tissue) and dynamic (comparative) information. There are multiple methods of parallel analysis, from traditional RNA gel blots and quantitative reverse transcription PCR (qRT-PCR) to the more comprehensive differential display, serial analysis of gene expression (SAGE) and microarrays. Microarrays allow the analysis of expression patterns of thousands of genes within a single experiment, using the same interactions between complementary strands of DNA principles as Southern blot assays.

Solid glass substrates, accurate robotics and fluorescence-based identification methodologies have increased the accuracy, rapidity and scale of expression analyses. Microarrays can be constructed using either PCR-amplified cDNAs or oligonucleotides. Arrays based on amplified ESTs are used for microspotting. ESTs are usually generated by sequencing methods that generate 300–900 bp cDNAs. EST sequence and homology information provides a distinct and obvious advantage in expression studies over anonymous clones, since it is important to make direct functional interpretations.

Important new resources that are available for tomato include substantial sequence information, the EST database and microarray technology (http://solgenomics.net/). To create microarrays, unique DNAs are printed onto chemically coated glass microscope slides. Glass has low inherent fluorescence, minimizing the intrinsic background level, and its non-porous surface prevents the diffusion of deposited samples, minimizing the required hybridization volumes.

These tools are allowing tomato and Solanaceae researchers to answer previously unexplored biological questions. A cDNA microarray was constructed recently to examine fruit development and ripening. A time course of ten intervals, during fruit development from 7 days post-anthesis to 15 days past the breaker stage, represents biologically significant stages in fruit development. Initially, cultivar 'Ailsa Craig' was used to establish a baseline of normal gene expression. The investigation was performed with several ripening-related mutants to identify their specific functions and effects. Probes were constructed for each stage and used in step-wise dual hybridizations in different replications. These included 'dye-swap' experiments to compensate for variability in signal intensity due to the characteristics of an individual fluorochrome. The obtained data functionally analyse subsets of genes during fruit development. There were specific genes differentially expressed for each developmental stage and particularly an increased number of genes were induced at the beginning of ripening (Moore *et al.*, 2002). These genes are involved in ethylene synthesis, carotenoid accumulation and cell-wall modifications. As these studies progress, it may become possible to predict functions for ESTs with little or no known homology, based on expression patterns and relationships to better-known genes. In particular, comparisons of expression profiles with the tomato proteome hold the promise of identifying underlying genetic and molecular events contributing to fruit development and ripening.

18.6 A Functional Genomics Case Study: Gene Regulation in Transgenic Parthenocarpic Tomato Fruit

In tomato, seeds contribute to flavour and are not considered undesirable in fresh market fruits. However, seedless fruits would facilitate tomato processing. Seed formation is an integral component of fruit development: developing seeds promote cell expansion via the synthesis of auxin and other unknown molecules.

Parthenocarpy is the formation of fruits without seeds. In some species, fruit can develop without fertilization, indicating that it is possible to separate fruit formation from seed development. This phenomenon has a genetic basis (Vardy *et al.*, 1989) and can be induced artificially with hormone applications of gibberellins, auxins and cytokinins. Raspberry fruit were genetically transformed with a DefH9–iaaM gene construct to induce parthenocarpic fruits (Mezzetti *et al.*, 2004). Seedlessness is highly desired by consumers in fresh citrus and grapes.

The INNER NO OUTER (INO) gene is needed for ovule development in Arabidopsis. Expression is observed only at the initiation site and developing outer (abaxial) cell layer of the ovule outer integument (Meister *et al.*, 2004). INO can induce parthenocarpy and significantly reduce seed number (Martinelli *et al.*, 2009). Transgenic tomato plants were obtained for each of four promoter–gene combinations (INO or DefH9 promoter with iaaM or rolB) to produce parthenocarpy (see Plates 9 and 10). Among the transgenic tomato lines expressing iaaM, close to one-third of the fruits showed no seeds, while slightly more than one-third had a few seeds. The remaining fruits had greatly reduced seed production. Similar results were obtained for transgenic tomato plants for rolB gene. The promoters of INO and DefH9 regulated expression of iaaM or rolB similarly and produced similar proportions of seedless and reduced-seed lines. Seedlessness did not appear to affect fruit shape or quality, although in DefH9–rolB transgenic tomatoes a decreased number of seeds was linked to altered soluble solids.

The Affymetrix tomato array was employed to compare the breaker stage of wild-type and transgenic fruits on a large transcriptomic scale and to determine changes directly caused by transgene expression. At this stage, the fruit has a yellow colour. Although direct changes induced by genetic transformation in both the transcriptome and metabolome are likely to occur at earlier developmental stages, the high numbers of gene expression changes at the breaker stage emphasized the differences due to transformation. Pairwise comparison (one-way analysis of variance, $P < 0.001$) of the highly expressed genes showed hundreds of downregulated genes in rolB- and iaaM-transformed plants (0.96 and 0.98% of all genes represented on the microarray, respectively). Fewer upregulated genes were observed (0.13 and 0.21% of the represented genes, respectively).

There was a large overlap in the expression of 1748 differentially expressed genes between DefH9- and INO-transformed plants, suggesting similar effects of these promoters on gene expression (see Plate 11). This strongly supports the hypothesis that the Arabidopsis thaliana AtINO promoter induces ovule-specific expression in tomato. Functional analysis of the differently regulated genes using MapMan software showed several transcriptomic differences in pathways involved in minor carbohydrate metabolism, DNA synthesis, transcription factors, secondary metabolism and hormones. Key differentially regulated genes were also related to transport: a cation exchanger, two sugar transporters and the nitrate transporter NRT1-3. Important transcriptional effects were observed in ethylene- and indole-3-acetic acid (IAA)-related pathways.

Possible interactions between auxin and ethylene pathways (biosynthesis, metabolism and perception) are also of interest. Ethylene-responsive element binding proteins (EREBPs) are both transcriptional activators and repressors (Fujimoto *et al.*, 2000), and represent an important gene family involved in fruit qualitative aspects. As some EREBPs induce ripening and others are repressed, EREBPs may affect fruit ripening using antagonistic mechanisms (Fei *et al.*, 2004). Seedless fruit is associated with a longer shelf-life than seeded fruit because seeds produce hormones that cause senescence.

Gene set enrichment analysis showed that IAA-responsive genes were downregulated in all transgenic fruits compared with seedless or seeded controls, implying that IAA and RolB downregulate other auxin-associated genes independently of seeds. It is possible that these regulators induce parthenocarpy in a similar way to the normal downregulation of SlARF7 ovary transcripts after pollination in tomato (*Solanum lycopersicum*) (de Jong *et al.*, 2009).

For metabolomic analysis, the concentrations of >400 metabolites were evaluated in transgenic and control fruits. The large-scale data were compared to identify differences and similarities among transgenic seedless fruits and seeded control fruits at the breaker stage. Few metabolomic changes were induced by IAA or RolB: principal component analyses could not separate the transgenic and control lines (Martinelli *et al.*, 2009). Thus, these gene/promotor combinations could induce parthenocarpy in tomato without large changes to the transcriptome or metabolome.

18.7 Functional Genomics to Study Fruit Development and Ripening in Olive

Olive (*Olea europaea* L.) is an evergreen species commonly grown in the Mediterranean basin. The oil extracted from the fruit is a predominant component of the 'Mediterranean diet', which has documented heart- and cancer-protective benefits. These derive from the lipid composition and from biologically active molecules that accumulate during olive fruit development. The oil can reach up to 30% of the total fruit fresh weight at full ripening and is highly present in the mesocarp and at lower levels in seeds. Oil increase reaches a plateau in pulp after veraison. A marked triacylglycerol accumulation in seed and pulp occurs after endocarp lignification, when about 40 mg of oil per fruit per week can be synthesized. The fatty acid profile of the oil accumulating in the fruit is important in relation to its nutritional properties. The main fatty acid is oleic acid (C18:1), which represents about 75% of total fatty acids, followed by linoleic (C18:2), palmitic (C16:0), stearic (C18:0) and linolenic (C18:3) acids. It is known that important metabolites accumulate during olive fruit development. These are chlorophylls, carotenoids, polyphenols, sterols and terpenoids, all important from an olive oil qualitative, technological and nutritional perspective. Information regarding the genetic regulation of these metabolic processes in olive is still very limited. Only a few genes involved in fatty acid metabolism have been functionally studied. Among these, a sugar transporter (*OeMST2*), expressed during maturation, has been cloned (Conde *et al.*, 2007). A gene encoding a geranylgeranyl reductase (*OeCHLP*) has been identified and its involvement in organ development and stress response was shown (Bruno *et al.*, 2009). Knowledge regarding molecular regulation mechanisms in important pathways such as polyphenol and triterpenoid metabolism is scarce, as well as the mechanisms involved in olive fruit development and ripening.

The elucidation of gene regulatory networks based on the regulation of key metabolic pathways during fruit growth and development is essential for improving olive oil quality and nutritional value. One olive transcriptomic analysis used the cultivar 'Leccino', a popular Italian variety with a short, highly synchronized fruit developmental cycle (Galla *et al.*, 2009). The suppression subtractive hybridization approach identified 1132 differentially expressed gene sequences at three stages: initial fruit set (30 days after flowering (DAF)), completed pit hardening (90 DAF) and veraison (130 DAF). The analysis identified 642 differentially regulated sequences. Among these, 89 (14%) corresponding to 61 key genes were further analysed by real-time PCR, which confirmed expression patterns for up to 69% of the results. The bioinformatic annotation of all gene sequences allowed insight into the metabolic pathways and elucidated specific regulatory networks.

These data are a significant contribution to the elucidation of control of carbohydrates, fatty acids, transcription factors, secondary metabolites, hormones and responses to environmental stress at the transcript level. Of particular interest are data showing the complexity of the role played by hormones in olive fruit development and ripening. These molecular and bioinformatic data represent a first step towards the elucidation of gene functions and regulatory networks active in olive fruit biochemical and morphological processes.

18.8 Tomato Mutants with Modified Light Signal Perception

While the hormone ethylene is required to complete ripening in climacteric fruit, the impact of light is specific to the regulation of pigment accumulation (Alba *et al.*, 2000). Tomato high-pigment mutations (*hp1* and *hp2*) accumulate more carotenoids and flavonoids due to greater light sensitivity without changing other ripening processes (Peters *et al.*, 1989). Ripe fruit pigments like carotenoids and flavonoids have antioxidant properties that neutralize the effects of photo-oxidation and are important human nutrients. Because mutations in the light signalling pathway increase pigmentation of ripe fruit, the light signalling pathway is a potential target for efforts to engineer increased fruit nutrition. Although carotenoid content in edible parts has been changed by altering the content of biosynthetic enzymes (e.g. Golden Rice), the results of such approaches did not follow expectations due to misunderstandings of the molecular mechanisms and/or undesirable side effects on non-target metabolites of the modified pathway (Beyer *et al.*, 2002). Engineering an existing signal transduction network that regulates flux through the carotenoid synthesis pathway in a biologically viable way could represent an alternative to enhance carotenoids in fruit.

Regulating expression of HY5 and COP1 involved in signal transduction using transgenic approaches, it is possible to modify fruit carotenoid content. MADS-box genes are present in eukaryotes and are linked to floral determination and development in plants. MADS-box proteins form heterodimers and higher-order multimers, implying that MADS-box genes might play a key role in ripening. Indeed, several MADS-box genes expressed in ripening tomato fruit could be good candidates for functional analysis of fruit ripening. Orthologous genes from agriculturally important fruit species are being targeted to enhance fruit quality and shelf-life.

Two putative transcription factors regulating ripening and fruit development in tomato by inducing climacteric ethylene biosynthesis and through ethylene-independent processes have been determined as an important first step in controlling fruit ripening. Isolation of the *Colorless non-ripening* (*Cnr*) locus will hopefully elucidate the developmental component of ripening regulation. Understanding the relationships among the *Cnr*, *ripening-inhibitor* (*rin*) and *non-ripening* (*nor*) gene products will follow.

Emerging genomics tools like ESTs and expression arrays will also help the identification of additional novel ripening regulators and homologous genes in other species, with evolutionary conservation established via comparative genomics. A recent comparison of ripening-related gene expression in non-climacteric grape with those of the climacteric tomato identified ripening-related transcription factor sequences from families not previously associated with ripening. EST content analysis was used to determine gene expression levels, and subsets of ripening-related genes from both species were compared to predict peptide homology and identify homologous genes with parallel expression patterns. Twenty ripening-related putative transcription factor sequences were identified in each species. The three common transcription factor sequences were members of the MADS-box, basic leucine zipper domain (bZIP) and zinc-finger families; bZIP and zinc-finger proteins have not previously been associated with ripening. Functional characterization of these genes and other regulatory candidates from ongoing genomics-based experiments will identify broadly conserved and species-specific genetic regulators of ripening in the near future.

18.9 Citrus Response to Huanglongbing Disease

Huanglongbing (HLB) or 'citrus greening' is a highly destructive citrus disease caused by phloem-limited bacteria of the genus *Candidatus* Liberibacter. Symptoms include blotchy, mottled and variegated leaf chlorosis, followed by tree decline. Infected leaves become upright, with leaf drop and twig dieback at later stages (Bové, 2006). Zinc, magnesium or iron deficiency cause similar symptoms, making diagnosis difficult.

Understanding host responses to pathogen infection at the molecular level will help develop novel strategies for early disease detection and therapy. Next-generation sequencing was used to examine the differential expression of a higher number of transcripts than is possible with microarrays. Next-generation sequencing also allowed a deeper analysis of different applications such as the study of gene isoforms, splice variants and microRNA. RNA sequencing provides direct counts of mRNA from expressed sequences rather than inferring expression based on hybridization of fluorescent molecules. Next-generation sequencing data can be used to create specific transcriptome assemblies for annotating genomes and differentially regulated genes and proteins analysed with any 'omics' technique.

Early host responses of citrus to infection with *Candidatus* Liberibacter asiaticus (CaLas) were examined using next-generation sequencing (Martinelli *et al.*, 2012). The deep mRNA profile was obtained from fruit peel of healthy and HLB-affected fruit, followed by pathway and protein–protein network analysis qRT-PCR validation of a subset of HLB-regulated genes. A deep gene regulatory network was constructed. Gene set enrichment analysis identified several pathways significantly affected by HLB, including the metabolism of starch, sucrose and α-linolenic acid and the synthesis of phenylpropanoids, flavonoids, terpenoids and anthocyanins. Plastid genes involved in photosynthetic light reactions were upregulated in symptomatic fruit. The resultant oxidative stress was linked to activation of protein degradation and misfolding. Transcripts for heat-shock proteins were repressed at all stages of disease, resulting in further protein misfolding. HLB strongly affected pathways involved in source-sink communications such as sucrose and starch metabolism and hormone biosynthesis and signalling. Transcription of several genes for synthesis and signal transduction of cytokinins and gibberellins was down-regulated, but ethylene pathways were induced. CaLas infection seemed to cause an induction of salicylic acid and jasmonic acid pathways and to enhance the transcript levels of several members of the WRKY family of transcription factors. A picture of the main changes in gene regulatory networks in response to HLB in the fruit was constructed (see Plate 12).

This study identified several genes differentially expressed before symptoms appear that could help disease detection at the primary stages of infection. Obviously, it will be important to determine that these genes are not also induced by *Citrus tristeza* virus, *Xylella fastidiosa* or *Xanthomonas axonopodis* infections, other diseases of citrus. In fruit peel, HLB induced altered levels of transcript abundance in hormone and isoprenoid pathways and in sucrose and starch metabolism. WRKY transcription factors seemed to regulate the defence responses to CaLas in the fruit. Treatments with small-molecule hormones could represent a short-term strategy to reduce the enormous negative effect of this disease.

18.10 Functional Genomics for Qualitative Improvement of Rosaceous Crops

The Rosaceae family contains some 3000 species in more than 100 genera and includes economically important crops grown in temperate environment (Dirlewanger *et al.*, 2002). The Rosaceous tree genera, including *Malus*, *Pyrus* and *Prunus*, derive mainly from an ancient *Malus* progenitor that gave rise to the cultivated apple (*Malus* × *domestica*), while domesticated European and Asian pears are derived from different species. Rosaceae species have been studied using molecular-assisted breeding, genetic engineering, and functional and structural genomics.

18.10.1 RNA interference (RNAi)

Computational analysis of ESTs in public databases identified seven conserved plant microRNA (miRNA) families and structures of precursor miRNAs (Schaffer *et al.*, 2007). Ten distinct sequences were classified into seven conserved plant miRNA families (Gleave *et al.*, 2008). A candidate gene approach has been linked with RNA RNAi silencing to elucidate the role of some genes in resistance to bacterial fire blight (Norelli *et al.*, 2007). Bioinformatics identified ESTs either specifically linked to fire blight or to *Pseudomonas syringae* pv. *tomato* infection. Genetic engineering was employed to upregulate a single EST-silencing gene and select apple RNAi mutants. Additional candidate ESTs are currently being identified using different biotechnological techniques (suppression subtractive hybridization and cDNA-amplified fragment length polymorphism analyses). Neither a mutant phenotype nor a gene sequence by itself explains the molecular function of a gene. Therefore, modern functional genomics consists of high-throughput methods of different 'omics' technologies followed by bioinformatics analysis for detailed functional genomics, while genetic engineering provides the means to validate gene function *in vivo*.

18.10.2 Functional genomics approaches

Apple is a food crop and a source of pectin, used to thicken jams and laboratory culture media. The fruits are processed into sauces, slices, sweets, alcoholic beverages, vinegar and juice. A large collection of apple ESTs has been used to produce microarrays for gene expression analyses using platforms like NimbleGen, Invitrogen and Affymetrix custom apple arrays.

These transcriptional approaches solved important issues in plant sciences. Proteomics studies in the Rosaceae have been more limited. Proteomic analyses of apple pseudocarp tissue have combined two-dimensional gel electrophoresis with matrix-assisted laser desorption/ionization-time of flight mass spectrometry and liquid chromatography/electrospray ionization mass spectrometry (Guarino *et al.*, 2007). Although many pseudocarp proteins remain unidentified, this study highlights the link between proteomics and functional genomics by linking identified proteins to their associated genes.

A proteomics approach was also employed to determine flesh browning in stored Conference pears (Pedreschi *et al.*, 2007), to identify novel isoforms of major cherry allergens (Reuter *et al.*, 2005) and to examine the role of dehydrins in cold temperature stress responses (Renaut *et al.*, 2008).

Rosaceous plants are rich in specialized metabolites important for human health and nutrition. Metabolomics involves global analysis and interrogation of metabolic networks. This technique has been used to study the metabolic transition from immature to ripe fruit (Aharoni and O'Connell, 2002), and the effects of UV/white light irradiation and cold storage on primary and secondary pathways, ethylene synthesis, acid metabolism, flavonoid pigment synthesis and fruit texture. Metabolomics can help identify novel gene functions in primary and secondary metabolism and model metabolic networks that regulate human health-promoting metabolites.

18.10.3 Marker-assisted breeding

Marker-assisted breeding is the genetic improvement of crops using information generated by molecular marker technology. It is particularly useful for perennial tree crops like apple, as many important traits are expressed only after several years of field cultivation. Marker-assisted breeding allows marker-assisted introgression of important and/or favourable genes from wild species into cultivated ones to improve breeding material (Lecomte *et al.*, 2004). Several apple genetic maps have been developed by positioning genetic markers linked to genes of interest. Such markers are used for genetic mapping, localization of major genes and QTL detection. High-quality, accurate, high-density genetic linkage maps allow genetic markers to be linked to desired traits and both to be localized on the chromosome.

18.10.4 Genome-wide single-nucleotide polymorphism (SNP) arrays in apple

Recent progress in high-throughput sequencing for genome-wide assays of single-nucleotide mutations have helped link phenotypic variation with the underlying DNA variation. Genomics tools can greatly help breeders to improve important agronomic traits and clarify their genetic structure. SNP screening is composed of detection, validation and final selection for marker development. During detection, a large pool of SNPs is detected in the crop using high-resolution melting or resequencing techniques. Validation informs SNP assay development by increasing the number of functional polymorphic markers in the genome. Because of the high costs, the validation can be performed only for a subset of SNPs. For genome-wide SNP assays, adequate genome coverage is essential. For cultivated species that have a sequenced genome and high-dense genetic maps, genome coverage can be based on physical and/or genetic factors, as preferred. Once SNP sets are available, screening uses highly parallel techniques for analysing germplasm needed for the specific research purposes. High-throughput technologies have increased the efficiency of SNP genotyping and several platforms for large-scale analysis can now genotype up to 1 million SNPs at the same time. The International RosBREED SNP Consortium

(IRSC) used the Illumina Infinium II system to produce high-throughput SNP screening methods for genome-wide evaluation of allelic variation in apple (*Malus* × *domestica*) germplasm. The whole-genome sequence from 'Golden Delicious' was resequenced with 27 apple cultivars (Chagné *et al.*, 2012). More than 2 million SNPs were detected – equivalent to one SNP for every 288 bp of genome – and a subset of 144 SNPs was validated in 160 apple accessions. A total of 7867 apple SNPs were used to develop the IRSC apple 8K SNP array v1, of which 5554 were polymorphic after evaluation in segregating families and a germplasm collection. This publicly available genomics resource will allow unprecedented resolution of SNP haplotypes and enable structural and functional genomics studies and marker-locus-trait association discovery in apple and other Rosaceaous crops.

18.10.5 Peach functional genomics

Fruits from the genus *Prunus* are drupes in which seeds are enclosed in a hard, lignified endocarp (the stone) surrounded by an edible mesocarp. The genus includes cultivated species like *Prunus persica* (peach, nectarine), *Prunus domestica* (European or prune plum), *Prunus salicina* (Japanese plum), *Prunus cerasus* (sour cherry), *Prunus avium* (sweet cherry), *Prunus armeniaca* (apricot) and *Prunus amygdalus* (almond). Peach is self-compatible, which allows breeding of cultivars with lower genetic variability than other *Prunus* crops. Peach is the genetic and genomic reference species for the genus *Prunus* because it is both economically valuable and has a relatively short juvenile period. Its self-compatibility allows the development of F2 progenies, and homozygous doubled haploids are available (Pooler and Scorza, 1997). However, efficient transformation protocols have yet to be developed.

An important resource for marker-assisted breeding is the Genome Database for Rosaceae (http://www.rosaceae.org/), from which the reference *Prunus* map was created. This consensus map was constructed using an interspecific almond × peach F2 population (Texas 3 Earlygold) and has hundreds of transportable markers and several major morphological, quality and agronomic characters with simple Mendelian inheritance and QTLs. The peach doubled haploid 'Lovell' was selected by the US Department of Energy Joint Genomics Institute's Community Sequencing Program for shotgun sequencing (83-fold genome coverage). The Italian ESTree database (http://www.itb.cnr.it/estree/) was created by the ESTree Inter-university Centre to develop functional genomics in drupaceous species. The Centre produced four cDNA libraries from *P. persica* mesocarps from three different cultivars at different developmental stages (postfertilization, endocarp hardening, preclimateric and postclimateric/final maturation) and used them to generate thousands of ESTs. An automated pipeline was used to mine EST sequences using Perl scripts. A web interface allowed database queries. To create this important peach genomic resource, sequences were assembled into contig consensus sequences and annotated against public primary databases. The resulting database is a comprehensive tool to link genome sequences with peach EST sequences, allowing data to be obtained rapidly for each sequence/contig.

References

Adams-Phillips, L., Barry, C. and Giovannoni, J. (2004) Signal transduction systems regulating fruit ripening. *Trends in Plant Science* 9, 331–338.

Aharoni, A. and O'Connell, A.P. (2002) Use of DNA microarrays in the identification of genes involved in strawberry flavor formation. *Analysis of Taste and Aroma* 21, 7–28.

Aharoni, A., Keizer, L.C.P., Bouwmeester, H.J., Sun, Z., Alvarez-Huerta, M., Verhoeven, H.A., Blaas, J., van Houwelingen, A.M., De Vos, R.C., van der Voet, H., Jansen, R.C., Guis, M., Mol, J., Davis, R.W., Schena, M., van Tunen, A.J. and O'Connell, A.P. (2000) Identification of the SAAT gene involved in strawberry flavour biogenesis by use of DNA microarrays. *Plant Cell* 12, 647–661.

Alba, R., Cordonnier-Pratt, M.M. and Pratt, L.H. (2000) Fruit-localized phytochromes regulate lycopene accumulation independently of ethylene production in tomato. *Plant Physiology* 123, 363–370.

Alba, R., Fei, Z., Payton, P., Liu, Y., Moore, S.L., Debbie, P., Cohn, J., D'Ascenzo, M., Gordon, J.S., Rose, J.K., Martin, G., Tanksley, S.D., Bouzayen, M., Jahn, M.M. and Giovannoni, J. (2004) ESTs, cDNA microarrays, and gene expression profiling: tools for dissecting plant physiology and development. *Plant Journal* 39, 697–714.

Beyer, P., Al-Babili, S., Ye, X., Lucca, P., Schaub, P., Welsch, R. and Potrykus, I. (2002) Golden rice: introducing the β-carotene biosynthesis pathway into rice endosperm by genetic engineering to defeat vitamin A deficiency. *Journal of Nutrition* 132, 506S–510S.

Bové, J.M. (2006) Huanglongbing: a destructive, newly-emerging, century-old disease of citrus. *Journal of Plant Pathology* 88, 7–37.

Bruno, L., Chiappetta, A., Muzzalupo, I., Gagliardi, C., Iaria, D., Bruno, A., Greco, M., Giannino, D., Perri, E. and Bitonti, M.B. (2009) Role of geranylgeranyl reductase gene in organ development and stress response in olive (*Olea europaea*) plants. *Functional Plant Biology* 36, 370–381.

Chagné, D., Crowhurst, R.N., Troggio, M., Davey, M.W., Gilmore, B., Lawley, C., Vanderzande, S., Hellens, R.P., Kumar, S., Cestaro, A., Velasco, R., Main, D., Rees, J.D., Iezzoni, A., Mockler, T., Wilhelm, L., Van de Weg, E., Gardiner, S.E., Bassil, N. and Peace, C. (2012) Genome-wide SNP detection, validation, and development of an 8K SNP array for apple. *PLOS One* 7, e31745.

Conde, C., Agasse, A., Silva, P., Lemoine, R., Delrot, S., Tavares, R. and Gerós, H. (2007) OeMST2 encodes a monosaccharide transporter expressed throughout olive fruit maturation. *Plant and Cell Physiology* 48, 1299–1308.

de Jong, M., Wolters-Arts, M., Feron, R., Mariani, C. and Vriezen, W.H. (2009) The *Solanum lycopersicum* auxin response factor 7 (SlARF7) regulates auxin signaling during tomato fruit set and development. *Plant Journal* 57, 160–170.

Dirlewanger, E., Cosson, P., Tavaud, M., Aranzana, J., Poizat, C., Zanetto, A., Arús, P. and Laigret, F. (2002) Development of microsatellite markers in peach *Prunus persica* (L.) Batsch and their use in genetic diversity analysis in peach and sweet cherry (*Prunus avium* L.). *Theoretical and Applied Genetics* 105, 127–138.

Fei, Z., Tang, X., Alba, R.M., White, J.A., Ronning, C.M., Martin, G.B., Tanksley, S.D. and Giovannoni, J.J. (2004) Comprehensive EST analysis of tomato and comparative genomics of fruit ripening. *Plant Journal* 40, 47–59.

Frary, A., Nesbitt, T.C., Grandillo, S., Knaap, E., Cong, B., Liu, J., Meller, J., Elber, R., Alpert, K.B. and Tanksley, S.D. (2000) *fw2.2*: a quantitative trait locus key to the evolution of tomato fruit size. *Science* 289, 85–88.

Fujimoto, S.Y., Ohta, M., Usui, A., Shinshi, H. and Ohme-Takagi, M. (2000) Arabidopsis ethylene-responsive element binding factors act as transcriptional activators or repressors of GCC box-mediated gene expression. *Plant Cell* 12, 393–404.

Galla, G., Barcaccia, G., Ramina, A., Collani, S., Alagna, F., Baldoni, L., Cultrera, N.G., Martinelli, F., Sebastiani, L. and Tonutti, P. (2009) Computational annotation of genes differentially expressed along olive fruit development. *BMC Plant Biology* 9, 128.

Giovannoni, J. (2001) Molecular biology of fruit maturation and ripening. *Annual Review of Plant Physiology and Plant Molecular Biology* 52, 725–749.

Gleave, A.P., Ampomah-Dwamena, C., Berthold, S., Dejnoprat, S., Karunairetnam, S., Nain, B., Wang, Y.-Y., Crowhurst, R.N. and MacDiarmid, R.M. (2008) Identification and characterisation of primary microRNAs from apple (*Malus domestica* cv. Royal Gala) expressed sequence tags. *Tree Genetics and Genomes* 4, 343–358.

Guarino, C., Arena, S., De Simone, L., D'Ambrosio, C., Santoro, S., Rocco, M., Scaloni, A. and Marra, M. (2007) Proteomic analysis of the major soluble components in Annurca apple flesh. *Molecular and Nutrition and Food Research* 51, 255–262.

Hall, A.E. and Bleecker, A.B. (2003) Analysis of combinatorial loss-of-function mutants in the Arabidopsis ethylene receptors reveals that the *ers1 etr1* double mutant has severe developmental defects that are EIN2 dependent. *Plant Cell* 15, 2032–2041.

Lecomte, L., Duffé, P., Buret, M., Servin, B., Hospital, F. and Causse, M. (2004) Marker-assisted introgression of five QTLs controlling fruit quality traits into three tomato lines revealed interactions between QTLs and genetic backgrounds. *Theoretical and Applied Genetics* 109, 658–668.

Martinelli, F., Uratsu, S.L., Reagan, R.L., Chen, Y., Tricoli, D., Fiehn, O., Rocke, D.M., Gasser, C.S. and Dandekar, A.M. (2009) Gene regulation in parthenocarpic tomato fruit. *Journal of Experimental Botany* 60, 3873–3890.

Martinelli, F., Uratsu, S.L., Albrecht, U., Reagan, R.L., Phu, M.L., Britton, M., Buffalo, V., Fass, J., Leicht, E., Zhao, W., Lin, D., D'Souza, R., Davis, C.E., Bowman, K.D. and Dandekar, A.M. (2012) Transcriptome profiling of citrus fruit response to Huanglongbing disease. *PLOS One* 7, e38039.

Meissner, R., Jacobson, Y., Melamed, S., Levyatuv, S., Shalev, G., Ashri, A., Elkind, Y. and Levy, A. (1997) A new model system for tomato genetics. *Plant Journal* 12, 1465–1472.

Meissner, R., Chague, V., Zhu, Q., Emmanuel, E., Elkind, Y. and Levy, A.A. (2000) A high throughput system for transposon tagging and promoter trapping in tomato. *Plant Journal* 22, 265–274.

Meister, R.J., Williams, L.A., Monfared, M.M., Gallagher, T.L., Kraft, E.A., Nelson, C.G. and Gasser, C.S. (2004) Definition and interactions of a positive regulatory element of the *Arabidopsis INNER NO OUTER* promoter. *Plant Journal* 37, 426–438.

Mezzetti, B., Landi, L., Pandolfini, T. and Spena, A. (2004) The *defH9-iaaM* auxin-synthesizing gene increases plant fecundity and fruit production in strawberry and raspberry. *BMC Biotechnology* 4, 4.

Moore, S., Vrebalov, J., Payton, P. and Giovannoni, J. (2002) Use of genomics tools to isolate key ripening genes and analyse fruit maturation in tomato. *Journal of Experimental Botany* 53, 2023–2030.

Norelli, J.L., Borejsza-Wysocka, E., Baldo, A.M., Aldwinckle, H.S., Bassett, C.L., Farrell, R.E., Malnoy, M., Lalli, D.A., Korban, S.S., Gasic, K. and Wisniewski, M.E. (2007) Functional genomic analysis of apple (*Malus*) ESTs associated with fire blight (*Erwinia amylovora*). *Phytopathology* 97, 5185.

Park, S., Sugimoto, N., Larson, M.D., Beaudry, R. and van Nocker, S. (2006) Identification of genes with potential roles in apple fruit development and biochemistry through large-scale statistical analysis of expressed sequence tags. *Plant Physiology* 141, 811–824.

Pedreschi, R., Vanstreels, E., Carpentier, S., Hertog, M., Lammertyn, J., Robben, J., Noben, J.P., Swennen, R., Vanderleyden, J. and Nicolaï, B.M. (2007) Proteomic analysis of core breakdown disorder in Conference pears (*Pyrus communis* L.). *Proteomics* 7, 2083–2099.

Peters, J.L., van Tuinen, A., Adamse, P., Kendrick, R.E. and Koornneef, M. (1989) High pigment mutants of tomato exhibit high-sensitivity for phytochrome action. *Journal of Plant Physiology* 134, 661–666.

Pooler, M.R. and Scorza, R. (1997) Irradiation and heat affect peach pollen germination and fertility. *Hortscience* 32, 290–291.

Renaut, J., Hausman, J.-F., Bassett, C., Artlip, T., Cauchie, H.-M., Witters, E. and Wisniewski, M. (2008) Quantitative proteomic analysis of short photoperiod and low-temperature responses in bark tissues of peach (*Prunus persica* L. Batsch). *Tree Genetics and Genomes* 4, 589–600.

Reuter, A., Fortunato, D., Garoffo, L.P., Napolitano, L., Scheurer, S., Giuffrida, M.G., Vieths, S. and Conti, A. (2005) Novel isoforms of Pru av 1 with diverging immunoglobulin E binding properties identified by a synergistic combination of molecular biology and proteomics. *Proteomics* 5, 282–289.

Ronen, G., Cohen, M., Zamir, D. and Hirschberg, J. (1999) Regulation of carotenoid biosynthesis during tomato fruit development: expression of the gene for lycopene epsilon-cyclase is down-regulated during ripening and is elevated in the mutant *Delta*. *Plant Journal* 17, 341–351.

Ronning, C.M., Stegalkina, S.S., Ascenzi, R.A., Bougri, O., Hart, A.L., Utterbach, T.R., Vanaken, S.E., Riedmuller, S.B., White, J.A., Cho, J., Pertea, G.M., Lee, Y., Karamycheva, S., Sultana, R., Tsai, J., Quackenbush, J., Griffiths, H.M., Restrepo, S., Smart, C.D., Fry, W.E., Van Der Hoeven, R., Tanksley, S., Zhang, P., Jin, H., Yamamoto, M.L., Baker, B.J. and Buell, C.R. (2003) Comparative analyses of potato expressed sequence tag libraries. *Plant Physiology* 131, 419–429.

Rounsley, S. and Briggs, S. (1999) The paradigm shift of genomics – a complement to traditional plant science. *Current Opinion in Plant Biology* 2, 81–82.

Rudd, S. (2003) Expressed sequence tags: alternative or complement to whole genome sequences? *Trends in Plant Science* 8, 321–329.

Schaffer, R.J., Friel, E.N., Souleyre, E.J., Bolitho, K., Thodey, K., Ledger, S., Bowen, J.H., Ma, J.H., Nain, B., Cohen, D., Gleave, A.P., Crowhurst, R.N., Janssen, B.J., Yao, J.L. and Newcomb, R.D.

(2007) A genomics approach reveals that aroma production in apple is controlled by ethylene predominantly at the final step in each biosynthetic pathway. *Plant Physiology* 144, 1899–1912.

Strassner, J., Schaller, F., Frick, U.B., Howe, G.A., Weiler, E.W., Amrhein, N., Macheroux, P. and Schaller, A. (2002) Characterization and cDNA-microarray expression analysis of 12-oxophytodienoate reductases reveals differential roles for octadecanoid biosynthesis in the local versus the systemic wound response. *Plant Journal* 32, 585–601.

Van der Hoeven, R., Ronning, C., Giovannoni, J., Martin, G. and Tanksley, S. (2002) Deductions about the number, organization, and evolution of genes in the tomato genome based on analysis of a large expressed sequence tag collection and selective genomic sequencing. *Plant Cell* 14, 1441–1456.

Vardy, E., Lapushner, D., Genizi, A. and Hewitt, J. (1989) Genetics of parthenocarpy in tomato under a low-temperature regime: I. Line RP 75/59. *Euphytica* 41, 1–8.

Index